Linear Programs and Related Problems

This is a volume in
COMPUTER SCIENCE AND SCIENTIFIC COMPUTING

Werner Rheinbolt and Daniel Siewiorek, editors

Linear Programs and Related Problems

Evar D. Nering

Microcomputer Resource Facility
Department of Mathematics
Arizona State University
Tempe, Arizona

Albert W. Tucker

Department of Mathematics
Princeton University
Princeton, New Jersey

ACADEMIC PRESS, INC.
Harcourt Brace Jovanovich, Publishers
Boston San Diego New York
London Sydney Tokyo Toronto

This book is printed on acid-free paper. ∞

ACADEMIC PRESS, INC.
1250 Sixth Avenue, San Diego, CA 92101

United Kingdom edition published by
ACADEMIC PRESS LIMITED
24–28 Oval Road, London NW1 7DX

Library of Congress Cataloging-in-Publication Data

Nering, Evar D.
Linear Programs and related problems / Evar D. Nering, Albert W. Tucker.
p. cm. — (Computer science and scientific computing)
ISBN 0-12-515440-2 (alk. paper)
1. Linear programming. I. Tucker, Albert W. (Albert William).
Date– II. Title. III. Series.
T57.74.N46 1990
519.7'2—dc20 89–17804
CIP

Printed in the United States of America
92 93 94 95 BC 9 8 7 6 5 4 3 2 1

Contents

Foreword

I had the singular good fortune in early 1948, three years after the end of World War II, to meet Albert Tucker on the occasion of a trip from the Pentagon to Princeton to visit the world famous mathematician John von Neumann, the originator of game theory. I was seeking mathematical help to solve an important problem that had its origins in the planning and scheduling activities of our nation during the war.

Soon Tucker and his students David Gale and Harold Kuhn began to make history by laying the mathematical foundations of the linear programming field and exploring its relations to game theory and nonlinear programming. Under his leadership, his group became one of three main centers of the field. The mathematical world learned from Princeton that a new field of mathematics was evolving. At the University of Chicago, another group under the leadership of Tjalling Koopmans was busy relating linear programming developments to economics and paving the way to transform economic theory into a practical quantitative science. At the Pentagon, my branch in collaboration with the National Bureau of Standards was concentrating on the algorithms, the computers, and on the applications to real world problems.

To those of us engaged in these activities way back when, it was an exciting new world of great expectations. Computers had just been invented and it was only a matter of time before they would be perfected and available. Combinatorial problems nobody believed could ever be solved now were within reach.

The contributions of Al Tucker and his group at Princeton were many. They published rigorous proofs of the famous duality theorem conjectured by von Neumann. Theirs was an independent discovery. They were responsible for the equally famous K–T or Kuhn–Tucker conditions of nonlinear programs. These two theorems tell us how to check if a feasible solution to a linear or nonlinear program is optimal.

Al Tucker's greatest contribution to the field is as an expositor. His proofs and illustrative examples are remarkable for their clarity, elegance, and simplicity. Given a choice, he prefers a constructive proof over an existence proof because not only are constructive proofs often shorter and easier to understand, but most importantly they are algorithms. This book, which he has jointly authored with Evar Nering, is Tucker at his finest!

George B. Dantzig
Stanford University
July 10, 1992

Preface

George B. Dantzig invented the simplex algorithm in 1947. From that fertile seed the field of mathematical programming grew and flourished. Dantzig was motivated to create the simplex algorithm because he knew that it could be used for problems that he had been working on during World War II. These problems were important but it was hard to work them by hand. In military parlance, the planning activities he was engaged in were called "programming." It is from this usage that linear programming derives its name, not from computer programming.

He knew that computers were important in our ability to use the simplex algorithm, but it is difficult to believe that he or anyone could have predicted that computer programs to solve linear programming problems would become the major application in scientific computing. Today, linear programming is unique in that among mathematical subjects that could be deemed non-elementary, it is the most widely used.

It is gratifying that linear programming beyond being useful is also a very beautiful mathematical subject. In addition to the usefulness of the facts and methods developed for linear programming, we believe that the structure of the subject is rewarding on its own merits. Beauty in mathematics, as in art, is in the eye of the beholder. It is not possible for us to insist that you should see beauty in the same way that we do. But we can describe the sorts of things that we see as beautiful in any mathematical subject.

We are pleased when the proofs and arguments to support the conclusions are graceful and appropriate, when they are no more sophisticated than the subject warrants, when they contribute to insight, and when they place powerful tools in our hands. A mathematical exposition that is a bundle of tricks might amaze as a tour de force, but we are more satisfied when ideas are used in a consistent and intuitive way that suggests usefulness in other contexts. Part I develops powerful tools within the motivation of linear programming. The whole of Part II moves these ideas to a variety of related problems.

The principal idea is that of duality. This is an ancient concept in mathematics. Accidents of changes in mathematical curricula have resulted in a virtual absence of this idea from elementary mathematics. While linear programming may be the most advanced widely useful subject in mathematics, it is also among the most elementary in which duality can be fully developed.

There is more than enough material here for a one-semester course in linear programming. At a very minimum, one should study Chapters 2–5. It is in these chapters that the ideas and tools mentioned above are developed. One could skip Sections 2.8, 3.6, 3.8, and 5.5 without undermining support for any later chapter. Chapter 1 discusses three typical linear programming problems and sets the stage for many of the exercises in the following four chapters.

Chapter 6 discusses some problems that arise when one implements any algorithm in a computer program. We also discuss some recent techniques for solving linear programming problems. Since these new methods are sophisticated and outside the scope of this book, our discussion is mainly descriptive.

Part II discusses several problems that are closely related to linear programming. The topics in Chapters 8–11 are network problems. They could be developed independently from one another. However, to achieve this independence some ideas would have to be repeated in each chapter. Thus, we have organized Sections 8.1, 8.2, 8.3, and 8.5 to be prerequisites for Chapters 9–11.

Each network problem has its own algorithm. Each could be solved by using the simplex algorithm, but the network structure allows the use of algorithms in which the arithmetic steps do not require multiplication or division. The motivation for including them is to show the power and appropriateness of the ideas developed in Part I. This is particularly well demonstrated in Chapters 8 and 9.

Chapter 12 can be read independently of Chapters 7–11. There is a duality theory for nonlinear programs, but it does not have the same degree of symmetry that the linear problems in this book display. The special nonlinear programs in Chapter 12 do have dual problems that allow us to establish the Karush–Kuhn–Tucker conditions as sufficient conditions.

The answer set is complete. The answers also include much detail about reasoning or methodology involved in the exercises. Don't look at the answer set as a last resort. Look at the answers as being similar to the documentation or help support for a computer program. We put tools in your hands in the text—the answers are intended to help you use those tools.

We think that linear algebra should be a prerequisite for a course based on this book. Actually, we develop all the linear algebra we need within the text. However, it is linear algebra in a form suited to our needs and it is not expository material in linear algebra.

We prefer giving definitions in context rather than in formal, numbered definitions. Such defined terms are italicized and indexed.

Years ago, in formal mathematics books, "Q.E.D." designated the end of a proof. This Latin abbreviation means "what was to be proved." The use of this abbreviation has fallen from favor. In any case, the symbol we use, □, means something slightly different. It means that we are not going to offer any further argument to support the theorem. In some cases where the argument precedes the statement of the theorem or where we expect the reader to complete the proof, this symbol might even follow the statement immediately.

There are many people in addition to George Dantzig to whom we owe words of gratitude. Harold W. Kuhn of Princeton University contributed the most. He taught a course in linear programming almost every year during the time we were writing this book. Harold is an excellent teacher and he chose to present the material in his own way and to change his presentations with time, but preliminary copies of this text were made available to his students in library reserve. He also shared with us the extensive collection of classroom handouts that he gave his students. So many ideas moved in both directions that it is impossible to identify the individual ideas for which we owe him credit.

Torrence Parsons was an assistant to one of us (Tucker) when this material was going through many changes. He also used an early version of this text in his classes at Pennsylvania State University and gave us advice based on this experience. Stephen Maurer of Swarthmore University was an assistant teaching this course at a later date and also contributed to its development. Robert Singleton of Wesleyan University read an early version of this book and gave us many valuable suggestions, both in form and substance. Cynthia Paul, a student at Princeton University, read the manuscript when it was near its final form. She went through all the mathematics in the text and caught errors when we thought we had eliminated them. Despite our care, she saved us much embarrassment. She also made many suggestions for changes in wording. She is such a good student that anything that appeared less than clear to her would undoubtedly be obscure to most students.

Finally, we owe many thanks to Jenifer Swetland and Elizabeth Tustian at Academic Press who were helpful and encouraging and kept the preparation of the manuscript on track.

The final version for publication was prepared with TeX and LaTeX. All the graphical material, including the lines and overlays on the tableaux, was done in PostScript.

Tempe, Arizona
Princeton, New Jersey
June 1992

Part I

Linear Programs

Introduction

One of the oldest questions in mathematics is to find, or characterize, a "best" solution to a problem. For example, a straight line is characterized as the shortest curve joining two points. Of all rectangles with a given perimeter, the square encloses the largest area. Problems more in keeping with the problems discussed in this book would be to plan the production schedule of an industrial plant to use its available resources for maximum profit, or to schedule a truck line to meet its commitments at minimum cost.

Problems of this type are called *optimization problems* . To cast an optimization problem in mathematical form it is necessary to identify a variable whose value we wish to make as large as possible or as small as possible. This variable is called the *objective variable.* It is also necessary to state how this variable is limited in the values it can take. The range of values of the objective variable is usually constrained implicitly by placing limitations on other variables on which the objective variable depends. These restrictions, however they may be described, are called the *constraints* of the problem. Thus, an optimization problem is to maximize (or minimize) an objective variable subject to specified constraints.

The class of problems considered in this text fits the above description. It is relatively easy to establish conditions under which the problems have solutions and to show that the solutions have certain characteristics. However, finding the solution is another matter. The principal source of difficulty is sheer complexity. In managing a manufacturing plant or operating a truck line, the variables of the problem can number in the thousands, and it may be next to impossible to identify all of them!

The best method to use to solve a problem of *constrained optimization* depends on the way the constraints are given and the way the objective variable depends on the other variables. However, a purely mathematical approach to such problems often provides a solution in principle only. That is, if a particular problem is somewhat complicated it may be quite impractical to obtain numerical answers.

The kind of optimization problem considered in Part I is called a *linear programming problem* or, more briefly, a *linear program*. The programming problems we consider are linear because the constraints are expressed as linear equations and linear inequalities, and the objective variable depends linearly on the variables whose values we must choose.

Many problems in linear programming involve an enormous number of variables and equations. Our ability to solve such large scale problems effectively dates from the invention of the simplex algorithm by George B. Dantzig in 1947. The intervening years have seen some slight modifications in the details of the simplex algorithm, some adaptations to problems with special structures, and a few proposed alternate algorithms. In recent years, two alternate algorithms have received much publicity. One, published in 1979 by the Russian mathematician L. G. Khachiyan, is now known as the ellipsoidal algorithm. It is theoretically "better" than the simplex algorithm, but it has not displaced the simplex algorithm. It does not seem to offer advantages for the types of problems that arise in practical applications. The second, published in 1984 by N. Karmarkar, is reported to be faster than the simplex algorithm. It has been implemented in practical, commercially available computer programs. Initial marketing has targeted large, expensive applications, and at the time this book is being written it is too early to tell what its eventual role will be. The performance reports have also spurred the development of faster implementations of the simplex algorithm and additional research into methods derived from the Karmarkar algorithm. We will discuss these two algorithms further in Chapter 6.

Neither of these new algorithms is amenable to hand calculation. Neither sheds light on the structure of linear programming problems as clearly as does the simplex algorithm. We use the simplex algorithm as much to develop the theory of linear programming as we do to solve problems. In particular, it allows us to give simple constructive proofs for the fundamental theorems of linear programming. Backed by years of practical experience the simplex algorithm remains the pre-eminent tool in the field of linear programming.

The success of the simplex algorithm has lead to a mathematical subject area, *mathematical programming*. This term includes a wide variety of optimization problems that involve algorithmic methods. Over the years and continuing today there is research into the theory underlying programming problems, linear and nonlinear, research into other optimization problems that will yield to algorithms, research into practical applications of mathematical programming, and research into the peculiar problem of using computers to solve mathematical programs.

In this text we are concerned primarily with the theory of linear and nonlinear programming and a number of closely related problems, and with algorithms appropriate to these problems. In mathematics an *algorithm* is a procedure with clearly defined steps, usually in which some steps are repeated cyclically. After each cycle of steps the results are examined to make a decision. The decision could be to make an adjustment and go through the cycle again or to go on to something else, or to stop.

A properly constructed algorithm should terminate after a finite number steps with a solution or a determination that a solution does not exist. Hopefully, this finite number of steps is a small finite number of steps.

Students have encountered many algorithms in the course of their studies, but often they are not identified by such a formal title. The process that we learn in grade school that we call long division is an example of an algorithm. Algorithms have been known and used in mathematics for thousands of years, but the ascendancy of the computer in recent years has given algorithms a new importance. A human working through an algorithm is likely to regard it as rather boring, but the repetition of the cycle of steps makes an algorithm very attractive for use with a computer. The specific instructions that the computer must be given are reduced to the steps of one occurrence of each cycle in an algorithm even though the computer may perform this cycle of steps many times. The use of algorithms with computers has also meant that the algorithms have to be more carefully crafted since the judgment of a human mind cannot intervene.

One should not regard an algorithm solely as a way to compute answers. It is a powerful method of proof, and we make use of algorithms in this way throughout the book. Many proofs in mathematics are indirect and nonconstructive. They may show that a certain result is true or that a solution exists, but provide no method for actually constructing it or computing it. In some cases no constructive proofs are known, but where a constructive proof is available it should be the preferred proof. A large number of constructive proofs based on algorithms are now known in mathematical programming and it is an attractive feature of the subject.

Linear programming is now a highly developed mathematical subject. It has an attractive symmetry made possible through the concept of duality. Many students seem to regard duality as a mysterious extension of linear programming. We think it is an integral part of linear programming, and a natural and easy concept if introduced at the beginning and used to develop the subject. It does much to reveal the structure of the problems and the algorithms used to solve them.

Duality is the unifying concept in this book.

Chapter 1

Sample Linear Programs

1.1 A Production Problem

The "programs" we consider in this book are plans for economic optimization, not programs for computing machines. For example, in a manufacturing plant decisions have to be made about how much of its various products to manufacture from the resources available. An oil refinery must decide how to allocate the raw materials it has on hand to refine a variety of products with desired characteristics. A livestock feeder must decide how much of the various types of feed to purchase to meet the nutritional requirements of his stock. A trucking company must decide how to schedule its trucks to meet its commitments.

The decisions that must be made are usually designed to maximize the profits or minimize costs. The manufacturing plant or oil refinery seeks to maximize profits. The livestock feeder or trucking company seeks to minimize costs.

What makes these problems interesting is that the stakes are high and the problems are usually complex, primarily because of the large numbers of variables. We shall begin by analyzing some simple problems involving just a few variables. These problems are simple enough that one could probably combine intuition and guesswork to obtain best solutions. However, we are looking for a method, a systematic procedure, that will work for problems of any complexity. Thus, we shall treat these little problems as though they were big problems. A realistic big problem can involve several thousand variables.

The examples we give are intended to sound realistic, but the data are chosen to make the calculations convenient and are only roughly reasonable.

As an example, suppose we are managing a plant that produces plastic products. Though the operation involves many products, we shall consider just three: (1) a vinyl–asphalt floor covering, the output of which is measured in boxed lots, each covering a certain area, (2) a pure vinyl counter top, measured in linear yards, and (3) a vinyl–asphalt wall tile, measured in "squares," each covering 100 square feet.

Of the many resources needed to produce these plastic products, we shall consider only four: vinyl, asphalt, labor, and time on a trimming machine. Suppose a box of floor covering requires 30 pounds of vinyl, a yard of counter top requires 10 pounds of vinyl, and a square of wall tile requires 50 pounds of vinyl. Also, suppose we have a supplier that can provide up to 1500 pounds of vinyl per day for this particular part of the enterprise. If this were the only restriction we could produce 50 boxes of floor covering, or 150 yards of counter top, or 30 squares of wall tile per day, or some combination of these products.

Rather than write out these and other similar conditions in verbal form, as in the preceding paragraph, it is more convenient to display this information in a *data table*.

(1.1)

	floor cover	counter top	wall tile	available	
vinyl	30	10	50	1500	pounds
asphalt	5		3	200	pounds
labor	0.2	0.1	0.5	12	man-hours
machine	0.1	0.2	0.3	9	machine-hours
profit	10.00	5.00	5.50		dollars
	per box	per yard	per square	per day	

This data table should be interpreted in the following way. The number 0.1 at the intersection of the labor row and the counter-top column gives the number of man-hours required to produce one yard of counter top. The 5 in the basement row is the profit in dollars per yard of counter top produced. The 12 in the last column is the amount of labor available per day, measured in man-hours. Analogous interpretations apply to the other numbers in the data table. Where an entry is absent it is taken to be zero. That is, "blanks are zeros."

We must decide how much of each product to produce. Let x denote the number of boxes of floor covering, y the number of yards of counter top, and z the number of squares of wall tile that will be produced each day. These variables are the *decision variables* of the problem.

The values of the decision variables cannot be chosen freely. In the first place, we cannot produce a negative quantity of anything. Thus, the values of these variables must be nonnegative. Also, the available supplies, manpower, and machine time limit the amounts that can be produced. A set of values for the decision variables that meets these conditions is called a *feasible solution.*

A convenient way to account for the effect of a limited supply is to introduce a variable for the amount of each supply that will be unused for a particular solution, and require that the value of that variable be nonnegative. These variables are called *slack variables.*

Let p denote the slack variable for vinyl. Then from the data table 1.1 we can see that the following equation must hold.

$$30x + 10y + 50z + p = 1500 \tag{1.2}$$

In a similar way, we can let q, r, and s denote the slack variables for the other three supplies and obtain the equations

$$\begin{aligned} 5x + 3z + q &= 200 \\ 0.2x + 0.1y + 0.5z + r &= 12 \\ 0.1x + 0.2y + 0.3z + s &= 9 \end{aligned} \tag{1.3}$$

The basement row of the data table 1.1 gives the information needed to compute the profit for a particular solution. The profit is

$$10x + 5y + 5.5z = f \tag{1.4}$$

measured in dollars per day.

There is a convenient way to get these equations directly from the data table 1.1: Erase the identifying information written on the margins of the data table and replace it with algebraic information, as in the following tableau.

(1.5)

x	y	z	-1	
30	10	50	1500	$= -p$
5	0	3	200	$= -q$
0.2	0.1	0.5	12	$= -r$
0.1	0.2	0.3	9	$= -s$
10	5	5.5	0	$= f$

The tableau 1.5 generates equations in the following way. For each row, multiply each entry in the row by the corresponding variable or number on the top margin, add these products, and set the sum equal to the entry on the right margin. For example, the first row of the tableau 1.5 generates the equation

$$30x + 10y + 50z - 1500 = -p \tag{1.6}$$

This equation is equivalent to equation 1.2. In a similar way, the next three rows generate equations equivalent to the equations in 1.3. The basement row generates the profit function 1.4.

Expressed in terms of the tableau alone, the problem we have posed is to

> **find nonnegative values of x, y, z, p, q, r, s which satisfy the equations implied by tableau 1.5 and maximize the value of f.**

In this example, one of the options is to do nothing. That is, we can take $x = 0$, $y = 0$, and $z = 0$. Then we compute $p = 1500$, $q = 200$, $r = 12$, and $s = 9$. Since these numbers are nonnegative, this is a feasible set of values for the variables. Of course, the profit is also zero. This is a starting point and we can consider increasing the production rate for one of the products. Since all the coefficients in $f = 10x + 5y + 5.5z$ are positive, increasing the production rate for any one of the products will increase f. Suppose, for example, that we consider increasing x, the amount of floor cover produced.

The amount that x can be increased is limited by the supplies available. These conditions can be seen by taking $y = 0$ and $z = 0$ on the top margin of tableau 1.5 and writing down the equations that are generated. We obtain

$$\begin{aligned} 30x - 1500 &= -p \\ 5x - 200 &= -q \\ 0.2x - 12 &= -r \\ 0.1x - 9 &= -s \end{aligned} \tag{1.7}$$

The condition that each slack variable be nonnegative requires that $1500 - 30x$, $200 - 5x$, $12 - 0.2x$, and $9 - 0.1x$ be nonnegative. The first condition means, for example, that x can be no more than $1500/30 = 50$. The second condition means that x can be no more than $200/5 = 40$. The maximum increase permitted is obtained by computing the ratios $1500/30 = 50$, $200/5 = 40$, $12/0.2 = 60$, and $9/0.1 = 90$, and taking the minimum ratio.

Thus, we can increase x to the value 40. Suppose we do this. This reduces the slack variable q to zero. Since it is the second equation that limits (controls) the value of x, we solve for x in this equation to obtain

$$0.2q + 0.6z - 40 = -x \tag{1.8}$$

In this equation we have interchanged the roles of x and q. Conceptually, this means that we are now thinking of controlling the value of the variable q instead of controlling the value of x.

We use equation 1.8 to eliminate x in the other equations. We obtain

$$\begin{aligned} -6q + 10y + 32z - 300 &= -p \\ -0.04q + 0.1y + 0.38z - 4 &= -r \\ -0.02q + 0.2y + 0.24z - 5 &= -s \\ -2q + 5y - 0.5z + 400 &= f \end{aligned} \tag{1.9}$$

The tableau corresponding to these new equations is

(1.10)

q	y	z	-1	
-6	10	32	300	$= -p$
0.2	0	0.6	40	$= -x$
-0.04	0.1	0.38	4	$= -r$
-0.02	0.2	0.24	5	$= -s$
-2	5	-0.5	-400	$= f$

This tableau 1.10 has the same form as the tableau 1.5 that we started with. We have cast the old problem in a new tableau in which q, y, and z are the decision variables. "Doing nothing" in this context means taking $q = 0$ (use all the asphalt), $y = 0$, $z = 0$. This gives $x = 40$ and $f = 400$, which is an improvement.

In this new tableau, the entries in the basement row are the coefficients that express the objective variable as a function of the variables q, y, and z. Since the coefficients of q and z are negative, increasing either of them would have the effect of decreasing the profit. Since the coefficient of y is positive, let us consider increasing the value of y.

To find out how much y can be increased, we compute the ratios $300/10 = 30$, $4/0.1 = 40$, and $5/0.2 = 25$. Since the coefficient of y in the second equation is zero, the second equation does not limit the value of y. Thus, we see that we can increase y to 25, which will also reduce the slack variable s to zero. We solve the fourth equation for y to obtain

$$-0.1q + 5s + 1.2z - 25 = -y \tag{1.11}$$

Use this equation to eliminate y from the other equations and obtain

$$
\begin{aligned}
-5q - 50s + 20z - 50 &= -p \\
0.2q + 0.6z - 40 &= -x \\
-0.03q - 0.5s + 0.26z - 1.5 &= -r \\
-1.5q - 25s - 6.5z + 525 &= f
\end{aligned}
\tag{1.12}
$$

The new tableau corresponding to these equations is

(1.13)

q	s	z	-1	
-5	-50	20	50	$= -p$
0.2		0.6	40	$= -x$
-0.03	-0.5	0.26	1.5	$= -r$
-0.1	5	1.2	25	$= -y$
-1.5	-25	-6.5	-525	$= f$

Again, we can regard this tableau 1.13 as a new presentation of the same problem. In this case, doing nothing means taking $q = 0$, $s = 0$, $z = 0$. This time the coefficients of q, s, and z that appear in the basement row are negative. Thus, the profit would decrease if q, s, or z were increased. These are the only changes that can be considered since we cannot operate with negative values for q, s, or z. For these values we get $x = 40$ and $y = 25$. We should produce per day 40 boxes of floor cover, 25 yards of counter top, and no wall tile. Since $q = 0$ and $s = 0$, this production schedule will use all the available asphalt and machine time. Since $p = 50$ and $r = 1.5$, 50 pounds of vinyl and 1.5 man-hours will be unused each day. The maximum profit under these conditions is \$525 per day.

To check, substitute $x = 40$, $y = 25$, $z = 0$ in tableau 1.5 to get $p = 50$, $q = 0$, $r = 1.5$, $s = 0$, and $f = 525$.

A maximum linear program may be presented in any of a variety of different forms. If the information about the problem can be cast in a data table, like table 1.1, then it is easy to obtain a tableau representation by erasing the label information on the margins and writing in the appropriate variables. In particular, that is the way tableau 1.5 was obtained from table 1.1.

Quite often a maximum linear program is presented in the form:

Maximize

$$f = c_1x_1 + \cdots + c_nx_n \tag{1.14}$$

subject to the conditions

$$\begin{aligned} a_{11}x_1 + \cdots + a_{1n}x_n &\leq b_1 \\ a_{21}x_1 + \cdots + a_{2n}x_n &\leq b_2 \\ &\vdots \\ a_{m1}x_1 + \cdots + a_{mn}x_n &\leq b_m \end{aligned} \tag{1.15}$$

and

$$x_1 \geq 0, \ldots, x_n \geq 0 \tag{1.16}$$

These inequalities can be converted to equalities by inserting a nonnegative slack variable on the left side of each inequality. In this way we obtain the equivalent problem:

Maximize

$$f = c_1x_1 + \cdots + c_nx_n \tag{1.17}$$

subject to the conditions

$$\begin{aligned} a_{11}x_1 + \cdots + a_{1n}x_n + y_1 &= b_1 \\ a_{21}x_1 + \cdots + a_{2n}x_n + y_2 &= b_2 \\ &\vdots \\ a_{m1}x_1 + \cdots + a_{mn}x_n + y_m &= b_m \end{aligned} \tag{1.18}$$

and

$$x_1 \geq 0, \ldots, x_n \geq 0, y_1 \geq 0, \ldots, y_m \geq 0 \tag{1.19}$$

When the constraints are presented in the form of 1.18, we can obtain an equivalent form that can be directly converted to a tableau representation. Exchange the slack variables (y_i) and the constants (b_i) to obtain the linear program:

Maximize

$$f = c_1x_1 + \cdots + c_nx_n \tag{1.20}$$

subject to the conditions

$$\begin{aligned} a_{11}x_1 + \cdots + a_{1n}x_n - b_1 &= -y_1 \\ a_{21}x_1 + \cdots + a_{2n}x_n - b_2 &= -y_2 \\ &\vdots \\ a_{m1}x_1 + \cdots + a_{mn}x_n - b_m &= -y_m \end{aligned} \tag{1.21}$$

and

$$x_1 \geq 0, \ldots, x_n \geq 0, y_1 \geq 0, \ldots, y_m \geq 0 \tag{1.22}$$

Finally, this form can be converted to the tableau form

$$
\begin{array}{|ccc|c|l}
\multicolumn{1}{c}{x_1} & \multicolumn{1}{c}{\cdots} & \multicolumn{1}{c}{x_n} & \multicolumn{1}{c}{-1} & \\
\hline
a_{11} & \cdots & a_{1n} & b_1 & = -y_1 \\
a_{21} & \cdots & a_{2n} & b_2 & = -y_2 \\
\vdots & & \vdots & \vdots & \vdots \\
a_{m1} & \cdots & a_{mn} & b_m & = -y_m \\
\hline
c_1 & \cdots & c_n & 0 & = f \\
\hline
\end{array}
\tag{1.23}
$$

subject to

$$x_1 \geq 0, \ldots, x_n \geq 0, y_1 \geq 0, \ldots, y_m \geq 0 \tag{1.24}$$

In mathematics, the words "standard," "normal," and "canonical" are often used to denote something that is standard in a defined sense. In linear programming, any one of the three equivalent representations in equality form is called a *canonical linear program*. To qualify as a canonical program, if there are slack variables,

1. **the representation must be in equality form,**
2. **there must be one and only one slack variable in each equation, and**
3. **all the variables (including the slack variables) must be nonnegative.**

If there are no slack variables,

1. **the constraints must be in inequality form, and**
2. **all the variables must be nonnegative.**

In the tableau representation, the fact that the objective variable, f, is to be maximized is implied by the form of the representation and it does not have to be stated explicitly. Later we will deal with linear programs in which all the variables are not required to be nonnegative. In such cases, the conditions that must be satisfied will have to be stated explicitly. For any of these representations, we will either state the nonnegativity conditions explicitly, or say that the linear program is canonical.

1.2 A Diet Problem

A typical minimization problem is a "minimum cost" problem of meeting given requirements at least cost. Suppose, for example, that we are to supply some feed to livestock at minimum cost. The livestock require weekly at least 450 pounds of protein, 400 pounds of carbohydrates, and 1050 pounds of roughage. A bale of hay supplies 10 pounds of protein, 10 pounds of carbohydrates, and 60 pounds of roughage, and costs \$3.80. These and other facts about the requirements and costs are given in the following *data table*.

(1.25)

	protein	carbohydrate	roughage	cost	
hay	10	10	60	3.80	per bale
oats	15	10	25	5.00	per sack
pellets	10	5	55	3.50	per sack
sweet feed	25	20	35	8.00	per sack
requirements	450	400	1050		per week
	pounds per week	pounds per week	pounds per week	dollars per week	

The entries in the data table 1.25 are interpreted in a way similar to the interpretation of the data table 1.1. For example, the 25 in the fourth row tells us that the sweet feed supplies 25 pounds of protein per sack. The numbers in the basement row give the minimum requirements for each nutriment. The numbers in the last column give the costs, in dollars, per unit of feed purchased.

We can proceed from the data table to a *tableau* by erasing labels outside the box and writing variables and numbers on the margins, as follows.

(1.26)

h	10	10	60	3.80
k	15	10	25	5.00
m	10	5	55	3.50
n	25	20	35	8.00
-1	450	400	1050	0
	$= u$	$= v$	$= w$	$= g$

Here, each column generates an equation in the following way. Multiply each entry in the column by the corresponding variable or number on the left margin, accumulate the sum of these products, and set the sum equal to the variable on the lower margin. For example, the first column generates the equation

$$10h + 15k + 10m + 25n - 450 = u \tag{1.27}$$

The variable u represents the amount by which the protein requirement is oversupplied. The minimum requirement of 450 pounds is enforced by requiring that u be nonnegative. In this example h, k, m, and n are the *decision variables*. The variables u, v, and w are called *slack variables*.

Similar equations are generated by the other columns. The last column generates the equation

$$3.8h + 5k + 3.5m + 8n = g \tag{1.28}$$

for the total cost g of buying the feeds. The objective of the problem is to meet the requirements, or more, at least cost.

Expressed in terms of the tableau 1.26 alone, the problem we have posed is to

> **find nonnegative values of h, k, m, n, u, v, and w which satisfy the equations implied by tableau 1.26 and minimize the value of g.**

The reasons for using the rows of a tableau to represent the conditions for a maximization problem and using the columns of a tableau to represent the conditions for a minimization problem are not expected to be obvious yet. There are good reasons for this difference in formulation and it is a major goal of the next few chapters to make these reasons clear and natural.

Some differences and some similarities between the maximization problem of the previous section and this minimization problem should be noticeable immediately. A slack variable in the maximization problem represents an underuse of a resource, and a slack variable in the minimization problem represents an oversupply of a requirement. Each should be nonnegative. The slack variables for the maximization problem appear on the right margin with coefficient -1, and the slack variables for the minimization problem appear on the lower margin with coefficient $+1$. Also, doing nothing (setting $x = y = z = 0$) in tableau 1.5 satisfies the conditions imposed on the problem; it is just not optimal. Doing nothing (setting $h = k = m = n = 0$) in tableau 1.26 does not meet the conditions imposed. If it did meet the nutritional requirements it would be optimal since the coefficients in the cost function 1.28 are positive and an increase in the value of any of the variables would increase the cost.

Suppose we decide to meet at least one of the requirements, say the protein requirement. To meet this requirement solely with hay would require $450/10 = 45$ bales which would cost $450(3.80/10) = 171$ dollars. Similarly, to meet this requirement solely with oats would cost $450(5/15) = 150$ dollars. With pellets it would cost $450(3.50/10) = 157.50$ dollars, and with sweet feed it would cost $450(8/25) = 144$ dollars. In each of these expressions the quantity in parentheses is the cost per pound for the protein. It is the cost per pound, not the total protein requirement, that determines which feed will be the least expensive.

We see that the protein requirement can be met most inexpensively with sweet feed (a mixture of various grains and molasses). As with the maximization problem we solve the equation

$$10h + 15k + 10m + 25n - 450 = u \tag{1.29}$$

for n and obtain

$$-0.4h - 0.6k - 0.4m + 0.04u + 18 = n \tag{1.30}$$

Conceptually, we are considering controlling the value of u, the amount of over-satisfaction of the protein requirement, rather than controlling n.

We eliminate n from the other equations to obtain

$$\begin{aligned} 2h - 2k - 3m + 0.8u - 40 &= v \\ 46h + 4k + 41m + 1.4u - 420 &= w \\ 0.6h + 0.2k + 0.3m + 0.32u + 144 &= g \end{aligned} \tag{1.31}$$

The tableau corresponding to these new equations is

(1.32)

h	-0.4	2	46	0.60
k	-0.6	-2	4	0.20
m	-0.4	-3	41	0.30
u	0.04	0.8	1.4	0.32
-1	-18	40	420	-144
	$= n$	$= v$	$= w$	$= g$

We can regard the tableau 1.32 as representing a new formulation of the problem we started with. Setting $h = 0$, $k = 0$, $m = 0$, $u = 0$ will still not meet the requirements because then $v = -40$ and $w = -420$. That is, we still have not met the carbohydrate and roughage requirements.

The positive numbers in the last column tell us that increasing any of the variables h, k, m, or u will increase the cost. This means doing nothing would be optimal if it met requirements. Suppose we consider trying to meet the carbohydrate requirement. First, notice that the negative entries in the second column mean that increasing k or m would actually move us farther from meeting the carbohydrate requirement. The costs of the other choices are $40(0.60/2) = 12$ dollars for h, and $40(0.32/0.8) = 16$ dollars for u. Thus, we choose to increase h.

With this choice we solve the equation

$$2h - 2k - 3m + 0.8u - 40 = v \tag{1.33}$$

for h to obtain

$$0.5v + k + 1.5m - 0.4u + 20 = h \tag{1.34}$$

Conceptually, we are now controlling the values of the variables v, k, m, and u.

We eliminate h from the other equations to obtain

$$\begin{aligned} -0.2v - k - m + 0.2u + 10 &= n \\ 23v + 50k + 110m - 17u + 500 &= w \\ 0.3v + 0.8k + 1.2m + 0.08u + 156 &= g \end{aligned} \tag{1.35}$$

The new tableau is

(1.36)

v	-0.2	0.5	23	0.30
k	-1	1	50	0.80
m	-1	1.5	110	1.20
u	0.2	-0.4	-17	0.08
-1	-10	-20	-500	-156
	$= n$	$= h$	$= w$	$= g$

For this tableau, taking $v = 0$, $k = 0$, $m = 0$, $u = 0$, we get $n = 10$, $h = 20$, and $w = 500$. That is, all requirements will be met. Furthermore, the positive entries in the last column tell us that increasing any of the values of v, k, m, or u would increase the cost. Since these are the only changes that are possible we must now have an optimal solution. The minimum cost is \$156. We should buy $h = 20$ bales of hay and $n = 10$ sacks of sweet feed. This will oversupply the roughage by $w = 500$ pounds. Substitute these values in tableau 1.26 to check against errors.

A minimum linear program may be presented in any of a variety of different forms. If the information about the problem can be cast in a data table, like table 1.25, then it is easy to obtain a tableau representation by erasing the label information on the margins and writing in the appropriate variables. In particular, that is the way tableau 1.26 was obtained from table 1.25.

Quite often a minimum linear program is presented in the form:

Minimize

$$g = b_1 v_1 + \cdots + b_m v_m \tag{1.37}$$

subject to the conditions

$$
\begin{aligned}
v_1a_{11} + \cdots + v_ma_{m1} &\geq c_1 \\
v_1a_{12} + \cdots + v_ma_{m2} &\geq c_2 \\
&\vdots \\
v_1a_{1n} + \cdots + v_ma_{mn} &\geq c_n
\end{aligned}
\tag{1.38}
$$

and

$$v_1 \geq 0, \ldots, v_m \geq 0 \tag{1.39}$$

These inequalities can be converted to equalities by inserting a nonnegative slack variable on the right side of each inequality. In this way we obtain the equivalent problem:

Minimize

$$g = b_1v_1 + \cdots + b_mv_m \tag{1.40}$$

subject to the conditions

$$
\begin{aligned}
v_1a_{11} + \cdots + v_ma_{m1} &= c_1 + u_1 \\
v_1a_{12} + \cdots + v_ma_{m2} &= c_2 + u_2 \\
&\vdots \\
v_1a_{1n} + \cdots + v_ma_{mn} &= c_n + u_n
\end{aligned}
\tag{1.41}
$$

and

$$v_1 \geq 0, \ldots, v_m \geq 0, u_1 \geq 0, \ldots, u_n \geq 0 \tag{1.42}$$

When the constraints are presented in the form of 1.41, we can obtain an equivalent form which can be directly converted to a tableau representation. Move the term c_i from the right side to the left side to obtain the linear program:

Minimize

$$g = b_1v_1 + \cdots + b_mv_m \tag{1.43}$$

subject to the conditions

$$
\begin{aligned}
v_1a_{11} + \cdots + v_ma_{m1} - c_1 &= u_1 \\
v_1a_{12} + \cdots + v_ma_{m2} - c_2 &= u_2 \\
&\vdots \\
v_1a_{1n} + \cdots + v_ma_{mn} - c_n &= u_n
\end{aligned}
\tag{1.44}
$$

and

$$v_1 \geq 0, \ldots, v_m \geq 0, u_1 \geq 0, \ldots, u_n \geq 0 \tag{1.45}$$

Finally, this form can be converted to the tableau form

$$\begin{array}{c|ccc|c} v_1 & a_{11} & \cdots & a_{1n} & b_1 \\ v_2 & a_{21} & \cdots & a_{2n} & b_2 \\ \vdots & \vdots & & \vdots & \vdots \\ v_m & a_{m1} & \cdots & a_{mn} & b_m \\ \hline -1 & c_1 & \cdots & c_n & 0 \\ \hline & = u_1 & \cdots & = u_n & = g \end{array} \tag{1.46}$$

subject to

$$v_1 \geq 0, \ldots, v_m \geq 0, u_1 \geq 0, \ldots, u_n \geq 0 \tag{1.47}$$

To qualify as a canonical program, if there are slack variables,

1. **the constraints must be in equality form,**
2. **there must be one and only one slack variable in each equation, and**
3. **all the variables (including the slack variables) must be nonnegative.**

If there are no slack variables,

1. **the constraints must be in inequality form, and**
2. **all the variables must be nonnegative.**

In the tableau representation, the fact that the objective variable, g, is to be minimized is implied by the form of the representation and it does not have to be stated explicitly. Later we will deal with linear programs in which not all the variables are required to be nonnegative. Thus, that condition must be stated. For any of these representations, we will either state the nonnegativity conditions explicitly, or say that the linear program is canonical.

1.3 A Transportation Problem

Suppose a distributor of a commodity is faced with the problem of supplying several market locations from several origins where the commodity is available. For each route between a source of supply and a point of demand there is a rate for shipping a unit of the commodity. Assuming that the total supply is sufficient to meet the total demand, the problem is to devise a shipping schedule that will meet the demand at each market with the supply available at each origin and minimize the total shipping cost.

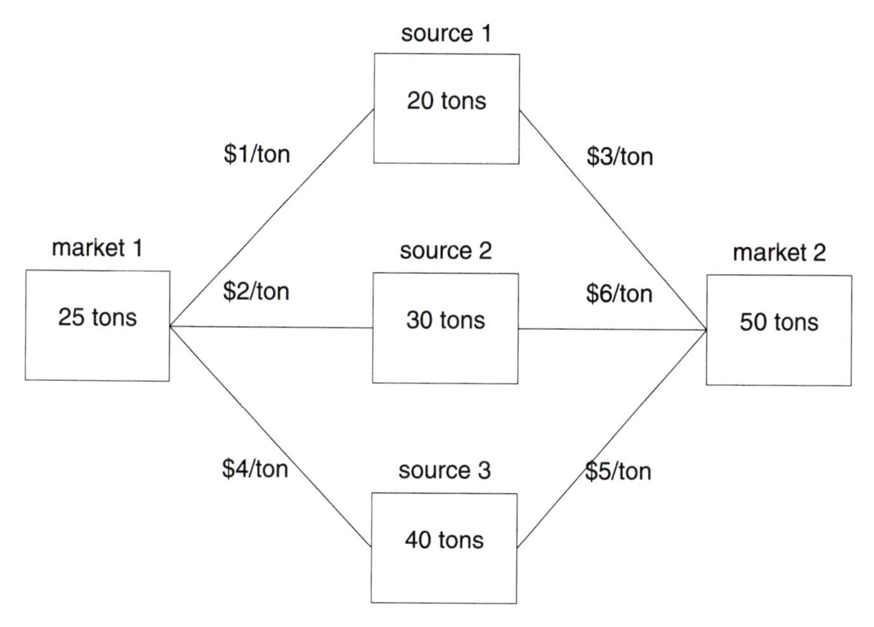

Figure 1.1: Initial transportation network.

As a specific example, suppose we have the situation illustrated in Figure 1.1. The origins are in the middle. In each box the quantity available is shown. The boxes on the left and right are the destinations, and the minimum requirement for each destination is shown in these boxes. The shipping rate for each route is shown adjacent to that route.

Let x_{ij} denote the amount to be shipped from origin i to destination j, and let y_i be the amount left unshipped at origin i. Then for the first origin we have an equation of the form

$$20 - x_{11} - x_{12} = y_1 \tag{1.48}$$

and we have a similar equation for each of the other origins. Let z_j be the amount by which destination j is oversupplied. At the first destination we have the equation

$$x_{11} + x_{21} + x_{31} - 25 = z_1 \tag{1.49}$$

and there is a similar equation for the other destinations. The total shipping cost is

$$g = x_{11} + 3x_{12} + 2x_{21} + 6x_{22} + 4x_{31} + 5x_{32} \tag{1.50}$$

The equations of the form 1.48, 1.49, and 1.50 can be represented in the tableau 1.51.

(1.51)

x_{11}	-1			1		1
x_{12}	-1				1	3
x_{21}		-1		1		2
x_{22}		-1			1	6
x_{31}			-1	1		4
x_{32}			-1		1	5
-1	-20	-30	-40	25	50	0
	$=y_1$	$=y_2$	$=y_3$	$=z_1$	$=z_2$	$=g$

Expressed in terms of the tableau alone, our problem is to

> **find nonnegative values of x_{11}, x_{12}, ... , x_{32}, y_1, y_2, y_3, z_1, z_2 which satisfy the equations implied by tableau 1.51 and minimize the value of g.**

If nothing is shipped, the variables $y_1 = 20$, $y_2 = 30$, $y_3 = 40$ are nonnegative, but $z_1 = -25$ and $z_2 = -50$ are not. We do not ship more from an origin than is available there, but we do not satisfy the demand at either destination.

Suppose we decide to supply one of the destinations by the cheapest route. The cheapest route into destination 2 is from origin 1. Supplying destination 2 means we increase the variable z_2, the amount by which this destination is oversupplied, to zero. We do this by solving the equation that involves z_2 for the variable x_{12}, the amount shipped from origin 1. We obtain

$$z_2 - x_{22} - x_{32} + 50 = x_{12} \qquad (1.52)$$

Only one other equation and the objective function contain x_{12}. When x_{12} is replaced in these expressions we obtain the equations represented by the tableau 1.53.

(1.53)

x_{11}	-1			1		1
z_2	-1				1	3
x_{21}		-1		1		2
x_{22}	1	-1			-1	3
x_{31}			-1	1		4
x_{32}	1		-1		-1	2
-1	30	-30	-40	25	-50	-150
	$=y_1$	$=y_2$	$=y_3$	$=z_1$	$=x_{12}$	$=g$

Taking the variables on the left margin to be zero amounts to the shipping schedule represented in Figure 1.2.

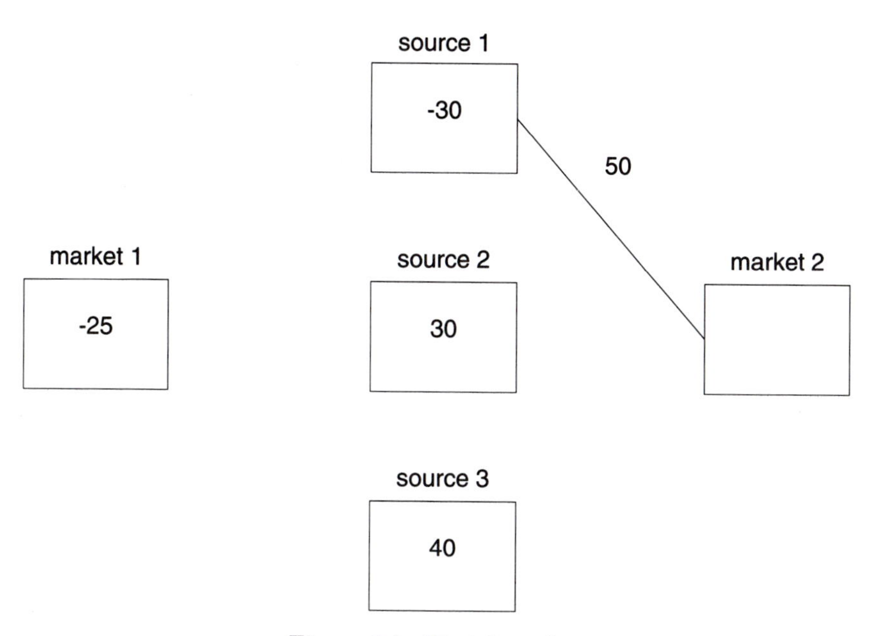

Figure 1.2: First iteration.

The -30 in the box for origin 1 means that the supply at that origin is overdrawn by 30 tons. The -25 in the box for destination 1 means that the demand there is under supplied by 25 tons. The blank in the box for destination 2 means that the demand there is met exactly.

The deficit at origin 1 will have to made up from other origins. The 3 in the fourth row of the cost column in 1.53 is the increase in the rate if a shipment is made from origin 2 instead of from origin 1. Similarly, the 2 in the sixth row of the cost column in 1.53 is the increase in the rate if a shipment is made from origin 3 instead of from origin 1. We choose to make up the deficit from origin 3 since that involves the least increase in the rate. To do this we solve the equation that contains y_1 for the variable x_{32}. We obtain

$$x_{11} + z_2 - x_{22} + y_1 + 30 = x_{32} \tag{1.54}$$

When we use this equation for x_{32} to eliminate x_{32} from the other equations, we obtain the equations represented by tableau 1.55.

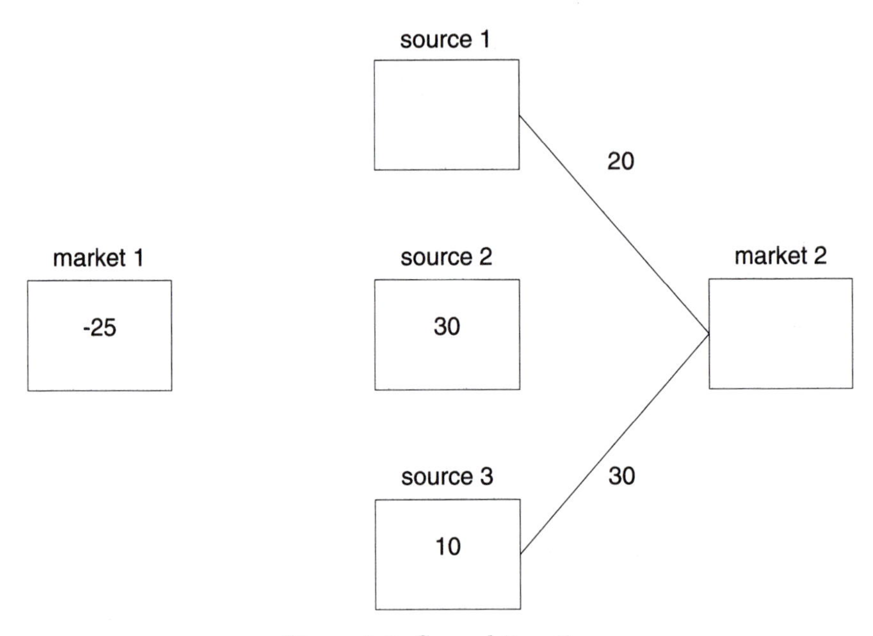

Figure 1.3: Second iteration.

(1.55)

x_{11}	1		−1	1	−1	3
z_2	1		−1			5
x_{21}		−1		1		2
x_{22}	−1	−1	1			1
x_{31}			−1	1		4
y_1	1		−1		−1	2
−1	−30	−30	−10	25	−20	−210
	$= x_{32}$	$= y_2$	$= y_3$	$= z_1$	$= x_{12}$	$= g$

Taking the variables on the left margin of 1.55 to be zero amounts to the shipping schedule represented in Figure 1.3.

Since the destination 1 is still under supplied, we want to ship material from an origin such that the resulting increase in cost will be minimal. The increases in rates are given in the cost column; an increase of \$3 per ton to draw from origin 1, an increase of \$2 per ton to draw from origin 2, and an increase of \$4 per ton to draw from origin 3. The cheapest alternative is to ship from origin 2. We do this by solving the equation involving z_1 for x_{21}, and substituting for x_{21} in the other equations. We obtain tableau 1.56.

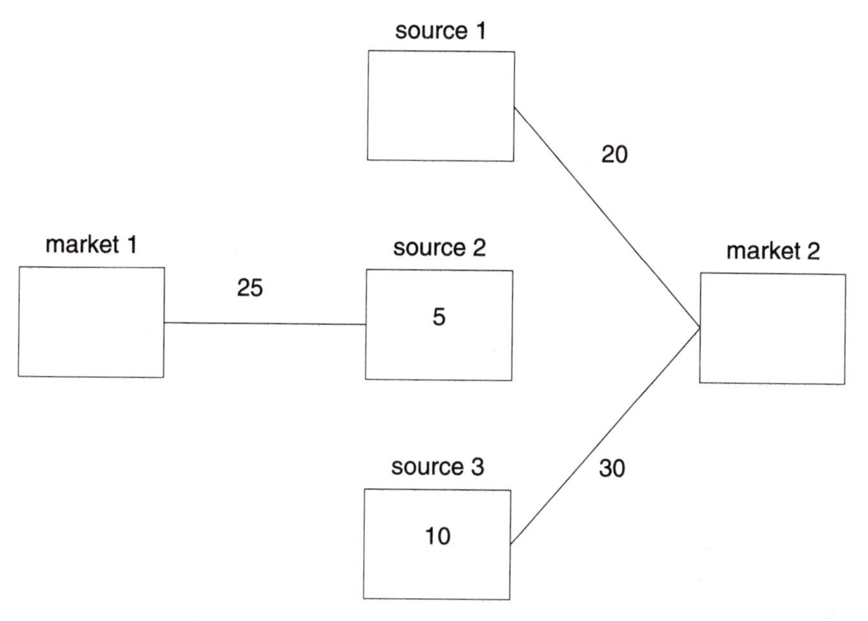

Figure 1.4: Transportation problem solution.

x_{11}	1	1	−1	−1	−1	1
z_2	1		−1			5
z_1		−1		1		2
x_{22}	−1	−1	1			1
x_{31}		1	−1	−1		2
y_1	1		−1		−1	2
−1	−30	−5	−10	−25	−20	−260
	$= x_{32}$	$= y_2$	$= y_3$	$= x_{21}$	$= x_{12}$	$= g$

(1.56)

The negative entries in the basement row mean that when zero values are given to the variables on the left margin, the variables on the bottom margin are positive. In this case all variables are nonnegative and we have a feasible solution.

The entries in the last column are the relative costs that would result from increasing any variable on the left margin. Since these entries are all positive, increasing any variable on the left margin would increase the cost. Thus, the schedule obtained by taking all variables on the left margin zero is optimal. The demands at the destinations are met exactly. We see that

5 tons remain unshipped from origin 2 and 10 tons remain unshipped from origin 3. The total cost is \$260. This shipping schedule is illustrated in Figure 1.4.

1.4 An Informal Algorithm

Despite the differences among the three problems we have considered in the previous sections, there is a common thread in the paths we have taken to their solutions. In each of the tableaux we have encountered, the entries in the right-hand column (other than the entry in the lower right corner) were positive. For the maximization problem, where the rows generated the equations, this meant that setting the variables on the top margin equal to zero resulted in positive values for the variables on the right margin. Thus, for this particular choice of values for the variables on the top margin the solution was feasible. For the minimization problems, where the columns generated the equations, the positive values of the entries in the right column meant that increasing the value of any variable on the left margin would cause an increase in the value of the objective variable, an increase in the cost.

In every case where an optimal solution was not yet achieved we selected a column with a positive entry in the basement row. For the maximization problem this identified a variable that could be increased to increase the value of the objective variable. For the minimization problems this identified a requirement or a demand that was not satisfied and, therefore, a variable that must be increased.

We achieved an optimal solution when the entries in the basement row (other than the entry in the lower right corner) were negative. For the maximization problem this meant that any increase in the value of any variable on the top margin would result in an decrease in the value of the objective variable. For the minimization problems this meant that the solution obtained by setting the variables on the left margin equal to zero was feasible.

In each of the problems we considered, the problem was cast in such a form that the right column of the first tableau we encountered was positive. Although the systems of equations in every case had many solutions, we directed our attention towards solutions obtained by setting the variables on the top and left margins equal to zero. For the maximization problem such a solution was feasible (because the entries in the right-hand column were positive) but not necessarily optimal (unless there was no positive entry in the basement row). For the minimization problems such solutions were not necessarily feasible (unless there was no positive entry in the basement row) but they would be optimal if feasible (because the entries in the right-hand column were positive).

In each case we obtained a sequence of tableaux in which the right-hand column was positive and we worked towards a tableau with a negative basement row. For the maximization problem, this preserved the feasibility of the special solutions (obtained by setting the variables on the top margin equal to zero) and worked for the condition of optimality for the special solutions (no positive entry in the basement row). For the minimization problems, this preserved the condition of potential optimality for the special solutions (obtained by setting the variables on the left margin equal to zero) and worked towards the feasibility of the special solutions (no positive entry in the basement row).

An *algorithm* for solving a problem is a set of specific instructions which, if followed, will lead to a solution of the problem. For linear programming problems the *simplex method*, or *simplex algorithm*, is such a set of instructions. First, the problem is represented in a form to which the instructions of the algorithm can be applied. For a linear programming problem we shall give the instructions in terms of a tableau.

We assume the tableau has a positive (or at least nonnegative) right-hand column. If the entries in the basement row are negative (or at least nonpositive) the special solution is feasible and optimal. If the basement row has at least one positive entry, we choose any column with a positive basement entry.

For the next step we choose a positive entry (other than the basement entry) in that column. If there is no such entry the problem would have no solution. For the maximization problem this would mean there is a product whose production we could increase, and increase the profit, while using less of the resources. For the minimization problems this would mean there is a requirement that cannot be satisfied no matter how much we buy.

If there is at least one positive entry (other than the basement entry) in the selected column, we must select one such that the next tableau we get will have a nonnegative last column. For the maximization problem we made this choice by determining which resource was the most restrictive on how much we could increase production. For the minimization problem we made this choice by determining which supply could meet the selected requirement at least cost.

When these choices are made, choosing the column and then choosing the row, the process of obtaining the next tableau is purely computational. In the examples we considered we carried this step out algebraically. In the next chapter we shall routinize this step, the pivot exchange.

We have described the essential features of the simplex method; expressed in terms of a tableau, they are the condition for the optimality and the rules for selecting a row and column for the pivot exchange. In the remainder of Part I of this book we shall refine and extend the description of the simplex algorithm and establish the theory that underlies the simplex algorithm and linear programming in general.

1.5 Graphical Representation

It is very helpful to have a geometric interpretation of constraints imposed in a linear programming problem. From such geometric pictures we can appreciate the significance of the constraints, where and how to look for optimal solutions, and how the simplex algorithm works. It does not, however, provide a method for working any but the smallest linear programming problems.

In the plane, a linear expression of the form

$$2x + 3y - 12 = L \tag{1.57}$$

divides the coordinate plane into three parts, as illustrated in Figure 1.5, one part in which L is positive, one part in which L is negative, and one in which $L = 0$.

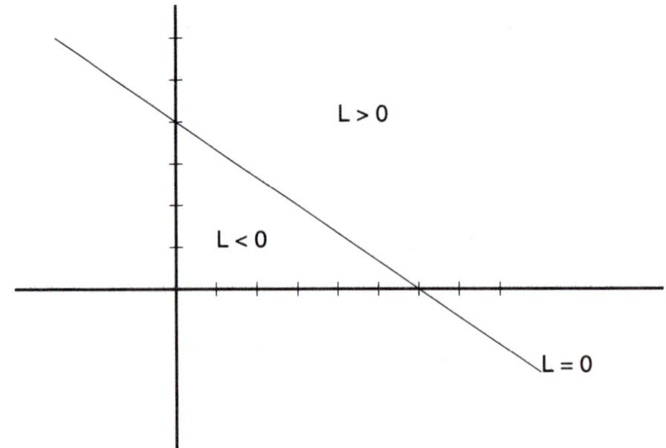

Figure 1.5: A linear expression divides space into three parts.

The region where $L \geq 0$ consists of two of these parts, the part where L is positive and the line itself. The region where $L \leq 0$ consists of the part where L is negative and the line itself.

Generally, in real spaces of any dimension the set of points that satisfy a linear inequality of the form $L \leq 0$, where L is a linear expression in the variables, consists of the line (plane or hyperplane) where $L = 0$ and one side of the line (plane or hyperplane) where L is negative.

Let us see how an example we have already considered can be cast in these terms. The maximization problem of Section 1.1 had three variables, x (the amount of floor cover), y (the amount of counter top), and z (the amount of wall tile). In the end, however, we found that the solution had $z = 0$. Thus, the situation can be represented in the plane $z = 0$, and we can take advantage of that fact so that we can draw the necessary figures in a plane. We drop z from the problem from the beginning. This will not destroy the spirit of the problem.

The first linear equation for the problem in Section 1.1 is

$$30x + 10y - 1500 = -p \tag{1.58}$$

The region where $p \geq 0$ is the shaded region together with the straight line $p = 0$ in Figure 1.6.

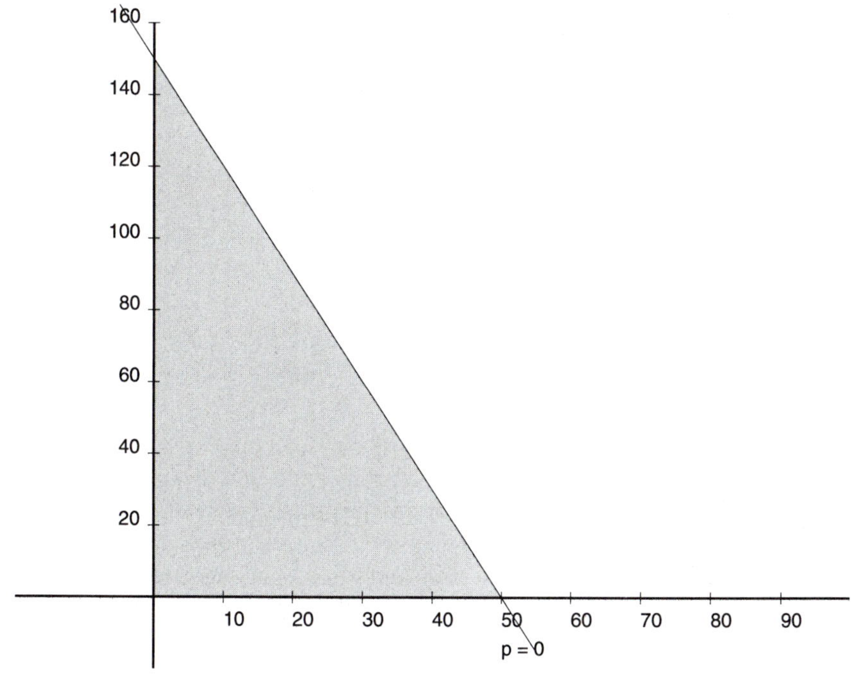

Figure 1.6: The feasible set for one linear expression.

In tableau 1.59, derived from 1.5 by deleting the z-column, there are six variables.

Each variable defines a constraint line and a half-plane on one side of that line. The set of points whose coordinates satisfy all the constraints is

$$
\begin{array}{|cc|c|l}
x & y & -1 & \\
\hline
30 & 10 & 1500 & = -p \\
5 & & 200 & = -q \\
0.2 & 0.1 & 12 & = -r \\
0.1 & 0.2 & 9 & = -s \\
\hline
10 & 5 & 0 & = f \\
\hline
\end{array}
\tag{1.59}
$$

precisely the intersection of the half-planes corresponding to each variable. The intersection for the constraints of tableau 1.59 is shown in Figure 1.7. The line for each variable is shown with an arrow pointing to the side of the line in which the variable is positive.

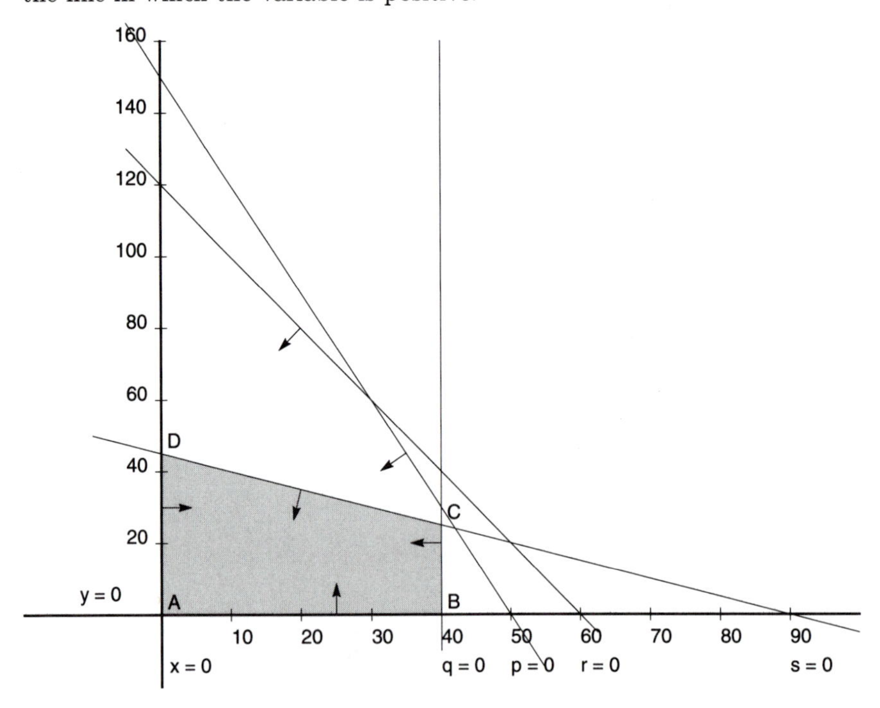

Figure 1.7: The feasible set for the production problem.

We can see that the constraints $p \geq 0$ and $r \geq 0$ are ineffective.

Each point in the intersection represents a feasible solution to the constraints and this set is called the feasible set. It is conceivable that the constraints might be inconsistent and the feasible set would then be empty. However, the effect of assuming that the entries in the right-hand column

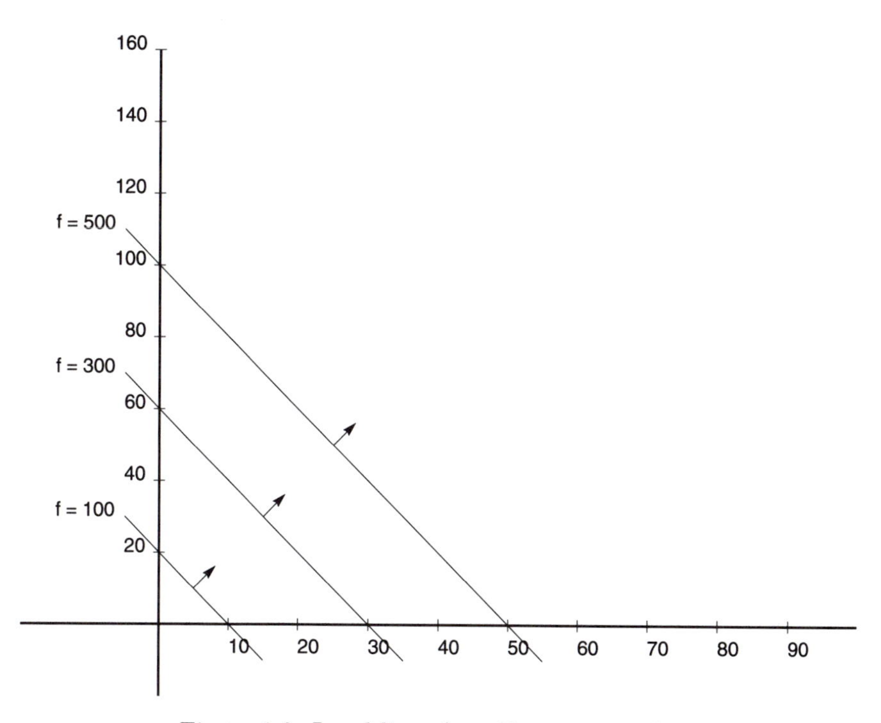

Figure 1.8: Level lines for a linear expression.

are positive, which they are for tableau 1.59, is that the origin is in the feasible set and the feasible set is not empty.

The objective function for the tableau 1.59 is

$$10x + 5y = f \tag{1.60}$$

For a constant value of f, the graph of 1.60 is a straight line. For distinct values of f these straight lines cannot intersect, so they are parallel. A family of straight lines corresponding to 1.60 is shown in Figure 1.8. The arrow points in the direction in which f increases.

For each straight line in the family of Figure 1.8, the intersection of the line with the feasible set gives the feasible points which yield that value of f. Since the values of f increase as the line is moved farther to the upper right, the maximum value of f for a feasible solution will be obtained at a point of the feasible set that is farthest towards the upper right. A comparison of Figure 1.8 with Figure 1.7 shows that the corner C, where $s = 0$ and $q = 0$, is the optimal solution. In Figure 1.9 we show the feasible set and the line $f = 525$ passing through the corner C.

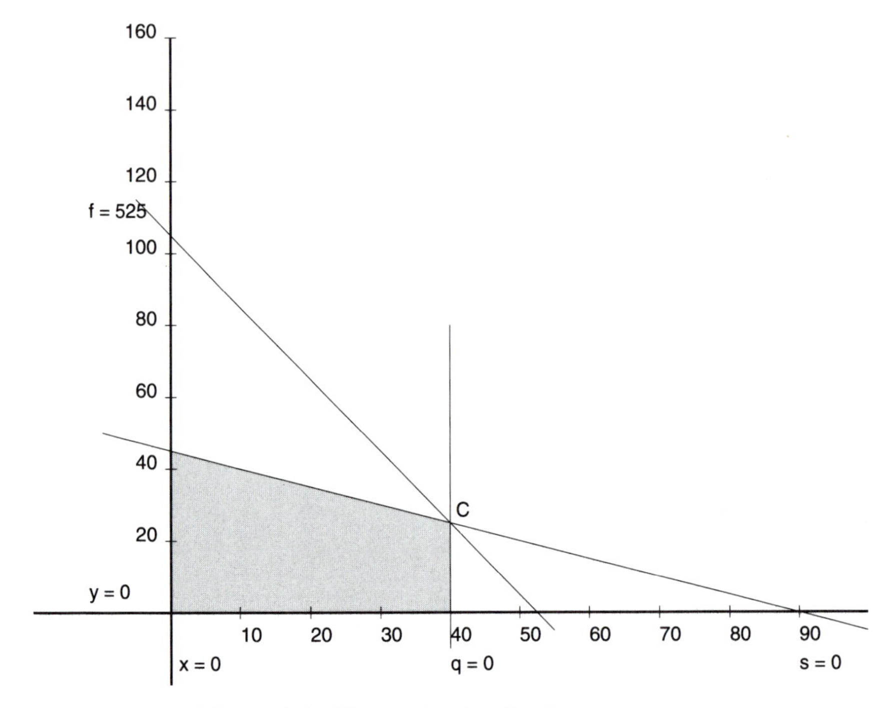

Figure 1.9: The optimal point is at a vertex.

Notice that the feasible set touches the line $f = 525$ and lies entirely on the negative side of that line. For any set, a line that touches the set and for which the set lies entirely on one side is called a *line of support* for the set.

Generally, the feasible set is the intersection of a finite number of half-spaces, each defined by a hyperplane. The feasible set could be empty or it could be unbounded. If it is nonempty and bounded, it is the n-dimensional generalization of a polyhedron. Since the objective function is linear, an optimal solution will be found at an extreme point of the feasible set where the hyperplane on which f is a constant passes through that point and is a hyperplane of support.

Theoretically, it is possible to solve a linear programming problem by drawing the feasible set and the family of hyperplanes for f. However, this procedure is quite impractical if the problem involves more than a few variables. The simplex algorithm is practical for problems that can involve several thousand variables if a computing machine is available, and it is practical for hand calculation for a dozen or so variables.

Setting the variables on the top margin of tableau 1.59 equal to zero gives the corner of the feasible set at the origin in Figure 1.7. The values of the other variables at that point can also be evaluated from the tableau. Any nonzero entry in the tableau 1.59, other than those in the last row or last column, can be used for a pivot exchange. For example, we can use the nonzero 0.1 in the third row to solve the equation

$$0.2x + 0.1y - 12 = -r \tag{1.61}$$

for y and obtain

$$2x + 10r - 120 = -y \tag{1.62}$$

We can then eliminate y from the other equations and obtain a tableau with x and r on the top margin. Setting the variables on the top margin of that tableau equal to zero would give us the point where the lines $x = 0$ and $r = 0$ in Figure 1.7 intersect.

In general, every variable of the problem corresponds to a line (or hyperplane). In particular, x =0 is the y-axis and $y = 0$ is the x-axis. Every point where two of these lines intersect is a point where the two variables corresponding to these lines vanish. This is represented in the tableaux in the following way. Starting with the tableau 1.59, every tableau that we can obtain by some sequence of pivot exchanges will have two variables on the top margin. Setting those two variables equal to zero will give the point of intersection of the two lines corresponding to those two variables. There is, in fact, a one-to-one correspondence between tableaux that can be obtained and points of intersections of pairs of lines from the network of lines. (If three lines intersect in one point this must be counted as three pairs of lines in this correspondence.) If two lines are parallel, no tableau can be obtained in which the corresponding variables are on the top margin. In particular, the lines $x = 0$ and $q = 0$ are parallel, and the zero in tableau 1.59 prevents the pivot exchange that would yield a tableau with x and q on the top margin.

These assertions are not expected to be obvious from this brief description. The necessary facts will be proved for tableaux in the next three chapters. However, an awareness of these facts may help motivate the tableau algebra and provide a basis for understanding the connection between the algebra and the geometry. It would be helpful if you would reread this section several times as you progress through the next three chapters.

The rules of the simplex algorithm as applied to this maximization problem are designed so that we examine only those intersection points which are vertices of the feasible set. The tableau 1.59 corresponds to examining the origin. The 10 in the basement row, being positive, means that we could increase the value of f by increasing x while keeping $y = 0$. In

Figure 1.7 this corresponds to moving along the edge $y = 0$ from vertex A towards vertex B. The 5 in the basement row means that we could increase f by increasing y while keeping $x = 0$. This corresponds to moving along the edge $x = 0$ towards the vertex D. In that direction the first intersecting line is the line $s = 0$.

In the sense of the discussion in Section 1.1, selecting a column amounts to selecting the line on which the variable at the head of the column vanishes. Selecting a column in which the basement entry is positive corresponds to selecting a line for which the segment of that line in common with the feasible set is in a direction in which the objective variable increases. The other lines that cross this line correspond to variables on the right margin for which the entries in its row are nonzero. The positive entries in that column correspond to intersecting lines in the direction in which the objective function increases, and negative entries in that column correspond to intersecting lines in the direction in which the objective function decreases. Choosing the minimum ratio (with a positive denominator) is the same as selecting the nearest intersecting line in the direction for which the objective function increases.

The arithmetic of performing the pivot exchange with tableau 1.59 is the same as that for tableau 1.5, except that the z-column is not involved. We obtain tableau 1.63.

(1.63)

q	y	-1	
-6	10	300	$= -p$
-0.2		40	$= -x$
-0.04	0.1	4	$= -r$
-0.02	0.2	5	$= -s$
-2	5	-400	$= f$

The 5 in the basement row of tableau 1.63, being positive, means that we can increase y while keeping $q = 0$ and increase the value of f. In Figure 1.7 this corresponds to moving along the edge where $q = 0$ towards the vertex C. The first line condition in this direction is the line where $s = 0$. This was also the most restrictive condition at this step when we worked through this example in Section 1.1. Thus, we exchange y and s by a pivot exchange and obtain tableau 1.64.

The negative entries in the basement row mean no further improvements are possible. This situation is represented geometrically in Figure 1.9. The edges of the feasible set where $q = 0$ or $s = 0$ are on the negative side of the line on which f is a constant. It is a line of support through C. Vertex C gives the optimal solution.

(1.64)

q	s	-1	
-5	-50	50	$= -p$
0.2		40	$= -x$
-0.03	-0.5	1.5	$= -r$
-0.1	5	25	$= -y$
-1.5	-25	-525	$= f$

The geometric interpretation of a minimization problem is similar to that of a maximization problem, but the simplex algorithm has a different interpretation and a separate description is worthwhile. The optimal solution for the minimization problem in Section 1.2 had $k = 0$ and $m = 0$. It will, therefore, simplify the description of the problem without changing the result if we delete those two rows from the tableau 1.26. The simplified tableau is then 1.65.

(1.65)

h	10	10	60	3.80
n	25	20	35	8.00
-1	450	400	1050	0
	$= u$	$= v$	$= w$	$= g$

The constraint lines are shown in Figure 1.10, and the feasible set is shown as a shaded polygon (extending to infinity to the upper right).

This time the origin is not a corner of the feasible set. The objective function is

$$3.8h + 8n = g \tag{1.66}$$

The lines where g is constant form a family of parallel lines. As g increases, the line moves to the upper right. Thus, the minimal value of g will be taken on at one of the corners on the lower left. By drawing a line representing 1.66 for a suitable value of g we could determine which corner is optimal. However, our purpose is to illustrate the working of the simplex algorithm, so we start with the tableau 1.65.

If we take $h = 0$ and $n = 0$, corresponding to the origin in Figure 1.10, we get $u = -450$ (a positive entry in the basement row). Since the feasible set must be on the positive side of the line $u = 0$, this means the line $u = 0$ separates the origin from the feasible set. Also, the feasible set is on the positive sides of the lines $h = 0$ and $n = 0$. To simplify the relevant details, Figure 1.11 shows the situation with the lines $v = 0$ and $w = 0$ deleted.

The feasible set lies somewhere in the shaded region of Figure 1.11. The line $g = 0$ is shown passing through the origin, our starting point

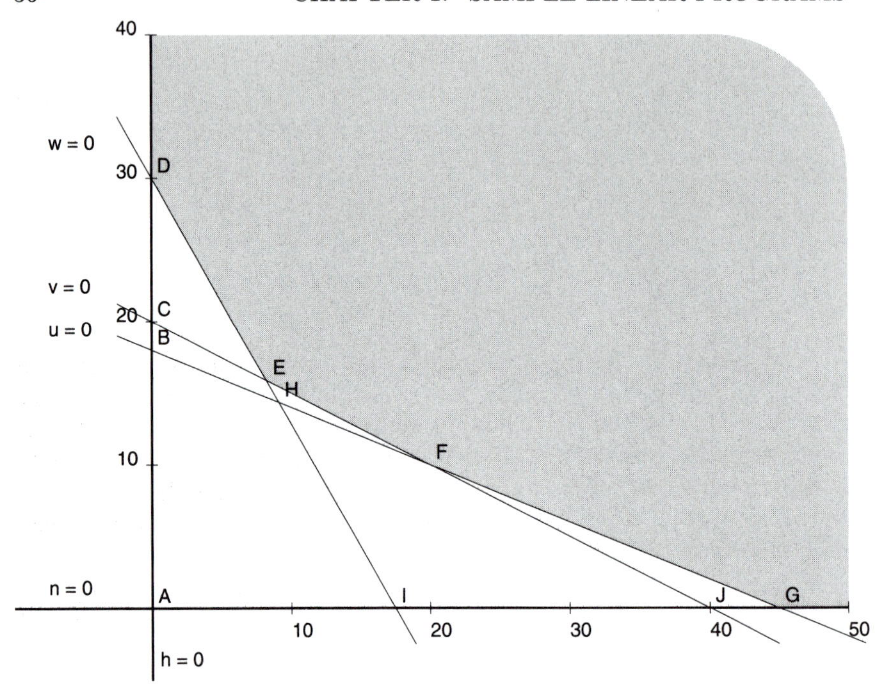

Figure 1.10: The feasible set for the diet problem.

corresponding to the tableau 1.65. Our choice of the u-column means that the pivot exchange we are considering will move the variable u to the left margin. That is, we will set $u = 0$ in the next trial solution. Since we will keep either $h = 0$ or $n = 0$, the two points under consideration are B and G. We do not want the line g =constant to go through the interior of the feasible set. The feasible set is smaller than the shaded region in Figure 1.11 (we did not show the lines $v = 0$ and $w = 0$). However, if we choose the point B we can be sure the new location of the objective function g =constant will not go through the interior of the feasible set. Thus, we are going to select the point B. We want to interpret this choice in terms of the numbers in the tableau.

Which point we choose depends on the relative slopes of the lines $g = 0$ and $u = 0$. This, in turn, depends on the relative sizes of the ratios $3.8/10$ and $8/25$. The ratio $8/25$ is smaller and we exchange the roles of u and n. After this pivot exchange we obtain the tableau 1.67.

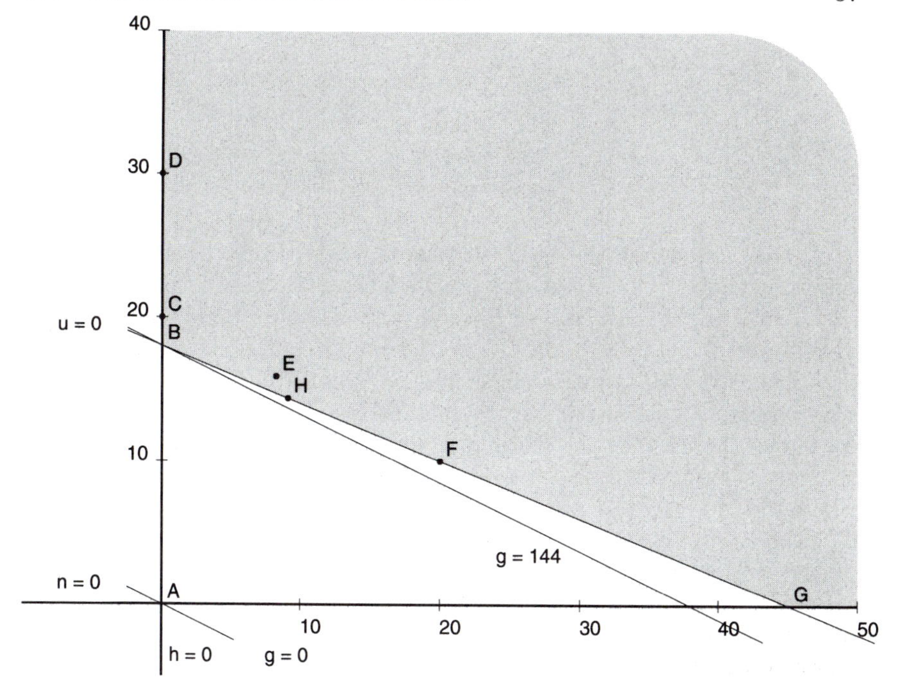

Figure 1.11: The level line for the optimum value of the objective function must not go through the interior of the feasible set.

					(1.67)
h	-0.4	2	46	0.6	
u	0.04	0.8	1.4	0.32	
-1	-18	40	420	-144	
	$= n$	$= v$	$= w$	$= g$	

Setting the variables on the left margin equal to zero corresponds to selecting the point B in Figure 1.10. In tableau 1.67 the v-column has a positive entry in the basement row. If we select that column for the next pivot exchange, then our next trial solution must be chosen from C and F. (If we were to choose the w-column our next choice would be between D and H.) The reasoning to decide which choice to make is the same as in the previous step. We decide to interchange v and h. The next tableau is then 1.68.

If we set the variables on the left margin to zero we obtain positive values for the variables on the bottom margin. That is, we obtain a feasible solution. This is the point F in Figure 1.10. It is the optimal solution.

v	-0.2	0.5	23	0.3
u	0.2	-0.4	-17	0.08
-1	-10	-20	-500	-156
	$= n$	$= h$	$= w$	$= g$

(1.68)

Let us summarize the ideas we have developed. Each variable of the problem defines a line (and this includes the variables of the coordinate axes.) In three dimensions we would be talking about planes, and in higher dimensions about hyperplanes. The feasible set is the intersection of the half-planes (or half-spaces) defined by these lines. The feasible set is possibly empty or unbounded. An optimal solution is obtained at one of the corners of the feasible set. (If the line for which the objective variable is a constant is parallel to one of the edges there might be more than one corner where the objective variable is maximal or minimal.)

If we start at a corner of the feasible set, we find an edge along which the objective variable increases (a column where the basement entry is positive) and move along that edge to the next corner of the feasible set. This is what determines the pivot row. When we reach a corner for which the objective variable decreases along edges out of that corner (negative entries in the basement row) we have obtained an optimal solution.

If we start at an intersection point that is not a corner of the feasible set, and for which the line on which the objective variable is a constant does not intersect the feasible set, we move along a line towards a next intersection point. We choose the direction so that the line on which the objective variable is a constant moves towards the feasible set but does not pass through the interior of the feasible set.

In the first case, we start with a line (objective variable a constant) and we move it in such a way that it continues to intersect the feasible set. We move it until it becomes a line of support of the feasible set. In the second case we start with a line that does not intersect the feasible set. We move it towards the feasible set so that when it finally touches the feasible set it is a line of support.

This is the essence of the simplex algorithm. However, for large problems these decisions must be made on the basis of numerical calculations.

Exercises

Solving a linear programming problem involves several steps. Depending on how the problem is posed, we have to interpret the information given and cast it in a form suitable for manipulation with the tools available. In the next four chapters we will develop the simplex algorithm, and that will be our principal tool. This will require the problem to be cast in the form of tableaux 1.23, 1.24 or 1.46, 1.47. This set of exercises deals with the steps that prepare for the application of the simplex algorithm. Thus, the answers should be in the form of one of the equivalent formulations, 1.14 through 1.24, or 1.37 through 1.47. In most cases you should first construct a data table or cast the problem in inequality form. Otherwise, it will be difficult to see the slack variables.

The exercises range from fanciful to tiny examples of serious and difficult problems. Some involve subtleties we have not prepared you for yet. Accept them as challenges for which help is available in the answers provided.

Some exercises are borrowed from other authors and, where this is the case, the sources are identified. They have been rephrased to fit the context of this book.

1. **The Furniture Maker's Problem**. A furniture maker has a line of four types of desks. They vary in the manufacturing processes and their profitability. The furniture maker has available 6000 hours of time in the carpentry shop each six months, and 4000 hours of time in the finishing shop. Each desk of type 1 requires 4 hours of carpentry and 1 hour of finishing. Each desk of type 2 requires 9 hours of carpentry and 1 hour of finishing. Each desk of type 3 requires 7 hours of carpentry and 3 hours of finishing. Each desk of type 4 requires 10 hours of carpentry and 40 hours of finishing. The profit is $12 for each desk of type 1, $20 for each desk of type 2, $28 for each desk of type 3, and $40 for each desk of type 4. How should the production be scheduled to maximize the profit?

(George B. Dantzig, *Linear Programs and Extensions*, 1963, p50)

2. **The President's Problem**. A certain President is concerned that his tax reform package is in trouble in the Senate. He has been told by his advisors that there are 12 Republicans and 16 Democrats who have not made up their minds. He decides to make telephone calls to convince the undecided Senators but finds that he only has time to make 20 calls before the vote in the Senate. Experience shows that a phone call to a Republican has a 0.9 chance of success while a phone call to a Democrat has a 0.6 chance of success. How many Senators in each party should he call to make best use of the calls he has time to make?

(Harold W. Kuhn, *unpublished class notes*)

3. **The Wyndor Glass Company's Problem.** This company produces a variety of products, but we shall restrict our attention to two, a high-quality door and a window. The company has some capacity available

at three plants: 4 units at Plant 1, 12 units at Plant 2, and 18 units at Plant 3. Plant 1 can produce a door with 1 unit of its capacity. It doesn't make windows. Plant 2 can produce a window with 2 units of its capacity. It doesn't make doors. Plant 3 can produce a door with 3 units of its capacity and a window with 2 units of its capacity. The company can sell as many of each as it can make at a profit of $3 for each door and $5 for each window. How should the company allocate its production capacities to maximize its profits?

(Frederick S. Hillier and Gerald J. Lieberman, *Operations Research*, p16)

4. **The Investment Manager's Problem.** An account manager in an investment firm has been assigned the task of investing $10,000 each month for a client. He has available many types of bonds, but we will assume that he has only three available. The return on each type varies from month to month. This month he can get 7% on the first type, 8% on the second type, and 8.5% on the third type. His instructions are that he need not invest the entire fund, but he can invest no more than 40% of the amount invested in any one type of bond. Within these conditions he is to maximize the return on the investment.

He considers it obvious how he must invest the money, but to give himself a mathematical justification for his decision he decides to cast his problem as a linear programming problem.

5. **The Investment Manager's Problem 2.** To minimize the risk in the investment portfolio, the investment manager is given the additional instruction that he must invest at least $2500 each month in the bond issue that is considered to be the safest. This month it is the bond with a 7% return. Formulate this altered form of the investment manager's problem.

6. **The Welfare Mother's Problem.** A welfare recipient decides to see if her knowledge of linear programming can help her with her financial problems. Peanut butter costs 20 cents an ounce and contains 1 unit of carbohydrate and 1 unit of protein. A small loaf of bread costs 12 cents and contains 1 unit of carbohydrate and no protein. A cup of milk costs 16 cents and contains 1 unit of protein and no carbohydrate. She estimates that she needs 0.9 units of carbohydrate and 0.6 units of protein per day. How can she get at least that much nutrient from these foods at least cost?

(Harold W. Kuhn, *unpublished class notes*)

7. **The Advertiser's Problem.** An advertising manager is assigned the task of putting together an advertising campaign for a client. The client has set the objectives of getting at least 160 million exposures, with at least

60 million of those exposures being persons with income of at least \$8000 per year and at least 80 million in the 18 to 40-year-old-age groups. Market studies indicate that an ad in a certain magazine will be seen by 8 million people of whom 3 million will be in the targeted income group and 4 million will be in the targeted age group. Each magazine ad will cost \$40,000. An ad on television will be seen by 40 million people of whom 10 million will be in the targeted income group and 10 million will be in the targeted age group. Each television ad will cost \$200,000. How can the objectives of this advertising campaign be achieved at least cost?

(William J. Baumol, *Economic Theory and Operations Analysis*, 1972, p98)

8. **The MaxMin Problem.** Find values for the variables x_1, x_2, x_3 that satisfy the inequalities

$$\begin{aligned} x_1 + 2x_2 + x_3 &\leq 16 \\ 4x_1 + x_2 + 3x_3 &\leq 30 \\ x_1 + 4x_2 + 5x_3 &\leq 40 \end{aligned}$$

for which the minimum value of x_1, x_2, and x_3, is as large as possible.

9. **The MinMax Problem.** Find values for the variables v_1, v_2, v_3 that satisfy the inequalities

$$\begin{aligned} v_1 + 2v_2 + v_3 &\geq 16 \\ 4v_1 + v_2 + 3v_3 &\geq 30 \\ v_1 + 4v_2 + 5v_3 &\geq 40 \end{aligned}$$

for which the maximum value of v_1, v_2, and v_3, is as small as possible.

10. **The Maintenance Manager's Problem.** An airline has five types of planes. Each requires certain routine maintenance operations periodically. The maintenance shop has adequate manpower and equipment to handle the planes whenever this operation is needed, but the hanger in which the planes must be parked while the operation is carried on will allow only three types of planes to be parked simultaneously. Types A, B, and D, types A, D, and E, types A, C, and E, types B, C, and D, and types B, C, and E can be hangered together. That is, the hanger must be in one of five different configurations at any time. The nature of the airline's equipment indicates that 9 planes of type A, 10 planes of type B, 14 planes of type C, 12 planes of type D, and 15 planes of type E will need servicing each week. The maintenance manager must decide on the portion of time the hanger is in each of the five configurations. The planes arrive for servicing at random

times. The objective is to have the maximum number of planes of any one type that arrive when they cannot be serviced as small as possible.

11. **The Gardener's Problem**. Three fertilizers have the following percentages of nitrogen and phosphorus, acidity ratings per pound, and costs per pound.

	% N	% P	acidity/lb	cost/lb
Tuf-Turf	25	10	2	0.20
Good-Gro	10	5	−1	0.08
Slik-Sod	25	5	1	0.22

Limestone has no nitrogen or phosphorus, has an acidity of −10 per pound, and costs 2 cents per pound.

Find the mixture of these four ingredients that contains at least 100 pounds of nitrogen and at least 50 pounds of phosphorus, has an acidity rating 0, and costs the least.

(Robert R. Singleton and William F. Tyndall, *Games and Programs*, 1974, p209)

12. **The Metallurgist's Problem**. A metallurgist requires an alloy that is 40% tin. He has available nine alloys with varying percentages of tin at various costs. These data are summarized in the following table.

alloy	A	B	C	D	E	F	G	H	I	desiredblend
% tin	80	60	10	10	40	30	50	10	50	40
cost/lb	4.10	4.30	5.80	6.00	7.60	7.50	7.30	6.90	7.30	min

How can the metallurgist blend an alloy that will produce what he needs at least cost?

(George B. Dantzig, *Linear Programming and Extensions*, 1963, p46)

Chapter 2

Tableau Algebra

2.1 Tableaux and the Duality Equation

In Section 1.1 we introduced a tableau 1.5 to represent the conditions for the maximization problem we were considering. In a setting more general than a maximization problem, tableaux can be used to represent systems of linear equations. Consider the tableau 2.1 below.

$$
\begin{array}{cccc|c|l}
x_1 & x_2 & \cdots & x_n & -1 & \\
\hline
a_{11} & a_{12} & \cdots & a_{1n} & b_1 & = -y_1 \\
a_{21} & a_{22} & \cdots & a_{2n} & b_2 & = -y_2 \\
\vdots & \vdots & & \vdots & \vdots & \vdots \\
a_{m1} & a_{m2} & \cdots & a_{mn} & b_m & = -y_m \\
\hline
c_1 & c_2 & \cdots & c_n & d & = f \\
\hline
\end{array}
\tag{2.1}
$$

For each row in the tableau 2.1, we multiply each entry in the row by the corresponding variable or number on the top margin, accumulate the sum of these products, and set the result equal to the variable on the right margin. Thus, for example, the first row of the tableau 2.1 generates the equation

$$a_{11}x_1 + a_{12}x_2 + \cdots + a_{1n}x_n - b_1 = -y_1 \tag{2.2}$$

In a similar way, each of the other rows of the tableau generates an equation.

The lines drawn within the box of the tableau serve only to visually subdivide the tableau into zones to allow for convenient reference. The *A-matrix* is the array of m rows and n columns in the upper left block. The *d-corner* is the entry in the lower right corner. The *b-column* is the part of

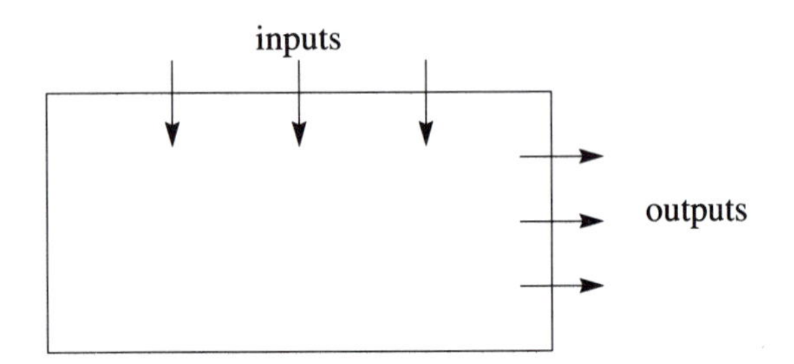

Figure 2.1: Max program input–output.

the last column that does not include the d-corner. The *c-row*, or basement row, is the part of the bottom row that does not include the d-corner.

Any set of values for the $m+n+1$ variables $x_1, x_2, \ldots, x_n, y_1, y_2, \ldots, y_m, f$ that satisfy all the equations generated by the tableau 2.1 is called a *solution of the tableau*. The variables on the top margin are called *independent variables*, regardless of the letters of the alphabet used to denote them. The variables of the right margin are called *dependent variables*, regardless of the letters of the alphabet used to denote them.

> **A particular solution is determined by specifying the values of the independent variables and computing the values of the dependent variables. See Figure 2.1.**

We look upon the tableau as representing an input–output device: The inputs at the top determine the outputs at the right.

In Section 1.2 we considered a minimization problem for which we used the columns of a tableau to express the constraint relations. In this spirit, consider the tableau 2.3 below.

$$
\begin{array}{c|cccc|c|}
\cline{2-6}
v_1 & a_{11} & a_{12} & \cdots & a_{1n} & b_1 \\
v_2 & a_{21} & a_{22} & \cdots & a_{2n} & b_2 \\
\vdots & \vdots & \vdots & & \vdots & \vdots \\
v_m & a_{m1} & a_{m2} & \cdots & a_{mn} & b_m \\
\cline{2-6}
-1 & c_1 & c_2 & \cdots & c_n & d \\
\cline{2-6}
\multicolumn{1}{c}{} & = u_1 & = u_2 & \cdots & = u_n & \multicolumn{1}{c}{= g}
\end{array}
\tag{2.3}
$$

For each column in tableau 2.3, we multiply each entry in the column by the corresponding variable or number on the left margin, accumulate the sum of these products, and set the result equal to the variable on the bottom margin. Thus, for example, the first column of the tableau generates the

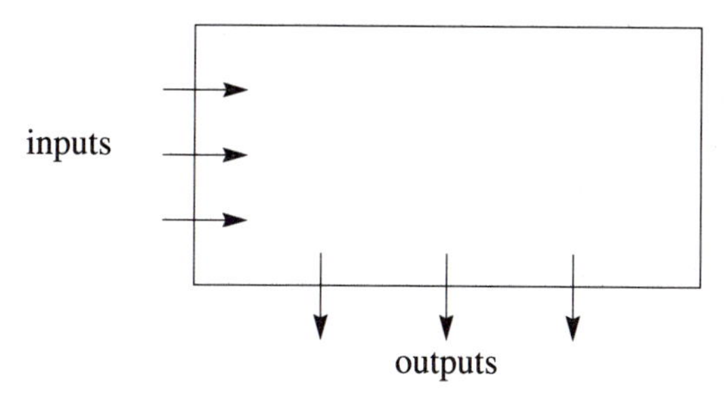

Figure 2.2: Min program input–output.

equation

$$v_1a_{11} + v_2a_{21} + \cdots + v_ma_{m1} - c_1 = u_1 \tag{2.4}$$

In a similar way each of the other columns of the tableau generates an equation.

Any set of values for the $m + n + 1$ variables u_1, u_2, $\ldots$, u_n, v_1, v_2, $\ldots$, v_m, g that satisfy all the equations generated by tableau 2.3 is called a *solution of the tableau.* The variables on the left margin are the *independent variables.* The variables on the bottom margin are the *dependent variables.*

A particular solution is determined by specifying the values of the independent variables and computing the values of the dependent variables. See Figure 2.2.

Again, we can look upon tableau 2.3 as an input–output device. The inputs at the left determine the outputs at the bottom.

Let A denote the A-matrix (an $m \times n$ matrix), b the b-column (a one-column matrix), c the c-row (a one-row matrix), and d the d-corner. Let x denote the one-column matrix of the x's on the top margin of tableau 2.1, and y the one-column matrix of the y's. Then the equations generated by the rows of tableau 2.1 can be written in matrix form.

$$Ax - b = -y \tag{2.5}$$

$$cx - d = f \tag{2.6}$$

In a similar way let v denote the one-row matrix of the v's on the left margin of tableau 2.3 and let u denote the one-row matrix of the u's on the bottom margin of tableau 2.3. The equations generated by tableau 2.3 can also be represented in matrix form.

$$vA - c = u \tag{2.7}$$

$$vb - d = g \tag{2.8}$$

Many authors consistently use columns where we have chosen to use row matrices. If that convention were used equation 2.7 would take the form $v^T A - c^T = u^T$, in which the superscript "T" denotes the transpose. Other than the extra symbols there is no disadvantage with either convention, as long as one is used consistently. We prefer the simpler notation. When we discuss duality, the distinction between the row equations and the column equations, where the variables of the row equations are columns and the variables of the column equations are rows, will prove to simplify the discussion.

It is important to note that the order of the rows and columns in tableau 2.1 is not significant. If we change the order of the rows and move the variables on the right margin to preserve the correspondence, the new tableau would represent the same row equations. Similarly, if we change the order of the columns and move the variables on the top margin to match, the new tableau would represent the same row equations. Thus, we shall consider such changes to be unessential. As long as each row is properly identified with the variable on the right margin and each column is properly identified with the variable on the top margin we know all we need to know about the tableau. If these identifications are properly maintained we shall permit rearrangements of the rows and columns and consider all such tableaux to be the same. Similar remarks apply to rearrangements of the rows and columns of tableau 2.3.

With this in mind, it is appropriate to identify a row or column by the variable with which it is associated, rather than by its position in the tableau. Thus, we can speak of the x_s-column or the y_r-row without prejudicing the actual number of the row or column that is indicated. In particular, notice that this is the situation in the tableau 1.56 that we used in the transportation problem example in Chapter 1. The variable y_1 might be associated with column 1 in the initial tableau and associated with a different column, or with a row, in another tableau. Within the tableau, we will use the subscripts to indicate the row and column numbers as it is usually done in matrix notation. Reference to a row by its row number or by its associated variable is acceptable and either will be used if it is clear what is intended.

The fact that tableau 2.1 and tableau 2.3 share the same entries suggests that the two tableaux are closely related. If we are given either tableau, we could easily write down the other one. In fact, there is no reason why we should not consider both tableaux at the same time in the form of tableau 2.9.

The rows of tableau 2.9 generate one system of equations for which the independent variables are on the top margin and the dependent variables

$$
\begin{array}{c|cccc|c|l}
 & x_1 & x_2 & \cdots & x_n & -1 & \\
\hline
v_1 & a_{11} & a_{12} & \cdots & a_{1n} & b_1 & = -y_1 \\
v_2 & a_{21} & a_{22} & \cdots & a_{2n} & b_2 & = -y_2 \\
\vdots & \vdots & \vdots & & \vdots & \vdots & \vdots \\
v_m & a_{m1} & a_{m2} & \cdots & a_{mn} & b_m & = -y_m \\
\hline
-1 & c_1 & c_2 & \cdots & c_n & d & = f \\
\hline
 & = u_1 & = u_2 & \cdots & = u_n & = g &
\end{array}
\tag{2.9}
$$

are on the right margin. The columns generate another system of equations for which the independent variables are on the left margin and the dependent variables are on the bottom margin. We refer to the first system as the *row system* of tableau 2.9 and the second as the *column system* of the tableau.

Whichever of these two systems is given, we can obtain the other system merely by supplying the appropriate variables on the margins. We call the given system the *primal system*, and the system obtained from it the *dual system*. Either one can be considered to be the primal system. The two systems are said to be *dual* to each other.

Using matrix notation, tableau 2.9 can be written more concisely in the form 2.10.

$$
\begin{array}{c|c|c|l}
 & x & -1 & \\
\hline
v & A & b & = -y \\
\hline
-1 & c & d & = f \\
\hline
 & = u & = g &
\end{array}
\tag{2.10}
$$

From the matrix equations generated by tableau 2.10 we can obtain

$$
ux + vy = (vA - c)x + v(-Ax + b) = vb - cx = g - f \tag{2.11}
$$

Written out in expanded form this equation is

$$
u_1x_1 + \cdots + u_nx_n + v_1y_1 + \cdots + v_my_m = g - f \tag{2.12}
$$

Equation 2.12 is known as the *duality equation*. The factors that are paired in each term of the duality equation are said to be *dual variables*. That is, u_i and x_i are dual variables and v_j and y_j are dual variables.

The duality equation is so fundamental in the theory of linear programming that its importance cannot be overemphasized. Yet it is interesting to note that it depends on only the simplest ideas in matrix algebra. The only property of matrices needed to prove the duality equation is the associative law for matrix multiplication: In equation 2.11 we need $(vA)x = v(Ax)$.

The duality equation can also be written in the form

$$ux - g = -vy - f \tag{2.13}$$

The left side of equation 2.13 is obtained by multiplying the variables (and numbers) on the margins at the opposite ends of each column, including any prefixed signs, and accumulating the sum of these products. The right side of equation (1.15) is obtained by multiplying the variables on the margins at opposite ends of each row, including any prefixed signs, and accumulating the sum of these products. Diagram 2.14 shows the pairings of the factors in each term of the duality equation.

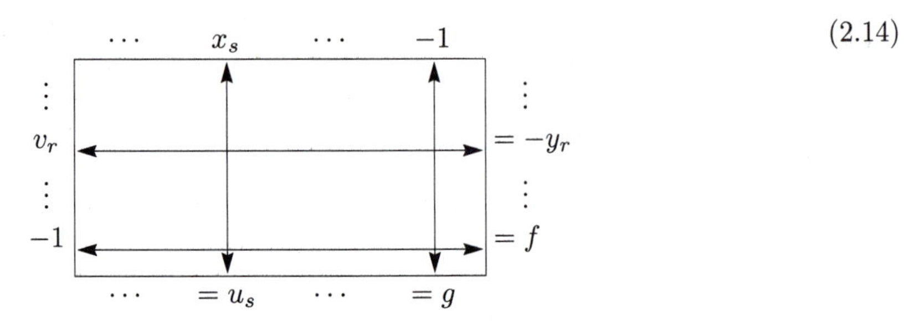

In this form the duality equation is easy to remember and write down regardless of the notation that is used.

Notice that equation 2.12 does not involve explicitly the entries in the interior of the tableau 2.9, only the variables and numbers on the margins. Also, we want to emphasize that

> **the duality equation is satisfied for any and every pair of solutions, one solution of the row system of equations and one solution of the column system.**

2.2 Pivot Exchange, Row Equations

In this section we shall describe an operation that is one step in the simplex algorithm. This step, the *pivot operation* or *pivot exchange*, consists in passing from one tableau with its dual pair of systems of linear equations to a new tableau representing the same systems of equations in a new form.

The simplex algorithm will consist of rules for selecting the pivot exchanges to be performed.

To simplify matters, let us consider first just the system of row equations for the tableau. That is, consider tableau 2.15.

$$\begin{array}{|cccc|c|l} \multicolumn{1}{c}{x_1} & x_2 & \cdots & \multicolumn{1}{c}{x_n} & \multicolumn{1}{c}{-1} & \\ \hline a_{11} & a_{12} & \cdots & a_{1n} & b_1 & = -y_1 \\ a_{21} & a_{22} & \cdots & a_{2n} & b_2 & = -y_2 \\ \vdots & \vdots & & \vdots & \vdots & \vdots \\ a_{m1} & a_{m2} & \cdots & a_{mn} & b_m & = -y_m \\ \hline c_1 & c_2 & \cdots & c_n & d & = f \\ \hline \end{array} \tag{2.15}$$

We wish to exchange the roles of the independent variable x_s and the dependent variable y_r. The y_r-row of the tableau generates the equation that involves these two variables. This equation has the form

$$\cdots + a_{rs}x_s + \cdots + a_{rj}x_j + \cdots - b_r = -y_r \tag{2.16}$$

To solve for x_s it is necessary and sufficient that $a_{rs} \neq 0$. Assuming this is the case we obtain

$$\cdots + \frac{1}{a_{rs}}y_r + \cdots + \frac{a_{rj}}{a_{rs}}x_j + \cdots - \frac{b_r}{a_{rs}} = -x_s \tag{2.17}$$

The y_i-row (where $i \neq r$) of the tableau generates the following equation.

$$\cdots + a_{is}x_s + \cdots + a_{ij}x_j + \cdots - b_i = -y_i \tag{2.18}$$

We substitute the expression in 2.17 for x_s in the equation 2.18 to obtain

$$\cdots - \frac{a_{is}}{a_{rs}}y_r + \cdots + (a_{ij} - \frac{a_{is}a_{rj}}{a_{rs}})x_j + \cdots - (b_r - \frac{a_{is}b_r}{a_{rs}}) = -y_i \tag{2.19}$$

We make a similar substitution for x_s in the equation generated by the basement row.

The structure of the formulas is much easier to see when they are displayed in a tableau. Furthermore, it is more convenient and expeditious to do the arithmetic of a pivot exchange directly on the tableaux without writing out the equations considered above. Consider the representative terms in equations 2.16 and 2.18 in the following tableau.

(2.20)

	$\cdots$	x_s	$\cdots$	x_j	$\cdots$	-1	
$\vdots$		$\vdots$		$\vdots$		$\vdots$	$\vdots$
pivot row	$\cdots$	a_{rs}^*	$\cdots$	a_{rj}	$\cdots$	b_r	$= -y_r$
$\vdots$		$\vdots$		$\vdots$		$\vdots$	$\vdots$
other row	$\cdots$	a_{is}	$\cdots$	a_{ij}	$\cdots$	b_i	$= -y_i$
$\vdots$		$\vdots$		$\vdots$		$\vdots$	$\vdots$
	$\cdots$	c_s	$\cdots$	c_j	$\cdots$	d	$= f$
	$\cdots$	pivot column	$\cdots$	other column	$\cdots$		

The pivot row generates equation 2.16 and the "other" representative row generates equation 2.18. The entry a_{rs} at the intersection of the *pivot row* and the *pivot column* is called the *pivot entry*, and it is indicated by an asterisk.

We can display the representative terms of equations 2.17 and 2.19 in a tableau and compare it with tableau 2.20.

(2.21)

	$\cdots$	y_r	$\cdots$	x_j	$\cdots$	-1	
$\vdots$		$\vdots$		$\vdots$		$\vdots$	$\vdots$
pivoted row	$\cdots$	$\frac{1}{a_{rs}}$	$\cdots$	$\frac{a_{rj}}{a_{rs}}$	$\cdots$	$\frac{b_r}{a_{rs}}$	$= -x_s$
$\vdots$		$\vdots$		$\vdots$		$\vdots$	$\vdots$
other row	$\cdots$	$-\frac{a_{is}}{a_{rs}}$	$\cdots$	$a_{ij} - \frac{a_{is}a_{rj}}{a_{rs}}$	$\cdots$	$b_i - \frac{a_{is}b_r}{a_{rs}}$	$= -y_i$
$\vdots$		$\vdots$		$\vdots$		$\vdots$	$\vdots$
	$\cdots$	$-\frac{c_s}{a_{rs}}$	$\cdots$	$c_j - \frac{c_s a_{rj}}{a_{rs}}$	$\cdots$	$d - \frac{c_s b_r}{a_{rs}}$	$= f$
	$\cdots$	pivoted column	$\cdots$	other column	$\cdots$		

The process of obtaining the tableau of the form 2.21 from tableau 2.20 is called a *pivot exchange*, or *pivot operation*. The pivot exchange consists of the following steps.

1. Select a pivot entry $a_{rs} \neq 0$.

2. Replace the pivot entry a_{rs} by $1/a_{rs}$.

3. In the pivot row, divide the other entries by a_{rs}. That is, replace a_{rj} (for $j \neq s$) by a_{rj}/a_{rs} and replace b_r by b_r/a_{rs}.

4. For the entries a_{ij} not in the pivot row or pivot column (for $i \neq r$ and $j \neq s$), replace a_{ij} by $a_{ij} - a_{is}a_{rj}/a_{rs}$. Similarly, replace b_i by $b_i - a_{is}b_r/a_{rs}$, replace c_j by $c_j - c_s a_{rj}/a_{rs}$, and replace d by $d - c_s b_r/a_{rs}$.

5. In the pivot column, divide the other entries by $-a_{rs}$. That is, replace a_{is} (for $i \neq r$) by $-a_{is}/a_{rs}$ and replace c_s by $-c_s/a_{rs}$.

6. Interchange the dependent variable y_r and the independent variable x_s.

While the steps in the pivot exchange can be performed in any order, we recommend that they be done in a systematic way and in the same order every time. We shall summarize these rules in an equivalent way that will provide such a systematic procedure. We shall use formulas in which we place a bar over a symbol to indicate that it is an entry in the new tableau.

1. Select a pivot entry $a_{rs} \neq 0$.

2. Set $\overline{a}_{rs} = 1/a_{rs}$.

3. For $j \neq s$, set $\overline{a}_{rj} = a_{rj}\overline{a}_{rs}$. Also, set $b_r = b_r\overline{a}_{rs}$.

Repeat the following two steps for each $i \neq r$.

4. For $j \neq s$, set $\overline{a}_{ij} = a_{ij} - a_{is}\overline{a}_{rj}$. Also, set $\overline{b}_i = b_i - a_{is}\overline{b}_r$. For the basement row set $\overline{c}_j = c_j - c_s\overline{a}_{rj}$ and $\overline{d} = d - c_s\overline{b}_r$.

5. Set $\overline{a}_{is} = -a_{is}\overline{a}_{rs}$ and $\overline{c}_s = -c_s\overline{a}_{rs}$.

6. Interchange y_r and x_s.

Notice in step 4 that an arithmetic step is saved by using the value of $\overline{a}_{rj}$ computed in step 3 rather than using the entries from the original tableau, as it is described in the first set of instructions for the pivot exchange. The order of steps described here also has the advantage that once an entry for the new tableau is computed, the corresponding entry in the old tableau is no longer needed. If steps 4 and 5 were performed in the reverse order, a_{is} (or a_{rs} or a_{rj}) would have to be saved in addition to $\overline{a}_{rj}$. In the second procedure, as each new entry is computed, it replaces the old entry in all future computational steps. These are important considerations for any procedure designed for use on a computer.

The new equations are obtained from the old equations by standard algebraic manipulations. Thus, any solution of the old system of equations is also a solution of the new system. If we pivot on a_{rs}, the entry that replaces it is $1/a_{rs}$ which is also nonzero. A pivot exchange in the new tableau on $1/a_{rs}$ will produce the first tableau again. Thus, any solution of the equations of the new tableau is also a solution of the equations of the old tableau, and conversely. That is,

a pivot exchange preserves the solution set of the system of linear equations generated by the rows of a tableau.

Let us see how these rules apply in an example. Consider tableau 1.5 from Section 1.1.

(2.22)

x	y	z	-1	
30	10	50	1500	$= -p$
5*	0	3	200	$= -q$
0.2	0.1	0.5	12	$= -r$
0.1	0.2	0.3	9	$= -s$
10	5	5.5	0	$= f$

We wish to exchange x and q, and the required pivot entry is indicated by an asterisk. By step 2 we obtain

(2.23)

0.2				

Using step 3, we multiply the other entries in the pivot row (including the entry in the b-column) of 2.22 by 0.2 ($= 1/a_{rs}$) and enter the results.

(2.24)

0.2	0	0.6	40

We now carry out steps 4 and 5 for the first row. We refer to the "30" in the pivot column of tableau 2.22, multiply it by the other entries in the pivot row of tableau 2.24 and, subtract the products from the other entries in the first row of tableau 2.22. We obtain

(2.25)

	10	32	300
0.2	0	0.6	40

Then we supply the entry in the pivot column of the first row by step 5.

(2.26)

−6	10	32	300
0.2	0	0.6	40

We continue to use steps 4 and 5 for the other rows, including the c-row and exchange the variables x and q on the margins.

(2.27)

q	y	z	-1	
−6	10	32	300	$= -p$
0.2	0	0.6	40	$= -x$
−0.04	0.1	0.38	4	$= -r$
−0.02	0.2*	0.24	5	$= -s$
−2	5	−0.5	−400	$= f$

This tableau is the same as tableau 1.10 that we obtained in Section 1.1. There we performed the pivot operation "bare handedly," without the formal rules of this section.

Notice that the routine for handling the entries in the b-column, the

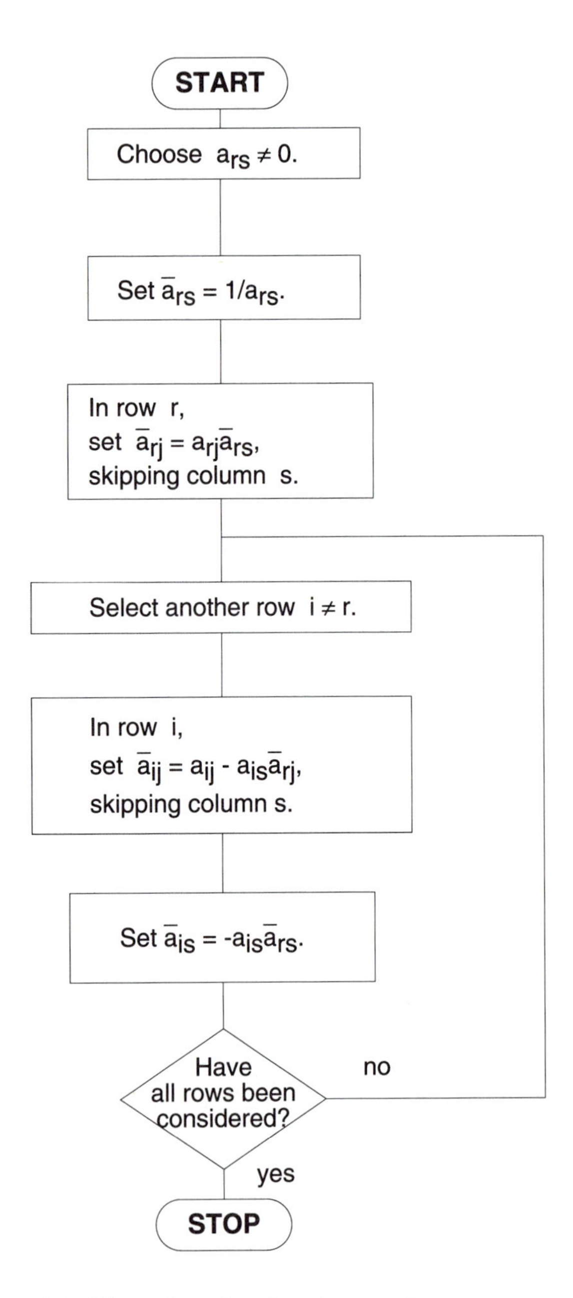

Figure 2.3: Flow chart for the pivot exchange.

c-row, and the d-corner is the same as that for the entries in the A-matrix. The only reason they appear with separate formulas is that different symbols were used to represent these entries.

The next pivot entry is indicated with an asterisk in tableau 2.27. The reader should use the rules of the pivot exchange to obtain tableau 1.13 of Section 1.1.

Exercises

Several levels of computational assistance are available for working linear programming problems, from hand calculators to sophisticated computer programs. In learning how linear programming problems are constructed and how they are solved, one should avoid using assistance for steps that are not yet understood. At this point you should perform the pivot exchange by hand calculations until you are confident that you know how it works. It is not necessary to become particularly fast or skilled, but it is essential that you be confident of your knowledge of the sequence of steps to be taken and can do the calculations accurately. Do not use assistance for any step you have not examined by making hand calculations.

This exercise set includes problems in which you can perform the pivot exchange. You may use a hand calculator if you wish, but it is not as much assistance as one might think. These exercises involve only rational numbers, and the pivot exchange will yield only rational numbers. With a hand calculator, the numbers are in decimal form and it is tedious to copy and re-enter numbers in that form. It is easier to use rational arithmetic when working these problems by hand. In each of these problems, perform a pivot exchange using the entry with the asterisk as the pivot entry.

1.

x_1	x_2	x_3	-1	
1^*	2	-3	4	$= -y_1$
2	3	1	5	$= -y_2$
-3	2	3	0	$= f$

2.

x_1	x_2	x_3	-1	
1	2	-3^*	4	$= -y_1$
2	3	1	5	$= -y_2$
-3	2	3	0	$= f$

3.

x_1	x_2	-1	
1	-2^*	-3	$= -y_1$
3	3	2	$= -y_2$
1	-3	3	$= -y_3$
4	5	4	$= f$

4.

x_1	x_2	x_3	-1	
1	-2^*	0	-3	$= -y_1$
3	3	4	2	$= -y_2$
1	-3	5	3	$= -y_3$
5	0	-3	-1	$= -y_4$
4	5	2	4	$= f$

5. Perform the two pivot exchanges indicated with asterisks. You can check your results against the calculations in Section 1.1.

x	y	z	-1	
30	10	50	1500	$= -p$
5^*		3	200	$= -q$
0.2	0.1	0.5	12	$= -r$
0.1	0.2^*	0.3	9	$= -s$
10	5	5.5	0	$= f$

6. Perform a pivot exchange on the starred entry.

x_1	x_2	x_3	x_4	-1	
4	9	7	10	6,000	$= -y_1$
1	1	3	40^*	4,000	$= -y_2$
12	20	28	40		$= f$

2.3 Pivot Exchange, Column Equations

At the beginning of Section 2.2 we suggested that the pivot exchange would preserve both of the dual systems of linear equations. Then we concentrated all our attention on the system of row equations. Let us examine the system of column equations, but in somewhat less detail than we devoted to the row system. Consider the tableau

$$
\begin{array}{c|cccc|c|}
\cline{2-6}
v_1 & a_{11} & a_{12} & \cdots & a_{1n} & b_1 \\
v_2 & a_{21} & a_{22} & \cdots & a_{2n} & b_2 \\
\vdots & \vdots & \vdots & & \vdots & \vdots \\
v_m & a_{m1} & a_{m2} & \cdots & a_{mn} & b_m \\
\cline{2-6}
-1 & c_1 & c_2 & \cdots & c_n & d \\
\cline{2-6}
\multicolumn{1}{c}{} & = u_1 & = u_2 & \cdots & \multicolumn{1}{c}{= u_n} & \multicolumn{1}{c}{= g}
\end{array}
\tag{2.28}
$$

We wish to exchange the independent variable v_r with the dependent variable u_s. The u_s-column of the tableau generates the equation that involves the variables v_r and u_s. This equation has the form

$$\cdots + v_r a_{rs} + \cdots + v_i a_{is} + \cdots - c_s = u_s \tag{2.29}$$

To solve for v_r it is necessary and sufficient that $a_{rs} \neq 0$. Assuming this is the case we obtain

$$\cdots + u_s \frac{1}{a_{rs}} - \cdots - v_i \frac{a_{is}}{s_{rs}} - \cdots - (-\frac{c_s}{a_{rs}}) = v_r \tag{2.30}$$

The u_j-column (where $j \neq s$) of the tableau generates the following equation.

$$\cdots + v_r a_{rj} + \cdots + v_i a_{ij} + \cdots - c_j = u_j \tag{2.31}$$

We substitute the expression for v_r in 2.30 in equation 2.31 and obtain

$$\cdots + u_s \frac{a_{rj}}{a_{rs}} + \cdots + v_i(a_{ij} - \frac{a_{is}}{a_{rj}} a_{rs}) + \cdots - (c_j - \frac{c_s a_{rj}}{a_{rs}}) = u_j \tag{2.32}$$

Compare equations 2.30 and 2.32 with the columns of tableau 2.21. We see that the rules for a pivot exchange in the system of column equations of a tableau give results the same as those obtained with the rules for pivoting in the system of row equations. In fact, this is the reason that the dependent variables of the row system are prefixed with a negative sign while the dependent variables of the column system are not. With this convention

> **the rules for the pivot exchange preserve the solution sets for both the row system of equations and the column system of equations.**

In Section 1.2 we considered a minimization problem and set up a tableau that represented the constraints and objective function in column form. In that example, the initial tableau 1.26 was

(2.33)

h	10	10	60	3.80
k	15	10	25	5
m	10	5	55	3.50
n	25*	20	35	8
-1	450	400	1050	
	$= u$	$= v$	$= w$	$= g$

We wish to exchange n and u and the required pivot entry is indicated by an asterisk. Apply the rules for a pivot exchange to this tableau and we obtain

(2.34)

h	-0.4	2	46	0.60
k	-0.6	-2	4	0.2
m	-0.4	-3	41	0.30
u	0.04	0.8	1.4	0.32
-1	-18	40	420	-144
	$= n$	$= v$	$= w$	$= g$

We obtained the same tableau with our bare hands in 1.32.

Since the dependent variables in the row system of a tableau are prefixed with negative signs while the dependent variables of the column system are not, it might appear that there is at least an operational difference between the two systems of equations. Actually, the relation between the two systems is symmetric and they are fully equivalent. Consider tableaux 2.35 and 2.36, both of which have dual systems of equations.

By comparing the equations generated by tableaux 2.35 and 2.36 we can see that each row equation of tableau 2.35 is a column equation of tableau 2.36, and conversely. Each column equation of tableau 2.35 is a row equation of tableau 2.36, and conversely.

(2.35)

	x_1	x_2	$\cdots$	x_n	-1	
v_1	a_{11}	a_{12}	$\cdots$	a_{1n}	b_1	$= -y_1$
v_2	a_{21}	a_{22}	$\cdots$	a_{2n}	b_2	$= -y_2$
$\vdots$	$\vdots$	$\vdots$		$\vdots$	$\vdots$	$\vdots$
v_m	a_{m1}	a_{m2}	$\cdots$	a_{mn}	b_m	$= -y_m$
-1	c_1	c_2	$\cdots$	c_n	d	$= f$
	$= u_1$	$= u_2$	$\cdots$	$= u_n$	$= g$	

(2.36)

	v_1	v_2	$\cdots$	v_m	-1	
x_1	$-a_{11}$	$-a_{21}$	$\cdots$	$-a_{m1}$	$-c_1$	$= -u_1$
x_2	$-a_{12}$	$-a_{22}$	$\cdots$	$-a_{m2}$	$-c_2$	$= -u_2$
$\vdots$	$\vdots$	$\vdots$		$\vdots$	$\vdots$	$\vdots$
x_n	$-a_{1n}$	$-a_{2n}$	$\cdots$	$-a_{mn}$	$-c_n$	$= -u_n$
$-l$	$-b_1$	$-b_2$	$\cdots$	$-b_m$	$-d$	$= -g$
	$= y_1$	$= y_2$	$\cdots$	$= y_m$	$= -f$	

The matrix of tableau 2.36 is the negative transpose of the matrix of tableau 2.35, and the matrix of tableau 2.35 is the negative transpose of the matrix of tableau 2.36. We extend this terminology to tableaux by saying that tableau 2.36, including the variables on the margins, is the *negative transpose* of tableau 2.35, and conversely.

We wish to emphasize that there is no intrinsic difference between the system of row equations and the system of column equations. Given any system of linear equations, we can write down a tableau for which the given system is the system of row equations, or we can write down a tableau for which the given system is the system of column equations. Furthermore, either tableau can be obtained from the other by taking the negative transpose. We can, in fact, take the negative transpose and then perform a pivot exchange, or perform a pivot exchange and then take the negative transpose with the same results in either case, provided in both cases the pivot exchange interchanges the same variables.

For example, the feed lot problem of Section 1.2 as given by tableau 2.33 can be transformed into tableau 2.37 by taking a negative transpose.

(2.37)

h	k	m	n	-1	
-10	-15	-10	-25	-450	$= -u$
-10	-10	-5	-20	-400	$= -v$
-60	-25	-55	-35	-1050	$= -w$
-3.80	-5.00	-3.50	-8.00	0	$= -g$

In this form, the tableau represents a maximum program. The objective is to maximize $-g$, which is equivalent to minimizing g. Then successive pivot exchanges to exchange n and u and to exchange h and v will result in tableau 2.38.

(2.38)

v	k	m	u	-1	
0.20	1.00	1.00	−0.20	10.00	$= -n$
−0.50	−1.00	−1.50	0.40	20.00	$= -h$
−23.00	−50.00	−110.00	17.00	500.00	$= -w$
−0.30	−0.80	−1.20	−0.08	156.00	$= -g$

By taking v, k, m, and $u = 0$, we get $-g = -156$. This is the same as the value obtained for g in tableau 1.36. Notice that if we take the negative transpose of tableau 2.38 we will get tableau 1.36.

Finally, let us give an example where both the row system and the column system of equations are preserved by a pivot exchange. Consider

(2.39)

	x_1	x_2	-1	
v_1	1*	2	3	$= -y_1$
v_2	3	4	5	$= -y_2$
-1	5	6	7	$= f$
	$= u_1$	$= u_2$	$= g$	

	y_1	x_2	-1	
u_1	1*	2	3	$= -x_1$
v_2	−3	−2	−4	$= -y_2$
-1	−5	−4	−8	$= f$
	$= v_1$	$= u_2$	$= g$	

Each tableau is obtained from the other by a pivot exchange on the starred entry. You should verify that the row systems of equations for both tableaux and the column systems of equations for both systems are equivalent.

Exercises

7. Perform two pivot exchanges on the pivot entries indicated by asterisks. Check your results against the calculations in Section 1,2.

h	10	10*	60	3.80
k	15	10	25	5.00
m	10	5	55	3.50
n	25*	20	35	8.00
-1	450	400	1050	0
	$= u$	$= v$	$= w$	$= g$

8. Perform three pivot exchanges on the pivot entries indicated by asterisks. The exchanges can be performed in any order, except that a pivot exchange cannot be performed where an entry is zero. Notice that

a zero is starred here, but it will be possible to perform a pivot exchange there after one or more of the other pivot exchanges. Check your results against the calculations in Section 1.3.

x_{11}	-1			1		1
x_{12}	-1				1*	3
x_{21}		-1		1*		2
x_{22}		-1			1	6
x_{31}			-1	1		4
x_{32}	*		-1		1	5
-1	-20	-30	-40	25	50	0
	$= y_1$	$= y_2$	$= y_3$	$= z_1$	$= z_2$	$= g$

9. Perform a pivot exchange on the starred entry.

v_1	1*	1	20
v_2	1		12
v_3		1	16
-1	0.9	0.6	
	$= u_1$	$= u_2$	$= g$

10. Perform a pivot exchange on the starred entry.

v_1	8	3	4*	50
v_2	40	10	10	200
-1	160	60	80	0
	$= u_1$	$= u_2$	$= u_3$	$= g$

11. Perform pivot exchanges on the three -1's in the following tableau.

v_1	1	4	1	-1	0	0	0
v_2	2	1	4	0	-1	0	0
v_3	1	3	5	0	0	-1	0
v_4	0	0	0	1	1	1	1
-1	16	30	40	0	0	0	0
	$= u_1$	$= u_2$	$= u_3$	$= u_4$	$= u_5$	$= u_6$	$= g$

2.4 Equivalent Tableaux

A concept that appears throughout mathematics is that of an equivalence relation. A relation is said to be defined in a set of objects if for any two objects in the set it can be decided that the relation holds (is true) or that the relation does not hold (is not true). Two familiar relations of the type we have in mind are the equality of numbers and the congruence of triangles. In each case whether the relation between two objects holds or does not depends on whether properties that are defined for the particular relation are or are not true.

Let us use the notation "$A \sim B$" to mean "A is related to B." The relation is an *equivalence relation* if it has the following three properties.

1. **Reflexivity:** For any object A in the set, $A \sim A$.

2. **Symmetry:** For any two objects A and B in the set, if $A \sim B$ then $B \sim A$.

3. **Transitivity:** For any three objects A, B, C in the set, if $A \sim B$ and $B \sim C$, then $A \sim C$.

For each equivalence relation that is defined one should prove that the definition implies the three properties given above. Our study of numbers and geometry may have occurred so early in our training that we might have been spared formal proofs of these three properties, but we should have no difficulty showing that the equality of numbers and the congruence of triangles are relations that are indeed reflexive, symmetric, and transitive. Other examples of equivalence relations abound throughout mathematics.

We say that two tableaux are *equivalent* if one can be obtained from the other by a sequence of pivot exchanges.

We do not want to get tied up in fine details but we do have to be rather precise about what is meant by this definition. First, to have this relation be reflexive we shall allow an empty sequence of pivot exchanges so that a tableau is equivalent to itself. Second, although one can pivot on any nonzero entry in a tableau, we have in mind restricting pivot exchanges to pivots on nonzero entries within the A-matrix. In some contexts it may be desirable to allow pivot entries outside the A-matrix. That would simply extend the number of equivalent tableaux that one could obtain. Third, a rearrangement of the rows or columns of a tableau can often be obtained by a sequence of pivot exchanges. However, we wish to regard a rearrangement as yielding the same tableau, not just an equivalent tableau.

We shall leave it to the reader to satisfy himself that the relation defined is symmetric and transitive.

Note that the variables on the margin are an integral part of a tableau. Thus,

(2.40)

x_1	x_2	-1	
1	0	1	$= -y_1$
0	2	1	$= -y_2$
0	1	3	$= f$

and

(2.41)

y_1	x_2	-1	
1	0	1	$= -x_1$
0	2	1	$= -y_2$
0	1	3	$= f$

are equivalent but different tableaux, since x_1 and y_1 are exchanged. However, we can omit either the variables of the row system or the variables of the column system if we wish since the variables of the dual system are always implied.

Later we shall show that an optimal solution to a linear programming problem, if one exists, can be found by examining the different equivalent tableaux. For this reason it is important to show that for each tableau there are only finitely many different tableaux equivalent to it. For a small tableau it is possible, if a little tedious, to obtain the entire set of different equivalent tableaux by exhausting all possibilities. Two such examples are shown in 2.44 and 2.45 below.

Suppose we start with a tableau for which the A-matrix is an $m \times n$ matrix. That is, there are n independent variables on the top margin and m dependent variables on the right margin. Every pivot exchange will produce a tableau with the same number of independent variables and the same number of dependent variables. Thus, the number of ways the variables can be divided between the top and right margins is at most equal to the number of ways that the $m+n$ variables can be divided into two sets, one with n variables and the other with m variables. This is the number of combinations of $m+n$ things m-at-a-time, or $(m+n)!/m!n!$.

It will turn out that sometimes not every potential division of the variables between the two margins can be obtained by a sequence of pivot exchanges. The second example given below illustrates this possibility. However, for each division of the variables that can be obtained there is only one tableau of the equivalence class with those variables on the margins. This is a consequence of the following theorem.

Theorem 2.1 *Equivalent tableaux are the same if their independent and dependent variables are the same.*

Proof. Let (2.42)

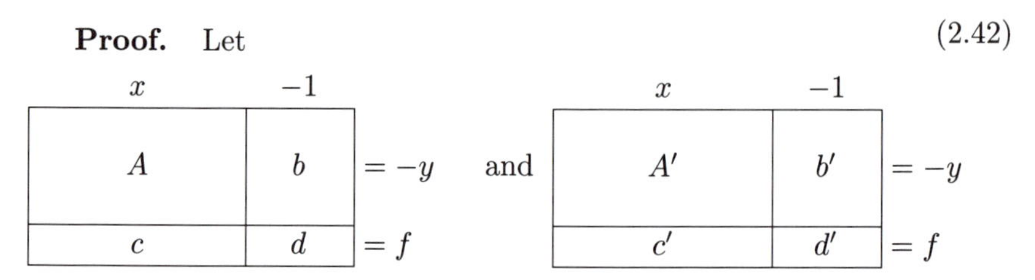

denote tableaux with the same independent and dependent variables. If they are equivalent, then their solutions are the same. Set $x = 0$ to obtain $b = y$ and $d = -f$ at the left and $b' = y = b$ and $d' = -f = d$ at the right.

Let A_k denote the k-th column of the A-matrix of the left tableau, and let A'_k denote the k-th column of the A-matrix of the right tableau. Then by setting $x_k = 1$ and $x_j = 0$ for all $j \neq k$, we obtain $A_k = b - y$ and $c_k = d + f$ at the left, and $A'_k = b' - y = b - y = A_k$ and $c'_k = d' + f = d + f = c_k$ at the right. We can conclude that $A = A'$, $b = b'$, $c = c'$, and $d = d'$. The two tableaux are the same. $\boxplus$

In Section 2.1 we associated the duality equation with a particular tableau. It is important to realize that the duality equation is preserved by the pivot exchange and that

all equivalent tableaux have the same duality equation.

For example, a pivot exchange with a_{rs} as the pivot entry will exchange the variables x_s and y_r for the row equations and it will exchange v_r and u_s for the column equations. Thus, v_r and y_r will be at opposite ends of a row before the pivot exchange and at opposite ends of a column after the exchange, while x_s and u_s will be at opposite ends of a column before the exchange and at opposite ends of a row after the exchange.

Here we shall give two examples of sets of equivalent tableaux. Consider first the tableau

(2.43)

	x_1	x_2	-1	
v_1	3	4	96	$= -y_1$
v_2	5	6	150	$= -y_2$
-1	300	384	0	$= f$
	$= u_1$	$= u_2$	$= g$	

There are six tableaux equivalent to tableau 2.43. They are shown in 2.44. None has a zero entry in the A-matrix. Thus, from each, four tableaux can be obtained by a single pivot exchange. Six different denominators appear, one in each tableau. The first tableau is the same as tableau 2.43 with 1's written in the denominators to make its identification similar to the others. For each nonzero entry, the integer in the numerator identifies, since it becomes the denominator, the tableau that is obtained if that entry is used as the pivot entry. The reader can verify that each possible pivot exchange produces one of the other tableaux in the set. Thus, this is the entire set of tableaux equivalent to tableau 2.43.

(2.44)

	x_1	x_2	-1	
v_1	3/1	4/1	96	$= -y_1$
v_2	5/1	6/1	150	$= -y_2$
-1	300	384	0	$= f$
	$= u_1$	$= u_2$	$= g$	

	y_1	y_2	-1	
u_1	$-6/2$	4/2	12	$= -x_1$
u_2	5/2	$-3/2$	15	$= -x_2$
-1	-60	-24	-9360	$= f$
	$= v_1$	$= v_2$	$= g$	

	x_2	y_1	-1	
u_1	4/3	1/3	32	$= -x_1$
v_2	$-2/3$	$-5/3$	-10	$= -y_2$
-1	-16	-100	-9600	$= f$
	$= u_2$	$= v_1$	$= g$	

	x_1	y_1	-1	
u_2	3/4	1/4	24	$= -x_2$
v_2	2/4	$-6/4$	6	$= -y_2$
-1	12	-96	-9216	$= f$
	$= u_1$	$= v_1$	$= g$	

	x_2	y_2	-1	
u_1	6/5	1/5	30	$= -x_1$
v_1	2/5	$-3/5$	6	$= -y_1$
-1	24	-60	-9000	$= f$
	$= u_2$	$= v_2$	$= g$	

	x_1	y_2	-1	
u_2	5/6	1/6	25	$= -x_2$
v_1	$-2/6$	$-4/6$	-4	$= -y_1$
-1	-20	-64	-9600	$= f$
	$= u_1$	$= v_2$	$= g$	

Since the number of combinations of 4 things 2-at-a-time is 6, this example yields the maximum number of different equivalent tableaux. The next example yields fewer than the maximum number of equivalent tableaux. The number of combinations of 5 things 2-at-a-time is 10 while the number of equivalent tableaux is just 8. The example also demonstrates a different way of indicating the network of possible pivot exchanges. In this example we use a simplifying convention that we first used in the transportation problem of Section 1.3. *Blanks denote zeros.*

(2.45)

A

x_1	x_2	-1	
1^B		1	$=-y_1$
1^C	1^E	1	$=-y_2$
	1^G	1	$=-y_3$
1	1		$=f$

B

x_2	y_1	-1	
	1^A	1	$=-x_1$
1^D	-1^C		$=-y_2$
1^H		1	$=-y_3$
1	-1	-1	$=f$

C

x_2	y_2	-1	
1^E	1^A	1	$=-x_1$
-1^D	-1^B		$=-y_1$
1^F		1	$=-y_3$
	-1	-1	$=f$

D

y_1	y_2	-1	
1^E		1	$=-x_1$
-1^C	1^B		$=-x_2$
1^F	-1^H	1	$=-y_3$
	-1	-1	$=f$

E

x_1	y_2	-1	
1^C	1^A	1	$=-x_2$
1^D		1	$=-y_1$
-1^F	-1^G		$=-y_3$
	-1	-1	$=f$

F

y_2	y_3	-1	
1^G	-1^E		$=-x_1$
	1^C	1	$=-x_2$
-1^H	1^D	1	$=-y_1$
-1		-1	$=f$

G

x_1	y_3	-1	
	1^A	1	$=-x_2$
1^H		1	$=-y_1$
1^F	-1^E		$=-y_2$
1	-1	-1	$=f$

H

y_1	y_3	-1	
1^G		1	$=-x_1$
	1^B	1	$=-x_2$
-1^F	-1^D	-1	$=-y_2$
-1	-1	-2	$=f$

Here each tableau is labeled with a capital letter. Each nonzero entry in an A-matrix has a superscript that indicates the tableau that will be obtained if that entry is chosen as the pivot entry. If we verify that the labels are correct then every possible pivot exchange will yield another tableau that is one of the eight shown in 2.45. Thus, no other tableau can be obtained and the eight shown are all the different equivalent tableaux that can be obtained.

The paths of possible pivot exchanges are illustrated in Figure 2.4. The nodes of the graph represent the various tableaux in the equivalence class of 2.45. The line segments show the tableaux that can be reached from a given tableau by a simple pivot exchange. Others can be reached with

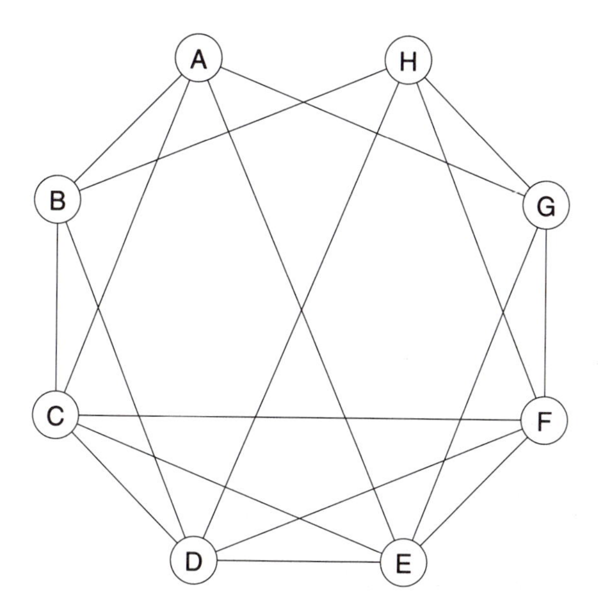

Figure 2.4: Paths of possible pivot exchanges.

more than one pivot exchange along the arcs shown.

Tableaux with x_1, y_1 or x_2, y_3 on the top margin do not exist here. A direct pivot exchange to one of these configurations from one of the eight tableaux shown in 2.45 is blocked by the zeros (blanks) in the A-matrices. In a more general setting we will have to show that if a particular tableau cannot be obtained directly because of the presence of a zero it also cannot be obtained by some long and devious sequence of pivot exchanges. Here the issue is settled because all possible pivot exchanges are accounted for.

These two examples will serve as resources for further discussions in the next few sections.

2.5 Basic Solutions

Consider a tableau in the form of 2.46.

$$
\begin{array}{c|cccc|c|l}
 & x_1 & x_2 & \cdots & x_n & -1 & \\
\hline
v_1 & & & & & b_1 & = -y_1 \\
v_2 & & & & & b_2 & = -y_2 \\
\vdots & & & & & \vdots & \quad \vdots \\
v_m & & & & & b_m & = -y_m \\
\hline
-1 & c_1 & c_2 & \cdots & c_n & d & = f \\
\hline
 & = u_1 & = u_2 & \cdots & = u_n & = g &
\end{array}
\tag{2.46}
$$

For the row system of equations, the variables on the right margin, the dependent variables, are called *basic variables*, and the variables on the top margin, the independent variables, are called *nonbasic variables*. Which variables are basic and which are nonbasic depends on the given tableau and will be different for different equivalent tableaux.

For the column system of equations, the basic variables are on the bottom margin and the nonbasic variables are on the left margin.

In each case a *basic solution* is obtained by setting the nonbasic variables equal to zero.

A basic solution is easy to obtain from the tableau since it does not depend on the entries in the A-matrix. For the row system we get $y_1 = b_1$, $y_2 = b_2, \ldots, y_m = b_m$, and $f = -d$. That is, a basic solution can be obtained by reading down the last column (and taking the signs into account). For the column system we get $u_1 = -c_1$, $u_2 = -c_2, \ldots, u_n = -c_n$, and $g = -d$. In this case the basic solution is obtained by reading across the bottom row (again, taking the signs into account).

The importance of the basic solutions stems from the fact that if a linear programming problem has an optimal solution an optimal solution can be found among the basic solutions. This is not obvious. Many treatments of linear programming state and prove this assertion as a theorem. They do this because they use this fact to establish the validity of the simplex method. We are going to establish the validity of the simplex method by other means. Since the optimal solution obtained by the simplex method is a basic solution, this assertion will then follow as a rather simple observation. Therefore, we need not prove this statement here.

In the previous section we showed that there are only finitely many different equivalent tableaux. It follows then that there are only finitely many different basic solutions. Thus, there are only finitely many places to look for an optimal solution.

For the moment let us concentrate our attention on the system of row equations. Each tableau equivalent to 2.46 yields a basic solution for the row system. Each basic solution has at least n of the variables equal to zero, the nonbasic variables. If no b_i is zero the corresponding basic solution is different from the basic solutions for other equivalent tableaux since at least one (nonzero) basic variable for the solution under consideration would be a (zero) nonbasic variable for the basic solution for any other equivalent tableau. Thus, if for all tableaux equivalent to 2.46 no b_i is zero, there is a one-to-one correspondence between tableaux and basic solutions. A similar remark applies to the column system if no tableau equivalent to 2.46 has a zero in the c-row.

On the other hand, if one of the b_i is zero, one of the basic variables has a zero value. Pivoting in the i-th row (which can be done unless the entire row is zero) will yield a tableau in which that variable becomes nonbasic. Then, setting the nonbasic variables equal to zero will yield the same solution. That is, we would get two equivalent tableaux for which the basic solutions were the same. The basic solutions are the same though the sets of basic variables are different.

For any tableau, when values are assigned to the nonbasic variables, the values of the basic variables are uniquely determined. Thus, for a basic solution, when it is determined which variables will be zeros, the values of the nonzero variables are uniquely determined. The following converse of this observation is also true.

Theorem 2.2 *For any set of variables for the row equations, if there is one and only one solution with those variables equal to zero, then that solution is a basic solution.*

Proof. Suppose we have any solution of the row system of 2.46 for which the variables $w_1, w_2, \ldots, w_r$ have zero values and that this is the only solution with these variables equal to zero. We try to perform pivot exchanges to move as many of the w_i as possible to the top margin. When the maximum number of the w_i on the top margin is obtained we have a tableau of the form

$$
\begin{array}{ccc|ccc|c|l}
w_1 & \cdots & w_s & x_{s+1} & \cdots & x_n & -1 & \\
\hline
 & & & & & & & = -y_1 \\
 & & & & & & & \vdots \\
 & & & & & & & = -y_s \\
\hline
 & & & 0 & \cdots & 0 & 0 & = -w_{s+1} \\
 & & & \vdots & & \vdots & \vdots & \vdots \\
 & & & 0 & \cdots & 0 & 0 & = -w_r \\
\hline
 & & & & & & & = f
\end{array}
\qquad (2.47)
$$

We must have a block of zeros (as illustrated), or the w-rows must be absent, or the x-columns must be absent. In any case, setting $w_1 = \cdots = w_s = 0$ gives $w_{s+1} = \cdots = w_r = 0$, satisfying the conditions of the hypothesis of the theorem. If any of the x-columns were present, we could assign any values we please to (say) x_n. This would contradict the assumed uniqueness of the solution with these zeros. Thus, the x-columns must be absent; that is, $s = n$. Then the given solution is a basic solution with w_1, $\cdots$, w_n the nonbasic variables. ◻

Theorem 2.2 does not require that the number of zero variables be equal to n (the number of nonbasic variables), but the number of zero variables will be at least n. A slightly sharper form of this theorem is also true.

Theorem 2.3 *If, for any solution of the row equations of a tableau 2.46, n of the variables are zeros for that solution and it is the only solution with at least those variables equal to zero, then that solution is a basic solution and the n variables involved are nonbasic variables.*

The proof of the previous theorem applies here without change. ◻

There are corresponding versions of these theorems for the system of column equations.

The first example 2.40 in the previous section shows a set of equivalent tableaux for which there is no zero entry in any of the b-columns or c-rows. The basic solutions for that example are

From the definition of basic and nonbasic variables we see that in every tableau, for every pair of dual variables, one of the pair is a basic variable for one of the systems of equations and the other is a nonbasic variable for the other system. Thus, for every pair of basic solutions for the same tableau one or the other in each pair is zero because it is a nonbasic variable. Some of the basic variables might also be zero, but let us postpone discussing the consequences of that possibility for the moment.

> **A pair of solutions, one for the row system and one for the column system of a tableau, is said to be** *complementary* **if for each pair of dual variables, one or the other is zero.**

In particular, a pair of basic solutions for the same tableau will always be complementary. This allows us to use our practice of regarding blanks as zeros advantageously in a display like Table 2.1. In each column of that table is a pair of basic solutions for the same tableau. Replace each listed value (always a zero) for a nonbasic variable by a blank. This leaves a space to write the value of the basic variable for the other (dual) solution.

Table 2.1: Basic Solutions (First Example)

	denominator					
	1	2	3	4	5	6
x_1	0	12	32	0	30	0
x_2	0	15	0	24	0	25
y_1	96	0	0	0	6	−4
y_2	150	0	−10	6	0	0
f	0	9360	9600	9216	9000	9600
u_1	−300	0	0	−12	0	20
u_2	−384	0	16	0	−24	0
v_1	0	60	100	96	0	0
v_2	0	24	0	0	60	64
g	0	9360	9600	9216	9000	9600

To distinguish between the two solutions we mark one of the solutions by writing those values in parentheses. In this way Table 2.1 can be rewritten in the form of Table 2.2.

Table 2.2: Complementary Basic Solutions (First Example)

	denominator					
	1	2	3	4	5	6
$x_1(u_1)$	(−300)	12	32	(−12)	30	(20)
$x_2(u_2)$	(−384)	15	(16)	24	(−24)	25
$y_1(v_1)$	96	(60)	(100)	(96)	6	−4
$y_2(v_2)$	150	(24)	−10	6	(60)	(64)
$f = g$	0	9360	9600	9216	9000	9600

Besides being shorter than Table 2.2 is easier to read and more informative. In Table 2.2 we have chosen to enclose numbers and variables for the column system in parentheses. The choice is arbitrary. For variables other than the objective variables we read the table by assigning values not in parentheses to variables not in parentheses (in the row system) and assigning values in parentheses to variables in parentheses (in the column system). The displaced variables are represented by blanks and are taken to be zeros. From the duality equation 2.12 we see that for a pair of complementary solutions the values of the two objective variables are equal. Thus, in the row of Table 2.2 where the values of f and g are given we do not have to distinguish between the two systems of equations.

Let us display the basic solutions for the set of equivalent tableaux in the second example of the previous section.

In this example sometimes both variables in a dual pair are zero, one because it is a nonbasic variable and the other because of a zero in the b-column or the c-row. In these cases we have chosen to show the zeros of the basic variables. That is, the blanks are the zeros that are forced for a basic solution while the indicated zeros are computed zeros.

Table 2.3: Complementary Basic Solutions (Second Example)

	tableau							
	A	B	C	D	E	F	G	H
$x_1(u_1)$	(−1)	1	1	1	(0)	0	(−1)	1
$x_2(u_2)$	(−1)	(−1)	(0)	0	1	1	1	1
$y_1(v_1)$	1	(1)	0	(0)	1	1	1	(1)
$y_2(v_2)$	1	0	(1)	(1)	(1)	(1)	0	−1
$y_3(v_3)$	1	1	1	1	0	1	1	2
$f = g$	0	0	1	1	1	1	1	2

We have shown earlier that if there is no zero in any b-column (or c-row) of a class of equivalent tableaux then there is a one-to-one correspondence between tableaux and basic solutions. Here, where zeros do occur in the b-column or c-row, we have multiple solutions. From Table 2.3 we see that x_2 is a computed zero (a basic variable) in tableau D while it is a zero nonbasic variable in tableaux A, B, and C. Also, u_2 is a computed zero in tableau C while it is a zero nonbasic variable in tableaux D, E, F, G, and H. As a result, the row system has identical basic solutions for tableaux B, C, and D and also identical basic solutions for tableaux E, F, and G. The column system has identical basic solutions for tableaux C, D, E, and F.

It is not difficult to see how multiple basic solutions can occur. Suppose there is a zero in the b-column. Any pivot exchange based on a pivot entry in the row containing this zero will exchange two variables, one of which is a zero basic variable and the other a zero nonbasic variable. After the exchange they will still both be zero, and the values of the other variables will be unchanged. A similar observation applies to the column system. A zero in the c-row will imply multiple basic solutions for the column system.

> **When a basic variable for a basic solution is zero the basic solution is said to be *degenerate*. Whenever any tableau in a class of equivalent tableaux has a degenerate basic solution, the class is said to be *degenerate*.**

It is easy to tell whether a basic solution is degenerate, but it is difficult to determine if a class of equivalent tableaux is degenerate. If one happens onto a degenerate tableau in a class, the class is degenerate. But if one believes that a class is *nondegenerate*, there is no practical way to establish that without examining every tableau in the class.

Degeneracy causes a number of problems in the theory of linear programming, of which the lack of a one-to-one correspondence between equivalent tableaux and basic solutions is a relatively minor example. (See Section 4.2 for further discussions of degeneracy and its implications.) However, a large amount of practical experience suggests that degeneracy is unlikely to cause difficulty in applied problems, though many of them are degenerate. Indeed, degeneracy has the possible advantage that it reduces the number of distinct basic solutions.

2.6 Inversion

We have described in Section 2.1 how a tableau could be viewed as an input-output device in which the nonbasic (independent) variables were inputs and the basic (dependent) variables were outputs. Inversion amounts to interchanging the inputs and the outputs. Consider the tableau 2.48.

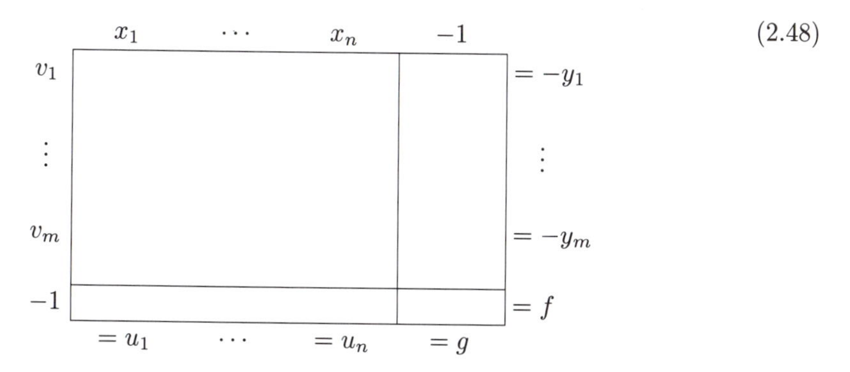

(2.48)

> **The row system of tableau 2.48 is *invertible* if a tableau in the form of tableau 2.49 exists, with the same solution set as tableau 2.48.**

(2.49)

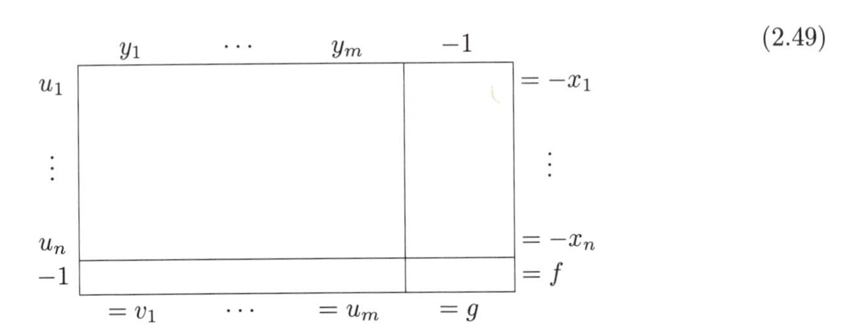

This definition of invertibility does not require that tableau 2.49 can be obtained from tableau 2.48 by a sequence of pivot exchanges. Our first objective is to prove that this is the case. When that is done it also will show that m and n must be equal.

The important thing about the form of tableau 2.49 is that the values of the independent variables (the y's) on the top margin can be chosen freely (arbitrarily) and independently, and once those values are assigned, the values of the dependent variables (the x's) are fixed (uniquely determined). This is what variables being independent and dependent implies.

A sequence of pivot exchanges is said to be *independent* if no two pivot entries in the sequence are in the same row or in the same column.

Theorem 2.4 *The row system of a tableau is invertible if and only if it can be inverted by a sequence of independent pivot exchanges.*

Proof. Starting with tableau 2.48 make as many pivot exchanges as possible, each time choosing the pivot entry in a new row and a new column (that is, independently), and each time exchanging a new pair of basic and nonbasic variables. We will be stopped when we run out of rows or columns, or when we encounter a block of zeros at the intersections of the remaining rows and columns. When that occurs, the tableau will look like tableau 2.50.

We assume that exactly r pivot exchanges were performed, and to simplify notation we assume the pivot entries were in the first r rows and first r columns and that any necessary rearrangement of the rows and columns is made so that the variables on the margins are indexed as shown. We allow $r = n$ or $r = m$, or both. Tableaux 2.48, 2.49, and 2.50 have the same solutions sets.

Set $y_1 = y_2 = \cdots = y_r = 0$ in tableau 2.50. Then, though the values of $x_{r+1}, \ldots, x_n$ are free to vary, the values of $y_{r+1}, \ldots, y_m$ are fixed. In tableau 2.49, when the values for the y's are chosen the values for the x's

$$
\begin{array}{c|ccc|ccc|c|l}
 & y_1 & \cdots & y_r & x_{r+1} & \cdots & x_n & -1 & \\ \hline
u_1 & & & & & & & & = -x_1 \\
\vdots & & & & & & & & \quad \vdots \\
u_r & & & & & & & & = -x_r \\ \hline
v_{r1} & & & & 0 & \cdots & 0 & & = -y_{r+1} \\
\vdots & & & & \vdots & & \vdots & & \quad \vdots \\
v_m & & & & 0 & \cdots & 0 & & = -y_m \\ \hline
-1 & & & & & & & & = f \\ \hline
 & = v_1 & \cdots & = v_r & = u_{r+1} & \cdots & = u_n & = g &
\end{array}
\qquad (2.50)
$$

are fixed. If n is larger than r then x_n is fixed in tableau 2.49 and free in tableau 2.50. This is a contradiction. If m is larger than r, then the value of y_m is determined by the values of $y_1, \ldots, y_r$ in tableau 2.50 and independent of the values of $y_1, \ldots, y_r$ in tableau 2.49. This is also a contradiction. Thus, $m = r = n$. This shows that tableau 2.48 must be square if it is invertible, and its inverted tableau 2.49 can be obtained by a sequence of n independent pivot exchanges. ◻

By a symmetrical argument we can establish similar results for the column system. Notice that when the row system is inverted by pivoting the column system is also inverted. One cannot invert the row system or the column system without inverting the other. Thus, we can speak of inverting the tableau rather than inverting one system or the other. Instead of giving a theorem for the column system corresponding to Theorem 2.4 we prefer the following statement, which deals with the row and column systems symmetrically.

Theorem 2.5 *If either the row or the column system can be inverted then*

1. *both can be inverted,*

2. *the tableau is square, i.e.,* $m = n$, *and*

3. *the inversion can be achieved by a sequence of* n *independent pivot exchanges.* ◻

It is easy to see by examples that it is possible that several different assignments of the independent variables yield the same values for the dependent variables. For example, in tableau 2.51 the value of x_2 does not affect the value of y_1.

(2.51)

x_1	x_2	-1	
1	0	1	$= -y_1$
1	1	0	$= f$

If different values of x determine different values of y in tableau 2.48 we say the mapping defined by the row system of the tableau is *one-to-one.*

We can also see by examples that it is possible that not all potential values for the dependent variables can be achieved. For example, in tableau 2.52 it is not possible for y_1 and y_2 to be equal.

(2.52)

x_1	-1	
1	1	$= -y_1$
1	2	$= -y_2$
1	0	$= f$

The set of values of the dependent variables that can be achieved is called the *range* of the row system of the tableau. If the range includes all conceivable values of y the mapping defined by the row system of tableau 2.48 is said to be *onto.*

The argument in the last paragraph of the proof of Theorem 2.4 can be used to shed some light on both these situations. Suppose, again, that we start with tableau 2.48 and make as many independent pivot exchanges as can be made to end up with a tableau in the form of tableau 2.50. Tableaux 2.48 and 2.50 have the same solution sets.

The range of the row system of tableau 2.48 can be found from tableau 2.50 by assigning values arbitrarily to $y_1, \ldots, y_r$ and computing the corresponding values of $y_{r+1}, \ldots, y_m$. The values of the x's that will yield these values for the y's can be found by choosing $x_{r+1}, \ldots, x_n$ (if $r < n$) and computing $x_1, \ldots, x_r$ from tableau 2.50. That is, the y's determined this way are in the range of the row system of the tableau. If $r < m$ then y_m is not independent of $y_1, \ldots, y_r$. That is, not all values of y can be obtained. Thus, if the mapping defined by the row system of tableau 2.48 is onto, then $r = m$. Also, if $r = m$, the y's in tableau 2.50 are independent and the mapping is onto.

If $r < n$ then x_n is still free in tableau 2.50. Thus, if the mapping defined by tableau 2.48 is one-to-one, then $r = n$. Also, if $r = n$, for each y in the range there is only one possible choice for x. That is, the mapping is one-to-one if and only if $r = n$.

The maximum number of independent pivots that are possible is r. This number is called the *rank* of the tableau. We have not shown that the rank is independent of the sequence of pivot exchanges used, though it is. We make little use of the concept of rank beyond those described in the theorems of this section, so we shall not go farther in proving properties associated with the rank of a matrix.

Let us summarize what we have observed so far.

The maximum number of independent pivot exchanges is r (the rank). The row system of equations is one-to-one if and only if $r = n$. The row system of equations is onto if and only if $r = m$.

Similar remarks can be made about the column system of equations. Just as the row equations in tableau 2.50 involving $y_{r+1}, \cdots, y_m$ are the equations defining the range of the mapping defined by the row equations, the column equations involving $u_{r+1}, \cdots, u_n$ are the equations defining the range of the mapping defined by the column equations. The column system of equations is one-to-one if and only if $r = m$. The column system of equation is onto if and only if $r = n$.

Theorem 2.6 *If any one of the following three statements holds in a specific tableau, all three hold.*

1. *The mapping defined by the row system is one-to-one.*

2. *The mapping defined by the column system is onto.*

3. $r = n$.

Theorem 2.7 *If any one of the following three statements holds in a specific tableau, all three hold.*

1. *The mapping defined by the row system is onto.*

2. *The mapping defined by the column system is one-to-one.*

3. $r = m$.

Proof. The equivalence of statements 1 and 3 in each theorem is the content of the discussion preceding Theorem 2.6. The equivalence of statements 2 and 3 in each theorem is the same argument applied to the column system. $\boxminus$

In summary:

$r = m$ is equivalent to the induced mapping defined by the row system being onto (and the column system being one-to-one), and

$r = n$ is equivalent to the induced mapping defined by the row system being one-to-one (and the column system being onto).

Since invertibility is equivalent to $r = m = n$, invertibility is equivalent to the condition that the induced mapping is both onto and one-to-one.

The ideas introduced here correspond to concepts in the more traditional language of linear algebra. We will not try to prove the exact connections between tableau algebra and linear algebra. To do so would require that we assume the reader already knows the corresponding definitions and theorems in linear algebra, or we would have to introduce and prove them. However, it is true that the rank of a tableau in tableau algebra is the same as the rank of the A-matrix in linear algebra; the mapping induced by the row system is onto if and only if the linear transformation represented by the A-matrix is onto; and the mapping induced by the row system is one-to-one if and only if the linear transformation represented by the A-matrix is one-to-one.

We will, however, say something about the connection between invertibility of a tableau and the invertibility of the A-matrix as defined in linear algebra. The connection is complicated by the fact that in tableau algebra we are free to rearrange the orders of the rows and columns. To discuss this connection, delete the basement row and the last column in tableaux 2.48, 2.49, and 2.50. The row and column equations in tableau 2.48 then become

$$Ax = -y \tag{2.53}$$

$$vA = u \tag{2.54}$$

If we let $\bar{A}$ denote the A-matrix of tableau 2.49 then the row and column equations in tableau 2.49 become

$$\bar{A}y = -x \tag{2.55}$$

$$u\bar{A} = v \tag{2.56}$$

The substitution of y in equation 2.53 for y in equation 2.55 gives

$$\bar{A}Ax = x \tag{2.57}$$

for all x. This statement requires that the variables on the rows of tableau 2.49 be in the same order as the variables on the columns of tableau 2.48.

The substitution of u in equation 2.54 for u in equation 2.56 gives

$$vA\bar{A} = v \tag{2.58}$$

for all v. This statement requires that the variables on the columns of tableau 2.49 be in the same order as the variables on the rows of tableau 2.48. Equations 2.57 and 2.58 together imply that

$$\bar{A}A = I = A\bar{A} \tag{2.59}$$

or

$$\bar{A} = A^{-1} \tag{2.60}$$

In linear algebra, a square matrix A is said to be *invertible* if there exist matrices B and C such that $AB = I$ and $CA = I'$ (not necessarily the same unit matrix). The argument above shows that B and C can both be taken to be $\bar{A}$. That is, if a tableau can be inverted, then the A-matrix has an inverse in the sense of linear algebra.

The converse is also easy to establish. If $CA = I$ and $AB = I$, then $B = (CA)B = C(AB) = C$. Then equation 2.53 becomes, by multiplying on the left by B, $By = -x$. Also, multiplying $By = -x$ by A on the left yields $Ax = -y$. That is, $Ax = -y$ and $By = -x$ have the same solution sets. In the sense of our tableau algebra, A is invertible.

Since tableau 2.49 can be obtained from tableau 2.48 by a sequence of n independent pivots, the inverse matrix A^{-1} can be obtained by pivoting, and rearranging rows and columns, if necessary. The independent pivots can be chosen arbitrarily in any order. If the inversion is blocked by a block of zeros, as in tableau 2.50, the matrix cannot be inverted since the mapping defined by the matrix will either fail to be onto or fail to be one-to-one. If n independent pivots can be found, the matrix can be inverted. To find the inverse matrix, however, it will usually be necessary to rearrange the rows and columns.

The ideas discussed here also have direct connections with theorems in linear algebra concerning the solvability of systems of linear equations.

Theorem 2.8 . *The matrix A is invertible if and only if the equation $Ax = b$ has a unique solution for every b.*

Proof. The statement that $Ax = b$ has a solution for every b is equivalent to saying that the linear mapping $Ax = y$ is onto. The statement that the solution is unique is equivalent to saying that the linear mapping is one-to-one. ◻

Theorem 2.9 *The matrix A is invertible if and only if the equation $vA = c$ has a unique solution for every c.* ⊟

If $Ax = b$ has a unique solution, the equation $Ax = 0$ has only the solution $x = 0$. In this case we say the columns of A are linearly independent. If $vA = c$ has a unique solution, the equation $vA = 0$ has only the solution $v = 0$. In this case the rows of A are linearly independent. If uniqueness fails in either case, the rows (or columns) are linearly dependent.

Finally, consider the system $Ax = b$ for any given $m \times n$ matrix A and $m \times 1$ matrix b. The question whether $Ax = b$ has a solution (i.e., whether the system is "consistent" or "inconsistent") is equivalent to the question as to whether the problem $Ax - b = -y$ has a solution with $y = 0$. This can be determined by attempting to invert the tableau. When we arrive at a tableau in the form of 2.50, set the y's on the top margin to zero. If any basic y is nonzero, there is no solution with $y = 0$. If all basic y's are zero, we can assign the nonbasic x's, if there are any, any values we wish. Those x's are the parameters of the solution set and the solution set is nonempty.

Which of the two alternatives described here holds can be determined by looking at the b-column of tableau 2.50. If the entries of the b-column in rows of basic y's are all zeros, set the nonbasic y's at the top all equal to zero, and the nonbasic x's to any arbitrary values, to get a basic solution of $Ax = b$. If any entry of the b-column in the row of a basic y is nonzero, set the u's and v's all equal zero at the left, except the v opposite the nonzero entry. Set the v in this row to 1 (say) to get a solution of $vA = 0$, $vb \neq 0$. This proves the following theorem.

Theorem 2.10 *(Theorem of Alternatives) Either $Ax = b$ has a solution or $vA = 0$, $vb \neq 0$ has a solution.* ⊟

The alternatives exclude one another because, with the basement row deleted, the duality equation is $ux + vy = g$. The condition $Ax = b$ means $y = 0$, and the condition $vA = 0$ means $u = 0$. But $g = vb \neq 0$.

By a symmetric argument one proves

Theorem 2.11 *Either $vA = c$ has a solution or $Ax = 0$, $cx \neq 0$ has a solution.* ⊟

Other terminology is often used for the concepts of onto, into, and one-to-one. An *epimorphism*, or a *surjection*, is an onto mapping. A *bimorphism*, or a *bijection*, is one-to-one and onto. A *monomorphism*, or an *injection*, is a one-to-one mapping onto a subset. A monomorphism or an injection does not quite correspond to an into mapping, which is not specific about being one-to-one.

2.7 Block Pivots

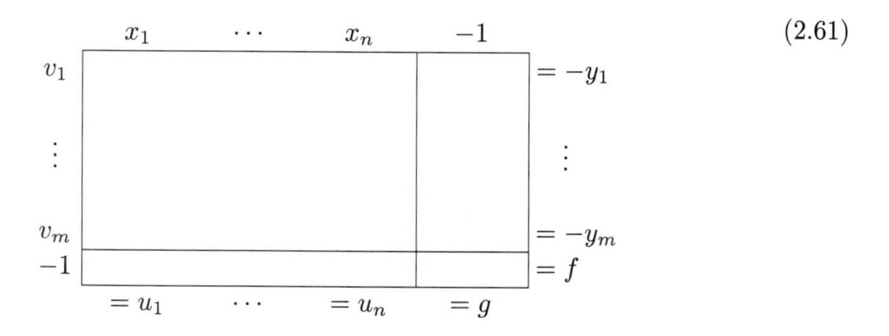

(2.61)

Suppose a tableau exists with the same variables as tableau 2.61 and with the same solution set, but with the variables of the row system redistributed between the top and right margins. To simplify the notation, assume this tableau has the following appearance.

(2.62)

y_1 $\cdots$ y_r	x_{s+1} $\cdots$ x_n	-1	
			$= -x_1$ $\vdots$ $= -x_s$
			$= -y_{r+1}$ $\vdots$ $= -y_m$
			$= f$

Notice, in particular, that we do not assume that the number of y's that have been moved to the top margin and the number of x's that have been moved to the right margin are equal. In both tableau 2.61 and tableau 2.62 set $x_{s+1} = \cdots = x_n = 0$ and ignore the rows involving $y_{r+1}, \ldots, y_m$. Setting the x's on the top margin equal to zero has the same effect as deleting them and their columns. This reduces both to tableaux that are inverses of each other. Since they have the same solution sets, $r = s$ and one can be obtained from the other by a sequence of independent pivot exchanges.

This shows only that a tableau with the same variables on the margins can be obtained with a sequence of r pivot exchanges entirely within the rows and columns of the exchanged variables. We have already shown in Section 2.4 that the variables on the margins determine the tableau. Thus, tableau 2.62 must be the same as the tableau obtained by the r pivot exchanges.

Theorem 2.12 *Any two tableaux with the same solution sets (either solutions to the row system or solutions to the column system) are equivalent. One can be obtained from the other by a sequence of independent pivot exchanges.* ◫

The argument above shows that the tableau that results from deleting the rows and columns of the variables that have not been exchanged is an invertible tableau. We can use this fact to simplify the notation involved. Let tableau 2.61 be rewritten in the form

(2.63)

	x'	x''	-1	
v'	P	Q	b'	$= -y'$
v''	R	S	b''	$= -y''$
-1	c'	c''	d	$= f$
	$= u'$	$= u''$	$= g$	

Here, x' denotes the one-column matrix of the variables $x_1, \ldots, x_r$ that are exchanged between 2.61 and 2.62, and x'' denotes the one-column matrix of the variables $x_{r+1}, \ldots, x_n$ that are not exchanged. Similarly, y' denotes the one-column matrix of the variables $y_1, \ldots, y_r$, and y'' denotes the one-column matrix of the variables $y_{r+1}, \ldots, y_m$. The rest of the tableau and the variables on the other margins are subdivided compatibly.

Using similar notation the tableau 2.62 can be rewritten in the form

(2.64)

	y'	x''	-1	
u'	$\bar{P}$	$\bar{Q}$	$\bar{b}'$	$= -x'$
v''	$\bar{R}$	$\bar{S}$	$\bar{b}''$	$= -y''$
-1	c'	c''	d	$= f$
	$= v'$	$= u''$	$= g$	

In matrix form the row equations of tableau 2.63 become

$$Px' + Qx'' - b' = -y' \tag{2.65}$$

$$Rx' + Sx'' - b'' = -y'' \tag{2.66}$$

$$c'x' + c''x'' - d = f \tag{2.67}$$

Since P is an invertible matrix, we can solve for x' in equation 2.65 to obtain

$$P^{-1}y' + P^{-1}Qx'' - P^{-1}b' = -x' \tag{2.68}$$

Substitute this expression for x' in equations 2.66 and 2.67.

$$-RP^{-1}y' + (S - RP^{-1}Q)x'' - (b'' - RP^{-1}b') = -y'' \tag{2.69}$$
$$-c'P^{-1}y' + (c'' - c'P^{-1}Q)x'' - (d - c'P^{-1}b') = f \tag{2.70}$$

Equations 2.68, 2.69, and 2.70 involve the same variables as the row equations of tableau 2.64. Since the entries within the tableau are determined by the variables on the margins, we have

$$\bar{P} = P^{-1} \tag{2.71}$$
$$\bar{Q} = P^{-1}Q \quad \text{and} \quad \bar{b}' = P^{-1}b' \tag{2.72}$$
$$\bar{S} = S - RP^{-1}Q \quad \text{and} \quad \bar{b}'' = b'' - RP^{-1}b' \tag{2.73}$$
$$\bar{c}'' = c'' - c'P^{-1}Q \quad \text{and} \quad \bar{d} = d - c'P^{-1}b' \tag{2.74}$$
$$\bar{R} = -RP^{-1} \quad \text{and} \quad \bar{c}' = -c'P^{-1} \tag{2.75}$$

We have merely duplicated the discussion of Section 2.2 in which we derived the formulas for the pivot exchange in the row equations. The formulas derived here are analogous. We have arranged the formulas above to suggest the analogy. Formula 2.71 corresponds to the reciprocal of the pivot entry. Formula 2.72 corresponds to the operations in the pivot row. Formulas 2.73 and 2.74 correspond to the operations in the field of a tableau. Finally, formula 2.75 corresponds to the operations in the pivot column. We could also duplicate the discussion of Section 2.3 with the column equations. We would, just as there, obtain the same formulas.

The exchange of a group of variables described here is called a *block pivot exchange.* The submatrix P in the rows and columns of the exchanged variables is called the *pivot block.* A proposed exchange of r basic variables and r nonbasic variables by a block pivot is possible if and only if the pivot block is an invertible matrix. The block pivot exchange can be achieved by the matrix formulas 2.71–2.75 or by a sequence of r independent simple pivot exchanges.

With this terminology, Theorem 2.12 becomes, with a slight rewording,

Theorem 2.13 *Two tableaux have the same solution sets if and only if they are equivalent and one can be obtained from the other by a block pivot exchange (subject to a suitable rearrangement).* □

Corollary 2.14 *There is a one-to-one correspondence between the tableaux of an equivalence class and the invertible submatrices of the A-matrix of any tableau in the class.* ◻

Let us take another look at the second example discussed in Section 2.4. The initial tableau A of that class is given here as tableau 2.76.

(2.76)

A	x_1	x_2	-1	
	1^B		1	$= -y_1$
	1^C	1^E	1	$= -y_2$
		1^G	1	$= -y_3$
	1	1		$= f$

There is one tableau obtainable from tableau A by a block pivot on the empty block, tableau A itself. Four can be obtained by simple block pivots on the nonzero entries indicated with superscripts. The two zero entries cannot be used as pivot blocks. The theorems above say that tableaux with x_1, y_1 or y_3, x_2 on the top margin cannot be obtained in the equivalence class by any other sequence of pivot exchanges. There are three 2×2 submatrices within the A-matrix. Each is invertible and each can be used for a block pivot to obtain another member of the equivalence class. This accounts for all eight different equivalent tableaux.

Several tableaux in this class have five nonzero entries. Each can be used for a simple block pivot. Since each is equivalent to exactly seven other tableaux, only two of the three 2×2 submatrices can be invertible.

The submatrix $S - RP^{-1}Q$ that appears in the tableau after a block pivot exchange is called the *Schur complement* of pivot block P in the A-matrix. Let us connect it with a topic that the reader has already encountered.

Consider the matrix product in 2.77, where S is a square matrix.

$$\begin{bmatrix} I & 0 \\ -RP^{-1} & I \end{bmatrix} \begin{bmatrix} P & Q \\ R & S \end{bmatrix} = \begin{bmatrix} P & Q \\ 0 & S - RP^{-1}Q \end{bmatrix} \tag{2.77}$$

The determinant of the first matrix in 2.77 is equal to 1. The determinant of the matrix on the right is the product of the determinant of P, the pivot matrix, and the determinant of $S - RP^{-1}Q$, the Schur complement of P.

Thus, the determinant of a square matrix is the product of the determinant of any invertible submatrix and the determinant of the Schur complement of that submatrix.

The observation of the previous paragraph yields a method for computing the determinant of a matrix. While it is not the most efficient known it is good enough for hand calculations with matrices of small orders. Suppose we can make a sequence of independent pivot exchanges down the main diagonal. The determinant is the product of the first pivot entry times the determinant of its Schur complement. The Schur complement has one less row and one less column. In turn, the determinant of this matrix is the product of the pivot entry times the determinant of its Schur complement. At each iteration the Schur complement gets smaller, eventually becoming of first order. The determinant of the original matrix is then the product of the successive pivot entries. If the pivot entries are not all in the main diagonal, the determinant is + or − the product of the pivot entries.

2.8 Expanded Tableaux

Most books on linear programming use a tableau with a format different from what we have been using. The format we have used and called a "tableau" is, where reference is made to it, called a *schema* or a *Tucker tableau.* To be able to make a distinction, we shall call the usual tableau an *expanded tableau.*

Consider a Tucker tableau

$$
\begin{array}{c|cccc|c|l}
 & x_1 & x_2 & \cdots & x_n & -1 & \\ \hline
v_1 & a_{11} & a_{12} & \cdots & a_{1n} & b_1 & = -y_1 \\
v_2 & a_{21} & a_{22} & \cdots & a_{2n} & b_2 & = -y_2 \\
\vdots & \vdots & \vdots & & \vdots & \vdots & \quad \vdots \\
v_m & a_{m1} & a_{m2} & \cdots & a_{mn} & b_m & = -y_m \\ \hline
-1 & c_1 & c_2 & \cdots & c_n & d & = f \\ \hline
 & = u_1 & = u_2 & \cdots & = u_n & = g &
\end{array}
\tag{2.78}
$$

The i-th row generates an equation

$$a_{i1}x_1 + \cdots + a_{in}x_n - b_i = -y_i \tag{2.79}$$

Transfer the basic variable on the right side of the equation to the left side and write the equation in the form

$$a_{i1}x_1 + \cdots + a_{in}x_n + y_i - b_i = 0 \tag{2.80}$$

Now put these equations back into a tableau of the form

$$
\begin{array}{c|cccccccc|c|l}
 & x_1 & \cdots & x_n & y_1 & y_2 & \cdots & y_m & -1 & \\
\hline
v_1 & a_{11} & \cdots & a_{1n} & 1 & 0 & \cdots & 0 & b_1 & = 0 \\
v_2 & a_{21} & \cdots & a_{2n} & 0 & 1 & \cdots & 0 & b_2 & = 0 \\
\vdots & \vdots & & \vdots & \vdots & \vdots & & \vdots & \vdots & \vdots \\
v_m & a_{m1} & \cdots & a_{mn} & 0 & 0 & \cdots & 1 & b_m & = 0 \\
\hline
-1 & c_1 & \cdots & c_n & 0 & 0 & \cdots & 0 & d & = f \\
\hline
 & = u_1 & \cdots & = u_n & = v_1 & = v_2 & \cdots & = v_m & = g &
\end{array}
\tag{2.81}
$$

In order to get the Tucker tableau from the expanded tableau, we need to have m columns of the type that are shown under the y_i in tableau 2.81. That is, it must be possible to rearrange the order of the columns so that they form an $m \times m$ identity matrix.

The expanded tableau appears in the literature in several slightly different variations. Usually, the variables and numbers we show in tableau 2.81 on the left, bottom, and right margins are absent, and a few additional columns or rows might be added for identification purposes. The variables that appear in an expanded tableau are usually not interpreted as playing an algebraic role. They are usually used primarily to identify the various parts of the tableau. Tableau 2.81 contains the essential features of an expanded tableau, and we shall use it to establish the correspondence between the expanded tableau and the Tucker tableau.

The columns of tableau 2.81 under the basic variables of the row equations are the *basic columns* or *basic vectors*. They are characterized within the tableau by the property that each column consists of zeros except for a single unit. If there should happen to be two identical such columns, one would be selected as a basic column and the other as a nonbasic column. However, if a tableau has two identical columns the problem probably should be recast to simplify the tableau.

Since the variables of the column equations are usually not shown in an expanded tableau, the v_i that we show on the left margin are replaced by their corresponding y_i, which are used to identify the basic columns.

The pivot exchange is usually called a *pivot operation* and it is described in terms of elementary row operations on the row equations of the expanded tableau. The elementary row operations on matrices are the following:

1. Multiply a row by a nonzero constant.

2. Add a multiple of one row to another row.

3. Interchange two rows.

In an expanded tableau, a pivot operation is a group of elementary row operations that produces a column of zeros, except for a single unit. In a column that has a nonzero entry this is achieved by dividing the row that contains the nonzero entry by that entry. Then subtract multiples of that row from the other rows to obtain zeros in the other positions in that column. The nonzero entry that becomes the single unit in that column is the pivot entry.

Suppose we perform a pivot operation with $a_{rs} \neq 0$ as the pivot entry. Consider representative entries of the expanded tableau in 2.82.

$$
\begin{array}{c|ccccccccc|c|c}
 & \cdots & x_s & \cdots & x_j & \cdots & y_r & \cdots & & -1 & \\
\hline
\vdots & & \vdots & & \vdots & & \vdots & & & \vdots & \vdots \\
v_r & \cdots & a_{rs}^* & \cdots & a_{rj} & \cdots & 1 & \cdots & & b_r & = 0 \\
\vdots & & \vdots & & \vdots & & \vdots & & & \vdots & \vdots \\
v_i & \cdots & a_{is} & \cdots & a_{ij} & \cdots & 0 & \cdots & & b_r & = 0 \\
\vdots & & \vdots & & \vdots & & \vdots & & & \vdots & \vdots \\
\hline
1 & \cdots & c_s & \cdots & c_j & \cdots & 0 & \cdots & & d & = f \\
\hline
 & \cdots & = u_s & \cdots & = u_j & \cdots & = v_r & \cdots & & = g &
\end{array}
\tag{2.82}
$$

Elementary row operations are now carried out to change the x_s-column into a basic column. That is, after the pivot operation the x_s-column will look like the y_r-column does before the pivot operation. To do this we multiply the v_r-row by a_{rs}^{-1}, then subtract a_{is} times the resulting row from the v_i-row. We obtain

$$
\begin{array}{c|ccccccccc|c|c}
 & \cdots & x_s & \cdots & x_j & \cdots & y_r & \cdots & & -1 & \\
\hline
\vdots & & \vdots & & \vdots & & \vdots & & & \vdots & \vdots \\
u_s & \cdots & 1 & \cdots & \dfrac{a_{rj}}{a_{rs}} & \cdots & \dfrac{1}{a_{rs}} & \cdots & & \dfrac{b_r}{a_{rs}} & = 0 \\
\vdots & & \vdots & & \vdots & & \vdots & & & \vdots & \vdots \\
v_i & \cdots & 0 & \cdots & a_{ij} - \dfrac{a_{is}a_{rj}}{a_{rs}} & \cdots & -\dfrac{a_{is}}{a_{rs}} & \cdots & & b_i - \dfrac{a_{is}b_r}{a_{rs}} & = 0 \\
\vdots & & \vdots & & \vdots & & \vdots & & & \vdots & \vdots \\
\hline
1 & \cdots & 0 & \cdots & c_j - \dfrac{c_s a_{rj}}{a_{rs}} & \cdots & -\dfrac{c_s}{a_{rs}} & \cdots & & d - \dfrac{c_s b_r}{a_{rs}} & = f \\
\hline
 & \cdots & = u_s & \cdots & = u_j & \cdots & = v_r & \cdots & & = g &
\end{array}
\tag{2.83}
$$

In tableau 2.83 we can recognize the y_r-column as the pivoted column in a Tucker tableau after the pivot exchange has been performed, and the x_j-column as a typical "other" column. The column headed by x_s becomes a basic column while the column headed by y_r becomes nonbasic. In an expanded tableau the pivot operation exchanges the roles of a basic and a nonbasic column. Otherwise, all arithmetic operations are identical for both tableau formats.

Let us see how the pivot operation in expanded tableau format comes out in an example we have worked in the Tucker tableau format. Tableau 2.84 is the tableau of the production problem discussed first in Section 1.1 and then in Section 2.2, with variables renamed, and tableau 2.85 is the corresponding expanded tableau.

(2.84)

x_1	x_2	x_3	-1	
30	10	50	1500	$= -x_4$
5*	0	3	200	$= -x_5$
0.2	0.1	0.5	12	$= -x_6$
0.1	0.2	0.3	9	$= -x_7$
10	5	5.5	0	$= f$

(2.85)

c_i		x_1	x_2	x_3	x_4	x_5	x_6	x_7	
0	x_4	30	10	50	1				1500
0	x_5	5*	0	3		1			200
0	x_6	0.2	0.1	0.5			1		12
0	x_7	0.1	0.2	0.3				1	9
$c_j - z_j$		10	5	5.5					0

The expanded tableau is shown with the left, bottom, and right margin information from tableau 2.81 deleted. We have added two columns, as is usually done, to supply the information that is lost by this deletion, but we defer explanation of these two columns. For the moment the reader should ignore them. Notice that the basement row contains more entries than the Tucker tableau. The extra entries are the (initially zero) coefficients of the initial basic variables.

Tableau 2.86 shows the effect of the pivot exchange on tableau 2.84 and tableau 2.87 shows the effect of the pivot operation on tableau 2.85.

(2.86)

x_5	x_2	x_3	-1	
−6	10	32	300	$= -x_4$
0.2	0	0.6	40	$= -x_1$
−0.04	0.1	0.38	4	$= -x_6$
−0.02	0.2*	0.24	5	$= -x_7$
−2	5	−0.5	−400	$= f$

(2.87)

c_i		x_1	x_2	x_3	x_4	x_5	x_6	x_7	
0	x_4		10	32	1	−6			300
10	x_1	1	0	0.6		0.2			40
0	x_6		0.1	0.38		−0.04	1		4
0	x_7		0.2*	0.24		−0.02		1	5
$c_j - z_j$			5	−0.5		−2			−400

Performing one more pivot step based on the starred entries in tableaux 2.86 and 2.87, we get tableau 2.88 from 2.86 and tableau 2.89 from 2.87.

(2.88)

x_5	x_7	x_3	-1	
−5	−50	20	50	$= -x_4$
0.2	0	0.6	40	$= -x_1$
−0.03	−0.5	0.26	1.5	$= -x_6$
−0.1	5	1.2	25	$= -x_2$
−1.5	−25	−6.5	−525	$= f$

(2.89)

c_i		x_1	x_2	x_3	x_4	x_5	x_6	x_7	
0	x_4			20	1	−5		−50	50
10	x_1	1		0.6		0.2		0	40
0	x_6			0.26		−0.03	1	−0.5	1.5
5	x_2		1	1.2		−0.1		5	25
$c_j - z_j$				−6.5		−1.5		−25	−525

The variables in the expanded tableaux 2.85, 2.87, and 2.89 serve as identifying labels. Those along the top identify columns—which are customarily regarded as "vectors." Those in the second column at the left identify vectors currently "in the basis." They are basic variables of the row equations of the corresponding Tucker tableau, not nonbasic variables of the column equations. The usual jargon refers to "basic vectors" and "nonbasic vectors." A basic vector is a column associated with our basic variable, and a nonbasic vector is a column associated with our nonbasic variable. The terminology stems from the fact that the basic columns are linearly independent and constitute a basis of the column space.

There is a column labeled "c_i" containing the c-entries from the initial tableau. Each c_i is the coefficient of the variable in the second column in the original objective function. That is, in tableau 2.89 the 10 in the c_i-column is c_1, the coefficient of x_1 in the objective function. The c_i all refer to the initial tableau. Notice that c_4 and c_6 are zero, and $c_2 = 5$. The c_i enter or leave as the variables in the second column enter or leave, and otherwise do not change value from iteration to iteration. In the initial tableau all c_i in the first column are zero.

In the lower left corner is a new entity, $c_j - z_j$. It suggests how the entries in the basement row can be calculated. Each z_j is the sum of products of the numbers in the c_i-column and the corresponding entries in column j. In tableaux 2.87 and 2.89 the entries in the basement row can be computed in either of two ways. They can be computed by pivoting, as we have done with the Tucker tableau, or they can be computed as $c_j - z_j$. Computing them both ways is often advocated as a check.

The reader should check that an entry in the basement row can be computed by multiplying each entry in the column of that entry by the entry in the c_i-column, accumulating the sum of these products, and subtracting that sum from the original basement row. This applies to the blanks and the d-corner as well as the nonzero entries.

Notice that every entry in a tableau of each type has its counterpart in the tableau of the other type. There is nothing that can be done with one type that cannot also be done with the other. There is nothing that can be observed in one type that cannot also be observed in the other, if you know where to look for it.

The expanded tableau offers the advantage that it is the traditional tableau and all commercially available computer programs to solve linear programming problems expect the data input to be in expanded tableau format. The Tucker tableau offers the advantage that it treats the row and column systems symmetrically. However, the most significant difference between the expanded tableau and the Tucker tableau has little to do with the format itself. It has to do with the way the entries and labels are interpreted.

As we have already mentioned, in the expanded tableau the columns and rows are labeled with vector notation. In this notation the rows and columns do not represent equations and the labels do not represent variables. Variables do appear implicitly in that the entries in the b-column are interpreted as the current values of the basic variables, just do with the

Tucker tableau. The value of the objective function of the row system is not computed from the basement row. It is computed by multiplying the entries in the c_i-column by the values of the basic variables in the b-column. The basement row of the expanded tableau is represented as $c_j - z_j$ and it is not easy to see the significance of these numbers in the dual system of equations.

If you use variables instead of labels it is quite easy to identify the column system in the expanded tableau. Refer to tableau 2.81, which is our first description of the expanded tableau. Both the row and the column system are there, constructed just as they are for the Tucker tableau. The duality equation is also constructed for the expanded tableau the same way it is constructed for the Tucker tableau.

Exercises

For each of the following three tableaux, determine the number of equivalent tableaux.

12.

	x_1	x_2	x_3	x_4	-1	
u_5	1	1	1	0	1	$= -x_5$
u_6	1	0	1	1	1	$= -x_6$
-1	1	1	1	1	0	$= f$
	$= u_1$	$= u_2$	$= u_3$	$= u_4$	$= g$	

13.

	x_1	x_2	x_3	-1	
u_4	1	2	3	4	$= -x_4$
u_5	4	5	6	7	$= -x_5$
u_6	7	8	9	10	$= -x_6$
-1	10	11	12	0	$= f$
	$= u_1$	$= u_2$	$= u_3$	$= g$	

14.

	x_1	x_2	x_3	x_4	-1	
u_5	0	1	0	0	1	$= -x_5$
u_6	0	0	0	1	1	$= -x_6$
u_7	1	0	0	0	1	$= -x_7$
u_8	0	0	1	0	1	$= -x_8$
-1	1	1	1	1	0	$= f$
	$= u_1$	$= u_2$	$= u_3$	$= u_4$	$= g$	

15. For the tableau given in Exercise 14, determine all the basic solutions.

16. Consider the class of tableaux equivalent to the tableau given in Exercise 13. How many of these tableaux have at least one zero entry, and how many zero entries does each of these tableaux have?

17. Determine whether the following tableau is invertible. If it is, find the inverse tableau and the inverse of the A-matrix.

x_1	x_2	x_3	-1	
2	0	-1	1	$= -y_1$
2	1	2	1	$= -y_2$
5	1	1	1	$= -y_3$
1	1	1	0	$= f$

18. Determine whether the following tableau is invertible. If it is, find the inverse tableau and the inverse of the A-matrix.

x_1	x_2	x_3	-1	
3	0	-1	1	$= -y_1$
2	1	2	1	$= -y_2$
5	1	1	1	$= -y_3$
1	1	1	0	$= f$

19. Determine whether the following tableau is invertible. If it is, find the inverse tableau and the inverse of the A-matrix.

	x_1	x_2	x_3	x_4	-1	
u_5	0	1	0	0	1	$= -x_5$
u_6	0	0	0	1	1	$= -x_6$
u_7	1	0	0	0	1	$= -x_7$
u_8	0	0	1	0	1	$= -x_8$
-1	1	1	1	1	0	$= f$
	$= u_1$	$= u_2$	$= u_3$	$= u_4$	$= g$	

20. Determine whether the system of linear equations

$$x_1 - 3x_2 + 5x_3 = 1$$
$$-2x_1 + 5x_2 - 2x_3 = 2$$
$$-3x_1 + 7x_2 + x_3 = 6$$

has a solution.

21. Use the formulas 2.71 through 2.75 to perform a block pivot exchange on the 2×2 submatrix in the upper left corner of the A-matrix.

x_1	x_2	x_3	-1	
1	1	-1	1	$= -y_1$
1	2	2	1	$= -y_2$
5	1	1	1	$= -y_3$
1	1	1	0	$= f$

22. Convert the following Tucker tableau into an expanded tableau.

x_1	x_2	x_3	-1	
1	1	-1	1	$= -y_1$
1	2	2	1	$= -y_2$
5	1	1	1	$= -y_3$
1	1	1	0	$= f$

23. Convert the following expanded tableau into a Tucker tableau.

c_i		x_1	x_2	x_3	x_4	x_5	x_6	
0	x_4	2	0	-1	1	0	0	2
0	x_5	2	1	-2	0	1	0	-2
0	x_6	5	1	1	0	0	1	3
		1	-2	4				0

24. Perform a pivot exchange on the starred entry in the following expanded tableau.

c_i		x_1	x_2	x_3	x_4	x_5	x_6	
0	x_4	2	0	-1	1	0	0	2
0	x_5	2	1	-2^*	0	1	0	-2
0	x_6	5	1	1	0	0	1	3
		1	-2	4				0

Questions

A mathematician practices his craft by looking at examples for insight and then making conjectures. When a reasonable conjecture is formulated, he tries to prove either that it is true or that it is false. The theorems you see in books and papers are conjectures that were true and could be proved. Not many false conjectures get published, at least not on purpose. Treat all the following problems as conjectures. They are not conjectures in the sense that mathematicians use the term since it is known whether they are true or false. We will call them Questions. Your mission, should you choose to accept it, is to prove or disprove these questions. If you don't choose to put that much effort into it, you can treat them as true-false questions.

Q1. If two tableaux are equivalent with an A-matrix with m rows and n columns, then either one can be obtained from the other with at most n pivot exchanges.

Q2. If for any $m \times n$ tableau, the set of tableaux equivalent to it contains $(m+n)!/(m!n!)$ tableaux, then every tableau in the set has only nonzero entries in its A-matrix.

Q3. For each basic solution for the primal problem, the dual problem has a basic solution that is complementary.

Q4. For each basic solution for the primal problem, the dual problem has a unique basic solution that is complementary.

Q5. The average of two solutions to a tableau is also a solution.

Q6. If a pair of solutions, one for the primal problem and one for the dual problem, are complementary, then both solutions are basic solutions.

Q7. The average of two basic solutions to a tableau is also a basic solution.

Q8. It is possible for a basic solution to be the average of two other solutions.

Q9. The average of two basic solutions cannot be a basic solution.

Q10. If a tableau is invertible, then every tableau equivalent to it is invertible.

Q11. If two equivalent tableaux have different basic solutions, then it is impossible to have the set of variables that are zero in one basic solution be a proper subset of the set of variables that are zero in the other.

Chapter 3

Canonical Duality

3.1 Canonical Dual Linear Programs

Consider the maximization problem discussed in Section 1.1 as a model for the following type of problem. We have a set of decision variables x_1, x_2, ... , x_n restricted to having nonnegative values. These variables are constrained by several linear inequalities of the form

$$a_{i1}x_1 + a_{i2}x_2 + \cdots + a_{in}x_n \leq b_i \tag{3.1}$$

We convert each inequality into an equation by supplying a slack variable

$$a_{i1}x_1 + a_{i2}x_2 + \cdots + a_{in}x_n + y_i = b_i \tag{3.2}$$

and requiring that the slack variable y_i be nonnegative. The equation 3.2 is more suitable for presentation in a tableau if we write it in the form

$$a_{i1}x_1 + a_{i2}x_2 + \cdots + a_{in}x_n - b_i = -y_i \tag{3.3}$$

We assume the *objective variable* f, the variable whose value is to be maximized, is given as a linear function of the decision variables.

$$f = c_1x_1 + c_2x_2 + \cdots + c_nx_n - d \tag{3.4}$$

The equations 3.3 and 3.4 can conveniently be represented by a tableau in the form

$$\begin{array}{cccc|c|l} x_1 & x_2 & \cdots & x_n & -1 & \\ \hline a_{11} & a_{12} & \cdots & a_{1n} & b_1 & = -y_1 \\ a_{21} & a_{22} & \cdots & a_{2n} & b_2 & = -y_2 \\ \vdots & \vdots & & \vdots & \vdots & \vdots \\ a_{m1} & a_{m2} & \cdots & a_{mn} & b_m & = -y_m \\ \hline c_1 & c_2 & \cdots & c_n & d & = f \end{array} \tag{3.5}$$

The problem of finding the maximum value of f for which the row equations of tableau 3.5 are satisfied and for which all variables (except the objective variable) are nonnegative is called a *maximum linear programming problem (or max program) in canonical form*.

To save words we shall describe such a problem as a *canonical maximum program*, or *canonical max program*. The term "canonical" implies that all variables (but f) are nonnegative.

Since the requirement that the y_i be nonnegative is equivalent to the inequalities of the form of 3.1, a canonical max program is equivalent to one in which the constraints are linear inequalities involving nonnegative variables. An advantage of presenting a canonical max program in terms of the tableau 3.5 is that it puts all variables, decision and slack variables, on an equal footing in equation form. When we get further involved with the concept of duality and noncanonical programs, other advantages will become apparent.

In Section 1.2 we discussed a minimization problem that we shall use as model for the following type of problem. We have a set of decision variables $v_1, v_2, \ldots, v_m$ restricted to having nonnegative values. These variables are constrained by several linear inequalities of the form

$$v_1 a_{1j} + v_2 a_{2j} + \cdots + v_m a_{mj} \geq c_j \tag{3.6}$$

We convert each inequality into an equation by supplying a slack variable

$$v_1 a_{1j} + v_2 a_{2j} + \cdots + v_m a_{mj} - c_j = u_j \tag{3.7}$$

and requiring that the slack variables u_j be nonnegative. We assume the objective variable g, the variable whose value is to be minimized, is given as a linear function of the decision variables.

$$g = v_1 b_1 + v_2 b_2 + \cdots + v_m b_m - d \tag{3.8}$$

The equations 3.7 and 3.8 can conveniently be represented by a tableau in the form

$$
\begin{array}{c|cccc|c|}
\cline{2-6}
v_1 & a_{11} & a_{12} & \cdots & a_{1n} & b_1 \\
v_2 & a_{21} & a_{22} & \cdots & a_{2n} & b_2 \\
\vdots & \vdots & \vdots & & \vdots & \vdots \\
v_3 & a_{31} & a_{32} & \cdots & a_{3n} & b_3 \\
\cline{2-6}
-1 & c_1 & c_2 & \cdots & c_n & d \\
\cline{2-6}
\multicolumn{1}{c}{} & \multicolumn{1}{c}{= u_1} & = u_2 & \cdots & = u_n & \multicolumn{1}{c}{= g}
\end{array}
\tag{3.9}
$$

The problem of finding the minimum value of g for which the column equations of tableau 3.9 are satisfied and for which all variables (except the objective variable) are nonnegative is called a *minimum linear programming problem (or min program) in canonical form*.

Given a canonical linear programming problem of either kind, we can immediately write down another canonical linear programming problem by supplying variables on the opposite margins and requiring that these variables be nonnegative. (The objective variable is never required to be nonnegative by implication of the word "canonical" and we shall usually not write out the statement that this variable is exempted from this requirement.)

The problem obtained in this way is called the *dual* of the given problem. When a linear program is described as a canonical linear program its dual program is always canonical.

Tableau 3.10 represents the dual systems of equations for a dual pair of canonical linear programs.

$$
\begin{array}{c|cccc|c|l}
\multicolumn{1}{c}{} & x_1 & x_2 & \cdots & x_n & \multicolumn{1}{c}{-1} & \\
\cline{2-6}
v_1 & a_{11} & a_{12} & \cdots & a_{1n} & b_1 & = -y_1 \\
v_2 & a_{21} & a_{22} & \cdots & a_{2n} & b_2 & = -y_2 \\
\vdots & \vdots & \vdots & & \vdots & \vdots & \;\vdots \\
v_m & a_{m1} & a_{m2} & \cdots & a_{mn} & b_m & = -y_m \\
\cline{2-6}
1 & c_1 & c_2 & \cdots & c_n & d & = f \\
\cline{2-6}
\multicolumn{1}{c}{} & \multicolumn{1}{c}{= u_1} & = u_2 & \cdots & = u_n & \multicolumn{1}{c}{= g} &
\end{array}
\tag{3.10}
$$

All variables (other than the objective variable) are required to be nonnegative.

In Section 2.3 we showed how taking the negative transpose of a tableau interchanged the pair of dual systems of equations. If we do this for tableau 3.10 we obtain

$$
\begin{array}{c|cccc|c|l}
 & v_1 & v_2 & \cdots & v_m & -1 & \\
\hline
x_1 & -a_{11} & -a_{21} & \cdots & -a_{m1} & -c_1 & = -u_1 \\
x_2 & -a_{12} & -a_{22} & \cdots & -a_{m2} & -c_2 & = -u_2 \\
\vdots & \vdots & \vdots & & \vdots & \vdots & \vdots \\
x_n & -a_{1n} & -a_{2n} & \cdots & -a_{mn} & -c_n & = -u_n \\
\hline
-1 & -b_1 & -b_2 & \cdots & -b_m & -d & = -g \\
\hline
 & = y_1 & = y_2 & \cdots & = y_m & = -f &
\end{array}
\tag{3.11}
$$

Maximizing f is equivalent to minimizing $-f$, and minimizing g is equivalent to maximizing $-g$. Thus, taking the negative transpose converts the canonical max program into an equivalent canonical min program and it converts the canonical min program into a canonical max program. There is, then, no intrinsic reason for casting a problem in the form of a min program in preference to a max program, or conversely. However, frequently there are practical computational reasons for casting a problem in one form or the other. We shall want to be free to convert any problem we work on from one form to the other at our convenience. We include the possibility of making such a conversion after some computation has been done.

A set of values for the variables that satisfy the row system or the column system for a tableau is called a *solution* for that system, whether it also satisfies the nonnegativity conditions or not.

A solution that satisfies the nonnegativity conditions is called a *feasible solution*.

A linear program is said to be *feasible* if it has at least one feasible solution. It is *infeasible* if no feasible solution exists. A linear program can fail to have an optimal solution in two ways; It can be infeasible, or it can be feasible but the objective variable can be unbounded. Otherwise, the linear program has at least one optimal solution. If it has an optimal solution, it may have a unique optimal solution or it may have an infinite number of optimal solutions. We will show that it is not possible to have a finite number of optimal solutions greater than one. We will also show that if a linear program is feasible but has no optimal solution, the objective variable must be unbounded.

It is the objective of this chapter and the following chapter to show that (1) infeasibility, (2) feasibility and at least one optimal solution, and (3) feasibility with an unbounded objective variable are the only possibilities;

to show how to determine which of these three cases holds; and to show how to find an optimal solution if one exists.

We shall refer to any conditions imposed outside the tableau itself (in this case the outside conditions are the nonnegativity conditions) as *feasibility specifications*. A value of the objective variable f, or g, for a feasible solution is called a *feasible value* of f, or g.

A linear program is *noncanonical* if some of the variables are unrestricted or restricted in a way other than being merely nonnegative. For all linear programs, canonical or noncanonical, the problem is to find the maximum, or minimum, feasible value of the objective variable.

3.2 Sufficient Conditions for Optimality

Consider a tableau for a pair of dual canonical linear programs.

(3.12)

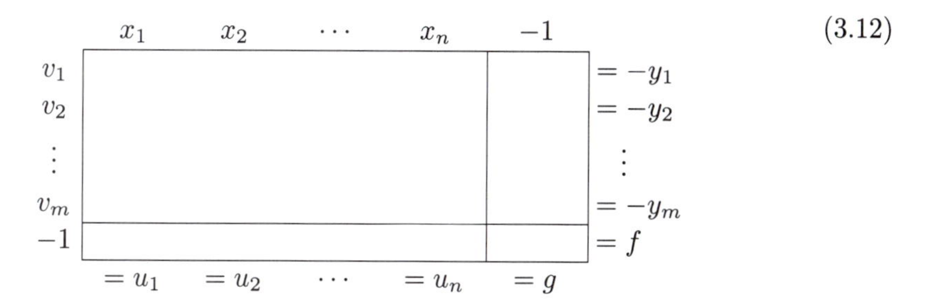

Theorem 3.15 *Let $(u, v, g) = (u_1, \ldots, u_n, v_1, \ldots, v_m, g)$ and $(x, y, f) = (x_1, \ldots, x_n, y_1, \ldots, y_n, f)$ be any feasible solutions for the minimum and maximum programs. Then*

$$f \leq g \tag{3.13}$$

Proof. In the duality equation

$$u_1x_1 + \cdots + u_nx_n + v_1y_1 + \cdots + v_my_m = g - f \tag{3.14}$$

each term on the left side, being the product of nonnegative factors, is nonnegative. Hence $g - f \geq 0$. ◻

It follows from this theorem that if $\bar{g}$ is any particular feasible value of g, then $\bar{g}$ is an upper bound for all feasible values of f. If $\bar{f}$ is any particular feasible value of f, then $\bar{f}$ is a lower bound for all feasible values of g.

Theorem 3.16 *(Sufficient Condition for Optimality) If we can find feasible solutions $(\bar{x}, \bar{y}, \bar{f})$ and $(\bar{u}, \bar{v}, \bar{g})$ for the maximum and minimum programs for which $\bar{f} = \bar{g}$, then each is optimal for its program.*

Proof. By the remark preceding the statement of this theorem, $\bar{f} = \bar{g}$ is an upper bound for all feasible values of f and, therefore, optimal. Similarly, $\bar{g} = \bar{f}$ is a lower bound for all feasible values of g and, therefore, optimal. ◻

We remind you that a pair of solutions, one for the row system and one for the column system, is complementary if for each pair of dual variables, one or the other is zero.

Theorem 3.17 *(Sufficient Condition for Optimality) Complementary feasible solutions are optimal.*

Proof. Because of the duality equation 3.14, for a complementary pair of solutions, feasible or not, we always have $f = g$. If both solutions are feasible, the sufficient condition for optimality is satisfied. ◻

Theorem 3.17 allows us to pose a problem which, if solved, provides optimal solutions for both of the dual problems.

> **Find feasible solutions for the row equations and the column equations that are complementary.**

As an example, let us consider again the canonical max program we discussed in Section 1.1. The tableau for that program is

(3.15)

	x	y	z	-1	
h	30	10	50	1500	$= -p$
k	5	0	3	200	$= -q$
m	0.2	0.1	0.5	12	$= -r$
n	0.1	0.2	0.3	9	$= -s$
-1	10	5	5.5	0	$= f$
	$= u$	$= v$	$= w$	$= g$	

We determined that $x = 40$, $y = 25$, $z = 0$, $p = 50$, $q = 0$, $r = 1.5$, and $s = 0$ was an optimal solution for the max program. In tableau 3.15 we have supplied variables for the canonical dual min program. The dual problem is sometimes called the *shadow problem*. Suppose $u = 0$, $v = 0$, $w = 6.5$, $h = 0$, $k = 1.5$, $m = 0$, and $n = 25$ is proposed as a solution of the min program. At the moment we are not concerned with what reasoning might be used to propose this particular solution. These numbers might have been obtained by guessing, or proposed by a friend. At least it can be tested for optimality. Enter these numbers in place of the variables in tableau 3.15.

(3.16)

	40	25	0	−1	
0	30	10	50	1500	$= -50$
1.5	5	0	3	200	$= 0$
0	0.2	0.1	0.5	12	$= -1.5$
25	0.1	0.2	0.3	9	$= 0$
−1	10	5	5.5	0	$= 525$
	$= 0$	$= 0$	$= 6.5$	$= 525$	

We can verify that the numbers on the margins of tableau 3.16 satisfy all the equations generated by tableau 3.15. The proposed solutions are nonnegative. This shows both solutions are feasible. Finally, both objective variables have the same value, $f = g = 525$.

Actually, there isn't any mystery about where the proposed solutions come from. Consider tableau 1.13 from Section 1.1, which we duplicate here with variables for the columns equations added.

(3.17)

	q	s	z	−1	
h	−5	−50	20	50	$= -p$
u	0.2		0.6	40	$= -x$
m	−0.03	−0.5	0.26	1.5	$= -r$
v	−0.1	5	1.2	25	$= -y$
−1	−1.5	−25	−6.5	−525	$= f$
	$= k$	$= n$	$= w$	$= g$	

The solutions given on the margins of tableau 3.16 are the basic solutions for tableau 3.17. The positive numbers in the b-column mean that the basic solution for the row program is feasible, and the negative numbers in the c-row mean that the basic solution for the column program is feasible. The two basic solutions for a tableau are complementary. Thus, if two basic solutions for the same tableau are both feasible, both are optimal.

If both programs are feasible the sets of feasible objective values for the two programs intersect in at most one point since any common point would be an optimal value for both programs. Actually, the two sets of feasible objective values intersect in exactly one point. This will be shown in Chapter 4. Also, in Chapter 4 we shall show that if the min program is infeasible and the max program is feasible, the feasible values of f are unbounded above and no optimal value exists. Similarly, if the max program is infeasible and the min program is feasible, the feasible values of g are unbounded below and no optimal value exists. Thus, the condition $f = g$ (for feasible values) will also turn out to be necessary for optimality. However, for the moment we shall use the condition $f = g$ for feasible values only as a sufficient condition for optimality.

3.3 Economic Interpretation of Duality

In Chapter 1 we discussed several linear programs with realistic interpretations. Now we see that every linear program has a dual program. What realistic interpretation can be given to these dual programs? To consider this question let us repeat here the tableau of the production max program of Section 1.1 with the variables of the dual program supplied.

(3.18)

	x	y	z	-1	
h	30	10	50	1500	$= -p$
k	5	0	3	200	$= -q$
m	0.2	0.1	0.5	12	$= -r$
n	0.1	0.2	0.3	9	$= -s$
-1	10	5	5.5	0	$= f$
	$= u$	$= v$	$= w$	$= g$	

The first thing that must be done is to determine what the variables of the dual program represent. A key to this interpretation can be found in the duality equation.

$$ux + vy + wz + hp + kq + mr + ns = g - f \tag{3.19}$$

In any sum, every term must be measured in the same units. Since the objective variable f is measured in dollars per day, so also must the objective variable g be measured in dollars per day. The term ux must also be measured in dollars per day. Since we know that x is measured in boxes (of floor cover) per day, we can determine the units in which u is measured. The simplest way to do this is to treat the dimensions as though they were algebraic quantities and put the measures after the word "per" in the denominator. Thus, "f dollars per day" is written "$f\frac{\text{dollars}}{\text{day}}$." Then for the ux term we have

$$(u\frac{\text{dollars}}{\text{box}})(x\frac{\text{boxes}}{\text{day}}) = ux\frac{\text{dollars}}{\text{day}} \tag{3.20}$$

The quantity u has the dimensions of a price. It is called a *shadow price*, a term due to the economist Paul A. Samuelson. In a similar way we see that the other variables of the min program have the dimensions of prices.

It is of interest to obtain the interpretation of the dual variables directly from the original data table of the problem rather than from the primal program. This places the two dual programs on a more symmetric footing. Table 3.21 repeats the data table of the plastics manufacturing problem from Section 1.1.

(3.21)

	floor cover	counter top	wall tile	available	
vinyl	30	10	50	1500	pounds
asphalt	5		3	200	pounds
labor	0.2	0.1	0.5	12	man-hours
machine	0.1	0.2	0.3	9	machine-hours
profit	10.00	5.00	5.50		dollars
	per box	per yard	per square	per day	

To simplify matters and still leave enough to illustrate how the variables of the two problems are related, let us ignore all but the first row (in vinyl) and the basement row, and the first column (involving the floor cover) and the last column.

(3.22)

	$\frac{\text{boxes}}{\text{day}}$	-1	
$\frac{\text{dollars}}{\text{pound}}$	$\frac{\text{pounds}}{\text{box}}$	$\frac{\text{pounds}}{\text{day}}$	$\frac{\text{pounds}}{\text{day}}$
-1	$\frac{\text{dollars}}{\text{box}}$	$\frac{\text{dollars}}{\text{day}}$	$\frac{\text{dollars}}{\text{day}}$
	$\frac{\text{dollars}}{\text{box}}$	$\frac{\text{dollars}}{\text{day}}$	

In each row, the numbers in the b-column are multiplied by -1, which has no dimensions. Thus, the numbers in the b-column and the corresponding variables on the right margin have the same dimensions. A similar remark applies to the numbers in the basement row and the variables on the bottom margin. Then for each entry in the tableau we have two dimensional relations, one for the row equation and one for the column equation. For the entry shown we have

$$\frac{\text{pounds}}{\text{box}}\frac{\text{boxes}}{\text{day}} = \frac{\text{pounds}}{\text{day}} \tag{3.23}$$

for the row equation and

$$\frac{\text{dollars}}{\text{pound}}\frac{\text{pounds}}{\text{box}} = \frac{\text{dollars}}{\text{box}} \tag{3.24}$$

for the column equation. This is sufficient information to establish the dimensional units for all the variables in both programs.

Now that we know how the variables of the dual program are dimensioned, how do we interpret them? Consider the effect of increasing any of the available supplies. Suppose, as an example of sensitivity to changes, we consider increasing the amount of machine time available. This should permit us to produce more and, perhaps, to increase the profit. To determine what the optimal solution would be under such changed circumstances we could change the appropriate numbers in tableau 3.18 and rework the problem. There is an easier way to see the sensitivity of the solution to such a change. Consider tableau 3.25.

$$\begin{array}{c|ccc|c|l} & x & y & z & -1 & \\ \hline h & & & & 1500 & = -p \\ k & & & & 200 & = -q \\ m & & & & 12 & = -r \\ n & & & & 9+e & = -s \\ \hline -1 & & & & 0 & = f \\ \hline & = u & = v & = w & = g & \end{array} \tag{3.25}$$

The quantity e is the "extra" machine time that is available. First, we move e to the right margin, as in tableau 3.26.

$$\begin{array}{c|ccc|c|l} & x & y & z & -1 & \\ \hline h & & & & 1500 & = -p \\ k & & & & 200 & = -q \\ m & & & & 12 & = -r \\ n & & & & 9 & = -(s-e) \\ \hline -1 & & & & 0 & = f \\ \hline & = u & = v & = w & = g - ne & \end{array} \tag{3.26}$$

We can now determine the corresponding change in any tableau equivalent to tableau 3.26. For example, consider tableau 1.13 for which the basic solution is optimal. We repeat the tableau here as 3.27 with s replaced by $s - e$.

(3.27)

	q	$s-e$	z	-1	
h	-5	-50	20	50	$= -p$
u	0.2		0.6	40	$= -x$
m	-0.03	-0.5	0.26	1.5	$= -r$
v	-0.1	5	1.2	25	$= -y$
-1	-1.5	-25	-6.5	-525	$= f$
	$= k$	$= n$	$= w$	$= g - ne$	

In the second column, the e on the top margin introduces a constant into each equation. The same result can be obtained by making a corresponding change in the b-column. In this way we get tableau 3.28.

(3.28)

	q	s	z	-1	
h	-5	-50	20	$50 - 50e$	$= -p$
u	0.2		0.6	40	$= -x$
m	-0.03	-0.5	0.26	$1.5 - 0.5e$	$= -r$
v	-0.1	5	1.2	$25 + 5e$	$= -y$
-1	-1.5	-25	-6.5	$-525 - 25e$	$= f$
	$= k$	$= n$	$= w$	$= g$	

Notice how the column equation for n is used to exactly remove the $-ne$ that was added to g in tableau 3.26. We see that the effect of all these manipulations is that when e is added to the b-entry in the s-row, the result in any equivalent tableau is to add e times the s-column to the b-column.

First, notice that only the last column is affected by this change. In the column system only the objective function is changed. The constraint equations are the same. Thus, any solution of the column equations of tableau 3.18 remains a solution, and any feasible solution remains a feasible solution. The situation is different for the row system. Consider just the basic solutions for tableau 3.28. The values of the basic variables are

$$\begin{aligned} p &= 50 - 50e \\ x &= 40 \\ r &= 1.5 - 0.5e \\ y &= 25 + 5e \end{aligned} \tag{3.29}$$

This basic solution will remain feasible as long as e does not exceed 1 machine-hour.

Thus, as long as e does not exceed 1, both basic solutions remain feasible and, therefore, optimal. Within this range the optimal value of the objective variable is

$$f = 525 + 25e \tag{3.30}$$

The coefficient 25 is the marginal value or marginal price of the machine time. It is neither the cost nor the value of an hour of machine time, though it is closely related to both.

In the primal problem the objective variable is the profit, and all costs have been accounted for in determining its functional dependence on the decision variables and the resources. A shadow price of \$25 means that an increase of an hour in the amount of machine time available will permit an adjustment in the operation of the plastics company that will increase its profit by \$25. The actual cost of the machine time is paid by the income realized from the products made and it may be relatively large or small compared with the profitability of the products.

In contrast, the values of h, the shadow price of vinyl, and m, the shadow price of labor, are both zero. That does not mean they have no value. The slack variables for vinyl and labor are both positive, meaning that these resources are not fully utilized. Increasing the supply of either would not allow an increase in production and, therefore, not allow an increase in profit. That's all a zero shadow price means.

To make the dual min program more vivid, let us suppose our plastics shop has the supplies, labor, and machine under contract and that they are the owner's to dispose of as he wishes. Suppose some entrepreneur needs these resources and is willing to pay the owner a premium to get access to them. Let h be the premium to be offered for a pound of vinyl, k the premium to be offered for a pound of asphalt, m for an hour of labor, and n for an hour of machine time. What conditions must these premium prices satisfy? Unless these prices satisfy the condition

$$30h + 5k + 0.2m + 0.1n \geq 10 \tag{3.31}$$

our owner can make a greater profit by using the resources to make floor cover in his own shop. Similar inequalities must be satisfied for the profitability of making counter top and wall tile.

The entrepreneur's offer must satisfy these conditions if it is to be considered, but he wants to pay as little as possible. The cost of this offer is

$$1500h + 200k + 12m + 9n = g \tag{3.32}$$

The entrepreneur faces a min program which is precisely the min program of tableau 3.18. It is the dual of the owner's max program.

The suggested relationship between owner and entrepreneur is probably too specialized to be of general interest, but we formulated the dual problem in terms of a second person for clarity. It is more realistic to consider the owner and entrepreneur to be the same person (or company), but that the two operations involved are different operations within the same company. These operations are competing for the available resources. It is from this point of view that the slack variables u, v, and w find an interpretation. The shadow prices are accounting costs that allocate the contributions of the various resources to the profitability of the items produced.

The positive value \$6.50 for w in the optimal solution to the shadow problem means that the value of the resources that would be required to produce wall tile exceeds the profitability of making wall tile by \$6.50 per square. The zero values for u and v mean that the sum of the contributions to the accounting costs of these products equals their profitability. This characterizes the products that are profitable.

We know that wall tile is unprofitable, but can anything be done to make it profitable? The most obvious point of attack is to try to raise the price. The shortfall of profitability is \$6.50 per square, so we would have to raise the price at least \$6.50 per square if that is the only action contemplated. Perhaps competition makes that impossible. We can consider changing production methods or the ingredients for wall tile. This amounts to changing the entries in the z-column.

For the optimal solution at hand, z is zero. Thus, changing the entries in the z-column will not change the feasibility of the current optimal solution for the row system. Reducing the contribution of labor will not affect the profitability of wall tile since the shadow price of labor is zero. We would have to attack the contributions of those ingredients that have positive shadow prices.

By setting $k = 1.5$ and $n = 25$ in tableau 3.18 we can see that the contribution of asphalt to the cost of wall tile is \$4.50. Thus, we could remove all the asphalt from wall tile and still not make it a profitable item. The contribution of machine time to the cost of wall tile is \$7.50. If there is any hope of making wall tile profitable something will have to be done about the amount of machine time used.

On an intuitive basis one might come to much the same conclusions. On a mathematical basis these conclusions are secure because for all the changes considered the solutions for the pair of dual programs remain feasible and complementary and, therefore, optimal.

Sensitivity analysis is the study of the effects that changes in the coefficients of a program have on the solution. We have been discussing the sensitivity of the solution to changes in the entries in the b-column. An extension of these ideas to a study of the effects of changes in the A-matrix is more complicated but carried out in a similar way.

Let us look at the feed lot min program of Section 1.2. What is the interpretation of the dual max program? As a start we write down the data table 1.25 and tableau 1.26 with the dual variables supplied.

(3.33)

	protein	carbohydrate	roughage	cost	
hay	10	10	60	3.80	per bale
oats	15	10	25	5.00	per sack
pellets	10	5	55	3.50	per sack
sweet feed	25	20	35	8.00	per sack
requirements	450	400	1050		per week
	pounds	pounds	pounds	dollars	

(3.34)

	x	y	z	-1	
h	10	10	60	3.80	$= -p$
k	15	10	25	5.00	$= -q$
m	10	5	55	3.50	$= -r$
n	25	20	35	8.00	$= -s$
-1	450	400	1050	0	$= f$
	$= u$	$= v$	$= w$	$= g$	

We see that the objective variables f and g are measured in dollars per week, and the other variables of the row system have the dimensions of prices.

The optimal solutions for both programs can be read off the tableau 1.36, which we repeat here as tableau 3.35 with the dual variables supplied.

(3.35)

	s	p	z	-1	
v	-0.2	0.5	23	0.30	$= -y$
k	-1	1	50	0.80	$= -q$
m	-1	1.5	110	1.20	$= -r$
u	0.2	-0.4	-17	0.08	$= -x$
-1	-10	-20	-500	-156	$= f$
	$= n$	$= h$	$= w$	$= g$	

Table 3.1 below has these solutions given in tabular form with their dimensions.

Table 3.1: Dual Optimal Solutions with Dimensions.

Min program	Max program
$u = 0 \ \dfrac{\text{pounds (of protein)}}{\text{week}}$	$x = 0.08 \ \dfrac{\text{dollars}}{\text{pound}}$
$v = 0 \ \dfrac{\text{pounds (of carbohydrate)}}{\text{week}}$	$y = 0.30 \ \dfrac{\text{dollars}}{\text{pound}}$
$w = 500 \ \dfrac{\text{pounds (of roughage)}}{\text{week}}$	$z = 0 \ \dfrac{\text{dollars}}{\text{pound}}$
$h = 20 \ \dfrac{\text{pounds (of hay)}}{\text{week}}$	$p = 0 \ \dfrac{\text{dollars}}{\text{bale}}$
$k = 0 \ \dfrac{\text{sacks (of oats)}}{\text{week}}$	$q = 0.80 \ \dfrac{\text{dollars}}{\text{sack}}$
$m = 0 \ \dfrac{\text{sacks (of pellets)}}{\text{week}}$	$r = 1.20 \ \dfrac{\text{dollars}}{\text{sack}}$
$n = 10 \ \dfrac{\text{sacks (of sweetfeed)}}{\text{week}}$	$s = 0 \ \dfrac{\text{dollars}}{\text{sack}}$

The prices for protein, carbohydrate, and roughage are marginal prices. For example, if the protein requirement is increased by one pound per week, the cost of the optimum solution would be increased by \$0.08. The marginal price of zero for roughage does not mean that the cost of roughage is zero. Since the roughage requirement is oversupplied by 500 pounds per week, this requirement could be increased by that much without increasing the cost of an optimal solution. Zero is not the cost of an extra pound purchased; it is the cost of an extra pound *required.*

A model for the max program could be envisioned in the following way. Suppose there is a supplier who can provide protein, carbohydrate, and roughage in pure form. Let x, y, and z be his prices per pound for these nutriments. If his prices do not satisfy the inequality

$$10x + 10y + 60z \leq 3.80 \tag{3.36}$$

the cattle feeder can obtain some of his feed requirements more cheaply by buying hay. Similar inequalities hold for the other feed available to the feeder. These give the four inequalities of the max program. The total income to the pure nutriments supplier would be

$$450x + 400y + 1050z = f \tag{3.37}$$

Since this supplier wishes to make as much money as possible, his problem is the max program of tableau 3.35.

The variables p, q, r, s are the slack variables for the max program. The variable q, for example, is the amount by which the cost of a sack of oats exceeds the value of the pure nutriments contained in it. In this case oats are overpriced by \$0.80 per sack.

If the price of sweet feed is increased from \$8.00 to \$8.40, the effect of this change is to add 0.4 times the s-column to the b-column. The new b-column in the optimal tableau 3.35 would be

(3.38)

	-1	
-0.2	0.22	$= -y$
-1	0.40	$= -q$
-1	0.80	$= -r$
0.3	0.16	$= -x$
-10	-160	$= f$

Oats become more competitive and are now overpriced by only \$0.40. We still buy the same quantities of each feed and the cost rises to \$160.

It is not very surprising to have the competitive position of a good improve when the price of another product increases. However, a curious thing happens if the price of hay goes up, say to \$4.00 a bale. The effect of this increase is to add 0.20 times the p-column of tableau 3.35 to the b-column. The new b-column is

(3.39)

0.2	-1	
0.5	0.40	$= -y$
1	1.00	$= -q$
1.5	1.50	$= -r$
-0.4	0	$= -x$
-20	-160	$= f$

The marginal price of carbohydrate rises to \$0.40 and the marginal price of protein falls to zero. The total cost of the optimal solution rises to \$160. (The former purchase schedule is still optimal because the basic solutions are still optimal for both programs.) Because oats and pellets have proportionately less carbohydrate than does hay, they become less competitive. They are overpriced by \$1.00 and \$1.50 per sack. A competing product increases its price and these products become less favorably priced! The answer to this seeming paradox lies in the fact that this price increase raises the value of an ingredient that they lack and reduces the value of an ingredient they have in abundance.

Now let us look at the transportation problem discussed in Section 1.3. We duplicate tableau 1.51 here.

(3.40)

	u_1	u_2	u_3	v_1	v_2	-1	
x_{11}	-1			1		1	$= -w_{11}$
x_{12}	-1				1	3	$= -w_{12}$
x_{21}		-1		1		2	$= -w_{21}$
x_{22}		-1			1	6	$= -w_{22}$
x_{31}			-1	1		4	$= -w_{31}$
x_{32}			-1		1	5	$= -w_{32}$
-1	-20	-30	-40	25	50	0	$= f$
	$= y_1$	$= y_2$	$= y_3$	$= z_1$	$= z_2$	$= g$	

A model for the dual max program can be described in the following way. Suppose another entrepreneur has a sufficient supply of the goods to be shipped and that he can produce the goods from a warehouse at any of the destinations. He offers to buy the goods at origin i at the price u_i and sell it at the destination j at the price v_j To be competitive, his prices must satisfy the inequalities

$$v_j - u_i \leq c_{ij} \tag{3.41}$$

where c_{ij} is the shipping rate on the route between origin i and destination j. If his prices do not satisfy these inequalities, the shipper will find it cheaper to ship his own goods. The net income to the entrepreneur is

$$25v_1 + 50v_2 - 20u_1 - 30u_2 - 40u_3 = f \tag{3.42}$$

Since the entrepreneur wants to maximize his income, his problem is the max program of tableau 3.40. The variables w_{ij} are the amounts by which the shipping rates are too high compared with the entrepreneur's price differentials. A route for which w_{ij} is positive will not be used in the optimal shipping schedule for the min program.

Two further examples with realistic interpretations, the one for profit maximization and the other for cost minimization, will be discussed in Section 4.1.

3.4 Heuristic Pivoting

For each linear program there are only finitely many equivalent tableaux representing the problem. If, for one of these tableaux, both basic solutions are feasible then each is optimal for its program. We shall show in the next chapter that this must occur for at least one of the equivalent tableaux, or neither program has an optimal solution. We could, in principle, search through the finite number of equivalent tableaux to find one in which both basic solutions are feasible.

For a linear program with a large number of variables, the number of tableaux that might have to be obtained in an unsystematic search is impractically large. The simplex algorithm is a method for making this search systematic and decisive. However, even without the simplex algorithm the search does not have to be haphazard. Let us consider small programs where it is reasonable to search for optimality without using an efficient algorithm. The purpose of this exercise is to observe what effect the choice of the pivot entry has on the entries of a tableau and how these observations can direct us to making good choices for the pivot entries.

Consider the tableau

5/2	0	−1	1
2	1	−2	−4
1	−1	0	−3
2	2	−2	4
21/2	3	−7	0

(3.43)

We have omitted the variables on the margins since, as far as choosing pivot entries and searching for basic optimality are concerned, the identities of the variables are irrelevant.

We would like to find a sequence of pivot exchanges leading to a tableau in which the b-column is nonnegative and the c-row is nonpositive. If such a tableau can be obtained, the dual basic solutions would be feasible and complementary. The sufficiency condition would assure us that both are optimal, and how we obtained that tableau would not matter. Neither condition is satisfied for tableau 3.43.

Let us examine the possible effects of a pivot exchange. Select $a_{rs} \neq 0$ as pivot, with b_r, c_s also nonzero. After a pivot exchange the new values of the b-entry in the pivot row and the c-entry in the pivot column are given by

$$\bar{b}_r = b_r \bar{a}_{rs} \tag{3.44}$$

$$\bar{c}_s = -c_s \bar{a}_{rs} \tag{3.45}$$

where $\bar{a}_{rs} = 1/a_{rs}$. Notice that either the b_r-entry or the c_s-entry must change sign, but not both. The c_s-entry will change sign if $a_{rs} > 0$, and the b_r-entry will change sign if $a_{rs} < 0$.

If there is a negative entry in the b-column, we might try to pivot on a negative entry in that row to change the sign of the b-entry. If there is a positive entry in the c-row, we might try to pivot on a positive entry in that column to change the sign of the c-entry. Of course, other entries will have their signs changed and we can lose ground as well as gain. Furthermore, there is always the possibility that an optimal solution does not exist. It turns out, however, that for small problems it is rather easy to settle the matter by well directed trial-and-error.

Without arguing whether it is a good strategy, let us concentrate on trying to get a nonnegative b-column without worrying, at first, about what happens to the entries in the c-row. Thus, we shall elect to pivot on a negative entry in one of the second or third rows of tableau 3.43. With this restriction, there are two choices for the pivot entry.

When a pivot exchange is performed with a_{rs} as the pivot entry, the new entries in the b-column (other than in the pivot row) will be of the form

$$\bar{b}_i = b_i - a_{is}\bar{b}_r = b_i - a_{is}\frac{b_s}{a_{rs}} \tag{3.46}$$

In other words, we subtract a multiple of the pivot column from the b-column (except in the pivot row).

(3.47)

0	1		1
1	−4		−7
−1*	−3		3
2	4		−2
3	0		−9

In tableau 3.47 we show to the right the new b-column that would be obtained if we pivot on the starred entry shown to the left. The new b-entry in the pivot row is $b_r/a_{rs} = 3$. The other entries in the b-column are obtained by subtracting this multiple, 3, times the pivot column from the b-column.

In the same way, if we pivot on the −2 in the third column we would have

(3.48)

	−1	1		3
	−2*	−4		2
	0	−3		−3
	−2	4		8
	−7	0		14

Again, the new b-column is shown on the right. Comparing the potential new b-columns in 3.47 and 3.48, we decide to pivot on the -2 in the third column of tableau 3.43. We obtain

$3/2$	$-1/2$	$-1/2$	3
-1	$-1/2$	$-1/2$	2
1	-1^*	0	-3
0	1	-1	8
$7/2$	$-1/2$	$-7/2$	14

(3.49)

In tableau 3.49 we want to change the sign of the -3 in the b-column. The reasonable choice for the pivot entry is the -1 in that row that is starred in tableau 3.49. We obtain

1^*	$-1/2$	$-1/2$	$9/2$
$-3/2$	$-1/2$	$-1/2$	$7/2$
-1	-1	0	3
1	1	-1	5
3	$-1/2$	$-7/2$	$31/2$

(3.50)

In tableau 3.50 we have achieved our first objective, a nonnegative b-column. Now the problem is to try to get a nonpositive c-row without getting a negative entry in the b-column. We wish to pivot on a positive entry in the column of a positive c-entry. There are two choices. Each one will result in subtracting a multiple of the pivot column from the b-column, as described above. If we pivot on the 1 in the first row, we will subtract $9/2$ times the pivot column from the b-column. If we pivot on the 1 in the fourth row, we will subtract 5 times the pivot column from the b-column. To avoid getting a negative entry in the b-column we subtract as little as possible. We pivot on the 1 which is starred in tableau 3.50.

1	$-1/2$	$-1/2$	$9/2$
$3/2$	$-5/4$	$-5/4$	$41/4$
1	$-3/2$	$-1/2$	$15/2$
-1	$3/2^*$	$-1/2$	$1/2$
-3	1	-2	2

(3.51)

In tableau 3.51 we succeeded in getting a negative entry in the first position in the c-row, but the second entry turned positive. We have to try again. There is only one positive entry in the second column. If we pivot on the starred entry in the second column we will succeed in getting a nonnegative b-column and a nonpositive c-row.

While this sort of heuristic procedure works rather well for a small problem it could be frustrating and confusing for a large problem. Later in this chapter we shall describe an effective method for finding a nonnegative b-column. In the following chapter we shall describe an effective method (the simplex algorithm) for getting a nonpositive c-row after obtaining a nonnegative b-column.

The essential feature of a systematic and effective method for solving linear programming problems is a rule or procedure for the choice of the pivot entries. The purpose of this section is to encourage exploratory pivoting so that the rules of the simplex algorithm will seem reasonable.

3.5 Optimal Solutions—Unique, Multiple, None

We are concerned here with the nature of the set of optimal solutions of canonical dual linear programs represented by a tableau in the form

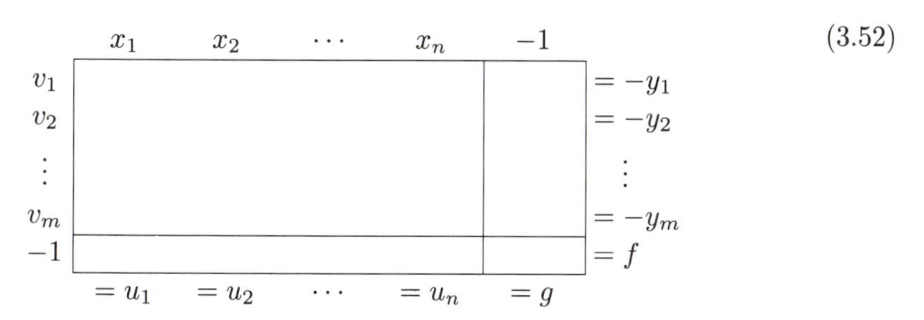

Several important conclusions about the structure of this set can be made from observations of patterns of signs that may be obtained by pivoting. For this purpose it is convenient to introduce notation that we will use frequently. Let

$+$ denote a positive number,
$\oplus$ denote a nonnegative number,
$-$ denote a negative number, and
$\ominus$ denote a nonpositive number.

Suppose we obtain a tableau equivalent to tableau 3.52 in which the pattern of signs shown in tableau 3.53 appears. For the basic solution of the max program, the basic variables have nonzero values and such a solution is said to be *nondegenerate*. If any basic variable in a basic solution is zero, the solution is *degenerate*.

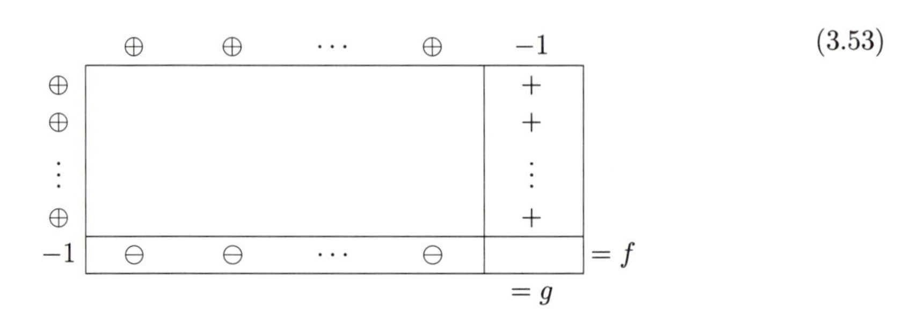

(3.53)

If all nonbasic variables in 3.53 are zero, we obtain complementary basic optimal solutions for the dual programs. If any nonbasic variable at the left is increased from zero, the value of g will be increased. Thus, the basic solution for the min program in tableau 3.53 is optimal and it is the unique optimal solution. A similar conclusion can be made if the entries in the c-row are negative and the b-column is nonnegative. Thus, we have

Theorem 3.18 *If a pair of dual linear programs can be represented by a tableau in which the b-column contains only positive entries and the c-row contains only nonpositive entries, then the optimal solution to the min program is unique. If the c-row contains only negative entries and the b-column is nonnegative, then the optimal solution to the max program is unique.* ◫

We shall describe some situations that indicate the presence of multiple optimal solutions. This information is obtainable by observing patterns of positive, negative, and zero terms in a tableau.

If one program has a degenerate basic optimal solution and the other has a nondegenerate basic optimal solution, then the degenerate optimal solution is unique and the program with the nondegenerate solution has an infinite number of optimal solutions. This is easy to establish and will be explained in the material following. When both optimal basic solutions are degenerate, one program or the other, or both, will have multiple optimal solutions. This is not so easy to establish. It will follow from the complementary slackness theorem, to be proved in Chapter 5. When both programs have degenerate optimal basic solutions, examples can be constructed where one program has multiple optimal solutions and where both programs have multiple optimal solutions.

Thus, both linear programs in a dual pair of linear programs have unique optimal solutions only when both have nondegenerate optimal basic solutions. When one program has a nondegenerate optimal basic solution and the other has a degenerate optimal basic solution, one program has multiple optimal solutions and the other, with the degenerate optimal solution, has a unique optimal solution.

We shall not attempt to describe the possibilities further from a theoretical point of view.

Suppose we have a tableau in which the entries of the c-row are nonpositive but at least one is zero. Any pivot exchange with a pivot entry taken from a column containing a zero entry in the c-row will not change the c-row. That is, we will obtain another tableau for which the basic solution for the column system is feasible. Actually, the two basic solutions are identical. The pivot exchange will exchange two variables that will have zero values in both basic solutions. One will be zero because it is nonbasic and the other will be zero because of the zero entry in the c-row. Also, if the b-column is nonnegative and there is at least one positive entry in the pivot column opposite a positive entry in the b-column, pivoting on that entry will produce a different optimal solution for the max program.

To see examples of what is being described, refer to the second example discussed in Sections 2.4 and 2.5, and to tableau 2.45 and table 2.2. Tableaux C, D, E, and F are tableaux in which both basic solutions are feasible. We duplicate these four tableaux in 3.54 for convenient reference.

(3.54)

C	x_2	y_2	-1	
u_1	1^{E}	1^{A}	1	$= -x_1$
v_1	-1^{D}	-1^{B}		$= -y_1$
v_3	1^{F}		1	$= -y_3$
-1		-1	-1	$= f$
	$= u_2$	$= v_2$	$= g$	

D	y_1	y_2	-1	
u_1	1^{E}		1	$= x_1$
u_2	-1^{C}	1^{B}		$= x_2$
v_3	1^{F}	-1^{H}	1	$= y_3$
-1		-1	-1	$= f$
	$= v_1$	$= v_2$	$= g$	

E	x_1	y_2	-1	
u_2	1^{C}	1^{A}	1	$= x_2$
v_1	1^{D}		1	$= y_1$
v_3	-1^{F}	-1^{G}		$= y_3$
-1		-1	-1	$= f$
	$= u_1$	$= v_2$	$= g$	

F	y_2	y_3	-1	
u_1	-1^{G}	1^{E}		$= x_1$
u_2		1^{C}	1	$= x_2$
v_1	1^{H}	-1^{D}	1	$= y_1$
-1	-1		-1	$= f$
	$= v_2$	$= v_3$	$= g$	

In all four tableaux of 3.54 the basic solutions for the column system of equations are feasible and identical. Each is obtained from the other by pivoting in the column of the zero c-entry. If we pivot in the row of the zero entry in the b-column we get a tableau with the same basic solution for the row system. Tableaux C and D form such a pair and so do tableaux

E and F. If we pivot in the row of a nonzero entry in the b-column we get a tableau with a different basic solution for the row system. In this example the new solutions are feasible, but generally the new solutions would be feasible only for suitable choices of the pivot entry.

Suppose we have a tableau

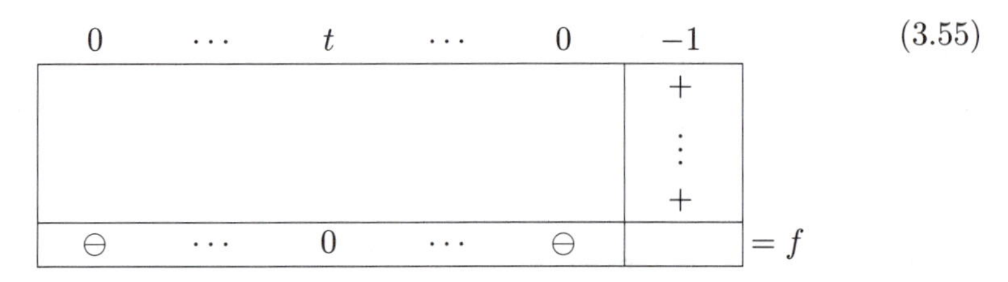

(3.55)

for which the b-column is positive and the c-row is nonpositive with at least one zero entry. We have set the basic variables at the top equal to zero except for one, denoted by t, that is above the zero entry in the c-row. If one or more of the entries in the t-column are positive, the basic variables in those rows will decrease as t increases. We increase t from zero until a basic variable at the right first becomes zero (much as we did in the production problem of Section 1.1). Since f does not change from its maximum value, we have generated an interval of optimal solutions for the max program. When we exchange t with the basic variable that became zero, we obtain a new tableau with a different basic optimal solution for the max program and the same basic optimal solution for the min program.

If the t-column in 3.55 contains no positive entry, we can increase t from zero through all positive values without changing the value of f or making any basic variable at the right nonpositive. Thus, we generate a half-line of optimal solutions for the max program.

We also want to discuss conditions under which we can see that a program is either infeasible or, though feasible, does not have an optimal solution. Suppose we obtain a tableau in which the following pattern of signs appears.

$$\begin{array}{cccc|c|l} \oplus & \oplus & \cdots & \oplus & -1 & \\ \hline \oplus & \oplus & \cdots & \oplus & - & = -t \end{array} \qquad (3.56)$$

In tableau 3.56 the basic variable t has a negative value, whatever nonnegative values the nonbasic variables at the top may have. This means that the max program has no feasible solution. We call a row with sign pattern as shown above an *infeasible row*, since it shows the row system to be infeasible. Similarly, the pattern of signs in tableau 3.57 indicates that the max program has no feasible solution. We call a column of that kind an *infeasible column*, since it shows the column system to be infeasible.

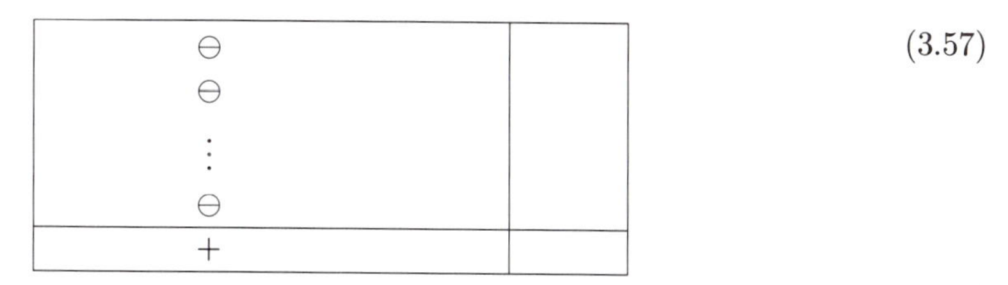

(3.57)

Tableau 3.56 tells us that the max program is infeasible but it does not tell us whether the min program is feasible or infeasible. However, in either case the min program does not have an optimal solution. Suppose, as indicated in tableau 3.58, that the min program has a feasible solution and that there is an infeasible row. At the left we indicate only the nonnegative value of the nonbasic variable heading the infeasible row because the nonnegative values of the other nonbasic variables will not change.

(3.58)

$+$	$\oplus$	$\oplus$	$\cdots$	$\oplus$	$-$
	$=\oplus$	$=\oplus$	$\cdots$	$=\oplus$	$=g$

If the nonbasic value $+$ at the head of the infeasible row is increased it will remain nonnegative and so will the basic values on the bottom margin. That is, the nonbasic value $+$ at the left can be increased indefinitely and the resulting solution will remain feasible. Then g will decrease without bound and the min program does not have an optimal solution.

If we have an infeasible column we cannot tell immediately whether the max program is feasible or infeasible. But if the max program is feasible its objective variable can be made arbitrarily large for feasible solutions so that an optimal solution for the max program does not exist.

Theorem 3.19 *An infeasible row implies that the max program has no feasible solution and that the min program has no optimal solution, either because the min program has no feasible solution or because it has feasible solutions with values of g unbounded below.* ⊟

Theorem 3.20 *An infeasible column implies that the min program has no feasible solution and that the max program has no optimal solution, either* *because the max program has no feasible solution or because it has feasible solutions with values of f unbounded above.* □

We can see that a great deal of information is available from the various patterns of signs that might be encountered.

> **If we obtain a nonnegative b-column, the max program is feasible. If we obtain a nonpositive c-row, the min program is feasible. If we obtain both in the same tableau, both basic solutions are optimal. If either basic solution is nondegenerate (no zero basic variables), then the other optimal solution is unique. If one optimal solution is nondegenerate and the other optimal solution is degenerate, then one is unique and the other is not.**
>
> **If we obtain an infeasible row or an infeasible column, neither program has an optimal solution. If we get an infeasible row, we know the max program is infeasible. We do not know whether the min program is feasible, but if it is, it is unbounded. If we get an infeasible column, the min program is infeasible. We do not know whether the max program is feasible, but if it is, it is unbounded. If we get both an infeasible row and an infeasible column, both programs are infeasible.**

3.6 The Geometric Picture

The purpose of this section is to describe the geometry of linear programming, approximately what a feasible set looks like, and the significance of basic feasible solutions. We are not interested in proofs unless they relate to the algebra of linear programming.

The setting is in an n-dimensional coordinate space, where n is the number of columns in the tableau. Since it is impossible to draw figures of high dimension on paper, our descriptions and illustrations will be of dimensions two and three. However, we will word our descriptions so that they work in higher dimensions as well. Be assured that as long as you consider linear problems, you can picture things in two and three dimensions and use these descriptions so that your reasoning will be reliable in any dimension.

The constraints in linear programming are linear. There may be many conditions, but each is a linear expression of the form

$$L(x) = a_1x_1 + a_2x_2 + \cdots + a_nx_n = b \tag{3.59}$$

The solution set is a line in two dimensions, a plane in three dimensions, and a hyperplane in n dimensions. Each linear expression divides the space into three parts: the hyperplane itself, one side where $L(x) < b$, and one side where $L(x) > b$. The regions that are of interest to us are the hyperplane, the region that satisfies $L(x) \leq b$, and the region where $L(x) \geq b$. These latter two regions are closed half-spaces (because they include the bounding hyperplane) as differentiated from the open half-spaces, which do not include the bounding hyperplane.

A *convex set* is a set that contains, with any pair of points, the entire line segment between the two points. A circle and rectangle in two dimensions, and a sphere and cube in three dimensions are examples of convex sets. The first quadrant in two dimensions is an unbounded convex set, as are the hyperplane and the closed and open half-spaces in n dimensions described above.

The intersection of any number of convex sets is also a convex set. Thus, the intersections of hyperplanes and closed half-spaces that are defined by each linear condition of the constraint conditions are also convex sets. That is,

the feasible set for a linear program is a convex set.

We want to connect three concepts: a basic feasible solution, an extreme point of the feasible set, and a vertex or corner of the feasible set. An *extreme point* of a set is a point that is not on a line segment between two other points in the set.

Let $U = (u_1, u_2, \ldots, u_n)$ and $V = (v_1, v_2, \ldots, v_n)$ be two points in an n-dimensional coordinate space. For each value of t the point

$$X = (1-t)U + tV = U + t(V - U) \tag{3.60}$$

is a point on the straight line through U and V. To see this notice that $X = U$ for $t = 0$, and $X = V$ for $t = 1$. For $0 < t < 1$, the point X is between U and V. Furthermore, for each coordinate x_i, if $u_i < v_i$.

$$u_i = (1-t)u_i + tu_i < x_i < (1-t)v_i + tv_i = v_i \tag{3.61}$$

That is, x_i is between u_i and v_i. If t is strictly between 0 and 1, the only way that x_i can be equal to either u_i or v_i is for u_i and v_i to be equal to each other.

The important observation here is that if $x_i = 0$ and either of u_i or v_i is positive, the other must be negative. We are now prepared to prove the following theorem.

Theorem 3.21 *The set of basic feasible solutions of a linear program is precisely the set of extreme points of the set of feasible solutions.*

Proof. The first part is easy. Let X be a basic feasible solution for a given linear program. Suppose U and V are feasible solutions for which there exists a t, $0 < t < 1$, such that $X = (1-t)U + tV$. For each coordinate where $x_i = 0$ we must have $u_i = v_i = 0$. Since X is a basic solution, it is the only solution with the given variables equal to zero. That is, $U = V = X$, and X is an extreme point.

The second part is not difficult, but it is a little fussy. Let X be an extreme point of the feasible set. We wish to show that X is a basic solution. To do this we will use Theorem 2.2 of Chapter 2. We must show that X is the only solution with a given set of variables equal to zero.

Let A be any point for which $x_i = 0$ implies $a_i = 0$. The other coordinates of A might be zero, positive, or negative. Let m be a positive number smaller than any of the absolute values of the nonzero coordinates in A or X, and let M be a positive number larger than any of the absolute values of the coordinates in A or X. Let

$$U = X + \frac{m}{2M}(A - X) \tag{3.62}$$

and

$$V = X + \frac{m}{2M}(X - A) \tag{3.63}$$

Now U and V satisfy every linear equation that X and A satisfy. Specifically, if $L(X) = b$ is a linear equation with $L(A) = b$, then

$$L(U) = b + \frac{m}{2M}(b - b) = b \tag{3.64}$$

and

$$L(V) = b + \frac{m}{2M}(b - b) = b \tag{3.65}$$

If $x_i = 0$, then $u_i = v_i = 0$.
If $x_i > 0$, then

$$u_i = x_i + \frac{m}{2M}(a_i - x_i) > x_i - \frac{m}{2M}M - \frac{m}{2M}M > x_i - m > 0 \tag{3.66}$$

and

$$v_i = x_i + \frac{m}{2M}(x_i - a_i) > x_i - \frac{m}{2M}M - \frac{m}{2M}M > x_i - m > 0 \tag{3.67}$$

That is, U and V are feasible solutions for the linear program. If $A \neq X$ then U and V are distinct from X. But $X = (U + V)/2$ is midway between U and V, and that contradicts the assumption that X is an extreme feasible point. Hence $A = X$. □

Generally, the feasible set for a canonical linear programming problem is an n-dimensional generalization of a convex polygon in two dimensions. It has hyperplanes as faces. It might be bounded or it might extend to infinity, like the first quadrant in two dimensions.

We are in the habit of thinking in terms of coordinate axes in a coordinate system, but it is also fruitful to think in terms of coordinate planes. For example, in three dimensions where points would be designated by coordinates (x, y, z), the coordinate plane where $x = 0$ is the plane containing the y- and z-axes. Similarly, the coordinate plane where $y = 0$ is the plane containing the x- and z-axes, and the coordinate plane where $z = 0$ is the plane containing the x- and y-axes.

The origin, then, is the point where the x, y, and z coordinate planes intersect. If we have a linear program with a tableau in the form of 3.68,

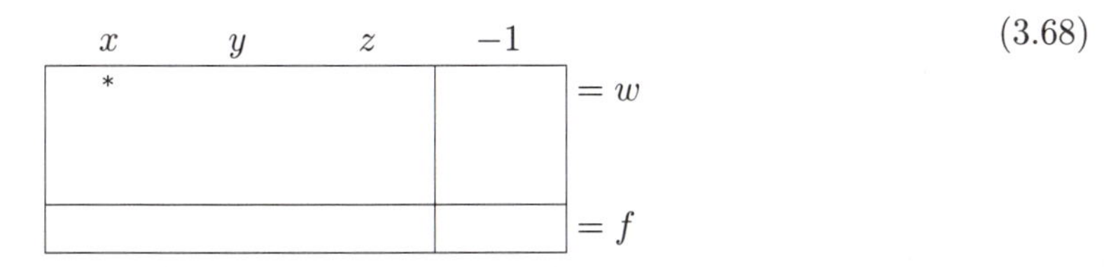

(3.68)

the basic solution corresponds to the origin. If the variables are canonical, the feasible set is contained in the first orthant (where the first orthant includes the coordinate planes). If the b-column is nonnegative, the origin is a feasible solution.

The set where $w = 0$ is another hyperplane. If we pivot on the starred entry in tableau 3.68, we obtain tableau 3.69.

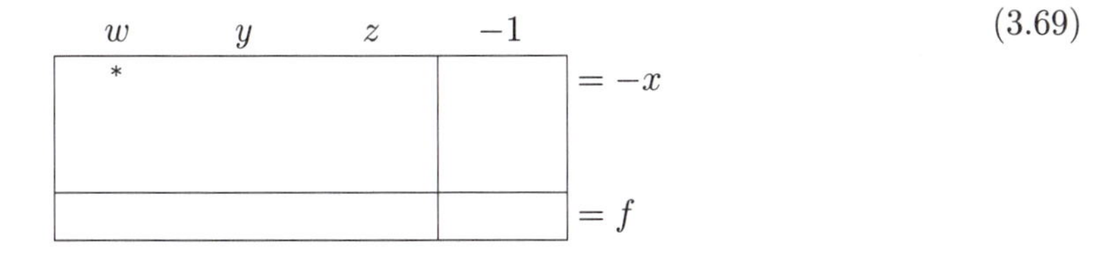

(3.69)

Now the basic solution represents the intersection of the planes $w = 0$, $y = 0$, and $z = 0$. It is fruitful to think of these as the new coordinate planes. Then all tableaux can be interpreted similarly, and each basic solution is the origin of a new coordinate system.

In three dimensions, a corner or vertex of a solid is a point where three planes intersect. However, we do not consider the intersection of three planes that have a common line as determining a vertex. A vertex is a point where the three intersecting planes are independent. This is equivalent to the condition that the variables corresponding to those planes can be brought to the top margin by a sequence of independent pivots. Accordingly, we define a vertex or corner of the feasible set to be a basic feasible solution for the linear program.

If the b-column in tableau 3.69 is nonnegative, then the new basic solution is feasible. That is, the new origin is a corner of the feasible set. If the tableau has n columns, the two basic solutions obtained before and after a pivot exchange will share $n-1$ basic variables, and the two corners will share $n-1$ hyperplanes. In two dimensions they share an edge. In three dimensions they share two faces and, therefore, an edge. In general, they are adjacent corners on the boundary of the feasible set. These corners share $n-1$ hyperplanes corresponding to the variables that are not involved in the pivot exchange. These $n-1$ hyperplanes determine the common edge. In a degenerate case this edge can collapse to a point.

If the value in the entry in the b-column of the first row is zero, then $w = 0$ at the first origin. That is, the pivot exchange does not produce a new point for the origin. Degeneracy for a basic feasible solution means that two (or more) corners collapse into one.

Now we turn our attention to the objective function. The set where $f = e$, where e is a constant, is another hyperplane. The family of hyperplanes obtained for different values of e is a family of parallel objective hyperplanes. For each objective hyperplane, the objective hyperplanes on one side correspond to smaller values of e and the objective hyperplanes on the other side correspond to larger values of e. That is, if we move the objective hyperplane in one direction, the values of e decrease. If we move the objective hyperplane in the other direction, the values of e increase.

There is a feasible solution that yields this value of the objective function if the hyperplane intersects the feasible set. If the goal is to maximize f, we move the objective hyperplane in the direction in which f increases as far as we can as long as it still intersects the feasible set. If the objective function is bounded, we will eventually reach a point where the objective hyperplane touches the feasible set and no point in the feasible set lies on the increasing side of the objective hyperplane.

Suppose that an objective hyperplane intersects the feasible set in a vertex of the feasible set. That vertex corresponds to a basic feasible solution, and there is a tableau for which that vertex is the basic feasible solution. Each basic variable represents an edge incident with that vertex. Look at the c-row. Positive entries in the c-row identify edges along which the objective function will increase, negative entries identify edges along which the objective function will decrease, and zero entries identify edges along which the objective function will not change. That is, the positive entries identify edges on the increasing side of the objective hyperplane, negative entries identify edges on the decreasing side of the objective hyperplane, and zero entries identify edges that lie in the objective hyperplane.

Given any set, a supporting hyperplane is a hyperplane that touches the set and for which the set lies entirely on one side of the hyperplane. If the feasible set is bounded, there are two supporting objective hyperplanes, one where the objective function is minimized and one where the objective function is maximized. Let us look more closely at the supporting objective hyperplane where the objective function is maximized.

All points of the feasible set that lie in the supporting objective hyperplane are optimal solutions. If the optimal solution is unique, only one feasible point lies in the supporting objective hyperplane. That point is a vertex of the polyhedron that defines the feasible set. Conversely, if the only point of contact with the supporting objective hyperplane is a vertex, the optimal solution is unique. In this case all edges of the feasible set incident with that vertex are on the decreasing side of the supporting objective hyperplane, and the c-row contains only negative entries.

If the supporting objective hyperplane intersects the feasible set in more than one point, it must intersect the feasible set in an entire edge or face. That is, there is more than one vertex in the supporting objective hyperplane. Some edges of the feasible set incident with this vertex lie in the supporting objective hyperplane, and some entries in the c-row are zeros.

3.7 Feasibility Algorithm

In Section 3.5 we have described sign patterns that indicate whether a linear program is feasible or infeasible. We know that a nonnegative b-column means that the row system of equations is feasible. We know that an infeasible row means that the row system is infeasible. We do not yet know that if the row system is feasible there must be an equivalent tableau for which the b-column is nonnegative. We do not know that if the row system is infeasible there must be an equivalent tableau with an infeasible row. What is lacking is the assurance that one or the other of those sign patterns must be obtained. In this section we shall describe a systematic method for resolving these questions.

The method to be presented is the first of several algorithms that will be described in this book. An algorithm is a prescription for a series of steps to be taken, and alternate choices to be made on the basis of the results of these steps, which will lead to an outcome that is sought or to the conclusion that the outcome sought cannot be obtained.

Ideally, the instructions contained in the algorithm should be unequivocal so that a computing machine could be programmed to carry out the steps of the algorithm. Often we shall not be completely explicit about which of several alternate choices to take (e.g., "choose a positive entry in the c-row") when any choice will lead to the same kind of outcome, if by a different route. But the algorithm described in this section will be explicit.

Many people think of an algorithm as a method for finding a solution to a problem for which one already knows, by some other means, that a solution exists. We look on algorithms as much more than that, and we use them for much more than that. In many cases an algorithm can supply the proof that the desired result exists and, wherever possible, constructive proofs are to be desired. The algorithms in this book are used to provide proofs as well as to obtain answers.

In 1976 Robert G. Bland introduced rules for selecting the pivot entries to avoid "cycling" or "circling" in the simplex algorithm (to be discussed in Chapter 4). The algorithm described here uses the Bland selection rules in a simpler context. An algorithm must be accompanied by a proof that the algorithm terminates with a decision in a bounded finite number of steps. The simpler context here allows a simple proof of the algorithm's termination.

The algorithm requires that all the variables of one of the linear programs be ordered. The particular ordering is arbitrary and the effect of different orderings is unpredictable. We assume the variables of the max program are $x_1, \ldots, x_{n+m}$, and we assume the dual variables of the min program are $u_1, \ldots, u_{n+m}$. The aim of the algorithm is to determine whether the max program is feasible or infeasible by obtaining a nonnegative b-column or an infeasible row. Since we are not initially concerned with the min program we shall suppress the c-row in discussing the algorithm.

Let the initial tableau be

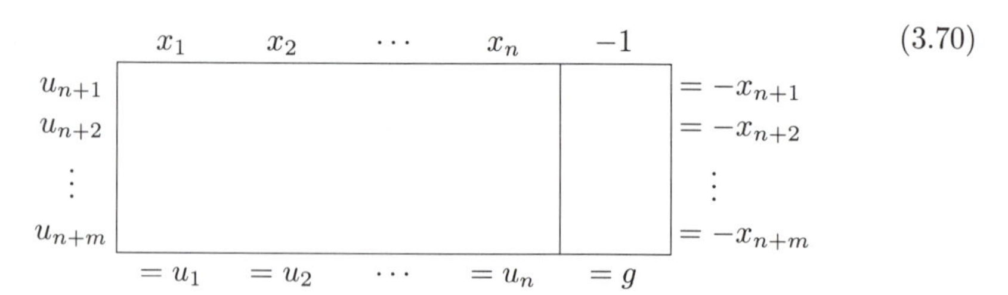

(3.70)

Feasibility Algorithm

1. If the b-column is nonnegative, STOP. If there is a negative entry in the b-column, select as the pivot row a row with a negative b-entry and for which the index of its basic variable on the right margin is least.

2. If the pivot row is an infeasible row, STOP. If there is a negative entry in the pivot row (other than in the b-column), select as the pivot column a column in which the entry in the pivot row is negative and for which the index of its nonbasic variable on the top margin is least.

3. Perform a pivot exchange with the pivot row and pivot column selected in 1 and 2. Return to 1.

If the algorithm stops by the first selection rule, we have a nonnegative b-column and the row system of tableau 3.70 is feasible. If the algorithm stops by the second selection rule, we have an infeasible row and the row system of tableau 3.70 is infeasible. We must show that we cannot perform an endless sequence of pivot exchanges without stopping by one rule or the other.

Theorem 3.22 *The feasibility algorithm with the Bland selection rules terminates in a finite number of steps with either a nonnegative b-column or an infeasible row.*

Proof. Suppose, for some initial tableau 3.70, the algorithm continues without stop, either by rule 1 or by rule 2. Then, sooner or later, it reaches a tableau that has occurred before, since the number of tableaux equivalent to 3.70 is finite. Between these two occurrences of the same tableau the algorithm circles through a loop which the unequivocal pivot selection rules will repeat again and again without stop.

During this loop some variables are pivoted from one margin to another. When the loop is closed, they must have been pivoted back (perhaps several times) to the margin from which they started. We use p as a generic index for these pivoted variables—except for those of greatest index, which we denote specifically by x_q and u_q. For variables (if any) not pivoted during the loop, attached to rows only or attached to columns only, we use r and s as generic indices. Then our loop must contain a tableau 3.71 in which x_q and u_q are attached to the pivot row and a tableau 3.72 in which x_q and u_q are attached to the pivot column. Since x_q and u_q are pivoted at least once, they must be pivoted in each direction at least once. (In 3.71 and 3.72 variables not needed later are omitted.)

Several pivot exchanges occur within the loop, either between 3.71 and 3.72 or between 3.72 and 3.71. These cause the arrangement of the x_p's in 3.71 to differ from the arrangement of the u_p's in 3.72, but do not affect the variables indexed by r and s. The entries $\oplus$ (nonnegative) and $-$ (negative) must occur in 3.71 and 3.72 as shown, with pivot entries starred. Otherwise

$$
\begin{array}{c|ccccccc|c|l}
 & (0) & \cdots & (0) & \cdots & (0) & (0) & \cdots & (0) & \\
 & x_p & \cdots & x_p & \cdots & x_p & x_s & \cdots & x_s & -1 \\
\hline
 & & & & & & & & & \oplus & = -x_p \\
 & & & & & & & & & \vdots & \vdots \\
 & & & & & & & & & \oplus & = -x_p \\
u_q & & & -^* & & & & & & - & = -x_q \\
\hline
 & & & & & & & & & & = -x_r \\
 & & & & & & & & & & \vdots \\
 & & & & & & & & & & = -x_r
\end{array}
\tag{3.71}
$$

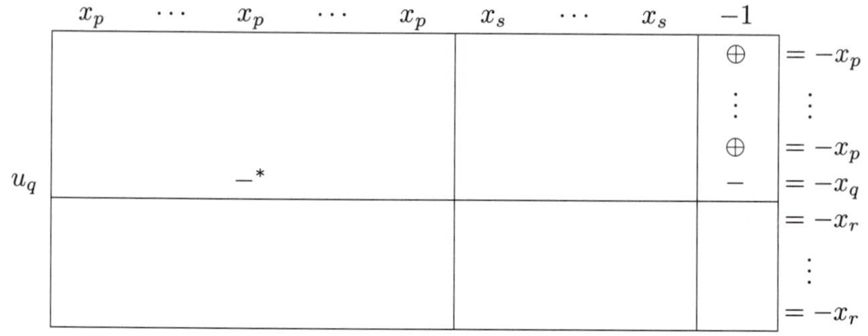

$$
\begin{array}{cc|cccc|ccc|c}
 & & & & & x_q & & & & \\
\hline
(0) & u_p & & & & & & & & \\
\vdots & \vdots & & & & & & & & \\
(1) & u_p & \oplus & \cdots & \oplus & -^* & & & & - \\
\vdots & \vdots & & & & & & & & \\
(0) & u_p & & & & & & & & \\
\hline
(0) & u_r & & & & & & & & \\
\vdots & \vdots & & & & & & & & \\
(0) & u_r & & & & & & & & \\
\hline
 & & = u_p & \cdots & = u_p & = u_q & = u_s & \cdots & = u_s & = g
\end{array}
\tag{3.72}
$$

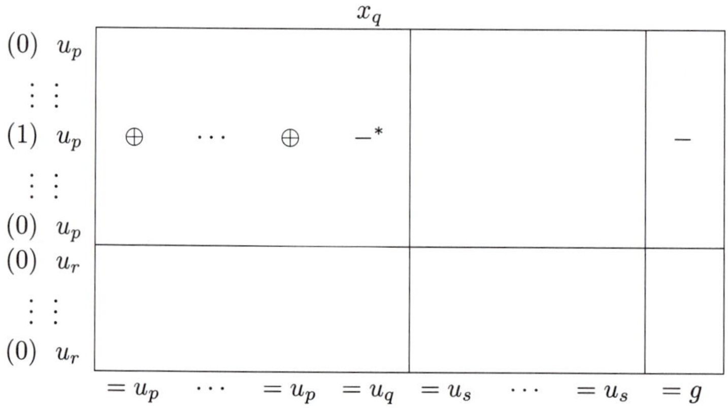

some earlier row in 3.71 would be its pivot row and some earlier column in 3.72 its pivot column.

On the top margin of 3.71 set the variables equal to 0, as indicated, to get a particular x-solution with each $x_p \geq 0$ (at right and top), $x_q < 0$, and each $x_s = 0$. On the left margin of 3.72 set the variables equal to 0 except $u_p = 1$ attached to the pivot row, as indicated, to get a particular u, g-solution with each $u_p \geq 0$ (at bottom and left), $u_q < 0$, each $u_r = 0$, and $g < 0$.

These two particular solutions make

$$
\begin{aligned}
u_p x_p &\geq 0 \\
u_q x_q &> 0
\end{aligned}
$$

$$\begin{aligned} u_r x_r &= 0 \\ u_s x_s &= 0, \\ &\text{and} \\ g &< 0 \end{aligned} \tag{3.73}$$

for all dual pairs of pivoted variables x_p and u_p, and all dual pairs of unpivoted variables x_r, u_r and x_s, u_s. On the other hand, the duality equation

$$\sum_p u_p x_p + u_q x_q + \sum_r u_r x_r + \sum_s u_s x_s = g \tag{3.74}$$

holds for any x-solution and any u, g-solution of the equivalent tableaux 3.71 and 3.72. Between 3.73 and 3.74 we have a clear contradiction!

We derived this contradiction from a chain of reasoning based on the assumption that for some initial tableau 3.70 the algorithm continues without stop. Therefore, we are able to conclude that the algorithm always does stop, either by rule 1 or by rule 2. This proves the theorem. The proof given here is due to Harold W. Kuhn (unpublished class notes.) ⊟

In previous discussions we have usually described a pair of dual solutions in terms of the same tableau. It should be emphasized, however, that the row equations of equivalent tableaux have the same solution sets, and the column equations of equivalent tableaux have the same solution sets. Thus, the duality equation must be satisfied even though one solution is described in terms of one tableau and the other is described in terms of another tableau.

For the column system there is a dual version of the feasibility algorithm with the Bland selection rules in which the pivot column is selected as the column with a positive c-entry and the least index, and the pivot row is selected as the row with a positive entry in the pivot column and the least index. In this dual form, the algorithm terminates with either a nonpositive c-row or an infeasible column, a column in which the c-entry is positive and the other entries are nonpositive.

The proof of the termination of the feasibility algorithm establishes the following important theorem.

Theorem 3.23 *(Theorem of the Two Feasibility Alternatives) In any class of equivalent tableaux there is either at least one tableau with a nonnegative b-column or at least one tableau with an infeasible row, but not both. In any class of equivalent tableaux there is either at least one tableau with a nonpositive c-row or at least one tableau with an infeasible column, but not both.* ⊟

Corollary 3.24 *If a linear program has a feasible solution, it has a basic feasible solution.* □

The example in Table 3.2 illustrates the feasibility algorithm in operation. It shows how the algorithm can terminate with feasibility and how it can terminate with infeasibility.

Table 3.2: A Run of the Feasibility Algorithm.

x_1	x_2	x_3	-1	(1)
-1^*	9	-9	-3	$= -x_4$
$-1/3$	2	-2	-2	$= -x_5$
2	-9	9	-2	$= -x_6$
$1/3$	-1	1	1	$= -x_7$

x_4	x_2	x_3	-1	(2)
-1	-9	9	3	$= -x_1$
$-1/3$	-1^*	1	-1	$= -x_5$
2	9	-9	6	$= -x_6$
$1/3$	2	-2	0	$= -x_7$

x_4	x_5	x_3	-1	(3)
2	-9	0	12	$= -x_1$
$1/3$	-1	-1	1	$= -x_2$
-1^*	9	0	-3	$= -x_6$
$-1/3$	2	0	-2	$= -x_7$

x_6	x_5	x_3	-1	(4)
2	9	0	6	$= -x_1$
$1/3$	2	-1	0	$= -x_2$
-1	-9	0	3	$= -x_4$
$-1/3$	-1^*	0	-1	$= -x_7$

x_6	x_7	x_3	-1	(5)
-1^*	9	0	-3	$= -x_1$
$-1/3$	2	-1	-2	$= -x_2$
2	-9	0	12	$= -x_4$
$1/3$	-1	0	1	$= -x_5$

x_1	x_7	x_3	-1	(6)
-1	-9	0	3	$= -x_6$
$-1/3^*$	-1	-1	-1	$= -x_2$
2	9	0	6	$= -x_4$
$1/3$	2	0	0	$= -x_5$

x_2	x_7	x_3	-1	(7)
-3	-6	3	6	$= -x_6$
-3	3	3	3	$= -x_1$
6	3	-6	0	$= -x_4$
1	1	-1^*	-1	$= -x_5$

x_2	x_7	x_5	-1	(8)
0	-3	3	3	$= -x_6$
0	6	3	0	$= -x_1$
0	-3	-6	6	$= -x_4$
-1	-1	-1	1	$= -x_3$

For the example in Table 3.2 the algorithm terminates with feasibility in seven pivot exchanges. If the x_3-column were deleted in the first tableau, the first six pivot selections would be unaffected but the pivot entry selected in the seventh tableau would be unavailable. In this case the algorithm would terminate with infeasibility. (That is, the original problem, without the deletion, has no feasible solution for which x_3 is zero.)

Notice that if we pivot on the -1 in the second column of tableau (6) to exchange x_7 and x_2, we would get the same basic and nonbasic variables that we have in tableau (1). The same sequence of pivot exchanges could then be repeated cyclically and endlessly. Cycling can be forced by choosing the pivot exchanges to make it occur, but it should not occur in a well-designed algorithm when one is following the rules of the algorithm. The principal problem in proving that the feasibility algorithm is effective is proving that it cannot cycle. Cycling is a theoretical problem in the simplex algorithm, and it will be discussed in more detail in the next chapter.

There is another expression of the theorem of the two feasibility alternatives that is equivalent to it. Consider the canonical tableau

<table>
<tr><td></td><td>x</td><td>-1</td><td></td></tr>
<tr><td>v</td><td>A</td><td>b</td><td>$= -y$</td></tr>
<tr><td>-1</td><td>0</td><td>0</td><td>$= f$</td></tr>
<tr><td></td><td>$= u$</td><td>$= g$</td><td></td></tr>
</table>

(3.75)

Notice the zeros in the c-row and the d-corner. The row equation is

$$Ax - b = -y \tag{3.76}$$

The value of the objective variable f is always 0. Thus, if the row system is feasible the optimal value is 0. The equation for the column system is

$$vA = u \tag{3.77}$$

The column system is feasible for every A-matrix.

Let us look at the possibilities in view of the theorem of the two feasibility alternatives. If the row system is feasible the feasible values of g are ≥ 0. Since 0 is a feasible value for g, the optimal value of g is 0. That is, g has no negative feasible values. If the row system is infeasible there is an equivalent tableau with an infeasible row. In that tableau, set the nonbasic variable associated with that row equal to 1 and set all other nonbasic variables equal to 0. Because the A-entries in the infeasible row are nonnegative the solution obtained this way is feasible. Since the b-entry in the infeasible row is negative this yields a negative feasible value for g. This establishes the following theorem.

Theorem 3.25 *(Theorem of the Two Feasibility Alternatives, Second Form) For any tableau either the row system $Ax - b = -y$ has a feasible solution or the column system $vA = u$ has a feasible solution with $vb = g < 0$.* ◻

If the row system is infeasible the solution described in the second alternative is found by applying the Bland feasibility algorithm. When the infeasible row is obtained, make the substitution for the nonbasic variables described above to obtain the required solution.

There is practical value associated with the ability to produce the feasible solution for the column equations as the second alternative. If someone comes to you with the claim that the row system is infeasible and presents a tableau with an infeasible row it would be a lot or work to verify that conclusion by checking the tableau. You would have to duplicate the pivoting. However, if you obtain the solution to the column equations, the second alternative can be verified by substitution in the original tableau.

3.8 Priority Pivoting

The *priority pivoting algorithm* is inspired by Bland's algorithm. The idea behind priority pivoting is to retain the features of the Bland algorithm that prevent cycling and to modify it so that it terminates with fewer iterations. Priority pivoting also introduces a completely new feature that provides an arrow for the algorithm, something that can be used to make choices to direct the algorithm to a conclusion.

To see how the feasibility algorithm with the Bland pivoting rules can be modified, let us examine the way the proof of termination is organized.

In Section 3.7 we showed that the Bland feasibility algorithm cannot cycle, because the variable x_q cannot be moved from margin to margin more than once if only variables with smaller indices are moved. If a variable with a larger index is moved after x_q is moved once, then x_q can be moved again. In particular, the variable with the largest index, x_n in the Bland algorithm, cannot be moved twice, while the variable with the smallest index, x_1, can be moved many times.

Notice that the Bland algorithm prevents cycling regardless of the indexing that is chosen for the variables, so long as the ordering remains fixed. It is easily seen that the number of iterations required by the Bland algorithm is dependent on the original indexing (just try a few examples), but it is not easy to see a pattern between the ordering of the indices and the number of iterations. The indices on the variables amount to a priority rating that selects one from several otherwise equal choices. The smaller indices have higher priorities.

We will assign labels that are equivalent to priorities. They will be independent of the indexing, and as few priorities will be assigned as are needed to make a decision about the pivot row and pivot column.

A variable that is not selected for the pivot row or pivot column may be passed over for one of two reasons, either it has a lower priority or it would not be selected anyway because the entry that would determine its selection is nonnegative. The idea is to make the second reason for being passed over the operative reason, if possible. That is, we try to assign high priorities to variables that will not be selected and low priorities to variables that are eligible for selection. We make as little commitment to the priorities as is needed to make the proof of noncycling work. We also allow the priorities to be changed as long as the change does not interfere with the property of the ordering that functions to prevent cycling.

For technical reasons, the exposition of the priority pivoting algorithm is more easily described if Bland's algorithm were described in terms of selecting the largest index at each step rather than the smallest index at each step. That is, in the priority pivoting algorithm, larger labels have higher priorities. The reason that this is more convenient is that the number of levels of priorities changes from step to step, and it is easier to have the fixed end at the low end, at level 1.

The priority pivoting algorithm avoids assigning priorities to those variables that are not involved in the choices, row variables with nonnegative b-entries and column variables with nonnegative entries in the pivot row. The substitute for the index used in the Bland algorithm is a priority level, which we call a label. In the initial description of the algorithm, the choice among variables with the same priority is made arbitrarily. Later we discuss refinements that will improve the speed of the algorithm.

We divide the choice of a pivot entry into two steps, the choice of a pivot row and the choice of a pivot column. We also regard the choice as being attached to the variable associated with the row or column. Those variables (rows or columns) that compete as choices at each step are assigned the same label, the same priority.

The algorithm proceeds by affixing labels and stars to the variables of the row system of equations. Labels and stars change twice at each pivot exchange, once when a pivot row is selected and once when a pivot column is selected. The labels are positive integers. One variable with each label is starred.

A basic variable is eligible to become a pivot variable if (a) the corresponding entry in the b-column is negative, and (b) no basic variable with a larger label (higher priority) corresponds to a negative b-entry. There may be several eligible basic variables.

When a pivot row is selected from among those eligible, a nonbasic variable is eligible as a pivot variable if (a) the corresponding entry in the pivot row is negative, and (b) no nonbasic variable with a larger label (higher priority) corresponds to a negative entry in the pivot row. There may be several eligible nonbasic variables.

Priority Pivoting Algorithm

1. At the start all variables are unlabeled. Label all eligible basic variables with a 1. Select one as pivot variable and star it. This determines the pivot row. Label all eligible nonbasic variables with a 2. Select one as pivot variable and star it. This determines the pivot column. Leave the remaining variables unlabeled. Perform a pivot exchange, exchanging the two pivot variables with stars and labels attached.

2. Determine the eligible basic variables. We shall prove that no eligible basic variable is currently starred. If no basic variable is eligible STOP—we have a nonnegative b-column. Otherwise, select an eligible basic variable as a new pivot variable and star it. It becomes a current pivot variable. Remove the labels and stars from all variables, both basic and nonbasic, with larger labels. Remove the labels and stars from all ineligible variables with the same label. The eligible variables and all variables with smaller labels retain their labels.

3. Determine the eligible nonbasic variables. We shall prove that no nonbasic variable currently starred is eligible. If no nonbasic variable is eligible STOP—we have an infeasible row. Otherwise, select an eligible variable as a new pivot variable. It becomes a current pivot variable. Remove the labels and stars from all variables with larger labels. Remove the labels and stars from all ineligible variables with the same label. The eligible variables and variables with smaller labels retain their labels.

4. Perform a pivot exchange, exchanging the two variables selected in 2 and 3. The labels and stars are attached to the variables and move with them. Go to step 1.

Lemma 3.26 *No starred variable is an eligible variable.*

Proof. If a variable x_q is currently starred, no variable with the same label or a smaller label has been selected and starred since x_q was starred, for such a selection would remove the star. The importance of this observation is that the labeling by labels smaller than or equal to the label on x_q remains unchanged. As long as x_q remains starred, the labels larger than the label on x_q may be changed, but only to other larger labels.

Suppose a variable x_q is already starred and becomes eligible at some stage. This means it was selected as a pivot variable once before, either when it was a basic variable or when it was a nonbasic variable. Consider the two pairs of tableaux shown below.

All tableaux represent the situation just before x_q is selected as a pivot variable. That is, in the first pair, x_q is first selected when it is a basic variable (at left). Then, when it is a nonbasic variable, x_k is selected and starred and x_q is eligible and starred. In the second pair, x_q is first selected when it is nonbasic (at left). Then, when it is a basic variable it is eligible and starred (at right). Most of the argument is the same for both cases. The variable x_k must be handled differently in the two cases. Except for the discussion of x_k, everything to follow applies to both cases.

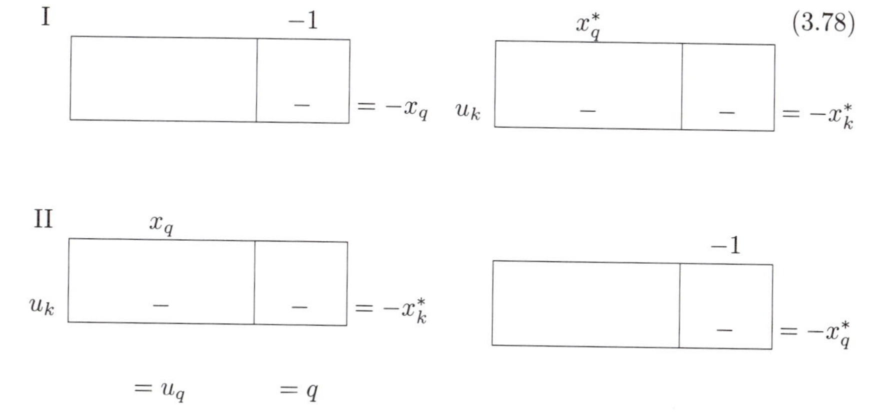

In each pair, we use a solution of the row system of equations for the tableau in which x_q is basic and a solution of the column system of equations for the tableau in which x_q is nonbasic.

In the tableau of each pair in which x_q is basic take the basic solution of the row system. Denote this solution by $\bar{x}$. Notice that $\bar{x}_q < 0$. In the tableau in which x_q is nonbasic, set $\bar{u}_k = 1$ and set all other nonbasic variables of the column system equal to 0. Determine this column solution and denote it by $\bar{u}$. Notice that $\bar{u}_q < 0$.

First, $\bar{u}_q < 0$, $\bar{x}_q < 0$, and $\bar{g} < 0$. Thus, $\bar{u}_q\bar{x}_q > 0$.

Second, consider variables with labels larger than the label on x_q.

For the solution to the row system: If x_i has a label larger than the label on x_q and x_i is nonbasic, $\bar{x}_i = 0$. If x_i has a label larger than the label on x_q and x_i is basic, $\bar{x}_i \geq 0$ by rule 2.

For the solution to the column system: If x_i has a label larger than the label on x_q and u_i is nonbasic, $\bar{u}_i \geq 0$ (including $\bar{u}_k$ if x_k has a larger label).

In all cases, for x_i with labels larger than the label on x_q, $\bar{u}_i\bar{x}_i \geq 0$.

Third, consider variables with labels less than or equal to the label on x_q.

It is in this connection that we must consider the possibility that x_k has a label less than or equal to the label on x_q. In the first pair of tableaux, x_k must have a label larger than the label on x_q. Otherwise, the star and label on x_q would have been erased when x_k was selected. For the second pair of tableaux, x_k will become a nonbasic variable after the pivot exchange is performed. If x_k has a label no larger than the label on x_q, x_k will remain a nonbasic variable for otherwise the star on x_q would be erased before the second tableau. In this case $\bar{x}_k = 0$.

Otherwise, for x_i with labels no larger than the label on x_q, x_i cannot be moved between the two tableaux without erasing the star on x_q. Thus either x_i is nonbasic in both of a pair of tableaux, or u_i is nonbasic in both of a pair of tableaux..

In all cases, for x_i with labels no larger than the label on x_q, $\bar{u}_i\bar{x}_i = 0$.

The arguments above show that $\bar{u}_i\bar{x}_i \geq 0$ for x_i with labels larger than the label on x_q, $\bar{u}_q\bar{x}_q > 0$, and $\bar{u}_i\bar{x}_i = 0$ for x_i with labels no larger than the label on x_q. Thus, $\bar{u}\bar{x} > 0$ and $\bar{g} < 0$. This contradicts the duality equation $\sum ux = g$. ◻

This lemma shows that the relabeling that follows the selection of a new pivot variable has the following effect on the labels. It does not change any smaller label. If there is a starred variable with the same label, it reduces by at least one the number of variables with that label since the starred variable is ineligible. If there is no starred variable with the same label (necessarily an unlabeled variable) it introduces a starred variable with that label. It also increases by one the number of different labels.

Theorem 3.27 *The algorithm terminates in a finite number of iterations.*

Proof. For the purpose of this proof, assign the smallest unused label to all unlabeled variables. For each label k, in any labeling, let h_k denote the number of variables with that label. For each labeling, $(h_1, h_2, \dots)$ is a *signature* of the labeling. Each pivot exchange (which involves two successive relabelings) results in a new signature that is lexicographically smaller. The signature can be decreased lexicographically only a finite number of times. ◻

To simplify implementation, either by hand or by machine, several things that are described in the algorithm can be skipped. The lemma shows that a starred variable is never an eligible variable. It would become eligible only after a subsequent pivot exchange erases both its label and its star. Thus, the star can be omitted when it is exchanged by the pivot exchange. The star only serves in the proof to show that the signature is decreased with each pivot exchange.

The only odd labels that appear on the top margin correspond to starred variables, and the only even labels that appear on the right margin correspond to starred variables. None of the unlabeled variables would have been starred. This class can appear on both margins.

The signature offers an attractive tool to hasten the termination of the algorithm. The following refinements are suggested by the possibility of making choices, where multiple choices are available, to reduce the signature more rapidly.

1. When a pivot row is selected from the eligible rows, the pivot column is going to be selected from a column with a negative entry in the pivot row. The signature can be reduced the most by selecting a row for which the eligible column variables have the smallest label.

2. When there are several pivot rows that meet the conditions of the rule 1 given above, the resulting signature will be smallest if the row with the smallest number of eligible column variables is selected.

3. When the pivot row is selected according to rules 1 and 2 given above, the b-column that is obtained after the next pivot exchange depends on the pivot column that is selected. For each positive entry in the pivot column, the corresponding entry in the b-column will be decreased. For each negative entry in the pivot column, the corresponding entry in the b-column will be increased. Select the column with the largest number of negative entries in the potential pivot column.

4. As a refinement of the above rule 3, select the column with the largest number of negative entries in the potential pivot column that are in rows with labels at least as large as the label on the pivot row.

5. We can change the definition of eligibility. A row is eligible, for a given label, if the b-entry is negative and the label on the row is at least the given label. Similarly, a column is eligible, for a given label, if the entry in the pivot row is negative and the label on the column is at least the given label.

   The signature will be reduced by selecting a pivot row/column if the number of eligible rows/columns, for the label of the selected row/column, is less than the number of rows/columns with the same label. The signature will be reduced the most by choosing the smallest label for which such a reduction is possible.

The rationale for refinement 5 is difficult to see without looking at an example. Suppose the signature is (5, 3, 4, ...) at a point where there are two basic variables with label 1 and two basic variables with label 3 that have negative entries in the b-column, and there are no other negative entries in the b-column. The priority pivoting rule would select one of the label 3 variables, which would produce a signature of (5, 3, 2, ...). However, if we select one of the label 1 variables the signature would be (4, ...). The refinement says there are two variables eligible for label 3 and four variables eligible for label 1. Selecting either will reduce the signature, but selecting the smallest label that will reduce the signature will effect the greatest reduction.

The only way the efficiency of a pivoting rule can be evaluated is to test it. Carl S. Ledbetter has run extensive tests to compare the various selection rules. He constructed a large number of tableaux with randomly generated entries. Without the refinements, the priority pivoting algorithm appears to do no better than the feasibility algorithm with the Bland selection rule. With refinements 1, 2, and 3, priority pivoting uses about 25% less computer time. The tests involved tableaux up to 20×20. It is likely that the savings would be greater with larger tableaux.

Refinement 5 seems to be unproductive. Subjective observations suggest that there is a structure that drives the algorithm to termination that is as important as the signature. If one prints out the patterns of signs (+, −, 0) that appear in the tableaux that are generated, striking patterns appear about midway between the initial tableau and the terminal tableau. Refinement 5 seems to break these patterns and they become more random. It takes several more iterations before noticeable patterns reappear. It seems that there is something about pivoting that is only dimly understood at this point.

Exercises

1. Write down the dual program for the following linear program. This canonical linear program is given in inequality form and what is required is the dual program in inequality form.

Maximize $f = 10x_1 - 11x_2 + 12x_3 - 13$ subject to

$$
\begin{aligned}
x_1 - 2x_2 + 3x_3 &\leq 4 \\
4x_1 + 5x_2 - 6x_3 &\leq 7 \\
-7x_1 + 8x_2 + 9x_3 &\leq 10
\end{aligned}
$$

and $x_1 \geq 0$, $x_2 \geq 0$, $x_3 \geq 0$.

For each of the following eight problems, determine whether the problem has optimal solutions and whether the solutions are unique or multiple. If a program has multiple optimal solutions, find at least two optimal solutions. These questions apply to both the max program and the min program in each case.

2.

	x_1	x_2	x_3	-1	
u_4	-1	1	2	1	$= -x_4$
u_5	0	1	1	2	$= -x_5$
u_6	-1	2	1	3	$= -x_6$
-1	-4	-5	-6	0	$= f$
	$= u_1$	$= u_2$	$= u_3$	$= g$	

3.

	x_1	x_2	x_3	-1	
u_4	-1	1	2	1	$= -x_4$
u_5	0	1	1	2	$= -x_5$
u_6	-1	2	1	3	$= -x_6$
-1	4	-5	-6	0	$= f$
	$= u_1$	$= u_2$	$= u_3$	$= g$	

4.

	x_1	x_2	x_3	-1	
u_4	-1	1	2	1	$= -x_4$
u_5	0	1	1	-2	$= -x_5$
u_6	-1	2	1	3	$= -x_6$
-1	-4	-5	-6	0	$= f$
	$= u_1$	$= u_2$	$= u_3$	$= g$	

5.

	x_1	x_2	x_3	-1	
u_4	-1	1	2	1	$= -x_4$
u_5	0	1	1	-2	$= -x_5$
u_6	-1	2	1	3	$= -x_6$
-1	4	-5	-6	0	$= f$
	$= u_1$	$= u_2$	$= u_3$	$= g$	

6.

	x_1	x_2	x_3	-1	
u_4	1	1	2	1	$= -x_4$
u_5	0	-1	-1	2	$= -x_5$
u_6	1	2	1	3	$= -x_6$
-1	0	-5	-6	0	$= f$
	$= u_1$	$= u_2$	$= u_3$	$= g$	

7.

	x_1	x_2	x_3	-1	
u_4	1	1	2	1	$= -x_4$
u_5	0	−1	−1	0	$= -x_5$
u_6	1	2	1	3	$= -x_6$
-1	0	−5	−6	0	$= f$
	$= u_1$	$= u_2$	$= u_3$	$= g$	

8.

	x_1	x_2	x_3	-1	
u_4	1	1	2	0	$= -x_4$
u_5	0	−1	−1	0	$= -x_5$
u_6	1	2	1	3	$= -x_6$
-1	0	−5	−6	0	$= f$
	$= u_1$	$= u_2$	$= u_3$	$= g$	

9.

	x_1	x_2	x_3	-1	
u_4	−1	1	2	0	$= -x_4$
u_5	0	−1	−1	0	$= -x_5$
u_6	1	2	1	3	$= -x_6$
-1	0	0	−6	0	$= f$
	$= u_1$	$= u_2$	$= u_3$	$= g$	

Several of the problems that were posed in the exercises of Chapter 1 are canonical problems, and we will discuss them in some detail here and in the following chapter. Some of the problems that were posed in Chapter 1 involve equality constraints. These are not in canonical form and we will postpone further discussion of them until Chapter 5.

10. In the spirit Section 3.3 formulate the dual of the Furniture Maker's Problem of Chapter 1.

11. Formulate the dual of the President's Problem.

12. Formulate the dual of the Wyndor Glass Co. Problem.

13. Formulate the dual of the Investment Manager's Problem.

14. Formulate the dual of the Welfare Mother's Problem.

15. Formulate the dual of the MaxMin Problem.

16. Formulate the dual of the MinMax Problem.

17. Use the Bland feasibility algorithm to determine whether the canonical row system in the following tableau is feasible.

	x_1	x_2	x_3	x_4	x_5	x_6	x_7	-1	
u_8	-1	0	0	0	-2	-2	-2	0	$= -x_8$
u_9	-2	-2	-2	1	-1	0	-2	-2	$= -x_9$
u_{10}	1	1	-1	1	-2	0	-1	-2	$= -x_{10}$
u_{11}	0	0	-2	1	0	-1	0	-2	$= -x_{11}$
u_{12}	-2	-1	1	-1	-1	1	0	-2	$= -x_{12}$
u_{13}	0	-2	-1	0	-2	0	-2	-1	$= -x_{13}$
u_{14}	-2	-2	-2	-2	0	1	0	-1	$= -x_{14}$
u_{15}	-2	-1	0	0	1	-1	-2	-1	$= -x_{15}$
-1	-2	-1	-1	-2	-1	-1	1	-1	$= f$
	$= u_1$	$= u_2$	$= u_3$	$= u_4$	$= u_5$	$= u_6$	$= u_7$	$= g$	

18. Use the Bland feasibility algorithm to determine whether the canonical row system in the following tableau is feasible.

	x_1	x_2	x_3	x_4	x_5	x_6	x_7	-1	
u_8	-2	-2	1	1	0	-2	-2	0	$= -x_8$
u_9	1	-2	-1	-2	-2	0	-1	-2	$= -x_9$
u_{10}	-1	-1	-1	0	1	-2	-1	1	$= -x_{10}$
u_{11}	1	1	0	-1	0	1	-1	-1	$= -x_{11}$
u_{12}	-2	1	1	1	0	1	-2	-2	$= -x_{12}$
u_{13}	-1	1	-1	0	1	-1	1	-1	$= -x_{13}$
u_{14}	1	0	1	0	1	1	-1	0	$= -x_{14}$
u_{15}	-2	0	-2	-1	-2	-2	0	0	$= -x_{15}$
-1	-1	0	-2	-1	-1	-1	1	0	$= f$
	$= u_1$	$= u_2$	$= u_3$	$= u_4$	$= u_5$	$= u_6$	$= u_7$	$= g$	

19. For any of the tableaux in the previous two problems for which the row system is infeasible, find the solution for the infeasible row described as the second alternative in the second form of the theorem of the two feasibility alternatives.

20.

x_1	x_2	x_3	-1	
-1	-4	-3	-1	$= -y_1$
-1	-9	-10	4	$= -y_2$
2	15	12	2	$= -y_3$

Pivot for feasibility using the following rule: Pivot in the first row with a negative entry in the b-column; pivot in the first column with a negative entry to the right of the negative entry last chosen. In this sense, "to the right" is interpreted cyclically. That is, if the last pivot is in the third column, the column to the right is the first column. Show that with this pivot rule pivoting will cycle. Use the Bland rule to illustrate that the Bland rule will work.

(Harold W. Kuhn, *unpublished class notes*)

Questions

Q1. If the b-column of a tableau is positive, the max program has an optimal solution.

Q2. If the b-column of a tableau is positive and the min program has an optimal solution, the optimal solution for the min program is unique.

Q3. If a feasible solution for the max program and a feasible solution for the min program are complementary, they are both optimal without regard as to whether either solution is a basic solution.

Q4. The two basic solutions for a given tableau (one for the row system and one for the column system) always give the same values for their respective objective variables, whether or not either basic solution is feasible.

Q5. If a solution for a linear programming problem is feasible but not optimal, no complementary solution can be feasible for the dual program.

Q6. If a solution for a linear programming problem is feasible but not optimal, no complementary solution can be optimal for the dual program.

Q7. If there is a feasible solution for the max program and a feasible solution for the min program that give the same values for their respective objective variables, the solutions are complementary.

Q8. A linear programming problem can have an optimal solution that is not a basic solution.

Q9. If the feasible set for a canonical max program with an optimal solution is enlarged without changing the objective function, the optimal value for the objective variable will decrease.

Q10. If the feasible set for a canonical max program with an optimal solution is enlarged without changing the objective function, the optimal value for the objective variable will increase.

Q11. If a tableau has a positive row the min program is feasible.

Q12. If all entries in a tableau are positive, both programs are feasible.

Q13. If all entries in a tableau are negative, both programs are feasible.

Q14. If a tableau has a nonnegative b-column, there is no tableau equivalent to it that has an infeasible row.

Q15. If the average of the entries in each column is positive, the column system is feasible.

Q16. It is possible for the objective variable for the max program to have a negative feasible value and the objective variable for the min program to have a positive feasible value.

Q17. If a max program has only one optimal basic solution then the optimal solution is unique.

Q18. If a program has nontrivial solutions for both of the pair of dual programs, then each optimal solution has at least one variable equal to zero. A program is nontrivial if each of the dual programs has an objective function that is not identically zero.

Chapter 4

The Simplex Algorithm

4.1 The Simplex Algorithm

In Section 3.7 we presented a fundamental algorithm that starts from any canonical tableau and moves systematically by pivot exchanges to reach a tableau with nonnegative b-column—unless an infeasible row bars the way. Here we present the simplex algorithm of G. B. Dantzig that starts from a canonical tableau with nonnegative b-column and moves systematically by pivot exchanges through tableaux with nonnegative b-columns and decreasing (or, at least, nonincreasing) d-entries to reach a tableau with nonpositive c-row also—unless an infeasible column bars the way or degeneracy gives trouble. Degeneracy, the problems it causes and how we can bypass these problems, will be discussed in following sections.

We have seen in the preceding chapter that if a canonical max program is feasible and we can find an equivalent tableau in which there is an infeasible column, then its objective variable is unbounded (and, therefore, the program does not have an optimal solution). If we can find an equivalent tableau for which the b-column is nonnegative and the c-row is nonpositive, then both programs have optimal solutions. We have not yet shown that we must necessarily obtain one or the other of these two alternatives.

The simplex algorithm is an algorithm that starts with a representation of the max program by a tableau with a nonnegative b-column and terminates only in one of these two alternatives. The idea behind the simplex algorithm is to select the pivot entry so that each new tableau has a nonnegative b-column (so that the basic solution for the max program is feasible) and for which the value of the objective variable (for a basic solution) increases. This is precisely what we did when we solved the production problem in Section 1.1.

First, let us see how the pivot entry should be selected to preserve basic feasibility (feasibility of the basic solution) in the max program. Select any column s for the pivot column. After the pivot exchange with $a_{rs} \neq 0$ as pivot, the new b-entry in the pivot row is b_r/a_{rs}. If b_r is positive, we must have $a_{rs} > 0$ to make this quotient positive. Even when $b_r = 0$ we shall want $a_{rs} > 0$. This choice changes the sign of the c-entry of the pivot column. The simplex algorithm will select as a pivot column a column in which the c-entry is positive, and we are trying to obtain an equivalent tableau with a nonpositive c-row. Thus, we shall consider only positive entries in the pivot column as potential pivot entries.

In the other positions in the b-column, the entry b_i is replaced by $b_i - a_{is}b_r/a_{rs}$. If $a_{is} \leq 0$, then $b_i - a_{is}b_r/a_{rs} \geq b_i \geq 0$. If $a_{is} > 0$, we need to have $b_i - a_{is}b_r/a_{rs} \geq 0$. This is equivalent to the condition

$$\frac{b_i}{a_{is}} - \frac{b_r}{a_{rs}} \geq 0, \text{ or } \frac{b_r}{a_{rs}} \leq \frac{b_i}{a_{is}} \tag{4.1}$$

These considerations suggest the following rule for selecting the pivot row.

In column s consider those a_{is} for which $a_{is} > 0$. For each such entry (if there are any) compute b_i/a_{is}. Select as pivot entry an a_{rs} for which b_r/a_{rs} is the minimum of these ratios. That is,

$$0 \leq \frac{b_r}{a_{rs}} \leq \frac{b_i}{a_{is}} \tag{4.2}$$

for all i with $a_{is} > 0$.

A pivot exchange for which the pivot entry is positive and the b-column is nonnegative both before and after the pivot exchange is called a *simplex pivot*. Thus, the rule described above determines a simplex pivot. The rule also shows that a simplex pivot is always possible in a tableau with a nonnegative b-column if the pivot column contains at least one positive entry. Thus, if a pivot exchange is a simplex pivot exchange the inverse pivot exchange is also a simplex pivot exchange. Note that this definition says nothing about the entry in the c-row.

How should the pivot column be selected? If the pivot row r is selected by the rule given above, the new entry in the d-corner will be $d - c_s b_r/a_{rs}$. Thus, for the new tableau, the new basic feasible value of the objective variable f will be

$$-(d - c_s b_r/a_{rs}) = -d + c_s b_r/a_{rs} \tag{4.3}$$

If we select s so that c_s is positive, the value of the objective variable will be increased (if $b_r > 0$) or be unchanged (if $b_r = 0$).

The Dantzig Simplex Algorithm

1. We assume that the initial tableau has a nonnegative b-column.

2. Examine the c-row. If the c-row is nonpositive, STOP. The basic solutions for both programs are optimal.

3. If the c-row has one or more positive entries, select as a pivot column any column s for which $c_s > 0$.

4. Examine the entries in the pivot column, other than the c-entry. If none is positive, STOP. We have an infeasible column.

5. If there are one or more positive entries in the pivot column, select a positive a_{rs} for which the pivot exchange on a_{rs} is a simplex pivot.

6. Pivot on a_{rs}. Go back to 2.

At this point the reader should reread the examples discussed in Sections 1.1–1.3, and verify that the pivot exchanges for those examples were selected according to these rules for the simplex algorithm.

Rule 1 is an assumption that the algorithm starts with a tableau with a nonnegative b-column. When the algorithm returns to step 2 with a simplex pivot, this assumption is again satisfied for the next iteration. If we are given a tableau for a program in which the b-column contains a negative entry, we could use the feasibility algorithm of Section 3.7 to obtain a nonnegative b-column or to show that a start with a nonnegative b-column is not possible. For small problems (those you can do by hand) the feasibility algorithm is suitable for this purpose. The standard computer program procedure for large problems is to use the phase I method, which we shall describe in Chapter 5

When we presented the feasibility algorithm in Section 3.7, we said that the choices made according to an algorithm should usually be unequivocal. There are two places where we have not been explicit in describing the simplex algorithm. The first is the choice of the pivot column. The second is the choice of the pivot row.

Rule 3 says to select any column for which $c_s > 0$. We shall show that the algorithm will terminate regardless of which column is selected in this way. However, a number of different rules are used for computer programs. The most frequently used is to select a column for which c_s is largest and, if there is more than one such column, the first largest. Another is simply to select the first column for which $c_s > 0$. Other much more complicated rules have been proposed.

The only way that one of these rules can be preferred over another is through experience. Experiments seem to indicate that all simple rules do approximately equally well, with the best performer doing about twice as well (in terms of the number of pivot exchanges required) as the poorest. There are complex rules that reduce the average number of pivot exchanges, but they require extensive computation to implement the rules.

The other place where we have not been explicit in describing the alternative to be chosen is in the choice of the pivot row when there is a tie for the minimum ratio. If there is a tie for the minimum ratio, the b-column in the following tableau will have zeros in the b-column (one less than the number of rows that are tied as candidates for the pivot row). That is, the resulting basic solution will be degenerate. We shall discuss the consequences of this possibility in some detail in the following sections and not go into it here.

Example. An interesting example of a linear programming application is described by Dorfman, Samuelson, and Solow, in *Linear Programming and Economic Analysis*, McGraw-Hill, 1958. A chemical firm processes a certain raw material by the use of two major types of equipment, called stills and retorts. Four different production processes are available to the firm. If Process 1 is used to treat 100 tons of the raw material, it will utilize 7% of the weekly capacity of the stills and 3% of the weekly capacity of the retorts. If 100 tons is treated by Process 1 the net profit to the firm is $60. The firm plans to process 1500 tons of raw material weekly.

The data table 4.4 gives the pertinent information for all four processes.

(4.4)

	production processes					
	(1)	(2)	(3)	(4)	available	
raw material	100	100	100	100	1500	tons
still capacity	7	5	3	2	100	%
retort capacity	3	5	10	15	100	%
profit	60	60	90	90		dollars
	per unit					

Tableau 4.5 displays the corresponding max program.

(4.5)

x_1	x_2	x_3	x_4	-1	
100	100	100	100	1500	$= -y_1$
7	5	3	2	100	$= -y_2$
3	5	10*	15	100	$= -y_3$
60	60	90	90		$= f$

Let us select as the pivot column the column with the first, largest positive value of c_i. The pivot row is selected by rule 4, and the chosen pivot entry is starred. After the pivot exchange we obtain

(4.6)

x_1	x_2	y_3	x_4	-1	
70*	50	−10	−50	500	$= -y_1$
6.1	3.5	−0.3	−2.5	70	$= -y_2$
0.3	0.5	0.1	1.5	10	$= -x_3$
33	15	−9	−45	−900	$= f$

The pivot column in 4.6 is the column containing the first, largest value of c_i. We pivot on the starred entry.

(4.7)

y_1	x_2	y_3	x_4	-1	
0.0143	0.714	−0.143	−0.714	7.14	$= -x_1$
−0.0871	−0.857	0.571	1.86	26.4	$= -y_2$
−0.00429	0.286	0.143	1.71	7.86	$= -x_3$
−0.471	−8.57	−4.29	−21.4	−1136	$= f$

Since the c-row of tableau 4.7 is negative, we stop according to rule 1. We have obtained an optimal solution. We see that $x_1 = 7.14$, $x_2 = 0$, $x_3 = 7.86$, $x_4 = 0$. Substituting these numbers in tableau 4.5 we get $y_1 = 0$, $y_2 = 26.4$, $y_3 = 0$, and $f = 1136$. Notice that more than 26% of the capacity of the stills is unused.

If it is required that we produce only integral units, we would have to consider $x_1 = 7$ and $x_3 = 8$, or $x_1 = 8$ and $x_3 = 7$. However, the first alternative is not feasible.

It is almost always informative to look at the solutions for the dual program. In tableau 4.7 we see that \$0.471 is the marginal price for the raw material. That means an additional 100 tons of the raw material, which would allow an increase in production of one unit, would allow the profit to increase by \$47.10. This is less than the profit that would accrue from processing this unit by any of the available processes. The reason for this apparent discrepancy can be seen in the first column of tableau 4.7.

The optimal operation to process this additional 100 tons would require an additional 1.43 units to be processed by Process 1, and 0.43 less units to be processed by Process 3. The capacity of the retort is already saturated. To allow the plant to process the additional 100 tons, some production will have to be shifted from the more profitable Process 3 to the less profitable Process 1.

Process 2 is unprofitable by a margin of \$8.57. Compared with Process 1, with which it seems comparable, it uses more of the capacity of the retort (which is in short supply) and less of the capacity of the still (which is not fully utilized). An additional 1% increase in the capacity of the retort would allow an increase in the weekly profit by \$4.29. This would be obtained by decreasing by 0.143 the number of units processed by Process 1 and increasing by 0.143 the number of units processed by Process 3.

Example. An interesting minimization problem is discussed by William J. Baumol in *Economic Theory and Operations Analysis*, 3rd edition, Prentice-Hall, 1972. An advertising firm wishes to minimize the cost of getting a total of 160 million "audience exposures." The company wants at least 60 million of these exposures to involve persons with family incomes over \$8000 per year, and at least 80 million of the exposures to involve persons between 18 and 40 years of age. The budget is to be divided between magazine and television advertising. Surveys give estimates for the size and composition of the audiences of the two media. The relevant information is contained in the following data table.

(4.8)

	total	income over \$8000	ages 18–40	cost	
magazine	4	3	8	40	per ad
television	40	10	10	200	per ad
requirement	160	60	80		per year
	Exposures in millions			dollars in thousands	

We convert the data table 4.8 to a tableau and obtain

(4.9)

v_1	4	3	8	40
v_2	40*	10	10	200
-1	160	60	80	
	$= u_1$	$= u_2$	$= u_3$	$= g$

The basic solution for the min program in tableau 4.9 is infeasible. However, since the b-column is positive, the basic solution for the max program associated with this tableau is feasible. The rules of the simplex algorithm are motivated in terms of a requirement to preserve the feasibility of the basic solution for the max program. However, the selection rules are expressed in terms of the tableau itself and not specifically in terms of the max program. We can apply the simplex algorithm to tableau 4.9 without writing in the variables of the max program. We select as pivot column the first column with a positive c-entry (to demonstrate a different way of selecting the pivot column). The pivot entry is indicated with an asterisk. We obtain

(4.10)

v_1	-0.1	2^*	7	20
u_1	0.025	0.25	0.25	5
-1	-4	20	40	-800
	$= v_2$	$= u_2$	$= u_3$	$= g$

Again, we select the first column with a positive c-entry. The pivot entry is indicated with an asterisk. Performing this pivot exchange we get

(4.11)

u_2	-0.05	0.5	3.5	10
u_1	0.0375	-0.125	-0.625	2.5
-1	-3	-10	-30	-1000
	$= v_2$	$= v_1$	$= u_3$	$= g$

In tableau 4.11 the basic solution for the min program is feasible. Since the basic solution for its dual program is also feasible, the solution is optimal. That is, the firm should place 10 magazine ads and 3 television ads at a cost of $1,000,000. The requirement to obtain at least 80 million exposures in the 18-to-40 age bracket will be oversupplied by 30 million exposures.

Again, the optimal solution for the dual program is informative. The marginal cost for the total exposure requirement is $2500 per million exposures. The marginal cost for the income-over-$8000 exposure requirement is $10,000. This suggests the savings that could be obtained by relaxing either of these requirements.

Notice that the simplex algorithm assumes a nonnegative b-column in the starting tableau. It does not matter whether we are interested in the maximum program or in the minimum program. It is the objective of this chapter to show that the simplex algorithm is decisive. If a solution can be obtained, the simplex algorithm will solve both the maximum and the minimum program. If the program has no optimal solution, neither program has an optimal solution. The simplex algorithm will also handle that case.

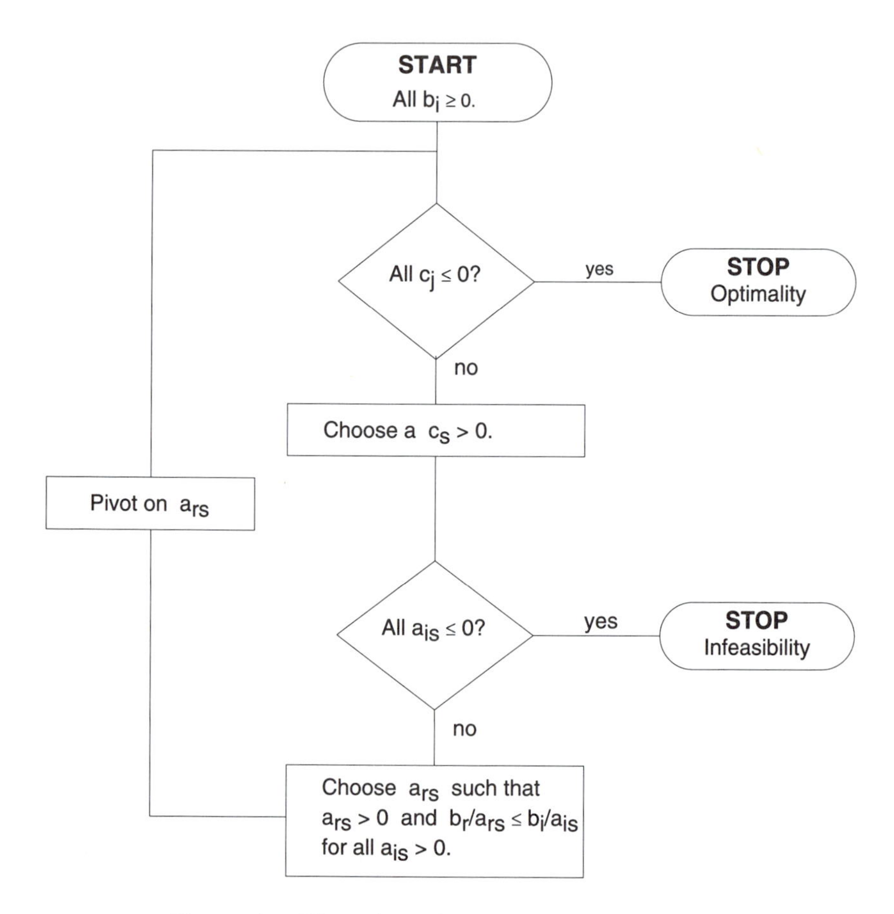

Figure 4.1: Flow chart for the simplex algorithm.

4.2 Degeneracy and Cycling

There are several ways to prove that an algorithm terminates. One is to use a variable that can be associated with each configuration the algorithm produces, to show that this variable can take on only a finite number of different values, and to show that this variable changes with each iteration of the algorithm and changes in only one direction. You might think of this variable as a pointer. The algorithm drives this pointer in one direction and, therefore, does not repeat any configuration.

The pointer that is usually used to establish termination of the simplex algorithm is the value of the objective variable f for basic solutions of the max program. Since the number in the d-corner is the negative of the value of the objective variable (for a basic solution), an increase in the value of the objective variable (for a basic solution) is equivalent to a decrease in the number in the d-corner.

If d is the number in the d-corner and a_{rs} is the pivot entry, after the pivot exchange the new entry in the d-corner is

$$d - c_s b_r / a_{rs} \tag{4.12}$$

The rules for the simplex algorithm require a_{rs} and c_s to be positive. If b_r is also positive, the new entry in the d-corner is decreased. There are only finitely many equivalent tableaux and, hence, only a finite number of different values for the entry in the d-corner. Thus, if the rules of the simplex algorithm always select a pivot row in which the b-entry is positive, the simplex algorithm will terminate in a finite number of iterations.

If a pivot exchange is made for which $b_r = 0$, the value of the entry in the d-corner will not change. If no tableau in the class of tableaux equivalent to the original tableau has a 0 in the b-column, the problem is said to be *nondegenerate.* For a nondegenerate problem, the entry in the d-corner always decreases after each pivot exchange. Thus, for a nondegenerate linear program, no tableau can be obtained twice in any sequence of pivot exchanges under the rules of the simplex algorithm. For a nondegenerate linear program the simplex algorithm will terminate in a finite number of pivot exchanges.

The usual way to ensure that a linear program is nondegenerate is to assume it. This assumption has a certain amount of rationale since a set of randomly generated linear constraints has a very high probability of being nondegenerate. However, the assumption is unrealistic from at least three points of view. First, it is impossible to check that a program presented in a tableau is nondegenerate without looking at all tableaux that can be obtained by simplex pivoting, or doing something equally time consuming. Second, one does not need to assume this much. All that is necessary is that the b-columns of those tableaux actually encountered in a run of the algorithm be nondegenerate. If the simplex algorithm is applied to a problem and no zero b-entry is encountered, that is all one will know. Third, a significant number of applied problems are degenerate.

If a linear program is degenerate, it is possible that an unending sequence of pivot exchanges can be made without changing the entry in the d-corner and without reaching termination. An endless loop of tableaux obtained repeatedly is called a *cycle* or a *circle.* Let us give an example of a degenerate linear program for which the simplex algorithm can cycle. Consider

(4.13)

x_1	x_2	x_3	x_4	x_5	-1	
3^*	3	0	0	1	6	$= -x_6$
1	10	1	-9	$-5/3$	2	$= -x_7$
1	2	$1/3$	-2	0	2	$= -x_8$
12	9	1	-6	4	0	$= f$

To make the selection rules definite, choose the pivot column with the first, largest entry in the c-row, and in the pivot column choose the first row that minimizes b_i/a_{is} for $a_{is} > 0$. In the following sequence of tableaux the pivot entry is starred in each case.

(4.14)

x_6	x_2	x_3	x_4	x_5	-1	
$1/3$	1	0	0	$1/3$	2	$= -x_1$
$-1/3$	9	1^*	-9	-2	0	$= -x_7$
$-1/3$	1	$1/3$	-2	$-1/3$	0	$= -x_8$
-4	-3	1	-6	0	-24	$= f$

(4.15)

x_6	x_2	x_7	x_4	x_5	-1	
$1/3$	1	0	0	$1/3$	2	$= -x_1$
$-1/3$	9	1	-9	-2	0	$= -x_3$
$-2/9$	-2	$-1/3$	1^*	$1/3$	0	$= -x_8$
$-11/3$	-12	-1	3	2	-24	$= f$

(4.16)

x_6	x_2	x_7	x_8	x_5	-1	
$1/3$	1	0	0	$1/3$	2	$= -x_1$
$-7/3$	-9	-2	9	1^*	0	$= -x_3$
$-2/9$	-2	$-1/3$	1	$1/3$	0	$= -x_4$
-3	-6	0	-3	1	-24	$= f$

(4.17)

x_6	x_2	x_7	x_8	x_3	-1	
$10/9$	4	$2/3$	-3	$-1/3$	2	$= -x_1$
$-7/3$	-9	-2	9	1	0	$= -x_5$
$5/9$	1^*	$1/3$	-2	$-1/3$	0	$= -x_4$
$-2/3$	3	2	-12	-1	-24	$= f$

(4.18)

x_6	x_4	x_7	x_8	x_3	-1	
$-10/9$	-4	$-2/3$	5	1	2	$= -x_1$
$8/3$	9	1^*	-9	-2	0	$= -x_5$
$5/9$	1	$1/3$	-2	$-1/3$	0	$= -x_2$
$-7/3$	-3	1	-6	0	-24	$= f$

(4.19)

x_6	x_4	x_5	x_8	x_3	-1	
$2/3$	2	$2/3$	-1	$-1/3$	2	$= -x_1$
$8/3$	9	1	-9	-2	0	$= -x_7$
$-1/3$	-2	$-1/3$	1^*	$1/3$	0	$= -x_2$
-5	-12	-1	3	2	-24	$= f$

(4.20)

x_6	x_4	x_5	x_2	x_3	-1	
$1/3$	1	$1/3$	1	0	2	$= -x_1$
$-1/3$	-9	-2	9	1^*	0	$= -x_7$
$-1/3$	-2	$-1/3$	1	$1/3$	0	$= -x_8$
-4	-6	0	-3	1	-24	$= f$

Tableau 4.20 is the same tableau 4.14, differing only in a rearrangement of the columns. The pivot entry in 4.20 is the same as the pivot entry in 4.14, and the same cycle of tableaux will be obtained ad infinitum.

If for any reason (a different original ordering of the variables, for example) the first pivot column chosen were the x_5-column, no degeneracy would have been encountered in the simplex algorithm and termination with optimality would have been achieved in two pivot exchanges.

Examples, such as the one given above, show that cycling can occur in a degenerate program. However, it is not easy to construct an example in which cycling occurs. The first examples were constructed by A. J. Hoffman and E. M. L. Beale.

Even where cycling does occur it is, apparently, very easily broken. The example given above was constructed with the given selection rules in mind. If one were to select the pivot columns and pivot rows at random when several choices are available, the cycle would eventually be broken. Generally, if ties were decided by a chance device, cycling would eventually be broken and termination would be obtained.

Consider the following example in which degeneracy occurs.

(4.21)

x_1	x_2	x_3	-1	
1	0	0	2	$= -y_1$
2	1	0	4	$= -y_2$
1	1	1	2	$= -y_3$
1	1	2	0	$= f$

If the x_1-column is selected as the pivot column, there is a three-way tie for the pivot row. In this case, the next tableau will contain two zeros in the b-column. However, it is not possible to obtain a cycle following the rules of the simplex algorithm no matter what choices are made when more than one is possible.

It is instructive to try to obtain a cycle starting with tableau 4.21. Sooner or later, the pivot column will have no positive entries in the rows of zeros in the b-column. This forces a change in the entry in the d-corner, and it is not possible to return to any tableaux obtained before that change.

Degeneracy is the source of more theoretical concern than practical difficulty. Over the years several systematic ways to prevent or break cycling have been developed. However, it has been part of the folklore of the subject that no one has ever encountered a practical problem that cycled. Apparently, that is not quite true, but it is such a rare event that it is considered to be more economical of computer time to make no provision for the possibility of cycling and to face the problem of doing something about it if it should occur.

4.3 Bland's Anticycling Rule

While it is possible to live with degeneracy and the possibility of cycling in the practical applications of linear programming, from the point of view of the theory of linear programming it is very desirable to show that the simplex algorithm terminates in a finite number of iterations. Otherwise, the proofs would be nonconstructive and would have to depend on mathematical concept more sophisticated than are otherwise required.

Several different methods for preventing cycling have been devised. Generally, they are refinements of the simplex algorithm in that they become operational only if the simplex algorithm, as given in Section 4.1, does not lead to a unique choice for the pivot row. That is, a refinement breaks ties. One method, indicated in 1951 by George B. Dantzig and developed by A. Charnes in 1952, is a perturbation method and alters the entries in the b-column slightly so that zeros will not occur. Another, proposed by George B. Dantzig, A. Orden, and P. Wolfe in 1954, introduces an auxiliary ordering, a lexicographic order, to break ties.

By far the simplest method (the easiest to state, the easiest to implement, and the easiest to prove finite) was proposed by Robert G. Bland in 1976. It is the one we describe here.

As long as the simplex algorithm chooses a pivot entry in a row with a positive b-entry, the number in the d-corner will be reduced after the pivot exchange is performed. This is enough to show that no tableau can be repeated. We must consider what might happen when zeros appear in the b-column, when degeneracy becomes apparent. If, after several pivot exchanges, a pivot entry is chosen in a row with a positive b-entry, the number in the d-corner will be reduced and we cannot return to any of the tableaux that were obtained previously.

Thus, we must consider the possibility of an infinite sequence of pivot exchanges where each pivot entry is in a row with a zero b-entry. For this purpose we can disregard the rows with positive b-entries. Also, the b-column and the d-corner will not change and we can disregard them. We are left with a tableau of the form

(4.22)

The rules of the simplex algorithm instruct us to choose a column with a positive c-entry as the pivot column. Then we choose a row with a positive entry in the pivot column as the pivot row. These choices are precisely the duals of the choices that we considered in the discussion of the feasibility algorithm in Section 3.7. The proof of the feasibility algorithm, in this context, shows that cycling will be avoided if the pivot rows and pivot columns are chosen by the Bland rule.

Thus, as a refinement of the simplex algorithm we propose the

Bland's Anticycling Rule

> Use the simplex algorithm in unchanged form as long as the simplex algorithm chooses a pivot entry in the row with a positive b-entry. When zeros appear in the b-column and the rules of the simplex algorithm would select a pivot entry in a row with a zero b-entry, switch to the Bland selection rule. That is,
>
> 1. If there is a positive entry in the c-row, select the column

with a positive c-entry and the least index for a variable associated with such a column.

2. If there is a positive entry in the pivot column with a zero b-entry in its row, select the row with a positive entry in the pivot column and the least index for a variable associated with such a row.

With this refinement, a sequence of pivot exchanges in rows with zero b-entries will terminate with one of the two alternatives illustrated in 4.23.

$$
\begin{array}{|ccc|c|}
\hline
 & & & 0 \\
 & & & \vdots \\
 & & & 0 \\
 & & & + \\
 & & & \vdots \\
 & & & + \\
\hline
\ominus & \cdots & \ominus & \\
\hline
\end{array}
\qquad
\begin{array}{|c|c|}
\hline
\ominus & 0 \\
\vdots & \vdots \\
\ominus & 0 \\
? & + \\
\vdots & \vdots \\
? & + \\
\hline
+ & \\
\hline
\end{array}
\tag{4.23}
$$

If the choice in rule 1 is not available, the c-row is nonpositive. The left tableau in 4.23 illustrates this possibility. In this case the simplex algorithm terminates with an optimal solution for both the max program and the min program.

If the choice in rule 2 is not available, the pivot column contains no positive entry in a row with a zero b-entry. The right tableau in 4.23 illustrates this possibility. If there is no other positive entry in the pivot column, the simplex algorithm terminates with the information that the max program is unbounded and the min program is infeasible. If there is a positive entry in some row with a positive b-entry, the rules of the simplex algorithm will choose such a row as the pivot row. The resulting pivot exchange will reduce the number in the d-corner and this will prevent any return to a previously obtained tableau.

We have described the simplex algorithm in its simplest form, in which the pivot column is selected arbitrarily from among all columns with a positive entry in the c-row. If there are several such columns this leaves a "tie" for the selection of the pivot column. Computers cannot make an arbitrary selection. Several suggestions have been made to make this choice specific. Among the simplest would be to choose the first (in any convenient ordering, usually whatever the computer uses to list the columns) column in which the c-entry is positive. Among the more complicated would be to choose a column that yields the largest change in the entry in the d-corner.

Even this choice would not always be unique.

The cost of using a complicated rule is that it takes computer processing time to implement. The most commonly used rule is to select the first column with the largest positive entry in the c-row. This seems to be the favored compromise between a simple rule that results in a large number of iterations and a complicated rule that reduces the number of iterations at the expense of computations involved in the rule itself. We have not included any favored rule in our definition of the simplex algorithm since we want the reader to experiment with different rules.

As originally proposed, the Bland rule was to break ties in the simple form of the simplex algorithm by selecting the row or column with the least index on the associated variable. In this form, the pivot column would be selected from all columns with a positive entry in the c-row. Selecting a column with the largest positive entry in the c-row would likely be in conflict with the Bland rule. However, as long as the rules of the simplex algorithm chose a pivot row with a positive b-entry, cycling was not possible in any event. Experiments seem to indicate that if the Bland rule is applied at all times the number of iterations will usually be larger than would be obtained, for example, by selecting the column with the largest positive c-entry.

It seems best, therefore, to use whatever favored rule one wishes for selecting the pivot column, as long as zeros do not appear in the b-column. When zeros appear in the b-column, switch to the Bland anticycling rule. This would also mean switching to the Bland rule for selecting the pivot column. As soon as a pivot exchange occurs in a row with a positive b-entry, switch back to whatever rule was being used.

Consider the following example.

(4.24)

x_1	x_2	x_3	x_4	-1	
9	1^*	-9	-2	0	$= -x_5$
1	$1/3$	-2	$-1/3$	0	$= -x_6$
-9	-1	9	2	1	$= -x_7$
-3	1	-6	0	0	$= f$

We shall use the common rule for choice of pivot entry with the simplex algorithm: Choose the first largest positive entry in the c-row to select the pivot column, and choose the first largest entry in the pivot column to select the pivot row. An application of this rule to tableau 4.24 leads to the choice indicated with the asterisk. Let us use this rule and continue to use it. From 4.24 we obtain

(4.25)

x_1	x_5	x_3	x_4	-1	
9	1	−9	−2	0	$= -x_2$
−2	−1/3	1*	1/3	0	$= -x_6$
0	1	0	0	1	$= -x_7$
−12	−1	3	2	0	$= f$

(4.26)

x_1	x_5	x_6	x_4	-1	
−9	−2	9	1*	0	$= -x_2$
−2	−1/3	1	1/3	0	$= -x_3$
0	1	0	0	1	$= -x_7$
−6	0	−3	1	0	$= f$

(4.27)

x_1	x_5	x_6	x_2	-1	
−9	−2	9	1	0	$= -x_4$
1*	1/3	−2	−1/3	0	$= -x_3$
0	1	0	0	1	$= -x_7$
3	2	−12	−1	0	$= f$

(4.28)

x_3	x_5	x_6	x_2	-1	
9	1*	−9	−2	0	$= -x_4$
1	1/3	−2	−1/3	0	$= -x_1$
0	1	0	0	1	$= -x_7$
−3	1	−6	0	0	$= f$

(4.29)

x_3	x_4	x_6	x_2	-1	
9	1	−9	−2	0	$= -x_5$
−2	−1/3	1*	1/3	0	$= -x_1$
−9	−1	9	2	1	$= -x_7$
−12	−1	3	2	0	$= f$

If the pivot exchange indicated in tableau 4.29 is performed we will get tableau 4.24, with the columns rearranged. That is, we will get cycling. The first time where the Bland anticycling rule would lead to a different choice is in tableau 4.28, where the second entry in the second column would have been chosen. However, this choice breaks the cycle and two more iterations would lead to termination with a nonpositive c-row.

4.4 Theorems of the Two Alternatives

In this section we prove two forms of the theorem of the two alternatives. The first is just a dual form of the Theorem of the two feasibility alternatives, proved as Theorem 3.23 in Section 3.7.

Let the initial tableau under consideration be

(4.30)

Let the variables on the left and bottom margins be ordered in any way whatever. Apply the dual form of the feasibility algorithm of Section 3.7. That is, select columns for which the c-entry is positive and, among these, select as pivot column the column for which the index of the variable associated with that column is least. If this selection is possible, select rows for which the entry in the pivot column is positive and, among these, select as pivot row the row for which the index of the variable associated with that row is least.

The b-column and the d-corner are carried along in the computations of the pivot exchange, but they do not enter into the rules for selecting the pivot column and the pivot row. We know that these selections cannot continue indefinitely. The process must terminate when one or the other of the indicated selections is impossible.

If the selection of a pivot column is impossible, we have a tableau in the form of

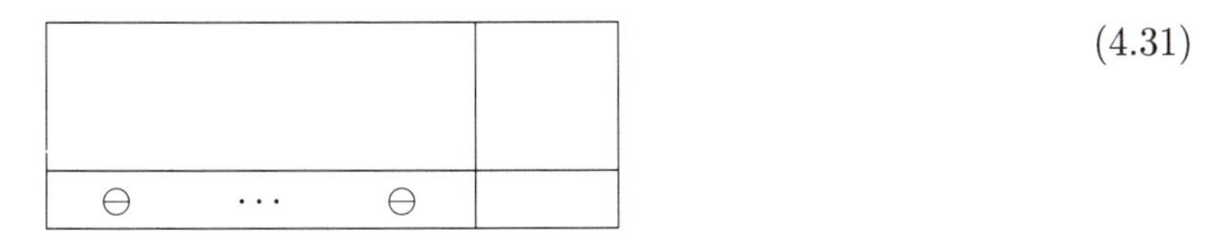

(4.31)

When the selection of a pivot row is impossible, we have a tableau in the form of

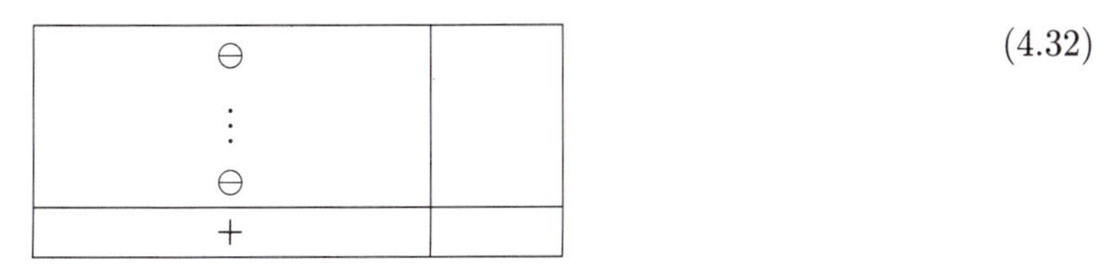

(4.32)

Tableaux 4.31 and 4.32 are inequivalent since tableau 4.31 has a non-negative solution to the column equations and tableau 4.32 does not. Thus, we have the *first theorem of the two alternatives.*

Theorem 4.28 *A tableau in the form of tableau 4.30 is equivalent to a tableau in the form of tableau 4.31 or to a tableau in the form of tableau 4.32, but not to both. Furthermore, the dual form of the feasibility algorithm will obtain one or the other in a finite number of iterations.* □

Suppose we start with a tableau in the form of tableau 4.30 in which the b-column is nonnegative. Now, apply the simplex algorithm with the Bland anticycling rule. Again, the iterations of the simplex algorithm cannot continue indefinitely. One or the other of the selection rules must eventually fail.

If we are unable to select a pivot column we have a tableau in the form of

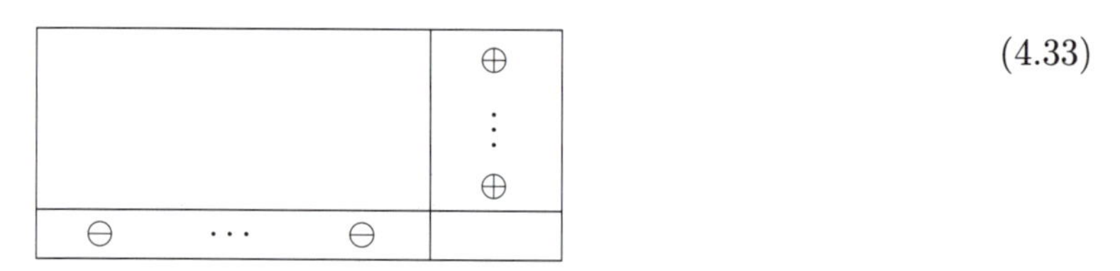

(4.33)

If we unable to select a pivot row we have a tableau in the form of

$\ominus$	$\oplus$
$\vdots$	$\vdots$
$\ominus$	$\oplus$
$+$	

(4.34)

Tableaux 4.33 and 4.34 are inequivalent since tableau 4.33 has a non-negative solution to the column equations and tableau 4.34 does not.

Thus, we have the *second theorem of the two alternatives.*

Theorem 4.29 *A tableau in the form of tableau 4.30 with a nonnegative b-column is equivalent to a tableau in the form of tableau 4.33 or to a tableau in the form of tableau 4.34, but not to both. Furthermore, the simplex algorithm with the Bland anticycling rule will obtain one or the other in a finite number of iterations.* □

The two theorems of the two alternative differ in the role of the b-column. The first theorem is indifferent to the nature of the b-column. The second starts with a nonnegative b-column and preserves the nonnegativity of the b-column.

The second theorem of the two alternatives has an immediate and important implication for canonical linear programs. If we start with a linear program whose constraints are represented by tableau 4.30 for which the b-column is nonnegative, then the simplex algorithm must terminate with a tableau in the form of tableau 4.33 or with a tableau in the form of tableau 4.34. The basic solutions for the row and column equations in tableau 4.33 satisfy the sufficient conditions for optimality. That is, we obtain optimal solutions for both the max program and the min program.

In tableau 4.34 the form of the tableau shows that the column equations are infeasible, and the row equations are feasible but the objective variable is unbounded. That is, we see that neither program has an optimal solution.

The simplex algorithm can terminate in only one of two ways, either with optimal solutions for both programs or with the information that neither program has an optimal solution. The important thing here is that the simplex algorithm, with the Bland anticycling rule, forces the termination in one of these two alternatives in a finite number of iterations. Furthermore, experience shows that it reaches this decision in a reasonably small number of iterations.

4.5 The Dual Simplex Algorithm

In Section 3.1 we showed that the max program and the min program represented by a tableau can be interchanged by taking the negative transpose of the tableau. The max program in one tableau is equivalent to the min program of the negative transpose tableau. If the initial tableau representing a dual pair of programs has a nonpositive c-row, we can obtain a tableau with a nonnegative b-column by taking the negative transpose.

Consider the effect of applying the simplex algorithm to the negative transpose tableau. When the simplex algorithm terminates we have optimal solutions to both the min program and the max program, or we know that neither program has an optimal solution. We can take the negative transpose of the terminal tableau and obtain a tableau equivalent to the initial tableau. That is, if we are confronted with a tableau with a nonpositive c-row, we can take its negative transpose, apply the simplex algorithm to the negative transpose, and take the negative transpose of the terminal tableau.

It is reasonable to ask if we can achieve the same ends without taking the negative transpositions twice. The answer is easy enough. All we need to do is to formulate the rules of another algorithm so that they would have the same effect if they were applied to the negative transpose. To this end we recast the simplex algorithm by interchanging the roles of the rows and columns, and changing the sign that controls the selection of an entry (and not changing the sign where a quotient of two entries is involved). In this way we obtain

The Dual Simplex Algorithm

1. We assume that the initial tableau has a nonpositive c-row.
2. Examine the b-column. If the b-column is nonnegative, STOP. The basic solutions for both programs are optimal.
3. If the b-column has one or more negative entries, select as a pivot row any row r for which $b_r < 0$.
4. Examine the entries in the pivot row, other than the b-entry. If none is negative, STOP. We have an infeasible row.
5. If one or more negative entries are in the pivot row, select a negative a_{rs} for which the ratio c_s/a_{rs} is minimal among all ratios c_j/a_{rj} in which a_{rj} is negative. Such a pivot exchange is a *dual simplex pivot.*
6. Pivot on a_{rs}. Go back to 2.

It is to be emphasized that there is no practical difference between the dual simplex algorithm and applying the simplex algorithm to the negative transpose. One or the other is appropriate when the initial tableau has a nonpositive c-row. For computational purposes the choice might be dictated by preference, convenience, or necessity (if, for example, a computer program is available for the simplex algorithm but not for the dual simplex algorithm). For theoretical purposes it is convenient to have both the simplex algorithm and the dual simplex algorithm available.

If an expanded tableau is used to represent a linear program, the symmetry between the simplex algorithm and the dual simplex algorithm is not so apparent. The wording must be changed, but the dual simplex algorithm can be cast so that it applies to expanded tableaux. The pivot exchange in a Tucker tableau is fully equivalent to a pivot exchange in an expanded tableau. To apply to the expanded tableau the dual simplex algorithm only needs to be formulated so that it leads to the same choices for the pivot entries.

The two theorems of the alternatives in the previous section follow from the fact that the dual feasibility algorithm and the simplex algorithm terminate in a finite number of iterations. In the same way there are dual forms of the theorems of the alternatives that follow from the finiteness of the feasibility algorithm and the dual simplex algorithm.

The first dual theorem of the two alternatives is just the theorem of the two feasibility alternatives, proved in Section 3.6. We start with a tableau in the form of

(4.35)

Theorem 4.30 *A tableau in the form of tableau 4.35 is equivalent to a tableau in the form of tableau 4.36 or to a tableau in the form of tableau 4.37, but not to both. Furthermore, the feasibility algorithm will obtain one or the other in a finite number of iterations.* ◻

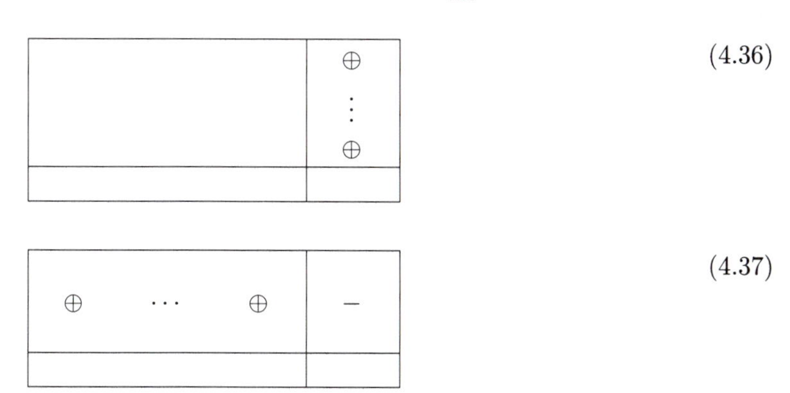

(4.36)

(4.37)

The second dual theorem of the two alternatives applies to a tableau in the form of tableau 4.35 in which the c-row is nonpositive.

Theorem 4.31 *A tableau in the form of tableau 4.35 with a nonpositive c-row is equivalent to a tableau in the form of tableau 4.38 or to a tableau in the form of tableau 4.39, but not to both. Furthermore, the dual simplex algorithm with the Bland anticycling rule will obtain one or the other in a finite number of iterations.* ◻

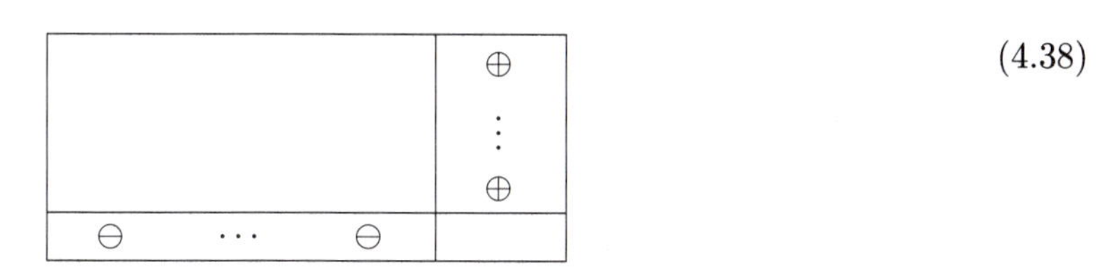

(4.38)

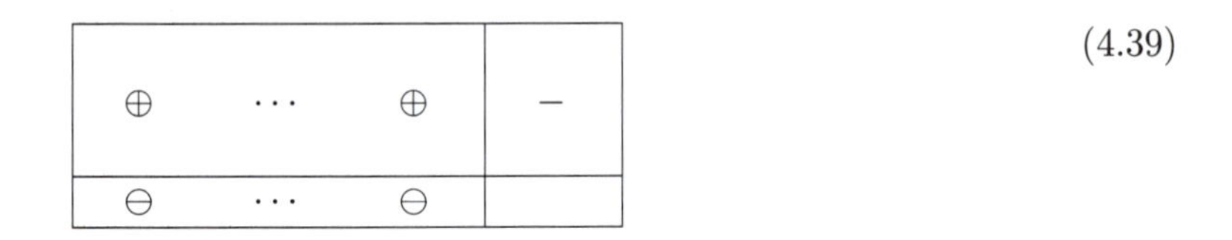

(4.39)

4.6 The Theorem of the Four Alternatives

By combining the results of the theorems of the two alternatives in their dual forms we can obtain the theorem of the four alternatives, which may be regarded as a fundamental theorem in the theory of linear programming.

Theorem 4.32 *(The theorem of the Four Alternatives) Given a tableau,*

A	b
c	d

(4.40)

it is equivalent to one and only one of the following inequivalent tableaux.

	$\oplus$ $\vdots$ $\oplus$
$\ominus \quad \cdots \quad \ominus$	

(4.41)

$\ominus$ $\vdots$ $\ominus$	$\oplus$ $\vdots$ $\oplus$
$+$	

(4.42)

$\oplus \quad \cdots \quad \oplus$	$-$
$\ominus \quad \cdots \quad \ominus$	

(4.43)

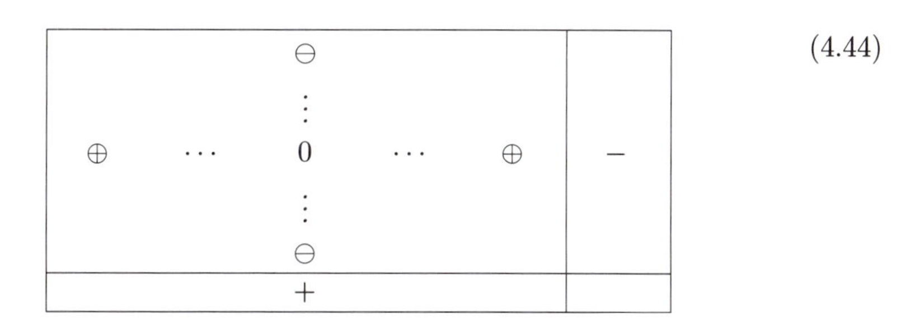
(4.44)

Proof. For canonical feasibility specifications, the row program is feasible only for tableaux 4.41 and 4.42 and the column program is feasible only for tableaux 4.41 and 4.43. Thus, no two of the four tableaux described in the theorem are equivalent.

We must show that the tableau 4.40 is equivalent to one of the four alternatives. The line of thinking that will establish these equivalences is illustrated in Figure 4.2, a flow chart for the theorem of the four alternatives. The initial tableau 4.40 is shown at the beginning of the flow chart.

The following notes apply to Figure 4.2.

Note 1. We apply the feasibility algorithm described in Section 3.7. We obtain a tableau either with a nonnegative b-column, the left alternative in Figure 4.2, or with an infeasible row, the right alternative. In the latter case we rearrange the rows so the infeasible row becomes the first row, and we rearrange the columns so that the positive entries in the infeasible row are in the first group of columns. This will simplify the discussion to follow.

Note 2. In the left alternative, where the b-column is nonnegative, we apply the simplex algorithm. At termination either we have a nonpositive c-row, the first of the four alternatives, or we have an infeasible column, the second of the four alternatives.

Note 3. In the right alternative, where we have an infeasible row, we apply the dual feasibility algorithm in which all pivot entries are selected from the columns which have a zero in the first row. Pivot exchanges selected in this way have no effect on the entries in the first row. At termination we obtain either the left branch with a nonpositive c-row, in this case only for the entries under the zeros in the first row, or the right branch in which we have an infeasible column. The right branch is the fourth of the four alternatives.

Note 4. For the left branch, if the remaining entries in the c-row are nonpositive we obtain the third alternative. If there is at least one positive entry we perform one more pivot exchange. We select the first row as the pivot row. Then we select the pivot column s so that c_s/a_{1s} is the maximum among all such quotients for which $a_{1j} > 0$. We then obtain a tableau with a nonpositive c-row, again the third alternative.

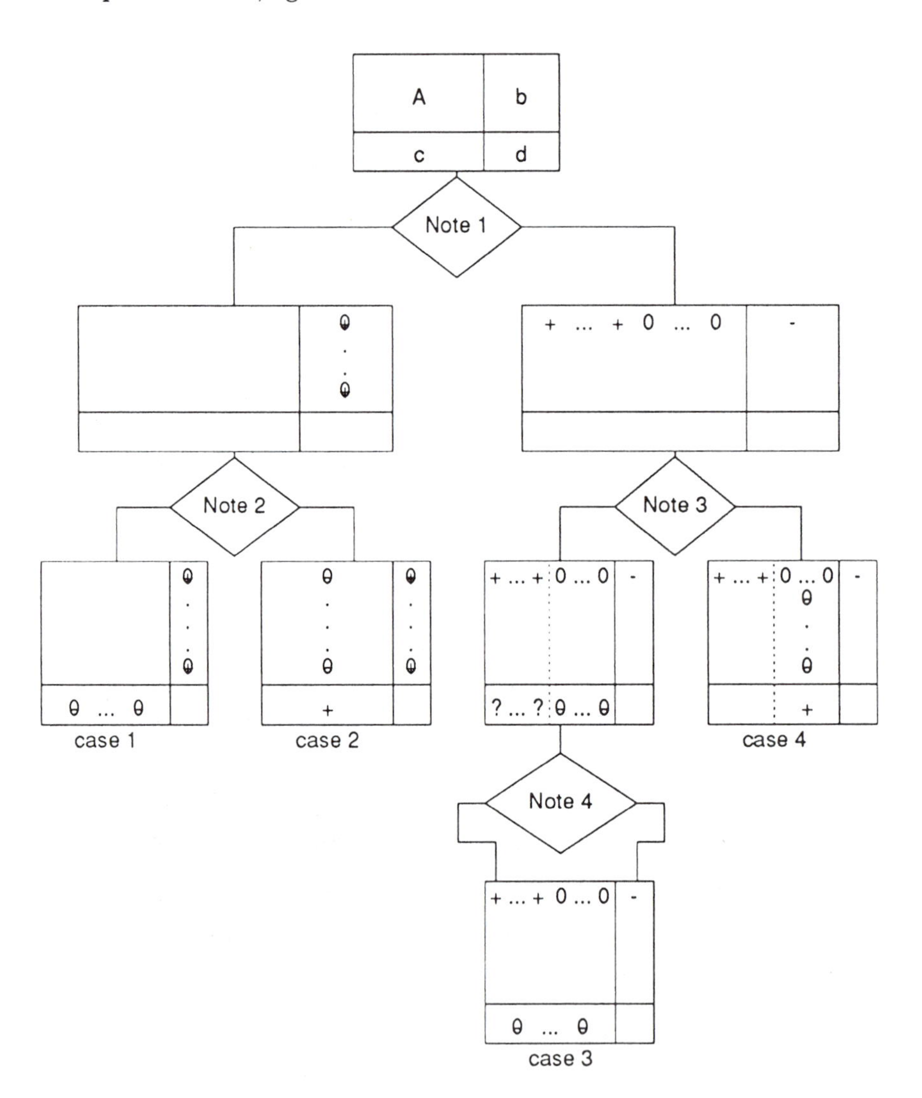

Figure 4.2: Flow chart for the theorem of the four alternatives.

It should be emphasized that the conclusion of this theorem does not depend on the particular algorithms used in the proof. If a random and haphazard sequence of pivot exchanges is used, one may still obtain one of the four terminal forms. The only one that can be obtained is the one to which the initial tableau is equivalent. The ones to which it is inequivalent cannot be obtained by any means. The use of specific algorithms is required only to assure ourselves that the process will terminate in a finite number of iterations.

Other algorithms may be used, or the same algorithms may be applied in different orders, or the choices may be totally ad hoc. The result will be the same. We do not mean that the terminal tableau is unique, only that the form of the terminal tableau is unique.

As an example, let us consider the tableau

(4.45)

x_1	x_2	x_3	-1	
3	-1	0	1	$= -y_1$
5	-3^*	-1	-3	$= -y_2$
5	-2	1	-1	$= -y_3$
4	-1	-1	0	$= f$

It is not necessary to rename and reindex the variables to apply the feasibility algorithm. We can regard any x_i as preceding any y_j, and use the prevailing indices to order the x's and y's among themselves. The rules of the feasibility algorithm lead to the starred entry in the second row, second column, as the first pivot entry. We obtain

(4.46)

x_1	y_2	x_3	-1	
$4/3$	$-1/3$	$1/3$	2	$= -y_1$
$-5/3$	$-1/3$	$1/3$	1	$= -x_2$
$5/3^*$	$-2/3$	$5/3$	1	$= -y_3$
$7/3$	$-1/3$	$-2/3$	1	$= f$

At this point we have determined that the row program is feasible. We know that the terminal form is either case 1 or case 2. To preserve the nonnegativity of the b-column we apply the simplex algorithm. The choice for the pivot entry will turn out to be unique at each step and the Bland anticycling rule does not have to be invoked. We obtain

(4.47)

y_3	y_2	x_3	-1	
$-4/5$	$1/5^*$	-1	$6/5$	$= -y_1$
1	-1	2	2	$= -x_2$
$3/5$	$-2/5$	1	$3/5$	$= -x_1$
$-7/5$	$3/5$	-3	$-2/5$	$= f$

(4.48)

y_3	y_1	x_3	-1	
-4	5	-5	6	$= -y_2$
-3	5	-3	8	$= -x_2$
-1	2	-1	3	$= -x_1$
1	-3	0	-4	$= f$

The first column of tableau 4.48 is an infeasible column and we see that we have case 2 of the theorem of the four alternatives.

In this example we used the algorithms mentioned in the proof of the theorem. At the point where we determined that the row program was feasible, we could have abandoned any attempt to preserve the nonnegativity of the b-column and concentrated on the column equations. We might, for example, have turned to the dual feasibility algorithm. Actually, there would be no advantage in doing that since experience indicates that the simplex algorithm will probably reach termination in fewer iterations than the dual feasibility algorithm. However, nothing is guaranteed in this respect.

Consider tableau 4.49.

(4.49)

1	-1	1	-1	b
-1	1	0	1	1
0	-1	1	1	-1
c	1	1	2	0

We have indicated the first entry in the b-column by a b, and the first entry in the c-row by a c. If b is positive the row program is more likely to be feasible than would be the case if b were negative. Similarly, if c is negative the column program is more likely to be feasible than would be the case if c were positive. Substitute 2 or -2 for b and 2 or -2 for c. Determine for each of the four cases that are produced which terminal form is equivalent to 4.49.

4.7 The Existence-Duality Theorem

For any pair of canonical dual linear programs there are four possibilities. (1) Both programs are feasible. (2) The max program is feasible and the min program is infeasible. (3) The max program is infeasible and the min program is feasible. (4) Both programs are infeasible.

This much is the most elementary consequence of the definitions. The important next step is to identify the existence or nonexistence of optimal solutions with each of these possibilities. From the theorem of the four alternatives we can see that each of those four alternatives corresponds to one of these possibilities. More important, from the terminal forms obtained in the theorem of the four alternatives we can make conclusions about the existence or nonexistence of optimal solutions for the linear programs.

In case 2, where we obtain a nonnegative b-column and an infeasible column, the max program does not have an optimal solution because the objective variable is unbounded and the min program does not have an optimal solution because it has no feasible solution. In case 3, where we obtain a nonpositive c-row and an infeasible row, the min program does not have an optimal solution because the objective variable is unbounded and the max program does not have an optimal solution because it has no feasible solution. In case 4, neither program is feasible and neither has an optimal solution.

In case 1, the terminal form provides complementary feasible solutions for the pair of dual linear programs. Thus, the sufficient condition for optimality is satisfied and both programs have optimal solutions. This is the only case in which either program has an optimal solution, the only case in which both programs have optimal solutions, and the only case in which both programs have feasible solutions. In addition, the optimal values for the objective variables for both programs are equal. These conclusions constitute what is known as the *existence-duality theorem of linear programming.*

Theorem 4.33 *For any pair of canonical dual linear programs, one and only one of the following alternatives holds.*

1. *Both linear programs are feasible. Both have optimal solutions. The optimal values for the objective variables for the two programs are equal. Both programs have basic optimal solutions and these basic solutions are complementary.*

2. *The minimum program is infeasible. The maximum program is feasible, and the set of feasible solutions contains a one-parameter subset for which the objective variable takes on arbitrarily large positive values.*

3. *The maximum program is infeasible. The minimum program is feasible, and the set of feasible solutions contains a one-parameter subset for which the objective variable takes on arbitrarily large negative values.*

4. *Neither program is feasible. Neither program has an optimal solution.*

   ⊟

Quite often the feasibility of a linear program can be established easily. For example, the sample programs in Chapter 1 had initial tableaux with nonnegative b-columns. This establishes the feasibility of the max program. Also, if any row consists entirely of positive entries, it is possible to take the nonbasic variable corresponding to that row large enough to ensure that all constraints of the min program are satisfied. This amounts to the observation that the nutritional requirements can be met by buying enough hay. This establishes the feasibility of the min program. In these cases we know the form of the outcome before a single computational step is taken.

From an intuitive point of view these conclusions are reasonable for the types of problems given in Chapter 1. The max program in Section 1.1 is a typical problem in which a variety of tasks utilize limited resources. Doing nothing uses no resource and is usually feasible. (Of course a legal, contractual, or licensing requirement to carry on a minimum level of activity would be another matter.) The positive entries in the A-matrix mean that the activity (the column) uses a positive amount of the resource (the row). This limits the level of activity and limits the profit so that an optimal value is expected to exist.

The min program in Section 1.2 is a typical problem in which some requirements must be met at minimum cost. A positive entry in the A-matrix means that a requirement (the column) can be met by some activity or resource (the row) if only enough of the activity or resource is committed. One expects the cost of each resource or activity to be positive, and this means that the b-column should be positive. Again, it is reasonable to expect that an optimal value exists.

The transportation problem in Section 1.3 is a little more specialized, but it is an important type of problem. If there are enough supplies in the combined origins and all origins are connected to all destinations, the problem of supplying the destinations is feasible. Since any method of supplying the destinations costs something, an optimal solution is expected.

Generally, in applied problems an intuitive grasp of the context of the problem should be sufficient to reach a conclusion as to whether the problem has an optimal solution. If the actual computation leads to a surprise, one should carefully examine the constraints to see that they correctly represent the problem.

In most treatments of linear programming, the existence-duality theorem is reached by proving the following theorem. It follows quite directly from the theorem of the four alternatives.

Theorem 4.34 *If a linear program has an optimal solution, there is a basic feasible solution that is optimal.*

Proof. According to the theorem of the four alternatives, the only case in which either program has an optimal solution is case 1. In this case, the basic solutions for the terminal tableau satisfy the sufficient condition for optimality. Thus, both programs have basic optimal solutions. □

While this theorem is not necessary in our development of the theory of linear programming, it has an interesting and important geometric interpretation. Generally, the constraint set is a convex set bounded by lines (in the plane), planes (in space), or hyperplanes (in general). The corners of this hyperpolyhedron correspond to the basic feasible solutions of the constraints. The theorem then says that if a linear program has an optimal solution, at least one corner will represent an optimal solution.

It is easily shown that if two points in the feasible set represent optimal solutions then the points on the line segment joining the two points also represent optimal solutions. One can also show that the set of optimal solutions is either a corner (if it is unique) or the smallest flat geometric object (edge, face, etc.) that contains those corners that are optimal.

If an optimal solution is obtained by the simplex algorithm, the basic optimal solution represents one of those corners. For the max program, the possibility of other optimal corners can be explored by continuing with simplex pivots in columns that have zeros in the c-row, if there are any. For the min program, the possibility of other optimal corners can be explored by continuing with dual simplex pivots in rows that have zeros in the b-column, if there are any.

Exercises

1. In the following tableau there are several entries on which a simplex pivot exchange may be performed. Identify all such entries.

	x_1	x_2	x_3	x_4	-1	
v_1	-1	2	3	4	1	$= -y_1$
v_2	0	3	5	5	2	$= -y_2$
v_3	2	4	9	6	3	$= -y_3$
-1	3	-1	3	0	0	$= f$
	$= u_1$	$= u_2$	$= u_3$	$= u_4$	$= g$	

2. The Production Problem, the Diet Problem, and the Transportation Problem, all discussed and solved in Chapter 1, were solved using steps consistent with the simplex algorithm. Check through the calculations there and verify that they were done with choices for the pivot exchanges made by the rules of the simplex algorithm.

We are going to use the simplex algorithm to solve the canonical problems that we set up in the exercises of Chapter 1. There may be several different sequences of pivot exchanges that lead to an optimal solution. When an optimal solution is obtained, there may be multiple optimal solutions, and there may be multiple optimal solutions to the dual program. Find all basic optimal solutions when there are more than one. Also, interpret the optimal dual solution in terms of the optimal solution of the primal problem. If you have not previously set up the tableaux for these problems you can obtain the initial tableaux in the Answers section on page 493.

3. Solve the Furniture Maker's Problem of Chapter 1.
4. Solve the President's Problem of Chapter 1.
5. Solve the Wyndor Problem of Chapter 1.
6. Solve the Investment Manager's Problem of Chapter 1.
7. Solve the Investment Manager's Problem 2 of Chapter 1.
8. Solve the Welfare Mother's Problem of Chapter 1.
9. Solve the Advertising Manager's Problem of Chapter 1.
10. Solve the MaxMin Problem of Chapter 1.
11. Solve the MinMax Problem of Chapter 1.

12. Determine which of the four alternatives the following tableau represents.

	x_1	x_2	x_3	x_4	-1	
v_1	-2	0	0	-1	-2	$= -y_1$
v_2	-2	-1	-1	-1	-1	$= -y_2$
v_3	1	-1	0	1	-2	$= -y_3$
v_4	-2	-2	-1	0	0	$= -y_4$
-1	-2	-1	-2	0	-2	$= f$
	$= u_1$	$= u_2$	$= u_3$	$= u_4$	$= g$	

13. Determine which of the four alternatives the following tableau represents.

	x_1	x_2	x_3	x_4	-1	
v_1	-1	-1	0	-1	0	$= -y_1$
v_2	1	0	-2	1	-1	$= -y_2$
v_3	0	-2	1	1	-2	$= -y_3$
v_4	1	-1	-2	1	0	$= -y_4$
-1	0	-1	1	1	-2	$= f$
	$= u_1$	$= u_2$	$= u_3$	$= u_4$	$= g$	

14. Determine which of the four alternatives the following tableau represents.

	x_1	x_2	x_3	x_4	-1	
v_1	0	0	2	2	3	$= -y_1$
v_2	-1	1	1	0	-2	$= -y_2$
v_3	0	2	-2	-1	-1	$= -y_3$
v_4	0	0	3	2	0	$= -y_4$
-1	-1	-1	2	-2	-2	$= f$
	$= u_1$	$= u_2$	$= u_3$	$= u_4$	$= g$	

15. Determine which of the four alternatives the following tableau represents.

	x_1	x_2	x_3	x_4	-1	
v_1	-2	-1	-2	0	-1	$= -y_1$
v_2	1	1	-1	0	-1	$= -y_2$
v_3	0	-2	-2	1	-1	$= -y_3$
v_4	-1	0	1	0	-1	$= -y_4$
-1	1	0	1	1	-2	$= f$
	$= u_1$	$= u_2$	$= u_3$	$= u_4$	$= g$	

Questions

Q1. If there are two equivalent tableaux, one with an infeasible row and the other with an infeasible column, then there exists an equivalent tableau with both an infeasible row and an infeasible column.

Q2. If there are two equivalent tableaux, one with a nonnegative b-column and the other with an infeasible column, then there exists an equivalent tableau with both a nonnegative b-column and an infeasible column.

Q3. If there are two equivalent tableaux, one with a nonpositive c-row and the other with an infeasible row, then there exists an equivalent tableau with both a nonpositive c-row and an infeasible row.

Q4. If there are two equivalent tableaux, one with a nonnegative b-column and the other with a nonpositive c-row, then there exists an equivalent tableau with both a nonnegative b-column and a nonpositive c-row.

Q5. If there are two equivalent tableaux, one with a positive row in the A-matrix and the other with a negative column in the A-matrix, then there exits an equivalent tableau with a nonnegative b-column and a nonpositive c-row.

Q6. The inverse pivot of a simplex pivot exchange is also a simplex pivot exchange.

Q7. Assume that the max program has a unique optimal solution. It is then possible to go from one basic feasible solution to any other basic feasible solution through a sequence of simplex pivot exchanges.

Q8. If the max program is feasible, the value of the objective variable is bounded from below.

Q9. The sufficient condition for optimality stated in Theorem 3.16 in Chapter 3 is also a necessary condition for optimality.

Q10. The sufficient condition for optimality stated in Theorem 3.17 of Chapter 3 is also a necessary condition for optimality.

Q11. If the max program is feasible and the feasible set is bounded, the min program has an optimal solution.

Q12. If the max program is feasible and the feasible set is unbounded, the min program cannot have an optimal solution.

Q13. If two successive pivot exchanges are performed under the rules of the simplex algorithm, it is not possible for a variable to switch from nonbasic to basic on the first exchange and then basic to nonbasic on the second.

Q14. If two successive pivot exchanges are performed under the rules of the simplex algorithm, it is not possible for a variable to switch from basic to nonbasic on the first exchange and then nonbasic to basic on the second.

Chapter 5

General Linear Programs

5.1 General Dual Linear Programs

Let us re-examine our earlier discussion of canonical linear programs with a view towards identifying the essential ideas that allowed us to establish the theory of linear programming. Many of these ideas are easily generalized and the proofs of more general theorems are either identical to those already given for canonical linear programs or follow from them directly.

Consider the initial tableau

(5.1)

	x	-1	
v	A	b	$= -y$
-1	c	d	$= f$
	$= u$	$= g$	

The relation that was used in virtually every development was the duality equation

$$u_1x_1 + \cdots + u_nx_n + v_1y_1 + \cdots + v_my_m = g - f \tag{5.2}$$

For nonnegative values of u, v, x, and y we were able to establish the inequality

$$\max f \leq \min g \tag{5.3}$$

from the duality equation. In turn, this leads to the sufficient condition for optimality, since a feasible value of f and a feasible value of g for which $f = g$ imply that this common value is optimal for both programs.

Our immediate objective is to relax the condition that the variables of the problems be nonnegative, but to do it in such a way that this line of conclusions remains valid. For each variable in one program, its dual variable is the one paired with it in the duality equation. Whatever conditions are imposed on a variable in one of the programs, we impose conditions on the dual variable in the following way.

1. If one variable is required to be nonnegative, we require its dual variable to be nonnegative. This is the *canonical condition.*

2. If one variable is required to be nonpositive, we require its dual variable to be nonpositive.

3. If one variable is *free* (i.e., unrestricted in sign), we require its dual variable to be zero. It may seem peculiar to consider a variable that can have only one value. However, it is convenient to allow such a variable to appear in the equations of a tableau and to specify its restriction as a separate (side) condition. A variable required to be zero is called an *artificial variable.*

4. If one variable is artificial, its dual variable is free.

These side conditions that the variables of a problem must satisfy are known as *sign specifications* or *feasibility specifications.* If we pose a general linear program with any feasibility specifications, we define the feasibility specifications for the dual problem so that the feasibility specifications of dual variables are related as described above. We call these relations *duality relations.* A solution of the linear equations of a linear programming problem is *feasible* if it satisfies the feasibility specifications.

With this extended meaning of the term "feasible" the statements of the theorems of Section 3.2 remain valid. In particular.

Theorem 5.35 *Let $(u, v, g) = (u_1, \ldots, u_n, v_l, \ldots, v_m, g)$ and $(x, y, f) = (x_1, \ldots, x_n, y_1, \ldots, y_m, f)$ be any feasible solutions for the minimum and maximum programs of tableau 5.1. If this pair of solutions satisfies feasibility specifications that are dually related, then*

$$f \leq g \tag{5.4}$$

Theorem 5.36 *(Sufficient Condition for Optimality) If we can find feasible solutions (x, y, f) and (u, v, g) for the maximum and minimum programs for which $f = g$, then each is optimal for its program.*

Theorem 5.37 *(Sufficient Condition for Optimality) Complementary feasible solutions are optimal.*

Proof. These three theorems have the same wordings as the corresponding canonical theorems 3.15, 3.16, and 3.17 of Section 3.2. They do not say quite the same things since the word "feasible" now means something different. However, the wordings for the proofs of the canonical theorems can be reread with the new interpretation of the word "feasible" and they will be found to be valid proofs of these general theorems. ◻

Just as for canonical linear programs, these sufficient conditions for optimality can be exploited to provide methods for solving general linear programs. We can try to find feasible solutions for the row and column equations which are complementary. The definitions of the duality relations already make artificial and free variables complementary. This, as we shall see in the following section, allows us to reduce the problem of solving general linear programs to that of solving canonical linear programs.

The second duality relation, where the variables are nonpositive, can always be rewritten in canonical form. Suppose, for example, that we have the dual feasibility conditions $v_r \leq 0$ and $y_r \leq 0$. An appropriate tableau, with emphasis on the v_r-row, is shown in 5.5.

$$
\begin{array}{c|cccc|c|l}
 & x_1 & x_2 & \cdots & x_n & -1 & \\
\hline
\vdots & \vdots & \vdots & & \vdots & \vdots & \vdots \\
v_r & a_{r1} & a_{r2} & \cdots & a_{rn} & b_r & = -y_r \\
\vdots & \vdots & \vdots & & \vdots & \vdots & \vdots \\
\hline
-1 & c_1 & c_2 & \cdots & c_n & & = f \\
\hline
 & = u_1 & = u_2 & \cdots & = u_n & = g &
\end{array}
\tag{5.5}
$$

The corresponding row equation and the j-th column equation are shown in 5.6 and 5.7.

$$a_{r1}x_1 + a_{r2}x_2 + \cdots + a_{rn}x_n - b_r = -y_r \tag{5.6}$$

$$v_1 a_{1j} + \cdots + v_r a_{rj} + \cdots + v_m a_{mj} - c_j = u_j \tag{5.7}$$

We can make the substitutions

$$v'_r = -v_r, y'_r = -y_r \tag{5.8}$$

Equations 5.6 and 5.7 then take the forms

$$(-a_{r1})x_1 + (-a_{r2})x_2 + \cdots + (-a_{rn})x_n - (-b_r) = -y'_r \tag{5.9}$$

$$v_1 a_{1j} + \cdots + v'_r(-a_{rj}) + \cdots + v_m a_{mj} - c_j = u_j \tag{5.10}$$

The interior of the tableau representing the equations with these new variables is the same as tableau 5.1, except that the new r-th row is the negative of the old one. The two new variables are canonically dual to each other

It is awkward, particularly when working a numerical problem, to have to make separate notations as to which variables are nonnegative, which are nonpositive, which are artificial, and which are free. This difficulty can be eased by making several notational agreements. For one thing, since the nonpositivity conditions can be converted to nonnegativity conditions, we shall avoid using nonpositive variables unless the situation requires it.

For dual variables related by the third and fourth duality relations, the situation is represented diagrammatically in tableau 5.11.

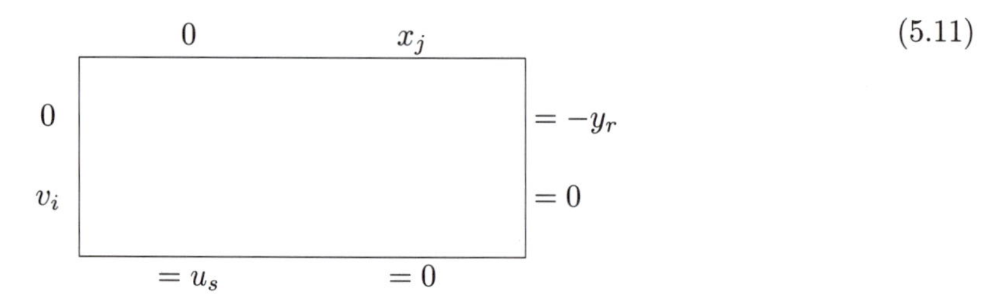

By implication of the duality relations, the dual variable of an artificial variable is free. Thus, the zeros shown in 5.11 are sufficient to indicate that the variables on the opposite margins are free. By avoiding the use of nonpositive variables, the feasibility status of all variables can be indicated by showing which variables are artificial. We can show the locations of the artificial variables by writing them as zeros, as above, or using some suggestive letter, like z for zero, to indicate artificial variables.

Artificial variables have been used in linear programming almost from the very beginning. The idea is usually used as a device for applying the simplex algorithm to a problem that does not have a nonnegative b-column in its initial tableau. (We shall discuss this technique in Section 5.3.) However, it has been the custom to delete a column or row headed by an artificial variable. Unfortunately, this also deletes the dual free variable. Sometimes this deletion causes no difficulty, but sometimes it destroys the interpretation of the dual program. Accepting and exploiting the duality of free and artificial variables turns out to make the theory of duality in linear programming quite graceful.

5.2 Reduction to Canonical Form

In this section we continue extending the ideas developed for canonical linear programs to noncanonical linear programs. Our main aim here is to obtain a general form of the existence-duality theorem of Section 4.7. In the preceding section the theorems and their proofs had the same wording for both the canonical and noncanonical cases. Here our theorem will have the same wording, but the proof will have to be different. It will depend directly on the corresponding theorem for canonical programs.

Suppose we have a linear programming problem in noncanonical form. If a variable is artificial, it is desirable to have it appear in the equations as an independent variable. We can then set it equal to zero. If a variable is free, it is desirable to have it appear as a dependent variable. It can then be computed after the values of the independent variables have been assigned.

Let us try, then, to perform pivot exchanges to move as many artificial variables as possible to the left and top margins, and as many free variables as possible to the right and bottom margins. That is, we will try to exchange a free or canonical nonbasic variable and an artificial basic variable, or a free nonbasic variable and an artificial or canonical basic variable. For the moment we are not interested in exchanging nonbasic and basic variables of the same feasibility type. When we can proceed no further we obtain a tableau in the form of 5.12.

(5.12)

	canonical			free			artificial				
	x_1	$\cdots$	x_s	x_{s+1}	$\cdots$	x_t	x_{t+1}	$\cdots$	x_n	-1	
v_1				0	$\cdots$	0				b_1	$= -y_1$
canonical		A_{11}		$\vdots$		$\vdots$		A_{13}		$\vdots$	canonical
v_r				0	$\cdots$	0				b_r	$= -y_r$
v_{r+1}	0	$\cdots$	0	0	$\cdots$	0				b_{r+1}	$= -y_{r+1}$
free	$\vdots$		$\vdots$	$\vdots$		$\vdots$		A_{23}		$\vdots$	artificial
v_q	0	$\cdots$	0	0	$\cdots$	0				b_q	$= -y_q$
v_{q+1}										b_{q+1}	$= -y_{q+1}$
artificial		A_{31}			A_{32}			A_{33}		$\vdots$	free
v_m										b_m	$= -y_m$
-1	c_1	$\cdots$	c_r	c_{s+1}	$\cdots$	c_t	c_{t+1}	$\cdots$	c_n	d	$= f$
	$= u_1$	$\cdots$	$= u_r$	$= u_{s+1}$	$\cdots$	$= u_t$	$= u_{t+1}$	$\cdots$	$= u_n$	$= g$	
	canonical			artificial			free				

For the column equations, all the entries in the third group of rows are multiplied by zeros, the artificial variables on the left margin. These entries play no role in the constraints of the min program and do not affect the value of the objective variable g. Thus, as far as the min program is concerned these rows could be deleted. For the row equations the variables on the right margin in the third group are free variables. We can file these equations away and try to solve the max program described by the other equations and the objective function. When that program is solved, if it can be solved, we can compute the values of $y_{q+1}, \ldots, y_m$.

In the same way the entries in the third group of columns do not enter into the max program and they determine equations in the min program that can be stored for a later computation. We are left with tableau 5.13.

(5.13)

	canonical			free				
	x_1	$\cdots$	x_s	x_{s+1}	$\cdots$	x_t	-1	
v_1				0	$\cdots$	0	b_1	$= -y_1$
canon		A_{11}		$\vdots$		$\vdots$	$\vdots$	canon
v_r				0	$\cdots$	0	b_r	$= -y_r$
v_{r+1}	0	$\cdots$	0	0	$\cdots$	0	b_{r+1}	$= -y_{r+1}$
free	$\vdots$		$\vdots$	$\vdots$		$\vdots$	$\vdots$	artificial
v_q	0	$\cdots$	0	0	$\cdots$	0	b_q	$= -y_q$
-1	c_1	$\cdots$	c_r	c_{s+1}	$\cdots$	c_t	d	$= f$
	$= u_1$	$\cdots$	$= u_r$	$= u_{s+1}$	$\cdots$	$= u_t$	$= g$	
	canonical			artificial				

The equations generated by the second group of rows have the form

$$-b_k = -y_k, \quad \text{for } k = r+1, \ldots, q \tag{5.14}$$

The system of row equations has no feasible solution unless $b_{r+1} = \ldots = b_q = 0$. Suppose one of these b's, say b_k, is nonzero and the minimum program is feasible. Assign the values of a feasible solution to the $v_1, \ldots, v_r$. The $v_{r+1}, \ldots, v_q$ can be assigned values arbitrarily without disturbing the feasibility of the resulting solution. Then the value of g can be made as negative as one wishes by taking appropriate values for v_k.

Similar remarks apply to $c_{s+1}, \ldots, c_t$. The column system of equations has no feasible solution unless all these c's are zero. If one of these c's is nonzero, then the min program is infeasible and the max program is either infeasible or unbounded.

We emphasize that so far we have not examined the first r row equations or the first s column equations. At this point we have determined one of the following three alternatives.

1. At least one artificial variable in each program cannot be zero. Thus, both programs are infeasible.

2. At least one artificial variable in one program cannot be zero. This program is infeasible. All artificial variables in the other program can be set equal to zero. This program is either infeasible or, if feasible, unbounded.

3. All artificial variables can be set equal to zero.

In the third case the rows in the second group are all zeros, and the columns in the second group are all zeros. We are left with tableau 5.15 in which all variables are canonical.

$$
\begin{array}{c|ccc|c|l}
 & x_1 & \cdots & x_s & -1 & \\
\hline
v_1 & & & & b_1 & = -y_1 \\
\vdots & & A_{11} & & \vdots & \quad \vdots \\
v_r & & & & b_r & = -y_r \\
\hline
-1 & c_1 & \cdots & c_s & d & = f \\
\hline
 & = u_1 & \cdots & = u_s & = g &
\end{array}
\tag{5.15}
$$

Tableau 5.15 represents a dual pair of canonical linear programs. We have reduced a general pair of linear programs to a canonical pair.

The theorem of the four alternatives and the existence-duality theorem apply to the reduced canonical problem. By combining the various alternatives given above with those given in the existence-duality theorem, we see that we can assert the general form of the existence-duality theorem.

Theorem 5.38 *(Existence-Duality Theorem) For any pair of general dual linear programs, one and only one of the following alternatives holds.*

1. *Both linear programs are feasible. Both have optimal solutions. The optimal values for the objective variables for the two programs are equal. Both programs have basic optimal solutions and these basic solutions are complementary.*

2. *The minimum program is infeasible. The maximum program is feasible, and the set of feasible solutions contains a one-parameter subset for which the objective variable takes on arbitrarily large positive values.*

3. *The maximum program is infeasible. The minimum program is feasible, and the set of feasible solutions contains a one-parameter subset*

for which the objective variable takes on arbitrarily large negative values.

4. *Neither program is feasible. Neither program has an optimal solution.* ⊟

As an example, consider the following tableau.

(5.16)

	x_1	x_2	x_3	x_4	-1	
t_1	1	2	-1	1^*	4	$= -z_1$
t_2	2	-2	3	4	19	$= -z_2$
v_3	1	-1	2	-1	1	$= -y_3$
-1	1	-10	7	1	0	$= f$
	$= u_1$	$= u_2$	$= u_3$	$= u_4$	$= g$	

We have used the notational convention described in the preceding section. The z's on the right margin are artificial variables. By implication, the t's on the left margin are free. All other variables, except the objective variables, are canonical. Our first goal is to perform pivot exchanges with pivot entries chosen in the first two rows to move the artificial variables to the top margin. The first choice is indicated with an asterisk. We obtain

(5.17)

	x_1	x_2	x_3	z_1	-1	
	u_4	1	2	-1	4	$= -x_4$
t_2	-2	-10	7^*	-4	3	$= -z_2$
v_3	2	1	1	1	5	$= -y_3$
-1	0	-12	8	-1	-4	$= f$
	$= u_1$	$= u_2$	$= u_3$	$= t_1$	$= g$	

(5.18)

	x_1	x_2	z_2	z_1	-1	
u_4	5/7	4/7	1/7	3/7	31/7	$= -x_4$
u_3	$-2/7$	$-10/7$	1/7	$-4/7$	3/7	$= -x_3$
v_3	$16/7^*$	17/7	$-1/7$	11/7	32/7	$= -y_3$
-1	16/7	$-4/7$	$-8/7$	25/7	$-52/7$	$= f$
	$= u_1$	$= u_2$	$= t_2$	$= t_1$	$= g$	

In tableau 5.18 the third and fourth columns are headed by artificial variables and could be set aside. They do not contribute further to the max program, but the free variables on the bottom margin are of interest to the min program. We prefer to carry them along, remembering not to choose a pivot entry in those columns.

The next pivot entry, chosen by the simplex algorithm from the first two columns, is shown with an asterisk in tableau 5.18. We obtain tableau 5.19.

(5.19)

	y_3	x_2	z_2	z_1	-1	
u_4	$-5/16$	$-3/16$	$3/16$	$-1/16$	3	$= -x_4$
u_3	$1/8$	$-9/8$	$1/8$	$3/8$	1	$= -x_3$
u_1	$7/16$	$17/16$	$-1/16$	$11/16$	2	$= -x_1$
-1	-1	-3	-1	2	-12	$= f$
	$= v_3$	$= u_2$	$= t_2$	$= t_1$	$= g$	

The basic solutions in tableau 5.19 assign nonnegative values to all canonical variables and zeros to all artificial variables. Thus they are feasible, complementary, and optimal. The positive entry in the c-row of tableau 5.19 does not disturb feasibility since the variable associated with that entry is free. The free variables of the min program have the values $t_1 = -2$ and $t_2 = 1$. Furthermore, $f = g = 12$ is the optimal value for both programs.

5.3 The Two-Phase Simplex Method

The simplex algorithm requires that the program to which it is applied be canonical and have a nonnegative b-column. By the method described in the preceding section we can reduce a noncanonical program to a canonical program. For the example given there we obtained a canonical program with a nonnegative b-column in tableau 5.18. If we had chosen to pivot in other columns we would not have been so lucky, and in general one should expect that the reduced canonical program will have negative entries in the b-column.

We could use the feasibility algorithm to handle a canonical program with negative entries in the b-column. However, that is not the way it is done in practice. A number of methods have been proposed for obtaining an equivalent canonical tableau with a nonnegative b-column to start the simplex algorithm. For every method that has been proposed it has been possible to devise an example for which the proposed method works rather poorly. Thus, it has not been possible to prove on a logical basis that any method is the best. The methods that have prevailed are those that work well for the types of problems that have been encountered in practice.

The Committee on Algorithms (COAL), a committee of the Mathematical Programming Society, maintains a collection of problems that it uses to test a proposed method. A new method would have to be more than marginally better to justify rewriting a computer program to incorporate it. However, that is a matter of judgment for those who maintain such programs. Not all commercial computer programs handle this problem the same way. There is a considerable amount of judgment exercised in designing the computer program.

The most widely used methods are based on the two-phase simplex method. Phase II is the method discussed in Chapter 4. It is the simplex algorithm applied to a tableau for which the variables are canonical and the b-column is nonnegative. Phase I is that part of the process where we are dealing with a tableau in which some nonbasic variables are free, or some basic variables are artificial, or the b-column contains some negative entries in the rows of canonical variables. In phase I we create an artificial problem that can be worked by the simplex algorithm for which the solution, if it exists, will provide a tableau in the proper form for phase II. Strictly speaking, phase I is not an algorithm or a method. It is a collection of devices to reformulate a problem so that the simplex algorithm can be used.

Let us first examine a device for handling a canonical program with at least one negative entry in the b-column. Consider, for example, tableau 5.20.

(5.20)

x_1	x_2	x_3	-1	
1	1	-1	1	$= -x_4$
2	1	-1	-2	$= -x_5$
1	-1	0	-3	$= -x_6$
3	0	-1	2	$= -x_7$
6	3	-5	0	$= f$

We have already remarked that a column with an artificial variable at the top can be deleted, at least as far as our interest in the row equations is concerned. This means we can freely introduce a new column if it is headed by an artificial variable. Introduce such a column in the tableau 5.20 as shown in tableau 5.21.

(5.21)

x_1	x_2	x_3	z	-1	
1	1	-1	0	1	$= -x_4$
2	1	-1	-1	-2	$= -x_5$
1	-1	0	-1^*	-3	$= -x_6$
3	0	-1	0	2	$= -x_7$
6	3	-5	0	0	$= f$

The column of the artificial variable z has a -1 for each negative entry in the b-column, and zeros otherwise. Pivot on the -1 in the z-column that is opposite the most negative entry in the b-column. This will always yield a tableau with a nonnegative b-column, as illustrated in tableau 5.22.

(5.22)

x_1	x_2	x_3	x_6	-1	
1	1	−1	0	1	$= -x_4$
1	2	−1	−1	1	$= -x_5$
−1	1	0	−1	3	$= -z$
3	0	−1	0	2	$= -x_7$
6	3	−5	0	0	$= f$

To clarify the discussion of what to do next, we shall make one more change in tableau 5.22. Consider 5.23.

(5.23)

x_1	x_2	x_3	x_6	-1	
1	1	−1	0	1	$= -x_4$
1	2	−1	−1	1	$= -x_5$
−1	1	0	−1	3	$= -z$
3	0	−1	0	2	$= -x_7$
6	3	−5	0	0	$= f$
−1	1	0	−1	3	$= -z$

We have simply copied the z-row into the basement row. We temporarily regard the z in the third row as a canonical variable, and the $-z$ in the basement row as an objective variable. The problem of maximizing $-z$ is equivalent to minimizing z. Since z (in the third row) is canonical, the maximum value of $-z$ (in the basement row) cannot be positive.

The simplex algorithm can be applied in tableau 5.23 to determine the minimum value of z. The f-row does not enter into the decision making, but it is carried along in the computations. If the minimum value of z is positive, the program in tableau 5.21, with z artificial, is infeasible and the original program in tableau 5.20 is also infeasible.

If the minimum value of z is zero, the simplex algorithm will terminate in one of two ways. It may terminate with z on the top margin, a nonbasic variable. In this case we can delete the z-column. The tableau that remains is equivalent to tableau 5.20 and has a nonnegative b-column. Phase I has terminated. We can apply the simplex algorithm to this canonical program. This is the beginning of phase II.

The simplex algorithm might also terminate with z on the right margin, a basic variable. In this case the b-entry of the z-row must be zero, since z has the value zero for the basic solution of the terminal tableau. We can pivot on any nonzero entry in the z-row. This will leave the b-

column undisturbed and, therefore, nonnegative. Again, the z-column can be deleted and phase II can begin. Actually, this case can be avoided since a zero will not appear in the b-entry of the z-row unless the z-row ties with some other row as a candidate for the pivot row. All we have to do is to give the z-row a higher priority as the pivot row if more than one choice is available.

Now let us look at the situation that we face when there are several artificial variables as basic variables. For an example, let us reconsider the example of the preceding section. We repeat tableau 5.16 here as tableau 5.24.

(5.24)

x_1	x_2	x_3	x_4	-1	
1	2	-1	1	4	$= -z_1$
2	-2	3	4	19	$= -z_2$
1	-1	2	-1	1	$= -y_3$
1	-10	7	1	0	$= f$

Construct a new objective row that is the sum of the rows for which the basic variables are artificial. In this way we get tableau 5.25.

(5.25)

x_1	x_2	x_3	x_4	-1	
1	2	-1	1	4	$= -z_1$
2	-2	3	4	19	$= -z_2$
1^*	-1	2	-1	1	$= -y_3$
1	-10	7	1	0	$= f$
3	0	2	5	23	$= -z$

We can see that

$$z = z_1 + z_2 \tag{5.26}$$

Again, we regard $-z$ as the objective variable that we want to maximize. We temporarily regard z_1 and z_2 as canonical variables. Since the b-column is nonnegative we can apply the simplex algorithm to determine the minimum value of z. Again, this is phase I. Since z is the sum of (temporary) canonical variables it is nonnegative, and if its minimum value is positive there is no feasible solution to the row equations in tableau 5.24 with $z_1 = z_2 = 0$. That is, the general max program in tableau 5.24 is infeasible.

If the minimum value of z is zero, there are two ways that phase I can terminate. It can terminate with all artificial variables as nonbasic variables, or it can terminate with some artificial variables as basic variables. In the latter case, all b-entries in the rows of basic artificial variables will be zeros. The basic artificial variables can be pivoted to the top margin without changing the b-column. (While the row of an artificial variable cannot be zero if it was nonzero in the initial tableau, such a row might have nonzero entries only in the columns of other artificial variables. Such rows can be deleted.) That is, in either case we can obtain a tableau in which all basic variables are canonical and the b-column is nonnegative. We can start phase II and apply the simplex algorithm.

Let us continue with the example started in tableau 5.25. To use a definite rule for choosing the pivot column, we shall use the order of the variables alphabetically and by index. The first pivot entry is shown in tableau 5.25. We obtain

(5.27)

y_3	x_2	x_3	x_4	-1	
-1	3^*	-3	2	3	$= -z_1$
-2	0	-1	6	17	$= -z_2$
1	-1	2	-1	1	$= -x_1$
-1	-9	5	2	-1	$= f$
-3	3	-4	8	20	$= -z$

(5.28)

y_3	z_1	x_3	x_4	-1	
$-1/3$	$1/3$	-1	$2/3^*$	1	$= -x_2$
-2	0	-1	6	17	$= -z_2$
$2/3$	$1/3$	1	$-1/3$	2	$= -x_1$
-4	3	-4	8	8	$= f$
-2	-1	-1	6	17	$= -z$

(5.29)

y_3	z_1	x_3	x_2	-1	
$-1/2$	$1/2$	$-3/2$	$3/2$	$3/2$	$= -x_4$
1	-3	8^*	-9	8	$= -z_2$
$1/2$	$1/2$	$1/2$	$1/2$	$5/2$	$= -x_1$
0	-1	8	-12	-4	$= f$
1	-4	8	-9	8	$= -z$

Finally, in tableau 5.29 we are forced to pivot in the last remaining row of an artificial variable. The next tableau will allow the start of phase II. Notice how the value of z in the lower right corner is driven down towards zero.

(5.30)

y_3	z_1	z_2	x_2	-1	
$-5/16$	$-1/16$	$3/16$	$-3/16$	3	$= -x_4$
$1/8$	$-3/8$	$1/8$	$-9/8$	1	$= -x_3$
$7/16$	$11/16$	$-1/16$	$17/16$	2	$= -x_1$
-1	2	-1	-3	-12	$= f$
0	-1	-1	0	0	$= -z$

Actually, tableau 5.30 represents the optimal solutions. Notice that the 2 in the c-row is acceptable since it is in the column of an artificial variable of the max program (it will be multiplied by 0) and its dual variable is free. The optimal solution is $x_1 = 2$, $x_2 = 0$, $x_3 = 1$, $x_4 = 3$, $z_1 = 0$, $z_2 = 0$, $y_3 = 0$, and $f = 12$. You should be able to obtain the solution for the dual min program.

5.4 Systems of Linear Inequalities

There are a number of important theorems concerning systems of linear inequalities that follow easily and directly from the general existence-duality theorem, Theorem 5.38. Consider the following tableau in matrix notation for a pair of general dual linear programs.

(5.31)

	x	-1	
v	A	b	$= -y$
-1	0	0	$= f$
	$= u$	$= g$	

Theorem 5.39 *One and only one of the following two alternatives holds. Either*

1. *there exists a feasible (x, y) for which $Ax - b = -y$, or*
2. *there exists a feasible (u, v) for which $vA = u$ and $vb < 0$.*

Proof. The column system of equations is feasible since $v = u = 0$ is a feasible solution, regardless of the nature of the feasibility specifications. If the row system of equations is feasible, then for any feasible (x, y) we have $f = 0$. In this case $vb = g \geq \min g = \max f = 0$. If the row system is infeasible, the value of g is unbounded below. In particular, there exists a feasible (u, v) for which $g < 0$. ◻

By particularizing the feasibility specifications we can obtain four theorems which are of more interest than Theorem 5.39 itself, even though they are just special cases of Theorem 5.39.

Theorem 5.40 *(Farkas's Lemma) One and only one of the following two alternatives holds. Either*

1. *there exists an $x \geq 0$ for which $Ax = b$, or*

2. *there exists a v (free) for which $vA \geq 0$ and $vb < 0$.* ◻

Theorem 5.41 *One and only one of the following two alternatives holds. Either*

1. *there exists an $x \geq 0$ for which $Ax \leq b$, or*

2. *there exists a $v \geq 0$ for which $vA \geq 0$ and $vb < 0$.* ◻

Theorem 5.42 *One and only one of the following two alternatives holds. Either*

1. *there exists an x (free) for which $Ax = b$, or*

2. *there exists a v (free) for which $vA = 0$ and $vb < 0$.* ◻

Theorem 5.43 *One and only one of the following two alternatives holds. Either*

1. *there exists an x (free) for which $Ax \leq b$, or*

2. *there exists a $v \geq 0$ for which $vA = 0$ and $vb < 0$.* ◻

In each of these theorems we are more interested in the first of the two alternatives than we are in the second. Consider the following four types of problems. Given the matrix A and the column b,

A. Find $x \geq 0$ for which $Ax = b$.

B. Find $x \geq 0$ for which $Ax \leq b$.

C. Find x (free) for which $Ax = b$.

D. Find x (free) for which $Ax \leq b$.

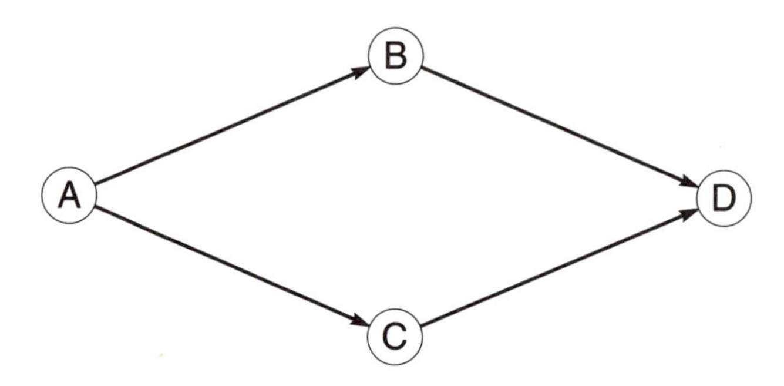

Figure 5.1: Feasibility implications for linear inequalities.

Consider all four types of problems with the same A-matrix and b-column. Any solution for a type-A problem is also a solution for a type-B problem and a type-C problem. Any solution for a type-B problem or a type-C problem is also a solution for a type-D problem. The possible implications of this kind are indicated in Figure 5.1.

Table 5.1 lists the various patterns of feasibility or infeasibility that can occur for these types of problems. A solidus is drawn through the letter representing the type of problem if that type of problem is infeasible. We use parentheses to indicate the types of problems for which the status follows by implication from the rest of the pattern.

Table 5.1: Feasibility patterns for a single data set.

1.	A	(B)	(C)	(D)
2.	A̸	B	C	(D)
3.	(A̸)	B̸	C	(D)
4.	(A̸)	B	C̸	(D)
5.	(A̸)	B̸	C̸	D
6.	(A̸)	(B̸)	(C̸)	D̸

Every one of these patterns of feasibility and infeasibility can occur. It would be very tedious to give an example for each of these possibilities. We shall content ourselves with an example to illustrate the second possibility. We will also consider the various possibilities in the exercises. Consider the following tableau.

(5.32)

	x_1	x_2	x_3	-1	
v_1	3	4	6	12	$= -y_1$
v_2	-3	2	3	-3	$= -y_2$
v_3	-3	-1	-3	-6	$= -y_3$
	$= u_1$	$= u_2$	$= u_3$	$= g$	

As a type-A problem or as a type-C problem, the y's are artificial. By implication, the v's are free. In either case the artificial variables should be pivoted to the top margin. This can be done, and we obtain

(5.33)

	y_1	y_2	y_3	-1	
u_1	$1/9$	$-2/9$	0	2	$= -x_1$
u_2	$2/3$	$-1/3$	1	3	$= -x_2$
u_3	$-1/3$	$1/3$	$-2/3$	-1	$= -x_3$
	$= v_1$	$= v_2$	$= v_3$	$= g$	

As a type-C problem, the solution $x_1 = 2$, $x_2 = 3$, $x_3 = -1$, $y_1 = y_2 = y_3 = 0$ is acceptable since the x's are free. Also, the column equations have no solution with the u's artificial and $g < 0$.

As a type-A problem, the solution given above is not acceptable since the x's must be canonical. The u's are also canonical and, by taking $u_3 > 0$, we can obtain a solution for the column equations that verifies alternative 2 in Theorem 5.40.

Now consider the row equations in tableau 5.32 as a type-B problem. All variables are canonical. We can use the feasibility algorithm and obtain the feasible solution $x_1 = 5/3$, $x_2 = 1$, $x_3 = 0$, $y_1 = 3$, $y_2 = y_3 = 0$.

Theorem 5.40 is perhaps the most interesting and famous among the four versions of Theorem 5.39. It is known as Farkas's lemma. Published by J. Farkas in 1902, it is the earliest known published theorem on linear inequalities of the general type discussed here. It is also a finite version of a theorem known as the *separating hyperplane theorem.*

A linear expression divides the space of variables into three parts: the part where the expression has the value zero, the part where the expression is positive, and the part where the expression is negative. The part where the expression is zero is the hyperplane defined by the expression, and the other two parts are on either side of the hyperplane. Thus, any set of points where the expression is of one sign is entirely on one side of the hyperplane.

To illustrate what the separating hyperplane theorem says, refer to Figure 5.2. A set is convex if whenever it contains two points P and Q it also contains all the points on the line segment connecting P and Q. In the plane a circular disk is convex while an annulus is not. In three dimensions a sphere is convex while a torus is not. The separating hyperplane theorem says that if C is a convex set and P is a point not in C, there is a hyperplane H for which C is on one side and P is on the other side of the hyperplane H. Figure 5.2 illustrates the situation for a circular disk C, a point P and a separating line H. In algebraic terms, this says there is a linear expression which has one sign at all points in the convex set C and the opposite sign at the point P.

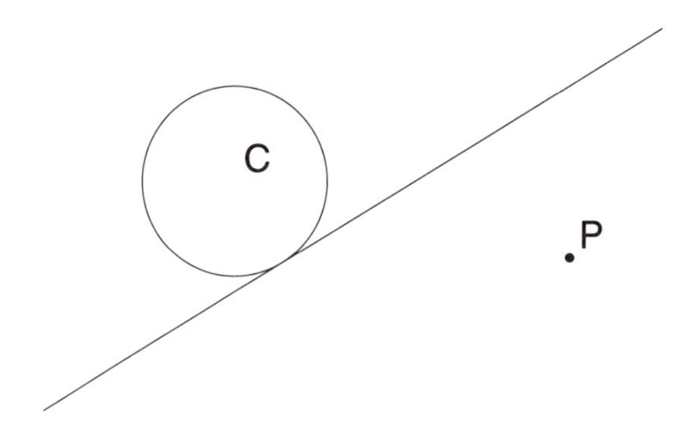

Figure 5.2: Line separating a convex set and a point.

We will not prove the separating hyperplane theorem with quite the generality implied in the previous paragraph and by Figure 5.2. In the plane, a circle and its interior, and a rectangle and its interior, are convex sets. The rectangle has four extreme points —points that are not located between two other points in the set. The circle has an infinite number of extreme points—all the points on the circumference. Similarly, in three dimensions, a cube would be a convex set with eight extreme points. A sphere would have an infinite number of extreme points. We will prove the separating hyperplane theorem in any finite-dimensional space for convex sets with a finite number of extreme points.

Let A_1, A_2, ... , A_n be a set of n points in an m-dimensional coordinate space. That is, each A_j is an m-tuple,

$$A_j = (a_{1j}, a_{2j}, \ldots, a_{mj}) \tag{5.34}$$

A point $b = (b_1, b_2, \ldots, b_m)$ is said to be a convex linear combination of the points $A_1, A_2, \ldots, A_n$ if

$$b_i = a_{i1}x_1 + a_{i2}x_2 + \cdots + a_{in}x_n \tag{5.35}$$

for each i, where the x_j's are nonnegative and

$$x_1 + x_2 + \cdots + x_n = 1 \tag{5.36}$$

Generally, if A_1 and A_2 are two points, then $tA_1 + (1-t)A_2$ represents the points on the line through A_1 and A_2. The points will be on the line segment joining A_1 and A_2 if both t and $(1-t)$ are nonnegative. The conditions that generalize to more than two points are the conditions that t and $(1-t)$ be nonnegative and $t + (1-t) = 1$. That is, the points on the line segment joining A_1 and A_2 are exactly the convex linear combinations of the two points.

In the two-point example discussed in the previous paragraph, the two points are extreme points of the line segment that joins them. If there are more than three points involved the situation is not quite so clear. To see what is involved, imagine a large number of points in a plane. Use a drawing board, and stick pins in the board to represent the position of each of the points. Then stretch a rubber band around the pins so that all pins are enclosed by the rubber band. The set enclosed will be the smallest convex set containing those points. It will be a convex polygon. Algebraically, the points in this set can be represented as convex linear combinations of the given points. The pins at the corners of that polygon will be the extreme points of the convex set. Every point in the convex polygon is between two of the extreme points (on an edge) or between an extreme point and a point on an edge. The smallest convex set containing any given set of points is called the *convex hull* of these points. The situation is suggested in Figure 5.3 for six points in a plane.

In a plane, the convex hull of a finite set of points is a convex polygon with some (not necessarily all) of the given points as the vertices and the remaining points in or on the boundary of the convex polygon. In three dimensions, the convex hull of a finite set of points is a convex polyhedron with some of the given points as the vertices. The algebraic interpretation of this situation is that a point is in the convex hull if it is a convex linear combination 5.35 of the points generating the convex hull. The separating hyperplane theorem says essentially the converse of this interpretation: that if a point cannot be represented as a convex linear combination of the given points then it lies outside this polyhedron. It does this by asserting the existence of a plane (or hyperplane) that separates the point from the convex polyhedron.

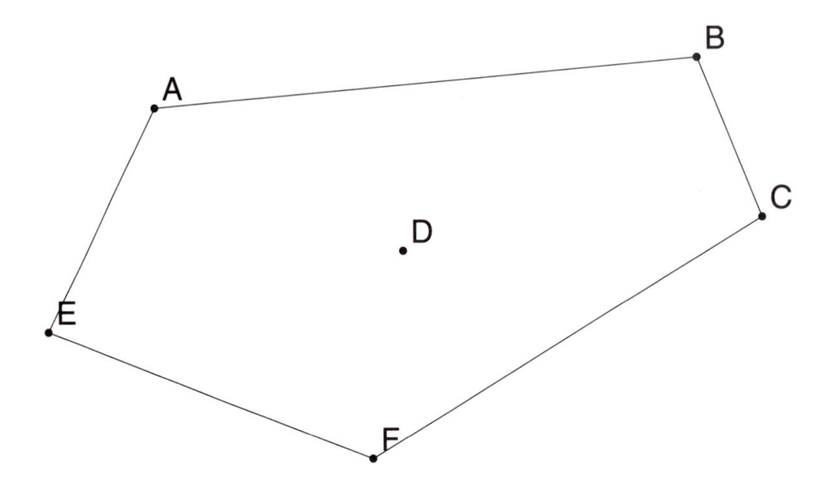

Figure 5.3: A convex set in the plane.

Let us show how the separating hyperplane theorem follows from Theorem 5.40. Consider the tableau 5.37.

(5.37)

	x_1	x_2	$\cdots$	x_n	-1	
v_1	a_{11}	a_{12}	$\cdots$	a_{1n}	b_1	$= -z_1$
v_2	a_{21}	a_{22}	$\cdots$	a_{2n}	b_2	$= -z_2$
$\vdots$	$\vdots$	$\vdots$		$\vdots$	$\vdots$	$\vdots$
v_m	a_{m1}	a_{m2}	$\cdots$	a_{mn}	b_m	$= -z_m$
v	1	1	$\cdots$	1	1	$= -z$
	$= u_1$	$= u_2$	$\cdots$	$= u_n$	$= g$	

Equations 5.35 and 5.36 for nonnegative x_j are equivalent to the row equations of tableau 5.37 with the x's canonical and the z's artificial. The point b is a convex linear combination of the points $A_1, A_2, \ldots, A_n$ if and only if the row equations have a feasible solution. If the row equations do not have a feasible solution, there exist $v_1, v_2, \ldots, v_m, v$ (free) for which all $u_j \geq 0$ and $g \geq 0$. Use these values of the v's to form a linear equation

$$v_1X_1 + v_2X_2 + \cdots + v_mX_m + v = 0 \tag{5.38}$$

The solution set of this linear equation is a hyperplane. Since

$$v_1b_1 + v_2b_2 + \cdots + v_mb_m + v = g < 0 \tag{5.39}$$

and

$$v_1 a_{1j} + v_2 a_{2j} + \cdots + v_m a_{mj} + v = u_j \geq 0 \tag{5.40}$$

for all j, the point b is on one side of this hyperplane and the points A_1, A_2, ... , A_m are on the other side of (or on) the hyperplane.

With a slight but significant change, Theorem 5.42 becomes a classical theorem in linear algebra.

Theorem 5.44 *Given a matrix A and an m-tuple b, one and only one of the following two alternatives holds. Either*

1. *there exists an x (free) for which $Ax = b$, or*
2. *there exists a v (free) for which $vA = 0$ and $vb = 1$.*

The change is in the second alternative. By changing the sign of v, which can be done since v is free, and changing the scale, the inequality can be replaced by an equality. The significance of this change is that Theorem 5.42 holds only in an ordered field, like the field of real numbers, while Theorem 5.44 has meaning for an unordered field, like the field of complex numbers. We must give a different proof of Theorem 5.44.

Proof. Start with an initial tableau in the form of tableau 5.31 with x and v free, and with y and u artificial. Use pivot exchanges to move as many artificial variables as possible to the positions of nonbasic variables, just as we did for the reduction of general to canonical form in Section 5.2. When no further progress in this direction is possible we will have a tableau in the form of 5.41.

(5.41)

	artificial	free	-1	
artificial	$*$	$*$	$*$	$= -$free
free	$*$	0	b	$= -$artificial
	= free	= artificial	$= g$	

The row equations are feasible if and only if the entries in the b-column in the rows of the artificial basic variables are zeros. If one of these entries, say b_k, is nonzero, set the corresponding free variable v_k equal to $1/b_k$ and set all other nonbasic variables of the column equations equal to zero. ☐

We include two theorems that follow directly from the 5.40–5.43 group of theorems.

Theorem 5.45 *Given any matrix A, one and only one of the following two alternatives holds. Either*

1. *there exists an $x \geq 0$ for which $Ax < 0$, or*
2. *there exists a $v \geq 0$, $v \neq 0$, for which $vA \geq 0$.*

Theorem 5.46 *Given any matrix A, one and only one of the following two alternatives holds. Either*

1. *there exists an x (free) for which $Ax < 0$, or*
2. *there exists a $v \geq 0$, $v \neq 0$, for which $vA = 0$.*

In each of these theorems, the "<" in the first alternatives means that every component of Ax is negative.

Proof. Take $b = (-1, -1, \ldots, -1)$ in Theorems 5.41 and 5.43. In alternative 1 of these theorems, $Ax \leq b$ implies $Ax < 0$. In alternative (2) of these theorems, $vb < 0$ implies $v \neq 0$. □

5.5 Complementary Slackness

Consider a pair of dual linear programming problems. We already know that for any pair of optimal solutions and any pair of dually related variables, one variable or the other in the dual pair must be zero. In this context the following questions seem natural. For each pair of dual variables is it possible to find optimal solutions for which one variable or the other is nonzero? Is it possible to find a pair of optimal solutions, one for each of the dual pair of programs, with the property that for every pair of dual variables one variable or its dual is nonzero?

For a pair of canonical dual variables the answer to the first question is affirmative. It is not necessarily true for a free/artificial pair of dual variables. The second question also has an affirmative answer for all pairs of dual canonical variables. It is the purpose of this section to establish these and other properties of complementary feasible solutions.

First, consider tableau 5.42 and its systems of homogeneous row and column equations. We allow any feasibility specifications for the two systems that are dually related.

$$
\begin{array}{c|c|l}
 & x & \\
\hline
v & A & = -y \\
\hline
 & = u &
\end{array}
\qquad (5.42)
$$

Theorem 5.47 *For any pair of feasible solutions, (x, y) for the row equations and (u, v) for the column equations, we have $ux + vy = 0$. For every pair of dually related canonical variables there exists a feasible solution (for one system or the other) in which the variable of the feasible system for that pair is positive.*

Proof. The first assertion is just the duality equation restated to set the context for the second assertion. However, it is worth emphasizing that this equation is just the associative law for matrix multiplication.

$$ux = (vA)x = v(Ax) = v(-y) = -vy \tag{5.43}$$

Consider a pair of dual canonical variables x_k on the top margin and u_k on the bottom margin. If x_k is positive in even one feasible solution for the row equations, $u_k = 0$ in every feasible solution for the column equations. Also, if x_k is positive in some feasible solution, there is a feasible solution for which $x_k = 1$ (since the equations are homogeneous and multiplying the entries in a solution by x_k^{-1} does not disturb feasibility).

Now, suppose the row equations have no feasible solution for which x_k is positive. Then the row equations in tableau 5.44 have no feasible solution.

$$\begin{array}{c|c|l} & 1 & \\ \hline v_1 & a_{1k} & = -y_1 \\ \vdots & \vdots & \quad\vdots \\ v_m & a_{mk} & = -y_m \\ \hline & = u_k & \end{array} \tag{5.44}$$

Tableau 5.44 differs from tableau 5.42 only in that we have set x_k equal to 1. Now change signs in the x_k-column to obtain tableau 5.45.

$$\begin{array}{c|c|l} & -1 & \\ \hline v_1 & -a_{1k} & = -y_1 \\ \vdots & \vdots & \quad\vdots \\ v_m & -a_{mk} & = -y_m \\ \hline & = -u_k & \end{array} \tag{5.45}$$

We now regard the x_k-column as playing the same role as the b-column in tableau 5.31. Since the row equations are infeasible, alternative 1 in Theorem 5.39 cannot hold. Thus, there exists a feasible solution for the column equations for which $-u_k < 0$. That is, $u_k > 0$.

We have shown that if there is not a feasible solution of the row equations with $x_k > 0$ then there is a feasible solution of the column equations with

$u_k > 0$. By taking the negative transpose of tableau 5.42 we obtain a tableau with equivalent systems of equations in which we can show that if the row system does not have a feasible solution with $v_j > 0$ then there is a feasible solution of the column equations with $y_j > 0$. ◻

A similar statement holds for a dual pair of nonpositive dual variables. However, a simple example shows that it does not hold for a free/artificial pair of dual variables. Consider the 1×1 tableau 5.46.

(5.46)

<table>
<tr><td></td><td>x</td><td></td></tr>
<tr><td>v</td><td>1</td><td>$= 0$</td></tr>
<tr><td></td><td>$= 0$</td><td></td></tr>
</table>

The variables x and v are both free, but neither system of equations has a nonzero feasible solution.

Theorem 5.48 *There exists a feasible solution (x, y) for the row system of equations in tableau 5.42 and a feasible solution (u, v) for the column system of equations such that for every pair of dual canonical variables, one variable or the other in the dual pair is positive.*

Proof. The row system of equations and the column system of equations for tableau 5.42 are both homogeneous. Thus, any sum of feasible solutions for either system of equations is also feasible. For each pair of dual canonical variables, determine a feasible solution (for one system or the other) in which the variable from that pair is positive. Sum all the solutions thus obtained for the row system of equations and sum all the solutions thus obtained for the column system of equations. (If either collection is empty, take this sum to be the solution for which all variables are zero.) The pair of solutions thus obtained have the property asserted in the statement of the theorem. ◻

Now consider a dual pair of general linear programs represented by tableau 5.47.

(5.47)

<table>
<tr><td></td><td>x</td><td>-1</td><td></td></tr>
<tr><td>v</td><td>A</td><td>b</td><td>$= -y$</td></tr>
<tr><td>-1</td><td>c</td><td>d</td><td>$= f$</td></tr>
<tr><td></td><td>$= u$</td><td>$= g$</td><td></td></tr>
</table>

We assume for the remainder of this discussion that both programs are feasible. Thus, both programs have optimal solutions. For convenience, let us add a constant to the objective functions (that is, to the d-corner) so that the common optimal value for the objective variables is zero. Then change the signs of the b-column and c-row to obtain tableau 5.48.

(5.48)

	x	1	
v	A	$-b$	$= -y$
1	$-c$	d	$= -f$
	$= u$	$= -g$	

Let us impose the condition $f \geq 0$ as a feasibility specification. Since the maximum value of f in tableau 5.47 is zero, a feasible solution for the row system in tableau 5.48 exists. Furthermore, there is no feasible solution for the row system in 5.48 in which $f > 0$ (since the maximum value of f is 0). Thus, every feasible solution for the row system in 5.48 is optimal for tableau 5.47.

Also, let us impose the condition $-g \geq 0$ as a feasibility specification. Since the minimum value of g in tableau 5.47 is zero, a feasible solution for the column system in tableau 5.48 exists. Since the row system has a feasible solution with the 1 on the top margin, every feasible solution for the column system has $g = 0$. That is, every feasible solution for the column system in tableau 5.48 is optimal for tableau 5.47.

Now make one further alteration in tableau 5.48.

(5.49)

	x	p	
v	A	$-b$	$= -y$
q	$-c$	d	$= -f$
	$= u$	$= -g$	

We require that p and q be nonnegative. Considered as a whole, the row and column systems of equations in tableau 5.49 are homogeneous. Also, q and f satisfy dual canonical feasibility specifications, and p and $-g$ satisfy dual canonical feasibility specifications. If we adopt the feasibility specifications for all other variables as given for tableau 5.47, we can apply Theorem 5.48 to tableau 5.49.

The row system of equations in tableau 5.49 has a feasible solution for which p is positive. Thus, $-g = 0$ for all feasible solutions for the column system. Similarly, the column system of equations in tableau 5.49 has a feasible solution in which q is positive. Thus, $f = 0$ for all feasible solutions for the row system.

Now let (x, y, p, f) and $(u, v, -g, q)$ be the pair of feasible solutions whose existence is asserted in Theorem 5.48. Since $f = 0$, q is positive and, since $-g = 0$, p is positive. Multiply every variable in the solution of the row system by p^{-1} and every variable in the solution of the column system by q^{-1}. In this way we obtain a pair $(\bar{x}, \bar{y}, 1, 0)$ and $(\bar{u}, \bar{v}, 0, 1)$ of feasible solutions for tableau 5.49. These solutions are also feasible for tableau 5.48 and, therefore, optimal for tableau 5.47. We have proved the following theorem.

Theorem 5.49 *(Complementary Slackness Theorem) For any dual pair of feasible linear programs there exists a pair of optimal solutions, one for each program, such that for every pair of dual canonical variables, one and only one variable in each pair is positive.* □

If a linear program is nondegenerate then the basic variables in a basic optimal solution will have positive values. If a linear program and its dual are both nondegenerate the conclusion of Theorem 5.49 is automatically satisfied. In case of degeneracy the complementary slackness theorem is both interesting and informative.

Consider, for example, a production problem of the type discussed in Section 1.1. We are to produce a variety of products which consume supplies that are limited, and we wish to do this with maximum profit. If x_k is the amount of an item we produce, its dual variable u_k is the amount by which the value of the materials in the product exceeds the price of the item. Suppose $x_k = 0$ in an optimal production schedule and $u_k = 0$ in an optimal solution of the dual program. We might want to have as many different items in our catalog as possible. Is it possible to produce this item without decreasing the profit?

The complementary slackness theorem tells us that either there is a production schedule in which x_k is positive, or there is a price schedule in which u_k is positive. In the former case we can produce the item. In the latter case x_k cannot be positive in any optimal solution. That is, the item cannot be produced without reducing the profit.

Exercises

In the exercises of Chapter 1 we set up several linear programming problems. Some of them were canonical programs and we have examined them in detail. Now we can look at those that are general linear programming problems.

Instead of writing down for each variable a note that the variable is canonical, free, or artificial we suggest a notation that will imply the type of variable. Use u and v for canonical variables for the column system and x and y for canonical variables for the row system. Use s for a free variable for the column system and t for a free variable for the row system. Use z or 0 for an artificial variable for either system.

The most common source of equality constraints involve a balance of input and output in a commercial enterprise. The second most common involve percentages, which are represented in the following exercises. All named problems are given in Chapter 1.

1. Set up the Maintenance Manager's Problem as a general linear programming problem.

2. Set up the Gardener's Problem as a general linear programming problem.

3. Set up the Metallurgist's Problem as a general linear programming problem.

4. Consider a general linear programming problem given in the following form.

Maximize

$$f = 50x + 60y - 10$$

subject to the following conditions:

$$\begin{aligned} 50x + 25y &\leq 200 \\ -6x + 5y &= 30 \\ 3x + 10y &\geq 10 \end{aligned}$$

where x is canonical and y is free. Formulate the dual program with a description in similar terms.

5. The Metallurgist's Problem described in Chapter 1 is the simpler of two similar problems posed by George Dantzig in *Linear Programming and Extensions*. The more complex problem imposes conditions on the percentages of lead and zinc as well as the percentage of tin. The pertinent data are contained in the following table. The metallurgist wants to produce an alloy that is 30% lead, 30% zinc, and 40% tin. How can he blend an alloy that will produce what he needs at least cost?

alloy	A	B	C	D	E	F	G	H	I	desiredblend
% lead	10	10	40	60	30	30	30	50	20	30
% zinc	10	30	50	30	30	40	20	40	30	30
% tin	80	60	10	10	40	30	50	10	50	40
cost/lb	4.10	4.30	5.80	6.00	7.60	7.50	7.30	6.90	7.30	min

6. Use the two-phase simplex method to solve the Maintenance Manager's Problem and its dual.

7. Use the two-phase simplex method to solve the Gardener's Problem and its dual.

8. Use the two-phase simplex method to solve the simpler Metallurgist's Problem and its dual.

9. Use the two-phase simplex method to solve the more complex Metallurgist's Problem and its dual.

10. Solve the following system of linear inequalities, or show that a solution does not exist. The x's are canonical and the z's are artificial.

x_1	x_2	x_3	-1	
2	−5	3	−1	$= -z_1$
1	−1	1	1	$= -z_2$
−4	1	−3	−7	$= -z_3$

11. Solve the following system of linear inequalities, or show that a solution does not exist. The x's are canonical and the z's are artificial.

x_1	x_2	x_3	-1	
3	4	6	12	$= -z_1$
−3	2	3	−3	$= -z_2$
−3	−1	−3	−6	$= -z_3$

12. Solve the following system of linear inequalities, or show that a solution does not exist. All variables are canonical.

x_1	x_2	x_3	-1	
3	4	6	12	$= -y_1$
−3	2	3	−3	$= -y_2$
−3	−1	−3	−6	$= -y_3$

13. Solve the following system of linear inequalities, or show that a solution does not exist. The t's are free and the z's are artificial.

t_1	t_2	t_3	-1	
3	4	6	12	$= -z_1$
−3	2	3	−3	$= -z_2$
−3	−1	−3	−6	$= -z_3$

14. Solve the following system of linear inequalities, or show that a solution does not exist. All variables are canonical.

x_1	x_2	x_3	-1	
−2	−2	1	1	$= -y_1$
2	2	3	0	$= -y_2$
−7	−6	−5	−1	$= -y_3$

15. Solve the following system of linear inequalities, or show that a solution does not exist. The t's are free and the z's are artificial.

t_1	t_2	t_3	-1	
−2	−2	1	1	$= -z_1$
2	−2	3	0	$= -z_2$
−7	−6	−5	−1	$= -z_3$

16. Solve the following system of linear inequalities, or show that a solution does not exist. All variables are canonical.

x_1	x_2	x_3	-1	
−3	2	1	−3	$= -y_1$
2	−2	3	−2	$= -y_2$
−7	−6	−5	2	$= -y_3$

17. Solve the following system of linear inequalities, or show that a solution does not exist. The t's are free and the z's are artificial.

t_1	t_2	t_3	-1	
−3	2	1	−3	$= -z_1$
2	−2	3	−2	$= -z_2$
−7	−6	−5	2	$= -z_3$

18. Solve the following system of linear inequalities, or show that a solution does not exist. All variables are canonical.

x_1	x_2	x_3	-1	
−3	−2	1	−5	$= -y_1$
2	2	3	3	$= -y_2$
−7	−6	−5	−10	$= -y_3$

19. Solve the following system of linear inequalities, or show that a solution does not exist. The t's are free and the z's are artificial.

t_1	t_2	t_3	-1	
-3	-2	1	-5	$= -z_1$
2	2	3	3	$= -z_2$
-7	-6	-5	-10	$= -z_3$

20. Solve the following system of linear inequalities, or show that a solution does not exist. The t's are free and the y's are canonical.

t_1	t_2	t_3	-1	
-3	-2	1	-5	$= -y_1$
2	2	3	3	$= -y_2$
-7	-6	-5	-10	$= -y_3$

21. Solve the following system of linear inequalities, or show that a solution does not exist. The t's are free and the y's are canonical.

t_1	t_2	t_3	-1	
-3	2	1	-4	$= -y_1$
2	2	3	3	$= -y_2$
-7	2	-7	-3	$= -y_3$

22. The Advertiser's Problem has a degenerate optimal solution and for every pair of dual basic optimal solutions there is at least one pair of dual variables for which both variables are zero. Find a pair of complementary optimal solutions for which one variable in every pair of dual variables is nonzero.

23. Use the two-phase simplex method to find an optimal solution for the following program, if an optimal solution exists. All variables are canonical, except for t_4 and t_6, which are free, and z_6, which is artificial.

	x_1	x_2	x_3	t_4	x_5	t_6	-1	
v_1	-2	0	0	-1	1	1	-2	$= -y_1$
v_2	-2	-1	-1	-1	-2	1	-1	$= -y_2$
v_3	1	-1	0	1	-2	0	-2	$= -y_3$
v_4	-2	-2	-1	0	0	-2	0	$= -y_4$
v_5	-1	-2	-1	0	-2	0	-2	$= -y_5$
s_6	-1	-1	0	-1	0	1	-2	$= -z_6$
-1	-2	-1	-2	0	-1	-1	-2	$= f$
	$= u_1$	$= u_2$	$= u_3$	$= z_4$	$= u_5$	$= z_6$	$= g$	

24. Use the two-phase simplex method to find an optimal solution for the following program, if an optimal solution exists. All variables are canonical, except for t_6, which is free, and z_5 and z_6, which are artificial.

	x_1	x_2	x_3	x_4	x_5	t_6	-1	
v_1	-2	0	0	-1	1	1	-2	$= -y_1$
v_2	-2	-1	-1	-1	-2	1	-1	$= -y_2$
v_3	1	-1	0	1	-2	0	-2	$= -y_3$
v_4	-2	-2	-1	0	0	-2	0	$= -y_4$
s_5	-1	-2	-1	0	-2	0	-2	$= -z_5$
s_6	-1	-1	0	-1	0	1	-2	$= -z_6$
-1	-2	-1	-2	0	-1	-1	-2	$= f$
	$= u_1$	$= u_2$	$= u_3$	$= u_4$	$= u_5$	$= z_6$	$= g$	

Questions

Q1. Consider a general linear max program for which an optimal solution exists. If a variable that was canonical is changed so that it is free, the optimal value of the objective variable might decrease.

Q2. Consider a general linear max program for which an optimal solution exists. If a variable that was canonical is changed so that it is free, the optimal value of the objective variable might increase.

Q3. Consider a general linear max program for which an optimal solution exists. If a variable that was canonical is changed so that it is free, the optimal value of the objective variable will necessarily increase.

Q4. Consider a general linear max program for which an optimal solution exists. If a variable that was canonical is changed so that it is free, the optimal value of the objective variable might remain unchanged.

Q5. Consider a general linear max program for which an optimal solution exists. If a variable that was canonical is changed so that it is artificial, the optimal value of the objective variable might decrease.

Q6. Consider a general linear max program for which an optimal solution exists. If a variable that was canonical is changed so that it is artificial, the optimal value of the objective variable might increase.

Q7. Consider a general linear max program for which an optimal solution exists. If a variable that was canonical is changed so that it is artificial, the optimal value of the objective variable will necessarily change.

Q8. Consider a general linear max program for which an optimal solution exists. If a variable that was canonical is changed so that it is artificial, the optimal value of the objective variable might remain unchanged.

Q9. A general linear program can fail to be feasible for either of two reasons. It may be impossible to make all artificial variables zero. It may be impossible to make all canonical variables nonnegative. In either case, if the dual program is feasible, its objective variable is unbounded.

Q10. If a general linear max program is feasible and the objective variable is bounded above for all feasible solutions, the program has an optimal solution.

Q11. If there exists a pair of complementary feasible solutions for a general linear program for which a pair of dual canonical variables are both zero, at least one of the two programs has multiple optimal solutions.

Q12. It is possible to have a dual pair of free and artificial variables zero in a pair of complementary feasible solutions and have both optimal solutions unique.

Q13. We have shown for canonical programs that when there are dual complementary optimal solutions that are both degenerate, both solutions can be multiple or one can be unique and the other multiple. It is never possible to have both optimal solutions unique for a canonical program when both are degenerate.

Chapter 6

Numerical Considerations

6.1 Numerical Considerations

We have been concerned with describing algorithms and proving that they work without worrying about how well they work. When it comes to finding real answers to real problems a number of practical considerations become important. How much work is required to get an answer and how accurate is it? The first question can be divided into two questions. How many iterations of an algorithm can we expect to have to do—on the average, or at worst? How much work is required in each iteration? The second question can also be divided into two questions. How does the accuracy of a result depend on the accuracy of the initial data and how does the accuracy depend on the computational steps taken?

To a large extent these questions take us into areas of study that are beyond the scope of this book, but we will try to give you some idea of what these problems are like and what is known about dealing with them. There is now a large amount of practical experience with the problems involved in doing linear computations on a computing machine, and to write successful programs one must be aware of these problems.

First, let us look at the amount of work expected using any algorithm that we might use to solve a linear programming problem. Generally, the amount of work required to solve a problem increases with the size of the problem. For a linear program, the size is measured by the number of rows and columns in the initial tableau. The size of the problem affects the amount of computation required and the amount of memory required to store the data. The amount of memory required is also affected by the amount of memory occupied by each number.

If we use the pivot exchange described in Section 2.2 in each iteration of the simplex algorithm, each pivot exchange will require $(m+1)(n+1)$ multiplications or divisions and mn additions or subtractions. Since there are approximately as many additive operations (addition or subtraction) as there are multiplicative operations (multiplication or division), we will count only the multiplicative operations as a relative measure of the amount of work required to perform the operation.

Furthermore, since we are mainly interested in how the work increases as the size of the problem increases, we will ignore all but the highest-degree terms in estimating the amount of work required. Thus, we will say that each pivot exchange requires approximately mn multiplicative operations. We are more interested in how the required work grows with the size of the problem, so we focus out attention on the term or factor that contributes most to the growth. For example, if a bound for the amount of work is $2n^2$ or $9n^2$, we will in either case refer to the n^2 factor as the growth factor. We say that these problems are "n^2 problems."

We consider reasonable problems to beproblems, and methods of solving them, for which the growth factor is a polynomial function of the size. Problems and methods for which the growth factor is exponential or worse we consider to be difficult problems.

We will look more closely at the question of the amount of work required in each iteration of the simplex algorithm when we discuss the revised simplex method later in this chapter. For the moment let us look at the number of iterations required in the simplex algorithm.

In 1972 Victor Klee and G. J. Minty published an example of a family of $n \times n$ linear programs for which the simplex algorithm required $2^n - 1$ iterations. We will give a modified version of their examples. Consider the tableau

$$
\begin{array}{c|cccc|c|l}
 & x_1 & x_2 & x_3 & x_4 & -1 & \\
\hline
v_1 & 1 & 0 & 0 & 0 & 1 & = -y_1 \\
v_2 & 4 & 1 & 0 & 0 & 8 & = -y_2 \\
v_3 & 8 & 4 & 1 & 0 & 64 & = -y_3 \\
v_4 & 16 & 8 & 4 & 1 & 512 & = -y_4 \\
\hline
-1 & 8 & 4 & 2 & 1 & 0 & = f \\
\hline
 & = u_1 & = u_2 & = u_3 & = u_4 & = g &
\end{array}
\tag{6.1}
$$

The rule for selecting the pivot column is: Select the first column with a positive entry in the c-row. This rule for selecting the pivot row is the usual simplex pivot rule. The problem is nondegenerate and the pivot row is always unique. With these rules the example requires 15 pivot exchanges. In fact, the final tableau could have been obtained in a single simplex pivot exchange if we had had the foresight to pivot in the last column at the start. Of the total number of equivalent tableaux, 16 are feasible and the selection rules trace a path through every one of them.

The Klee–Minty example of order 5 is

(6.2)

	x_1	x_2	x_3	x_4	x_5	-1	
v_1	1	0	0	0	0	1	$= -y_1$
v_2	4	1	0	0	0	8	$= -y_2$
v_3	8	4	1	0	0	64	$= -y_3$
v_4	16	8	4	1	0	512	$= -y_4$
v_5	32	16	8	4	1	4096	$= -y_5$
-1	16	8	4	2	1	0	$= f$
	$= u_1$	$= u_2$	$= u_3$	$= u_4$	$= u_5$	$= g$	

One should be able to see how the pattern generalizes to larger tableaux. The entries in the b-column increase by a factor of 8 as we go down the b-column. In the A-matrix and the c-row, the entries (except for those in the main diagonal) increase by a factor of 2 going from right to left. The $n \times n$ generalization of the problem has 2^n equivalent feasible tableaux and requires $2^n - 1$ iterations. That is, the amount of work required to solve this family of problems is exponential.

It is reasonable to ask if there might be a selection rule better than the one given. It seems that for every selection rule, if it is unequivocal, it is possible to provide an unfavorable example.

The publication of the Klee–Minty example left us in the following position. The example shows that there are problems for which the computational complexity of the simplex algorithm grows exponentially, but is this situation merely a property of the algorithm or is it a property of the linear programming problem itself? That is, will every algorithm for the linear programming problem also require exponential time or is there a different algorithm that can be used for which the growth of complexity is slower?

In 1979 L. G. Khachiyan published a paper that described a new algorithm for linear programming quite different from the simplex algorithm. This event was honored with a front page article in the *New York Times* declaring that the Khachiyan algorithm was a significant breakthrough. We will describe the algorithm later in this chapter. The algorithm does provide a method for solving linear programming problems for which the number of iterations required grows at a polynomial rate. This did answer one question. The exponential performance of the simplex algorithm is a property of the algorithm and not a property of the problem.

The amount of work required in each iteration of the Khachiyan algorithm and the number of iterations required for even small problems were very large and the algorithm never became useful.

In 1984 N. Karmarkar published another algorithm for which the complexity grows like a polynomial in the size of the problem. This event was also honored by front page treatment by the *New York Times*. The announcement of the Karmarkar algorithm was accompanied by much controversy. Important details about the algorithm were not divulged because of the potential monetary gain that might be available. The algorithm was regarded as proprietary. Enough is now known to say that the Karmarkar algorithm is practical. Another side of the story, however, is that the performance of the simplex algorithm is not nearly as bad as the Klee–Minty example would suggest. First, the Klee–Minty example is a contrived example, not at all the sort of problem that one might meet in a real applied situation. Furthermore, applied problems are not entirely random. If one constructed a problem with entries chosen at random in some way, very few of the entries would be zero.

In practice, in most applied problems almost all entries are zeros. There are several reasons for this. For one thing, a 1000×1000 problem would have one million entries. It would be a lot of work to provide all those numbers as initial data if each was meaningful. Usually, each datum is the contribution of some resource to the production of some product. Most products use only a few of the resources and many others are lumped together for convenience. Also, many equations have only zeros and ones as coefficients. These are equations that balance input and output in subsystems. Real applied problems are not random problems and experience seems to suggest that randomly generated problems behave quite differently from applied problems when it comes to running time required to solve them.

Estimating how much time it would take to solve an applied problem is not a question that yields to a theoretical analysis. It is primarily a matter of experience. This is a subject in which there is much folklore, and the folklore says that for most applied problems, over the range of problem sizes for which there is experience, the number of iterations grows linearly. Furthermore, the times required for phase I and phase II are approximately equal.

In summary: The linear programming problem is a polynomial time problem. The simplex algorithm is, in a worst case situation, an exponential time method. Two polynomial time algorithms have received prominent attention. The Karmarkar algorithm has been incorporated into a computer program marketed by AT&T. It also has inspired research into algorithms of a similar type, which are known generically as interior point algorithms. In addition, this activity has inspired closer examination of the simplex algorithm. There are now improved implementations of the simplex algorithm. Claims from those supporting the two approaches (the Karmarkar algorithm and the improvements of the simplex algorithm) are close to even, but comparisons are difficult since there has been little testing of the leading contenders on the same sets of problems. Reports of the performance of some newer interior point methods are even better.

6.2 The Revised Simplex Method

In the previous section we discussed complexity in terms of the number of iterations in a run of an algorithm. Here we wish to consider the amount of work required in each iteration. We mentioned in the previous section that for most applied problems the number of nonzero entries in the initial tableau was a small percentage. It is typically in the order of magnitude of 2 or 3 %. Of course, this significantly reduces the amount of calculation that needs to be done since the zeros are null operations in both addition and multiplication. However, after a few iterations, the tableaux begin to fill with nonzero terms and before long few entries in a tableau are zeros.

Still, this raises the question as to whether it is possible to take advantage of the large number of zeros in the initial tableau. The revised simplex method is designed to do just that. It is clear that to do this it is necessary to design the computation so that it uses the data from the initial tableau and as little as possible from tableaux that are generated later. The central idea is that to make a decision as to the pivot entry we need only know the entries in the c-row (to choose the pivot column) and the entries in the pivot column and b-column (to choose the pivot row).

Consider an expanded tableau in the form of 6.3 as an initial tableau.

$$
\begin{array}{c|ccc|cccc|c|c}
 & x_1 & \cdots & x_n & y_1 & y_2 & \cdots & y_m & -1 & \\
\hline
v_1 & a_{11} & \cdots & a_{1n} & 1 & 0 & \cdots & 0 & b_1 & = 0 \\
v_2 & a_{21} & \cdots & a_{2n} & 0 & 1 & \cdots & 0 & b_2 & = 0 \\
\vdots & \vdots & & \vdots & \vdots & \vdots & & \vdots & \vdots & \vdots \\
v_m & a_{m1} & \cdots & a_{mn} & 0 & 0 & \cdots & 1 & b_m & = 0 \\
\hline
-1 & c_1 & \cdots & c_n & 0 & 0 & \cdots & 0 & 0 & = f \\
\hline
 & = u_1 & \cdots & = u_n & = v_1 & = v_2 & \cdots & = v_m & = g &
\end{array}
\tag{6.3}
$$

It is simpler for the discussion that follows to represent the blocks that are shown in tableau 6.3 by single letters in matrix notation. Thus, tableau 6.3 can be represented in the form

(6.4)

	x	y	-1	
v	A	I	b	$= 0$
-1	c	0	d	$= f$
	$= u$	$= v$	$= g$	

Suppose, after several pivot exchanges, we obtain a tableau of the form

(6.5)

	x	y	-1	
w	A'	E	b'	$= 0$
-1	c'	D	d'	$= f$
	$= u$	$= v$	$= g$	

Since the rows in tableau 6.5 are linear combinations of the rows in tableau 6.4, the block matrices in tableau 6.5 can be obtained by multiplying the corresponding matrices in 6.4 by an appropriate matrix factor on the left. Since $E = EI$, we also have

$$A' = EA \tag{6.6}$$

$$b' = Eb \tag{6.7}$$

Also, from the column equations generated by tableaux 6.4 and 6.5 we can obtain

$$\begin{aligned} c' &= wA' - u = wEA - vA + c = (v + D)A - vA + c \\ &= DA + c \end{aligned} \tag{6.8}$$

Similarly, we get

$$d' = wb' - g = wEb - vb + d = Db + d \tag{6.9}$$

The point of obtaining the relations 6.6 – 6.9 is that it is sufficient to know the entries in the initial tableau 6.4 and just a few entries in tableau 6.5, namely E and D. Tableau 6.5 can then be computed from the relations 6.6 – 6.9.

Actually, it is sufficient to know just the E-matrix in tableau 6.5. Let u_0, v_0 be the basic solution obtained from tableau 6.4. That is, $u_0 = -c$ and $v_0 = 0$. The variables represented by w on the left margin of tableau 6.5 consist of m of the variables in u and v. Let w_0 denote the values of these variables in the basic solution for tableau 6.4. Since v consists of the nonbasic variables in tableau 6.4 and the variables on the bottom margin do not change in an expanded tableau with a pivot exchange, from tableau 6.5 we see that

$$D = w_0 E \tag{6.10}$$

For convenience and efficiency we are not going to use all these relations. For example, we do not need to know d until we get a tableau for which the basic solution is optimal.

The revised simplex algorithm differs from the simplex algorithm primarily in the way in which the pivot exchange is carried out. The rules for selecting the pivot column and pivot are unchanged.

The Revised Simplex Algorithm

We assume we have an initial expanded tableau in the form of tableau 6.3 with a nonnegative b-column.

1. At the start of each iteration of the revised simplex algorithm we assume the basic variables (and, therefore, the basic columns) are identified, and that the E-matrix, b-column, and c-row are known. Examine the D-row (which is just the part of the c-row under the E-matrix). If some $d_j > 0$, select as pivot column the column with the maximum d_s.

   If all d_j are nonpositive, go to step 5.

2. In the pivot column, if all $e_{is} \leq 0$, STOP. The column program is infeasible and the row program has no optimal solution.

3. For each $e_{is} > 0$, compute b'_i/e_{is}. Select r so that b'_r/e_{rs} is minimal among these ratios.

4. Pivot on e_{rs}. Compute only the new E-matrix, b-column, and D-row. Go to step 1.

5. Compute the c-row from equation 6.8. That is,

$$c'_j = d_1 a_{1j} + \cdots + d_m a_{mj} + c_j \tag{6.11}$$

   This computation need be done only for the nonbasic columns. The basement entry in a basic column is necessarily zero.

6. If all c_j are nonpositive, STOP. The basic solution for this tableau is optimal.

7. If some $c'_j > 0$, select as pivot column the column with the maximum c'_j.

8. Compute the other entries in the pivot column with equation 6.6. That is,

$$a'_{is} = e_{i1}a_{1j} + \cdots + e_{im}a_{mj} \tag{6.12}$$

9. If all a'_{is} are nonpositive, STOP. No optimal solution exists.

10. For all $a_{is} > 0$, compute b'_i/a'_{is}. Select r so that b'_r/a'_{rs} is minimal among these ratios.

11. Pivot on a'_{rs}. Compute only the new E-matrix, b-column, and D-row. Go to step 1.

To simplify programming the revised simplex algorithm we can bypass steps 1 through 4. Compute the c-row directly and select the maximum positive entry in the combined c, D-row. However, computing the c-row and the pivot column, if the pivot column is not within the E-matrix, significantly increases the number of computational steps in an iteration. The only way selecting a pivot column from the A-matrix when one is available from the E-matrix would be advantageous is if it can be shown that this selection method decreases the number of iterations.

In the initial tableau 6.3 there are n nonbasic variables and m basic variables. At each iteration, the variables w on the left margin identify the nonbasic variables of the minimum program and, therefore, the basic variables of the maximum program. Usually, when using an expanded tableau the variables of the minimum program are not shown. In that case we might write the variables of the maximum program on the left margin to identify the basic variables. These ideas can be adequately illustrated with examples.

Consider an expanded tableau of the type that we worked through in Section 2.8. For convenience we repeat that tableau here as tableau 6.13. We will omit the variables of the minimum program and write the basic variables of the maximum program on the left margin. The blanks are anticipated zeros.

(6.13)

	x_1	x_2	x_3	x_4	x_5	x_6	x_7	-1
x_4	30	10	50	1				1500
x_5	5*	0	3		1			200
x_6	0.2	0.1	0.5			1		12
x_7	0.1	0.2	0.3				1	9
	10	5	5.5					0

We select the column with the maximum positive entry in the c-row (since the D-row is nonpositive). This selects the x_1 column. Then the x_5-row is selected as the pivot row, as indicated by the asterisk. We perform the pivot exchange, but we compute only the entries in the E-matrix, the D-row, and the b-column. The d-corner is optional since it does not enter into the decisions. We obtain tableau 6.14.

(6.14)

	x_1	x_2	x_3	x_4	x_5	x_6	x_7	-1
x_4				1	-6			300
x_1	1				0.2			40
x_6					-0.04	1		4
x_7					-0.02		1	5
					-2			-400

Notice that the basic columns, the columns of the basic variables, are identified by the isolated 1's. We now compute the remaining entries in the c-row (since the D-row is nonpositive). The c-entry for the x_1-column does not have to be computed since x_1 is now a basic variable and the c-entry for a basic column will be zero. However, it is a good idea to compute that entry. Just as with the Tucker tableau, at any iteration of the simplex algorithm we will have inverted a submatrix of the initial A-matrix. This inverse appears in the revised simplex algorithm as a submatrix of the E-matrix. If that submatrix is accurate, we will get zeros in the c-row of the basic variables and an identity matrix with the columns rearranged. The extent to which those entries differ from what is expected is a measure of the inaccuracy of the E-matrix.

We use formula 6.11 to compute the c-entries. A convenient way to do this when computing by hand is to write the D-row in a column on the right edge of a paper in the form

(6.15)

0
-2
0
0
1

Place this column adjacent to each column of 6.13 successively and accumulate the sum of the products as the c-entry. We then obtain

(6.16)

	x_1	x_2	x_3	x_4	x_5	x_6	x_7	-1
x_4				1	-6			300
x_1	1				0.2			40
x_6					-0.04	1		4
x_7					-0.02		1	5
		5	-0.5		-2			-400

We select the x_2-column as the pivot column and use formula 6.12 to compute the remaining entries in that column. A convenient way to do this when computing by hand is to write the x_2-column of the initial tableau 6.13 on the top edge of a paper in the form

(6.17)

10	0	0.1	0.2

Place this row adjacent to each row of the E-matrix in tableau 6.16 successively and accumulate the sum of the products as the corresponding entry in the new x_2-column. We obtain

(6.18)

	x_1	x_2	x_3	x_4	x_5	x_6	x_7	-1
x_4		10		1	-6			300
x_1	1	0			0.2			40
x_6		0.1			-0.04	1		4
x_7		0.2*			-0.02		1	5
		5	-0.5		-2			-400

We select the pivot entry, as indicated with the asterisk, and obtain the new E-matrix, D-row, and b-column in tableau 6.19.

(6.19)

	x_1	x_2	x_3	x_4	x_5	x_6	x_7	-1
x_4				1	-5		-50	50
x_1	1				0.2		0	40
x_6					-0.03	1	-0.5	1.5
x_2		1			-0.01		5	25
			-6.5		-1.5		-25	-525

The basic variables are indicated on the left margin. For the revised simplex algorithm, we do not want to compute entries in the tableau unless forced to. The blanks in the columns of basic variables are anticipated zeros. Thus, we see that the c, D-row is nonpositive, and that tells us that an optimal solution has been obtained.

The algorithm we have described is known as the *explicit inverse form of the revised simplex algorithm*. The "explicit inverse" in the title is the matrix E in tableau 6.5.

We started this chapter with remarks that we were interested in computational efficiency. How many multiplicative operations are required for one iteration of the revised simplex algorithm, and how does it compare with a full pivot exchange? Let us assume we start an iteration with a tableau in the form of tableau 6.5. In the initial tableau 6.3, the m y's are basic variables and the n x's are nonbasic variables. Suppose at the step under consideration k of the y's have become nonbasic and, therefore, k of the x's have become basic variables. Then, as illustrated in tableau 6.19 for $k = 2$, the D-row has at most k nonzero entries. Furthermore, each row of the E-matrix has at most k nonzero entries, or at most k nonzero entries plus a unit.

Using formula 6.11 to compute each c'_j requires at most k multiplications, and using formula 6.12 to compute each a'_{is} requires at most k multiplications. Only $n - k$ of the c'_j need be computed. The number of multiplications and divisions required to compute the c-row, the pivot column, and the test ratios is

$$k(m + n - k) + m \tag{6.20}$$

After the pivot entry is selected we must update the E-matrix, the D-row, and the b-column. We might just as well update the d-corner. The number of multiplications required is either $(m + 1)k$, $(m + 1)(k + 1)$, or $(m + 1)(k + 2)$, depending on whether the number of basic columns in the E-matrix is decreased, stays the same, or is increased. Notice that if the D-row and b-column are computed from the data of the previous tableau instead of using formulas 6.10 and 6.7 each entry requires only one multiplication, which is better than using those formulas.

For the purpose of making comparisons it is desirable to make a few simplifying assumptions that will not destroy the spirit of the comparisons. Let us assume the E-matrix is "full." That is, all the columns of E are nonbasic columns. We are simply interested in evaluating the computation required in this "worst case" situation. Then $k = m$. Furthermore, we will not count the test ratios as these must be computed in any case. The number of multiplications required to update is then $(m + 1)^2$ and the number of multiplications required to compute the c-row and the pivot column is mn. Since n is usually significantly larger than m in most applications, if the pivot column is not in the E-matrix more than half the work is required to compute the c-row and the pivot column.

The revised simplex algorithm ordinarily calls for computing the c-row and choosing the largest entry in the combined c, D-row for the pivot column. The rationale for this choice is that it tends to reduce the number of iterations. On the other hand it is desirable to avoid computing the c-row, and this can be done by choosing the pivot column from the E-matrix if such a choice is available.

The number of multiplications required in one iteration of the regular simplex algorithm is

$$(n+1)(m+1) \tag{6.21}$$

and one iteration of the revised simplex algorithm takes

$$mn+(m+1)^2 \tag{6.22}$$

multiplications. An iteration of the revised simplex algorithm will take fewer multiplications if

$$n > m^2+m \tag{6.23}$$

This comparison does not make the revised simplex algorithm look like an efficient algorithm. However, in many problems involving a large number of equations and variables, most of the entries in the initial tableau are zeros. For example, a large manufacturing industry might make many products and use many supplies, but each product will use only a few of the different resources available. Some actual uses of resources are so small they are ignored or lumped into large categories like "overhead." Even if an initial tableau has a large number of zeros, after several pivot exchanges the tableaux tend to fill out and before long almost all entries are nonzero.

Often the large number of zeros in a tableau can be used to reduce the amount of computation that is required. The advantage of the revised simplex algorithm is that formulas 6.11 and 6.12 use entries from the initial tableau. If t is the proportion of nonzero entries in the initial tableau, then

$$tmn+(m+1)^2 \tag{6.24}$$

is an estimate of the number of multiplications required for each iteration of the revised simplex algorithm. In this case, the revised simplex algorithm will take fewer multiplications for each iteration if t satisfies the condition

$$t < \frac{n-m}{n} + \frac{n-m}{mn} \tag{6.25}$$

For $n = 2m$, the initial tableau should have less than 50% nonzero entries. Problems with less than 1% nonzero entries are common.

The form of the computation offers a possible way of formatting the data that might or might not be advantageous, depending on the particular computing equipment available. The fastest type of memory available in a computer is random access memory, or RAM. However, even more important than its speed is the fact that all data in RAM are available with equal speed. If a problem is so large that all that data cannot be stored in memory at the same time, part of the data must be stored externally, either on disks or on tape. Using such devices, the time it takes to access the data is more than is required to access data in RAM, and it will be further increased if successive items of data that are read or written are physically separated on the storage media.

Tape is essentially sequential. That is, the information in the middle of the tape cannot be read without reading over all the information preceding it. Disks are partially sequential in that the read/write head can be moved directly to a track in the middle of the disk. But still, a single item of data cannot be read from the disk without reading everything included with it in a sector of the disk.

With a tableau, one can arrange to have adjacent entries in a row stored in adjacent positions on the disk, or adjacent entries in a column stored in adjacent positions. But we cannot have it both ways. While improving computer technology is reducing this problem, for large problems the way data are stored is still a factor that must be considered.

We recommend that we store the initial tableau column-by-column and store the E-matrix row-by-row. The computation in formulas 6.11 and 6.12 will then use entries that appear in sequence in memory.

There is yet another form of the revised simplex algorithm that differs primarily in the way information about E is stored, and this offers advantages if E must be stored externally. With the large RAM memories available in today's computers, this is a vanishing problem. We will omit a discussion of the details of this form of the revised simplex method. This form of the revised simplex algorithm is known as the *product form of the revised simplex algorithm.*

6.3 Gaussian Elimination

If one must redo a sequence of pivot exchanges, as described in the previous section, it is desirable to reduce the amount of computation required and to choose pivot exchanges that preserve the numerical accuracy of the results. To achieve this we do not choose the pivot entries to preserve the nonnegativity of the b-column. Instead, we recommend that we use Gaussian elimination.

K. F. Gauss, while working on applied problems in astronomy and geodesy, devised a systematic method early in his great career for solving systems of linear equations that arose from the theory of "least squares." He devised the theory of least squares to allow him to handle data that were being accumulated during a survey of Germany. We will describe least squares in detail in Chapter 12. Later, the geodesist, Wilhelm Jordan, devised another method for solving linear equations. The first method is now know as *Gaussian elimination* and the second is known as *Gauss–Jordan elimination*.

For years, courses in linear algebra treated the two methods as roughly equivalent. Indeed, for problems of the size encountered in textbooks, the two methods take just about the same amount of work. In fact, Gauss–Jordan requires a little less copying. When computers allowed one to solve really large systems of linear equations, all methods for solving linear equations were examined with great care because there was a need for solving linear equations and large systems took distressingly long times to solve.

For a system of n linear equations in n variables that has a unique solution, Gaussian elimination takes approximately $n^3/3$ multiplications and Gauss–Jordan takes approximately $n^3/2$ multiplications. In both cases a careful count of the steps will provide terms of lower degree than three, but the cube terms are the dominant terms. The actual counts are $(n^3 + 3n^2 - n)/3$ and $(n^3 + n)/2$ in the two cases. For large systems of linear equations, Gauss–Jordan elimination requires about 50% more work than does Gaussian elimination.

A great deal of time and effort has gone into devising significantly more efficient methods for a variety of special kinds of systems of linear equations, but for systems of linear equations with no unusual patterns in the coefficients, Gaussian elimination remains as good as any. The description we shall give for Gaussian elimination differs from that given in most texts, but the arithmetic steps that are required are identical.

Consider a system of linear equations in the form

$$Ax = b \tag{6.26}$$

where A is an $m \times n$ matrix of coefficients, x is a one-column vector with n entries, and b is a one-column vector with m entries.

Choose any nonzero entry in the matrix A as the pivot entry. For convenience with the notation we take a_{11} as the pivot entry. In the example that follows we shall make another choice to illustrate other possibilities. Form a one-column matrix C_1 from the column containing the pivot entry. We get

$$C_1 = \begin{bmatrix} a_{11} \\ a_{21} \\ \vdots \\ a_{m1} \end{bmatrix} \tag{6.27}$$

Also, form a one-row matrix R_1 obtained by dividing the row containing the pivot entry by the pivot entry.

$$R_1 = \begin{bmatrix} 1 & a_{12}/a_{11} & \cdots & a_{1n}/a_{11} \end{bmatrix} \tag{6.28}$$

The product $C_1 R_1$ is an $m \times n$ matrix with the same pivot row and pivot column as the matrix A. Therefore,

$$A - C_1 R_1 \tag{6.29}$$

has at least one less nonzero row and at least one less nonzero column than the matrix A. If $A - C_1 R_1$ has a nonzero entry, choose another nonzero pivot entry and repeat the construction of a one-column matrix C_2 and a one-row matrix R_2 so that

$$A - C_1 R_1 - C_2 R_2 \tag{6.30}$$

has at least one less nonzero row and at least one less nonzero column.

Continue this process until

$$A - C_1 R_1 - C_2 R_2 - \cdots - C_r R_r \tag{6.31}$$

is a zero matrix. Then

$$A = C_1 R_1 + C_2 R_2 + \cdots + C_r R_r \tag{6.32}$$

Let c_{ik} denote the i-th entry in the column matrix C_k, and let r_{kj} denote the j-th entry in the row matrix R_k. Then, by 6.32, the entry a_{ij} in the matrix A is given by

$$\begin{aligned} a_{ij} &= c_{i1} r_{1j} + c_{i2} r_{2j} + \cdots + c_{ir} r_{rj} \\ &= \sum_{k=1}^{r} c_{ik} r_{kj} \end{aligned} \tag{6.33}$$

The sum in equation 6.33 is the type of sum that appears in a matrix product, and we can exploit this fact by forming an $m \times r$ matrix C and an $r \times n$ matrix R in which the k-th column of C consists of the entries in C_k and the k-th row of R consists of the entries in R_k. With these definitions,

$$A = CR \tag{6.34}$$

Before proceeding with the general discussion, let us illustrate this construction with an example.

Consider the 6×7 matrix

$$A_1 = A = \begin{bmatrix} -1 & 1 & -2 & -1 & 1 & 1 & -1 \\ 2 & -2 & 4 & 1 & 0 & -2 & 1 \\ -3 & 1 & -3 & -2 & 0 & 4 & -4 \\ -2 & 2 & -4 & -1 & 0 & 3 & -2 \\ 4^* & -2 & 5 & 3 & -1 & -4 & 4 \\ 3 & -1 & 3 & 1 & 2 & 0 & -1 \end{bmatrix} \tag{6.35}$$

We have indicated the selected pivot entry with an asterisk. There is nothing special here about the choice of the pivot entry, except that it must be nonzero. We will see in the following section that considerations of numerical accuracy may make some choices for the pivot entry more desirable than others. The one-column matrix C_1 is identical to the column containing the pivot entry in A, and the one-row matrix R_1 is

$$R_1 = \begin{bmatrix} 1 & -1/2 & 5/4 & 3/4 & -1/4 & -1 & 1 \end{bmatrix} \tag{6.36}$$

We have

$$C_1R_1 = \begin{bmatrix} -1 & 1/2 & -5/4 & -3/4 & 1/4 & 1 & -1 \\ 2 & -1 & 5/2 & 3/2 & -1/2 & -2 & 2 \\ -3 & 3/2 & -15/4 & -9/4 & 3/4 & 3 & -3 \\ -2 & 1 & -5/2 & -3/2 & 1/2 & 2 & -2 \\ 4^* & -2 & 5 & 3 & -1 & -4 & 4 \\ 3 & -3/2 & 15/4 & 9/4 & -3/4 & -3 & 3 \end{bmatrix} \tag{6.37}$$

Then

$$A_2 = A_1 - C_1R_1 = \begin{bmatrix} & 1/2 & -3/4 & -1/4 & 3/4 & 0 & 0 \\ & -1^* & 3/2 & -1/2 & 1/2 & 0 & -1 \\ & -1/2 & 3/4 & 1/4 & -3/4 & 1 & -1 \\ & 1 & -3/2 & 1/2 & -1/2 & 1 & 0 \\ & & & & & & \\ & 1/2 & -3/4 & -5/4 & 11/4 & 3 & -4 \end{bmatrix} \tag{6.38}$$

In A_2 the entries that must be zeros because of the construction are shown as blanks; the computed zeros are shown. For the next step we choose another nonzero entry as the pivot entry. Again, the choice is quite arbitrary. It is indicated in 6.38 with an asterisk. We get

$$R_2 = \begin{bmatrix} 0 & 1 & -3/2 & 1/2 & -1/2 & 0 & 1 \end{bmatrix} \tag{6.39}$$

and

$$C_2R_2 = \begin{bmatrix} 1/2 & -3/4 & -1/4 & -1/4 & 0 & 1/2 \\ -1^* & 3/2 & -1/2 & 1/2 & 0 & -1 \\ -1/2 & 3/4 & -1/4 & 1/4 & 0 & -1/2 \\ 1 & -3/2 & 1/2 & -1/2 & 0 & 1 \\ \\ 1/2 & -3/4 & 1/4 & -1/4 & 0 & 1/2 \end{bmatrix} \tag{6.40}$$

$$A_3 = A_2 - C_2R_2 = \begin{bmatrix} 0 & -1/2 & 1 & 0 & -1/2 \\ \\ 0 & 1/2 & -1 & 1 & -1/2 \\ 0 & 0 & 0 & 1 & -1 \\ \\ 0 & -3/2^* & 3 & 3 & -9/2 \end{bmatrix} \tag{6.41}$$

$$C_3R_3 = \begin{bmatrix} 0 & -1/2 & 1 & 1 & -3/2 \\ \\ 0 & 1/2 & -1 & -1 & -3/2 \\ 0 & 0 & 0 & 0 & 0 \\ \\ 0 & -3/2^* & 3 & 3 & -9/2 \end{bmatrix} \tag{6.42}$$

$$A_4 = A_3 - C_3R_3 = \begin{bmatrix} 0 & & 0 & -1 & 1 \\ \\ 0 & & 0 & 2^* & -2 \\ 0 & & 0 & 1 & -1 \\ \\ \end{bmatrix} \tag{6.43}$$

$$C_4R_4 = \begin{bmatrix} 0 & & 0 & -1 & 1 \\ \\ 0 & & 0 & 2^* & -2 \\ 0 & & 0 & 1 & -1 \\ \\ \end{bmatrix} \tag{6.44}$$

Finally, we see that $A_4 = C_4R_4$, so that

$$A = C_1R_1 + C_2R_2 + C_3R_3 + C_4R_4 \tag{6.45}$$

The only task that remains is to assemble C_1, C_2, C_3, C_4 into a single matrix C, and to assemble R_1, R_2, R_3, R_4 into a single matrix R. Note that the terms in the sum 6.32 commute. That is, the C_i and R_i could be indexed in any order. All that is required is that they be indexed so that C_i and R_i are corresponding factors in the sum. Using the indexing as given, we can construct

$$C = \begin{bmatrix} -1 & 1/2 & -1/2 & -1 \\ 2 & -1^* & & \\ -3 & -1/2 & 1/2 & 2^* \\ -2 & 1 & 0 & 1 \\ 4^* & & & \\ 3 & 1/2 & -3/2^* & \end{bmatrix} \tag{6.46}$$

$$R = \begin{bmatrix} 1^* & -1/2 & 5/4 & 3/4 & -1/4 & -1 & 1 \\ & 1^* & -3/2 & 1/2 & -1/2 & 0 & 1 \\ & & 0 & 1^* & -2 & -2 & 3 \\ & & 0 & & 0 & 1^* & -1 \end{bmatrix} \tag{6.47}$$

If the pivot entries are taken in order down the main diagonal, C will be a *lower triangular matrix* (all zeros above the main diagonal), and R will be an *upper triangular matrix* (all zeros below the main diagonal). In this case C is usually designated by the letter L and R is usually designated by the letter U. Then $A = LU$ is called a *lower-upper triangular factorization* of A, or an *LU decomposition* of A. We shall refer to the factorization $A = CR$ as an LU decomposition even when C and R are not in triangular form.

Notice that C has the same number of columns as R has rows, and both are equal to the number of iterations in the process before the zero matrix is obtained. This number also turns out to be the same for any sequence of choices for the nonzero pivot entries. It is the rank of A, but we will not prove that here.

For an $m \times n$ matrix of rank r, the number of multiplications required to obtain the LU decomposition is

$$[2r^3 - 3(m+n)r^2 + (6mn + 3m - 3n - 2)r]/6 \tag{6.48}$$

Since this expression is a little too complicated for the purpose of making comparisons, we usually take $m = n = r$. In this case the number of multiplicative steps required is

$$(n^3 - n)/3 \tag{6.49}$$

The cubic term dominates this expression for large values of n, and we usually say it takes $n^3/3$ multiplicative steps to obtain an LU decomposition.

After the LU decomposition is obtained very little more work is required to solve the system of linear equations,

$$Ax = b = CRx \tag{6.50}$$

We rewrite these equations in the form of two systems,

$$Cy = b \tag{6.51}$$

$$Rx = y \tag{6.52}$$

The coefficient matrices for these two systems of equations are triangular. The first system of equations can be solved for y with $n(n-1)/2$ multiplicative steps (if $m = n = r$ and if the system has a solution), and the second system of equations can be solved for x with $n(n+1)/2$ multiplicative steps. The second system can always be solved if the first one can. Thus, this method of solving a system of n linear equations of rank n in n variables requires

$$(n^3 + 3n^2 - n)/3 \tag{6.53}$$

multiplicative steps.

This is the number of multiplicative steps given earlier in this section for Gaussian elimination. Not only does the process require the same number of steps as required for standard Gaussian elimination, the same arithmetic steps are encountered in both methods. It is Gaussian elimination in a slightly different form. The reduced coefficient matrices obtained in Gaussian elimination can be obtained here by filling the blank rows in the matrices A_1, A_2, ... with the rows of R_1, R_2, ... The *forward course* of Gaussian elimination is completed when the equations 6.51 are solved. Solving 6.52 is the *backward course.*

There are two primary advantages in using the form of Gaussian elimination described here. If $Ax = b'$ is a new system of equations with the same coefficient matrix but with different constant terms, the new system can be solved with only n^2 additional multiplicative steps. A more important advantage in the context of this book is that the dual system of equations

$$vA = c \tag{6.54}$$

can also be solved with only n^2 additional multiplicative steps.

If the inverse matrix for A is known, the solution for equation 6.26 can be written in the form

$$A^{-1}b = x \tag{6.55}$$

This expression has some appeal from a theoretical point of view but it has nothing to offer from a computational point of view. The product that must be computed in 6.55 requires n^2 multiplicative steps, the same as is required to solve the systems 6.51 and 6.52 together. Even worse, computing A^{-1} is inefficient. If we form a tableau with A as the A-matrix, pivoting down the main diagonal to invert the matrix would require n pivot exchanges, each of which requires n^2 multiplicative steps. Any method you choose to use to find A^{-1} will do as badly. Even using the LU decomposition is no help. Finding the LU decomposition takes $(n^3 - n)/3$ multiplicative steps. Inverting C takes $(n^3+3n^2+2n)/6$ steps. Inverting R takes $(n^3-3n^2+2n)/6$ steps. Finally, computing $R^{-1}C^{-1} = A^{-1}$ takes $(n^3 - n)/3$ steps. All this adds up to n^3.

The bottom line is that one should not compute A^{-1} unless that is a specific objective. If one can work a problem by avoiding computing A^{-1}, one should.

We have already pointed out that a pivot exchange can be carried out without using more memory space than is required to store one tableau. The same thing can be accomplished in the process of obtaining an LU decomposition. The blank rows and columns in each A_k provide just enough space to store the C_k and R_k. Each pair C_k and R_k do overlap at the pivot entry, but the pivot entry for the R_k is always taken to be 1, so its value can be implied without storing it. The only extra records that need to be retained are the locations of the pivot entries and the order in which they were selected.

$$A_2 = A_1 - C_1R_1 = \begin{bmatrix} -1 & 1/2 & -3/4 & -1/4 & 3/4 & 0 & 0 \\ 2 & -1^* & 3/2 & -1/2 & 1/2 & 0 & -1 \\ -3 & -1/2 & 3/4 & 1/4 & -3/4 & 1 & -1 \\ -2 & 1 & -3/2 & 1/2 & -1/2 & 1 & 0 \\ 4^1 & -1/2 & 5/4 & 3/4 & -1/4 & -1 & 1 \\ 3 & 1/2 & -3/4 & -5/4 & 11/4 & 3 & -4 \end{bmatrix} \tag{6.56}$$

$$A_3 = A_2 - C_2R_2 = \begin{bmatrix} -1 & 1/2 & 0 & -1/2 & 1 & 0 & -1/2 \\ 2 & -1^2 & -3/2 & 1/2 & -1/2 & 0 & 1 \\ -3 & -1/2 & 0 & 1/2 & -1 & 1 & -1/2 \\ -2 & 1 & 0 & 0 & 0 & 1 & -1 \\ 4^1 & -1/2 & 5/4 & 3/4 & -1/4 & -1 & 1 \\ 3 & 1/2 & 0 & -3/2^* & 3 & 3 & -9/2 \end{bmatrix} \tag{6.57}$$

$$A_4 = A_3 - C_3R_3 = \begin{bmatrix} -1 & 1/2 & 0 & -1/2 & 0 & -1 & 1 \\ 2 & -1^2 & -3/2 & 1/2 & -1/2 & 0 & 1 \\ -3 & -1/2 & 0 & 1/2 & 0 & 2^* & -2 \\ -2 & 1 & 0 & 0 & 0 & 1 & -1 \\ 4^1 & -1/2 & 5/4 & 3/4 & -1/4 & -1 & 1 \\ 3 & 1/2 & 0 & -3/2^3 & -2 & -2 & 3 \end{bmatrix} \quad (6.58)$$

$$A_5 = A_4 - C_4R_4 = \begin{bmatrix} -1 & 1/2 & 0 & -1/2 & 0 & -1 & 0 \\ 2 & -1^2 & -3/2 & 1/2 & -1/2 & 0 & 1 \\ -3 & -1/2 & 0 & 1/2 & 0 & 2^4 & -1 \\ -2 & 1 & 0 & 0 & 0 & 1 & 0 \\ 4^1 & -1/2 & 5/4 & 3/4 & -1/4 & -1 & 1 \\ 3 & 1/2 & 0 & -3/2^3 & -2 & -2 & 3 \end{bmatrix} \quad (6.59)$$

In the matrices 6.56 through 6.59, consider the shaded entries as hidden so that the unshaded entries represent $A_1, \ldots, A_5$. The darkly shaded entries are in C, and the lightly shaded entries are in R. We would also record the sequence of pairs, $(5, 1)$, $(2, 2)$, $(6, 4)$, $(3, 6)$ as the locations of the pivot entries for the LU reduction. If these matrices are stored in a computer, they would not be shaded—that is done for our visual convenience. However, the sequence of pivot locations stored suffice to tell the computer which entries are the entries of the A_i. For example, in A_3, $(5, 1)$ and $(2, 2)$ are the pairs that have been recorded for the pivot entries up to that point. The entries of A_3 are those entries stored in all rows except rows 5 and 2, and in all columns except columns 1 and 2.

Finally, when the process terminates with A_5, we can construct C and R in the following fashion. The first pivot pair is $(5, 1)$. Store column 1 as the first column of C. Convert the pivot entry to a 1 and all other entries in column 1 to a 0. Store row 5 as the first row of R and convert all entries in that row to 0. The next pivot pair is $(2, 2)$. Record column 2 as the second column of C. Convert the pivot entry to a 1 and all other entries in that column to 0. Record row 2 as the second row of R and convert all entries in row 2 to 0. Proceed in that fashion until C and R are fully constructed.

6.4 Numerical Accuracy

The extent to which numbers in a computation are undependable is referred to as error. We are not talking here about blunders or mistakes, but about deviations of the computed numbers caused by factors inherent in the way the problem is given or the method used to obtain the solution.

There are three principal sources of errors. The numbers given in the original formulation of an applied problem are usually estimates or statistical observations or other records. Their accuracy is often uncertain. A long calculation on a computing machine usually leads to round-off errors which can, under certain circumstances, accumulate to considerable proportions. We assume most readers are familiar with the existence of errors of these types. Once the data are accumulated, there is nothing that can be done about the first kind of error. However, the methods used in the calculation can determine how much effect data errors, calculation errors, and the third type of error to be described below have on the reliability of the result.

There is another way that errors can become significant that is probably not so familiar. For linear calculations, there is a phenomenon known as the *condition* of the problem. The following *ill-conditioned problem* is an illustration of what can happen. Let us invert tableau 6.60.

(6.60)

x_1	x_2	-1	
3	1	6	$= -y_1$
1*	0.333	1.999	$= -y_2$

(6.61)

y_2	x_2	-1	
-3	0.001*	0.003	$= -y_1$
1	0.333	1.999	$= -x_1$

(6.62)

y_2	y_1	-1	
-3000	1000	3	$= -x_2$
1000	-3333	1	$= -x_1$

The calculations given above are exact, and the solution $x_1 = 1$, $x_3 = 3$ is exact. Actually, all linear calculations, if they are performed rationally, are exact. The problems occur when these rational numbers are stored as decimal approximations.

Using the same ideas we considered in Section 3.3, suppose we change the value of one of the entries in the b-column. Suppose, for example, we change the 1.999 in tableau 6.60 to 1.9995. If we repeat the sequence of calculations performed above, we obtain the solution $x_1 = 1.5$, $x_2 = 1.5$. This solution, for the altered system of equations, is also exact.

The numbers in the y_2-column of tableau 6.62 determine the sensitivity of the computed values in the b-column of that tableau to changes (or errors) in the value of the b-entry of the y_2-row of tableau 6.60. If the new b_2-entry is $1.999 + e$, the exact b-column of the computed tableau 6.62 would become

$$\begin{array}{c|c|l} & -1 & \\ \hline & 3 - 3000e & = -x_2 \\ & 1 + 1000e & = -x_1 \\ \hline \end{array} \tag{6.63}$$

This example was discussed as though 6.60 were the whole tableau. The same results would be obtained if the numbers in 6.60 were embedded within a tableau (that is, the entries in the b-column of 6.60 are merely entries in some other column). If any entry in the y_2-row is slightly in error, after the variables are exchanged the error in that entry will be magnified a thousand fold.

Intuitively, the difficulty is caused by the fact that the equations in tableau 6.60 are represented by two straight lines that intersect at a small angle. Changing a number in the b-column moves the straight line parallel to itself. However, when two lines intersect at a small angle, even a small displacement of one of the lines may move the intersection point a relatively large distance. Generally, in an example involving more variables, a basic solution corresponds geometrically to a vertex formed by the intersection of a number of hyperplanes (one for each variable set equal to zero). If some of the hyperplanes at that vertex meet at slight angles the coordinates of that vertex can be very sensitive to errors in the parameters of those hyperplanes. Such vertices are *ill-conditioned.*

The simplex algorithm involves obtaining a sequence of tableaux and, therefore, a sequence of basic solutions. Each basic solution corresponds to a vertex where the corresponding hyperplanes intersect. The simplex algorithm can be thought of as determining a path from one vertex to another until a vertex corresponding to an optimal solution is obtained. If this path leads through an ill-conditioned vertex any errors accumulated up to that point will be magnified and the magnified errors carried forward.

If the optimal solution corresponds to a vertex that is ill-conditioned, the difficulties caused by the condition of the vertex must be accepted. If the path of the simplex algorithm leads through an ill-conditioned vertex that is not the optimal vertex, the question is raised as to whether we can find another path to the optimal solution that does not go through an ill-conditioned vertex.

Before following this line of discussion any further, let us look at another example. Consider the tableau

(6.64)

x_1	x_2	-1	
0.0001	1	1	$= -y_1$
1	1	2	$= -y_2$

If this tableau is inverted rationally (exactly) we obtain the tableau

(6.65)

y_1	y_2	-1	
$-\frac{10000}{9999}$	$\frac{10000}{9999}$	$\frac{10000}{9999}$	$= -x_1$
$\frac{10000}{9999}$	$-\frac{1}{9999}$	$\frac{9998}{9999}$	$= -x_2$

In a computer, numbers are stored in memory in registers with a finite number of decimal (or binary or hexadecimal) places. If a computed number has too many places for the register, the representation of the numbers is chopped (the excess less-significant digits are dropped) or rounded (5 is added to the first nonretained place and the resulting number is chopped). For illustration, let us assume that our computer stores four significant digits and chops for storage. The tableau 6.65 would be represented in the computer by

(6.66)

y_1	y_2	-1	
-1.000	1.000	1.000	$= -x_1$
1.000	$-1.000E-4$	0.9998	$= -x_2$

We see that the basic solution for tableau 6.66 is

$$x_1 = 1.000, x_2 = 0.9998 \tag{6.67}$$

Now let us return to tableau 6.64 and perform the pivot exchanges with numbers the computer would be using. In this connection it is important and it makes a difference how the arithmetic is carried out. In particular, we are going to do the work just as it is described in Section 2.2, where we compute the numbers in the pivot row and overwrite the old entries in that row before we compute the new entries in other rows. If you wish to carry out the calculations by hand, remember to chop or round the number obtained for each arithmetic operation before performing the next one. The initial tableau 6.64 and the subsequent computation take the form

(6.68)

x_1	x_2	-1	
$1.000E{-4}^*$	$1.000E0$	$1.000E0$	$= -y_1$
$1.000E0$	$1.000E0$	$2.000E0$	$= -y_2$

(6.69)

y_1	x_2	-1	
$1.000E4$	$1.000E4$	$1.000E4$	$= -x_1$
$-1.000E4$	$-9.999E3^*$	$-9.998E3$	$= -y_2$

(6.70)

y_1	y_2	-1	
0.000	$1.000E0$	$1.999E0$	$= -x_1$
$1.000E0$	$-1.000E{-4}$	$9.998E{-1}$	$= -x_2$

The basic solution for tableau 6.70 agrees with 6.66 for the value of x_2, but it yields a value for x_1 that is 100% off.

This large relative error is caused by performing the first pivot exchange on the relatively small entry chosen in tableau 6.68. If any other entry had been chosen as pivot entry the result would be more satisfactory. Starting with the x_1, y_2 entry or the x_2, y_1 entry we would obtain the same tableau as 6.66, and starting with the x_2, y_2 entry we would obtain the same tableau with the entries rounded to integer values.

If instead of chopping we had used rounding, the error would have been smaller, but it still would be very large compared with the accuracy implied by the number of significant digits maintained in the decimal representation.

The two examples discussed above exhibit slightly different manifestations of the same phenomenon. For the two equations of tableau 6.60

$$\begin{aligned} 3x_1 + x_2 - 6 &= -y_1 \\ x_1 + 0.333x_2 - 1.999 &= -y_2 \end{aligned} \tag{6.71}$$

we can represent the various solutions on an x_1, x_2 coordinate system. Each choice for the values of (x_1, x_2) determines the computed values of y_1 and y_2. In Figure 6.1 the lines $x_1 = 0$ and $x_2 = 0$ are the coordinate axes. The line $y_1 = 0$ intersects these coordinate axes in the points $(2, 0)$ and $(0, 6)$. The line $y_2 = 0$ intersects the coordinate axes in the points $(1.999, 0)$ and $(0, 6.003)$. If we draw these lines on a scale of one unit equals one inch, they will differ (within the first quadrant) by less than the width of any reasonable ink or pencil drawn line. Mathematically, these are distinct lines and they have a mathematically defined intersection point. However, if one of these lines is slightly displaced the intersection point will move a great

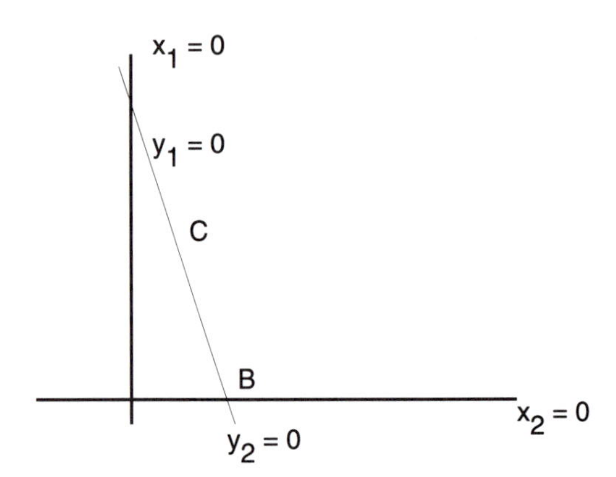

Figure 6.1: Two nearly identical lines.

distance. In the practical world of approximate numbers, the intersection point is not precisely defined.

Now look at the equations for our other example. The equations of tableau 6.64 are

$$\begin{aligned} 0.0001x_1 + x_2 - 1 &= -y_1 \\ x_1 + x_2 - 2 &= -y_2 \end{aligned} \tag{6.72}$$

Again, the lines $x_1 = 0$ and $x_2 = 0$ are the coordinate axes. The line $y_1 = 0$ intersects these axes in the points $(1000, 0)$ and $(0, 1)$. The line $y_2 = 0$ intersects the coordinate axes in the points $(2, 0)$ and $(0, 2)$. Again, it is impossible to draw the lines with an accuracy implied by the calculations. If the drawing were made to a scale that would include both the origin and the intersection point $(1000, 0)$ on the same sheet of paper, all intersection points would be contained within the width of either a pencil or an ink line.

In Figure 6.2, the solution is represented by the point E. Pivoting first on x_1, y_2 goes to E through the intermediate point D. Pivoting first on x_2, y_1 goes to E through the intermediate point B. Pivoting first on x_2, y_2 goes to E through the intermediate point C. The first pivot on x_1, y_1 is ill-conditioned and goes through the point $(1000, 0)$ far off the graph to the right. Again, the magnification of round-off errors is caused by the fact that the lines $x_2 = 0$ and $y_1 = 0$ intersect at a small angle. If the calculation is carried out rationally (exactly), there is no round-off and the solution is accurate.

In both of these examples, there is an ill-conditioned vertex. In the first

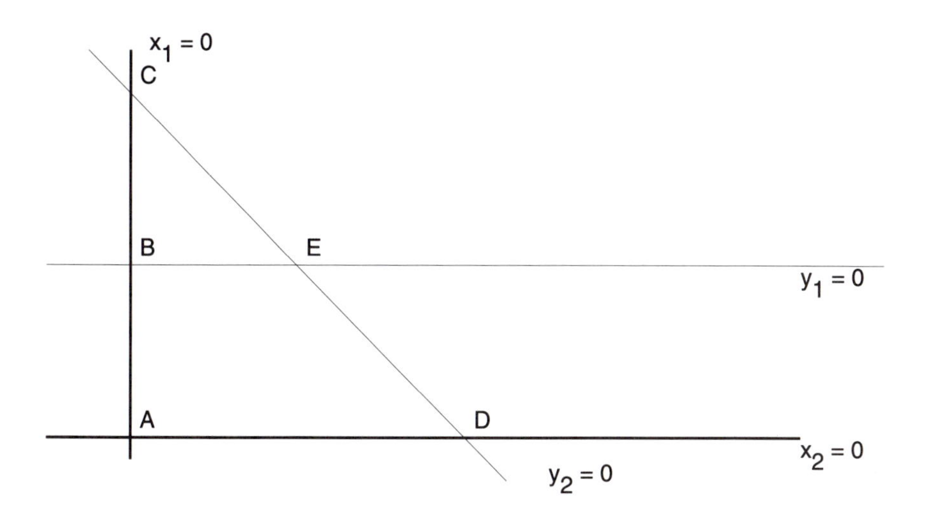

Figure 6.2: Two nearly parallel lines.

example this vertex is the one sought, and the inaccuracy in determining its location must be accepted. In the second case, the ill-conditioned vertex is not the one sought and a path to the desired vertex can be found that avoids the vertex that causes the problems with round-off error.

There has been much research in numerical analysis on how errors of the type discussed here are introduced and to what extent they can be estimated and controlled. There is no easy, definitive answer. It is clear that one must be careful in the choice of pivot entries. Most computing machine programs to solve linear equations include selection rules designed to reduce the likelihood of using an ill-conditioned pivot. The selection rules for the simplex algorithm are different from these selection rules but not inconsistent with them.

Generally, it is a bad practice to choose a pivot entry that is relatively small compared with the other entries in the matrix. Choosing a small entry for the pivot entry caused the problems with round-off error in the second example. However, the concept of "relative size" among the entries in a coefficient matrix is not very easy to pin down. If we multiply the entries in any one row by a nonzero constant, this has the effect of multiplying one of the equations of the system by that constant, and this has no effect on the solution set. If we multiply the entries in any one column by a nonzero constant, this has the effect of dividing the variable corresponding to that column by the same constant. As far as the equations are concerned these operations are considered to be inessential, but they can have a large effect on the relative sizes of the entries.

Multiplying the rows and/or columns of a matrix by nonzero constants is called *scaling*. For a given choice of a sequence of pivot entries, scaling has no effect on the relative errors in the computations. If you examine the calculations required you will see that if you multiply a row or column by a nonzero constant, that constant will appear as a common factor in two numbers in a quotient or as a common factor in two numbers in a difference, or it will not appear at all. However, if the rule for selecting the pivot entries depends on the relative sizes of the entries, scaling can make quite a difference in the choice of pivot entries.

Good computational practice calls for scaling the equations so that the maximum entry (in absolute value) in each row and column is 1, or some other arbitrary number. This is called *scaling for size*. A common rule for choosing the pivot entry is to pick the pivot entry from each column in order, starting with the left-most column, and picking the largest entry (in absolute value) in each column that is available. Another rule is to pick the largest entry available from the entire matrix. The first rule is called partial pivoting, and the second rule is called full pivoting. In numerical analysis the term "pivoting" refers to the rule by which the pivot entry is selected, not to the operation itself, as in linear programming. Experience seems to suggest that full pivoting is not worth the extra work involved in making the choice. At any rate, the intent of the rule is more to avoid disasters than to make the best choices. In the previous section, the worked example to obtain the LU decomposition used partial pivoting.

If an optimal solution is obtained by the simplex algorithm and this solution is located at an ill-conditioned vertex, the inherent inaccuracy may just have to be accepted. For example, if it is not known whether the second entry in the b-column of tableau 6.60 is accurate to four significant digits, the two solutions given will have to be considered as equally acceptable. The data, then, do not justify a more exact solution.

This is not an unreasonable or unacceptable conclusion. Suppose, for example, that tableau 6.60 represented the constraints for a production problem. (We ignore the objective variable for this discussion.) The numbers in the b-column, the 6 and the 1.999, represent the available quantities of two resources, say steel and copper, respectively. The exact solution, given in tableau 6.62, says we can make one unit of the first product and three units of the second product. If the amount of copper were increased to 1.9995, we could make 1.5 units of each product. However, each product uses about three times as much steel as it does copper. Unless the production methods are very precise the 0.025% increase in the supply of copper is not likely to be significant. Almost any production schedule would use about three times as much steel as copper, and a range of product mixes is reasonable and probably possible.

If an optimal solution is obtained that is not located at an ill-conditioned vertex but some intermediate tableau was represented by an ill-conditioned vertex, it may be worth reinverting from the initial tableau. No matter how many pivot exchanges were performed, and in what order, in the course of running the simplex algorithm, the initial tableau and the final tableau (for which the basic solution is presumed to be optimal, if accurate) differ in the exchange of a number of basic and nonbasic variables. If r variables have been exchanged we know that one tableau can be obtained from the other by exactly r independent pivot exchanges. For this purpose, the simplex algorithm can be ignored and the pivot entries can be selected by considerations of numerical accuracy. This means the pivot entries would be selected according to a rule like the partial or full pivoting rules.

If the purpose of a reinversion is to obtain the full tableau for the optimal solution, we might just as well perform the r independent pivot exchanges, selecting the pivot entries as described above. This would require $r(m + 1)(n + 1)$ multiplicative steps. However, it is very likely that we need to know only the basic solutions for the final tableau. In this case the amount of work required can be reduced by using Gaussian elimination, as described in the previous section.

Let us reindex the variables so that the nonbasic variables $x_1, \ldots, x_r$ in the initial tableau are exchanged in the final tableau. Then the initial tableau looks like 6.73.

$$
\begin{array}{c|ccc|ccc|c|l}
 & x_1 & \cdots & x_r & x_{r+1} & \cdots & x_n & -1 & \\
\hline
v_1 & & & & & & & & = -y_1 \\
\vdots & & A_{11} & & & A_{12} & & B_1 & \vdots \\
v_r & & & & & & & & = -y_r \\
\hline
v_{r+1} & & & & & & & & = -y_{r+1} \\
\vdots & & A_{21} & & & A_{22} & & B_2 & \vdots \\
v_m & & & & & & & & = -y_m \\
\hline
-1 & & C_1 & & & C_2 & & d & = f \\
\hline
 & = u_1 & \cdots & = u_r & = u_{r+1} & \cdots & = u_n & = g &
\end{array}
\tag{6.73}
$$

We have used single symbols to stand for blocks of entries within the tableau. Similarly, let

$$X_1 = \begin{bmatrix} x_1 \\ \vdots \\ x_r \end{bmatrix}, X_2 = \begin{bmatrix} x_{r+1} \\ \vdots \\ x_n \end{bmatrix}, Y_1 = \begin{bmatrix} y_1 \\ \vdots \\ y_r \end{bmatrix}, Y_2 = \begin{bmatrix} y_{r+1} \\ \vdots \\ y_m \end{bmatrix} \tag{6.74}$$

With this notation, the row equations for the tableau 6.73 can be written in the matrix form

$$\begin{aligned} A_{11}X_1 + A_{12}X_2 - B_1 &= -Y_1 \\ A_{21}X_1 + A_{22}X_2 - B_2 &= -Y_2 \\ C_1X_1 + C_2X_2 - d &= f \end{aligned} \tag{6.75}$$

Using similar notational conventions, the final tableau can be written in the form

(6.76)

	Y_1	X_2	-1	
U_1	$\bar{A}_{11}$	$\bar{A}_{12}$	$\bar{B}_1$	$= -X_1$
V_2	$\bar{A}_{21}$	$\bar{A}_{22}$	$\bar{B}_2$	$= -Y_2$
-1	$\bar{C}_1$	$\bar{C}_2$	$\bar{d}$	$= f$
	$= V_1$	$= U_2$	$= g$	

In matrix notation the row equations take the form

$$\begin{aligned} \bar{A}_{11}Y_1 + \bar{A}_{12}X_2 - \bar{B}_1 &= -X_1 \\ \bar{A}_{21}Y_1 + \bar{A}_{22}X_2 - \bar{B}_2 &= -Y_2 \\ \bar{C}_1Y_1 + \bar{C}_2X_2 - \bar{d} &= f \end{aligned} \tag{6.77}$$

For a basic solution in the final tableau we set $Y_1 = 0$ and $X_2 = 0$. This gives $X_1 = \bar{B}_1$ in 6.77 and

$$A_{11}\bar{B}_1 = B_1 \tag{6.78}$$

in 6.75. The system of equations in 6.78 can be solved for B_1 by any numerically sound method. Gaussian elimination will provide a solution with approximately $r^3/3$ multiplicative steps. Once $\bar{B}_1$ is obtained we can compute f from 6.75.

$$f = C_1\bar{B}_1 - d \tag{6.79}$$

Furthermore, $\bar{B}_2 = Y_2$ in 6.77 and, hence, from 6.75

$$\bar{B}_2 = B_2 - A_{21}\bar{B}_1 \tag{6.80}$$

Using similar notation for the column equations, $\bar{C}_1$ can be obtained by solving the system of equations

$$\bar{C}_1 A_{11} = -C_1 \tag{6.81}$$

If the system 6.81 is solved from scratch it would take approximately $r^3/3$ multiplicative steps. However, if the system 6.78 is solved by factoring A_{11} into a product of a lower triangular and an upper triangular matrix, the system 6.81 can be solved in approximately r^2 additional multiplicative steps. Then C_2 can be obtained by

$$\bar{C}_2 = C_2 + \bar{C}_1 A_{12} \tag{6.82}$$

The calculation in 6.80 requires $r(m-r)$ multiplicative steps, and the calculation in 6.82 requires $r(n-r)$ multiplicative steps. For large values of m, n, and r these are small numbers compared to $r^3/3$.

6.5 The Ellipsoidal Algorithm

The *ellipsoidal algorithm* does not deal with the linear programming problem directly. It solves an associated feasibility problem.

Theorem 6.50 *A canonical linear programming problem and a canonical feasibility problem are equivalent.*

Proof. In Chapter 5 we showed how a feasibility problem can be reduced to a linear program.

Consider a linear program represented by the following tableau.

(6.83)

	x	-1	
v	A	b	$= -y$
-1	c	d	$= f$
	$= u$	$= g$	

Consider the feasibility problem associated with the following tableau.

x	v^T	-1	
A	0	b	$= -y$
0	$-A^T$	$-c^T$	$= -u^T$
$-c$	b^T	0	$= -w$

(6.84)

If this problem has a feasible solution, we will obtain feasible solutions for $Ax = -y$ and $vA - c = u$ and also $vb - cx = b^T v^T - cx \leq 0$. Since $vb - cx \geq 0$ for feasible solutions we have $vb = cx$. Since this is a sufficient condition for optimality, we have optimal solutions for both of the dual pair of programs. □

An n-dimensional ellipsoid is a generalization of the familiar two-dimensional ellipses and three-dimensional ellipsoids. It has a center of symmetry and n planes of symmetry through the center.

We start with an ellipsoid large enough to contain points in the feasible set, if there are any. We examine the center of the ellipsoid to see if it is a feasible point. If the center is not a feasible point, there is at least one inequality in the system that is not satisfied at the center. The linear expression in that inequality increases on one side of the hyperplane on which it is constant and decreases on the other. Thus, the feasible set, if it is not empty, is on one side of that hyperplane through the center.

The next iteration of the algorithm involves finding the ellipsoid of least volume that contains the half of the ellipsoid described above. It turns out that that is not difficult to do. That ellipsoid is smaller, and it is smaller in volume by a predetermined ratio.

We examine the center of the new ellipsoid for feasibility and continue. The volumes of this sequence of ellipsoids decrease exponentially. We must either find a feasible point or the ellipsoids must decrease to the point where its dimensions are smaller than the accuracy of the representations of the numbers in the problem, in which case there is no feasible solution.

The estimated number of iterations of the ellipsoidal algorithm grows like the fifth power of the size of the problem. The amount of arithmetic also grows because the sizes of the numbers in the computation grow. The total growth size is estimated to be the eighth power.

The hopes raised by the publication of this algorithm have not been realized. Even small problems require enormous calculations and we cannot give a reasonable example that can be carried out by hand. However, this algorithm had very important consequences.

For some time it had not been known how the amount of work grew as the problem size grew. The publication of the Klee–Minty example showed that the simplex algorithm was an exponential algorithm. It was still not known whether the linear programming problem was or was not exponential. The ellipsoidal algorithm settled that. Among other things, it gave hope that a more efficient polynomial-time algorithm could be found.

6.6 The Karmarkar Algorithm

The *Karmarkar algorithm* uses a formulation of the linear programming problem that is different from the ones considered so far in this book. Consider a linear program represented by the following tableau.

(6.85)

x	-1	
A	b	$= -y$
c	d	$= f$

We take x to be canonical and y to be artificial. When y is artificial, the problem is said to be in *standard form.*

We assume the problem has an optimal solution and that we can make a reasonable estimate of a number M sufficiently large that it is an upper bound for the sum of the variables that might represent an optimal solution. That is,

$$\sum_j x_j \leq M \tag{6.86}$$

for an optimal solution. We are going to add this inequality as a constraint and the important thing about it is that it must not change the value of the optimal solution. We then change the scale of the coefficients in the problem so that we can take $M = 1$. When this constraint is added to the tableau we have

(6.87)

x	-1	
A	b	$= 0$
$\mathbf{1}$	1^*	$= -x_{n+1}$
c	d	$= f$

where $x_{n+1} \geq 0$.

The $\mathbf{1}$ in tableau 6.87 is a row of 1's. We now perform one pivot exchange indicated by the asterisk in tableau 6.87. We obtain

$$\begin{array}{cc|l} x & x_{n+1} & \\ \hline A' & b' & = 0 \\ \mathbf{1} & 1^* & = 1 \\ \hline c' & d' & = f \end{array} \tag{6.88}$$

If we include the new variable x_{n+1} as just another nonbasic variable, this tableau is of the form

$$\begin{array}{c|l} x & \\ \hline A & = 0 \\ \mathbf{1} & = 1 \\ \hline c & = f \end{array} \tag{6.89}$$

This is the form of the problem handled by the Karmarkar algorithm.

> **Minimize (or maximize) $f = cx$ subject to the constraints $Ax = 0$ and $\sum_j x_j = 1$ with all $x_j \geq 0$.**

We call this the *projective form* of the linear programming problem. Though it looks radically different, this formulation of the problem is equivalent to the standard form of the linear programming problem.

The set described by

$$\sum_j x_j = 1 \quad \text{with all } \; x_j \geq 0 \tag{6.90}$$

is called a *simplex*. (This is simply the term used for the geometric object and it does not imply any connection between the Karmarkar algorithm and the simplex algorithm.) In three dimensions, the simplex is a triangle in the first orthant satisfying 6.90. This gives a satisfactory geometric picture of a simplex. The transformations from 6.85 to 6.88 embed the feasible set for this problem in the simplex.

The point $\mathbf{e} = (\frac{1}{n+1}, \frac{1}{n+1}, \ldots, \frac{1}{n+1})$ is the *center of the simplex*. It is important that for each iteration of the Karmarkar algorithm that we have the center be a feasible point. This can be arranged in the following way. If $\mathbf{d} = (d_1, d_2, ..., d_{n+1})$ is any feasible point, then $\sum_j a_{ij} d_j = 0$. If we set

$$a'_{ij} = a_{ij} d_j \tag{6.91}$$

and

$$c'_j = c_j d_j, \tag{6.92}$$

we have a new problem,

> **Minimize (or maximize)** $f = c'x$ **subject to the constraints** $A'x = 0$, $\sum_j a'_{ij} = 0$, **and** $\sum_j x_j = 1$ **with all** $x_j \geq 0$.

If we take **d** to be a strictly interior point, all $d_j > 0$. Under these conditions there is a one-to-one correspondence between points that satisfy $Ax = 0$ and those satisfying $A'x = 0$.

Because the constraints are homogeneous, **e** satisfies $A'x = 0$. Since **e** is in the simplex, it is a feasible point for the revised constraints. Since this operation occurs in each iteration of the algorithm we will give it a name. We say that this operation *centers the constraints*.

Once the constraints are cast in this form the subsequent steps involve an interesting interplay between metric and projective geometry. At each step the method for finding the next feasible point uses metric geometry. Between steps, the method for constructing the next form of the problem uses projective geometry. Let us discuss the projective geometry part first.

Projective geometry was once a fashionable subject, but it is much neglected in the college curriculum today. Without going into a crash course in projective geometry let us summarize what is relevant here. An n-dimensional projective space is coordinatized with $n + 1$ coordinates. Two $(n+1)$-tuples represent the same projective point if they differ by a nonzero factor. The normalization $\sum_j x_j = 1$ merely selects one of these representations as the standard representation of a point. Some projective points cannot be represented this way, but they do not enter into the discussion to follow. A projective transformation is represented by an $(n+1) \times (n+1)$ matrix. Since coordinates that are nonzero multiples of each other represent the same point, matrices that are nonzero multiples of each other represent the same projective transformation. It is important to realize that projective transformations do not preserve distance or parallelism.

This part of the algorithm does not involve the objective function in an active way, so we will simply leave it out of the discussion for the moment. We start with the constraints posed in projective form. For this purpose it is convenient to let **1** denote the one-row matrix [1 1 ... 1]. The initial constraints are

$$Ax = 0 \tag{6.93}$$

$$\mathbf{1}x = 1 \tag{6.94}$$

$$x \text{ is canonical} \tag{6.95}$$

Let d_1 be any strictly interior feasible point. That is, the components of d_1 are positive. Let D_1 be the square matrix with the entries of d_1 down the main diagonal and zeros otherwise. That is, $D_1 = \text{diagonal}(d_1)$ and $d_1 = D_1\mathbf{1}^T$. Let $E_1 = D_1$.

We cast the first in a sequence of new constraints in the form

$$A_1 x = 0 \tag{6.96}$$

where $A_1 = AE_1$.

$$\mathbf{1}x = 1 \tag{6.97}$$

$$x \text{ is canonical} \tag{6.98}$$

$$\tag{6.99}$$

We note that

$$A_1\mathbf{1}^T = AE_1\mathbf{1}^T = AD_1\mathbf{1}^T = Ad_1 = 0 \tag{6.100}$$

The significance of condition 6.100 is that it means that the center of the simplex is a feasible point. That is, the constraints are centered.

Suppose we are at a point where we have centered constraints in the form

$$A_k x = 0 \tag{6.101}$$

where $A_k = AE_k$.

$$\mathbf{1}x = 1 \tag{6.102}$$

$$x \text{ is canonical} \tag{6.103}$$

$$A_k\mathbf{1}^T = 0 \tag{6.104}$$

The objective function for these constraints is

$$f = c_k x = cE_k x \tag{6.105}$$

Let d_{k+1} be any strictly interior feasible solution for these constraints and form $D_{k+1} = \text{diagonal}(d_{k+1})$. Then define $E_{k+1} = E_k D_{k+1}$ and $A_{k+1} = AE_{k+1}$. Then use A_{k+1} to define the next set of constraints. The first three conditions are merely definitions. The only thing that has to be proved is that the constraints are centered. That is,

$$A_{k+1}\mathbf{1}^T = AE_{k+1}\mathbf{1}^T = AE_k D_{k+1}\mathbf{1}^T = A^k d_{k+1} = 0 \tag{6.106}$$

We need to establish the relation between a feasible solution for any of these centered constraints and a solution of the original problem. Let

$$x_{k+1} = \frac{E_k d_{k+1}}{\mathbf{1}E_k d_{k+1}} \tag{6.107}$$

Then

$$Ax_{k+1} = \frac{AE_k d_{k+1}}{\mathbf{1}E_k d_{k+1}} = \frac{A_k d_{k+1}}{\mathbf{1}E_k d_{k+1}} = 0 \tag{6.108}$$

and

$$\mathbf{1}x_{k+1} = 1 \tag{6.109}$$

Finally, since all the diagonal entries in the E_k are positive, the division in 6.107 is possible and the entries in x_{k+1} are positive. That is, x_{k+1} is a feasible solution.

Now let us turn our attention to how the feasible solutions are chosen for each of the centered constraints.

In linear algebra any linear transformation T for which $T^2 = T$ is a *projection.* If T is a projection, $I - T$ is also a projection since $(I - T)^2 = I - T$. This gives us two projections, T and $I - T$. Each is a projection into the null space of the other, since $T(I - T) = T - T = (I - T)T = 0$. Our objective in this context is to find a projection into the feasible space in the simplex.

For each of the centered constraints constructed above, let B_k be the matrix formed from A_k by adjoining a row of 1's. We expect B_k to be a matrix with more columns than rows. We assume that the rank of B_k is equal to the number of rows. Under these circumstances, $B_k B_k^T$ is an invertible square matrix. Then $B_k^T(B_k B_k^T)^{-1}B_k$ is a projection since

$$[B_k^T(B_k B_k^T)^{-1}B_k][B_k^T(B_k B_k^T)^{-1}B_k] = B_k^T(B_k B_k^T)^{-1}B_k \tag{6.110}$$

As pointed out above, $I - B_k^T(B_k B_k^T)^{-1}B_k$ is also a projection. Finally, $B_k[I - B_k^T(B_k B_k^T)^{-1}B_k] = B_k - B_k = 0$. That is, $I - B_k^T(B_k B_k^T)^{-1}B_k$ is a projection into a subspace parallel to the feasible set. This gives us

$$[I - B_k^T(B_k B_k^T)^{-1}B_k](c_k)^T \tag{6.111}$$

as a vector pointing in the direction of maximum rate of change of the linear function $c_k x = cE_k x$ in the feasible set.

Let p be a unit vector in the direction of the vector obtained in 6.111. Then define

$$d_{k+1} = (\mathbf{1}^T + p) \tag{6.112}$$

We claim that d_{k+1} is a feasible solution of the centered constraints.

$$A_k d_{k+1} = (A_k \mathbf{1}^T + A_k p) = 0 \tag{6.113}$$

and

$$\mathbf{1}d_{k+1} = \frac{1}{n+1}(\mathbf{1}\mathbf{1}^T + \mathbf{1}p) = 1 \tag{6.114}$$

Finally, we must show that the entries in d_{k+1} are positive. The radius of the largest hypersphere that can be inscribed in the simplex is easy to compute. It is $\frac{1}{\sqrt{n(n+1)}} > \frac{1}{n+1}$. Since $\frac{1}{n+1}\mathbf{1}^T$ is the center of the simplex and $\frac{1}{n+1}p$ is of length smaller than the radius of the inscribed hypersphere, d_{k+1} is within the simplex. That is, all the entries in d_{k+1} are positive.

Now let us look at the change in the value of the objective variable with each iteration. Consider the smallest hypersphere with center at $\mathbf{e}$ that can be circumscribed about the simplex. The center of the simplex divides the line from a vertex of the simplex to the midpoint of the opposite face in the ratio $n : 1$. The ratio of the radius of the circumscribed hypersphere to the radius of the inscribed hypersphere is also $n : 1$.

It is a characteristic of linear functions that the change in the value of the function changes in any direction proportionally to the distance traveled. The optimal point is outside the inscribed hypersphere since an optimal solution will have at least one zero component. Since the optimal point is outside the inscribed hypersphere and inside the circumscribing hypersphere, the distance between the center and the next point chosen by the algorithm is at least $1/n$ of the distance between the center and the optimal point. Furthermore, the rate of change of the objective function is at least as large along the path to the new point as it is along a path to the optimal point. Thus, the difference between the optimal value and the value of the objective function is decreased by a factor of at most $(n-1)/n$ with each iteration.

The relationships between the initial problem and the derived problems constructed in the course of the Karmarkar algorithm are neither convenient nor predictable. However, the ratios between the approximations and the optimal solution decrease geometrically so that the algorithm converges. No calculation is difficult except for the calculation in formula 6.111. That calculation is daunting by hand and lengthy with a computer. However, the algorithm is practical and it has been implemented in a commercial program. At the time this is being written the only implementations are on very large and expensive computers.

At this time there is a great deal of interest and research in techniques inspired by the Karmarkar algorithm. They are known generally as *interior point methods.*

Let us go through one cycle of the Karmarkar algorithm with a small problem. Consider the canonical linear program represented by the tableau

(6.115)

x_1	x_2	-1	
1	0	4	$= -x_3$
0	2	12	$= -x_4$
3	2	18	$= -x_5$
3	5	0	$= f$

Whether one wishes to minimize or maximize does not affect the first few steps. The first step is to convert this problem into a standard linear program.

(6.116)

x_1	x_2	x_3	x_4	x_5	-1	
1	0	1	0	0	4	$= 0$
0	2	0	1	0	12	$= 0$
3	2	0	0	1	18	$= 0$
3	5	0	0	0	0	$= f$

The next step is to introduce the bound as an extra constraint. The optimal solution is $x_1 = 2$, $x_2 = 6$, $x_3 = 0$, $x_4 = 0$, $x_5 = 2$. Thus, 10 is an upper bound for the sum of the coordinates of an optimal solution. However, let us take $M = 50$ since we are not supposed to know what the optimal solution is. This gives us

(6.117)

x_1	x_2	x_3	x_4	x_5	-1	
1	0	1	0	0	4	$= 0$
0	2	0	1	0	12	$= 0$
3	2	0	0	1	18	$= 0$
1	1	1	1	1	50	$= -x_6$
3	5	0	0	0	0	$= f$

Then we have to scale the coefficients to reduce the bound to 1.

(6.118)

x_1	x_2	x_3	x_4	x_5	-1	
50	0	50	0	0	4	$= 0$
0	100	0	50	0	12	$= 0$
150	100	0	0	50	18	$= 0$
1	1	1	1	1	1^*	$= -x_6$
150	250	0	0	0	0	$= f$

Pivot on the 1 in the last column.

(6.119)

x_1	x_2	x_3	x_4	x_5	x_6	
46	−4	46	−4	−4	−4	= 0
−12	88	−12	38	−12	−12	= 0
132	82	−18	−18	32	−18	= 0
1	1	1	1	1	1	= 1
150	250	0	0	0	0	= f

One more step is required to start the first iteration of the algorithm. The center of the simplex is not a feasible solution to the constraints of tableau 6.119. We must center the constraints. We start with a feasible solution of tableau 6.115. The solution must represent an interior point. We can take $x_1 = 1$, $x_2 = 3$. When scaled, this gives

$$\mathbf{d}_1 = (0.02, 0.06, 0.06, 0.12, 0.18, 0.56) \tag{6.120}$$

as the initial feasible solution.

Following the calculations described above, we get

$$A_1 = \begin{bmatrix} 0.92 & -0.24 & 2.76 & -0.48 & -0.72 & -2.24 \\ -0.24 & 5.28 & -0.72 & 4.56 & -2.16 & -6.72 \\ 2.64 & 4.92 & -1.08 & -2.16 & 5.76 & -10.08 \end{bmatrix} \tag{6.121}$$

Since the constraints involved are homogeneous it is convenient to scale them so that the entries in A_k are integers. We obtain

$$A_1 = \begin{bmatrix} 92 & -24 & 276 & -48 & -72 & -224 \\ -24 & 528 & -72 & 456 & -216 & -672 \\ 264 & 492 & -108 & -216 & 576 & -1008 \end{bmatrix} \tag{6.122}$$

From this we obtain

$$A_1 = \begin{bmatrix} 92 & -24 & 276 & -48 & -72 & -224 \\ -24 & 528 & -72 & 456 & -216 & -672 \\ 264 & 492 & -108 & -216 & 576 & -1008 \\ 1 & 1 & 1 & 1 & 1 & 1 \end{bmatrix} \tag{6.123}$$

We have

$$E_1 = D_1 = \text{diagonal}(0.02, 0.06, 0.06, 0.12, 0.18, 0.56) \tag{6.124}$$

However, we can also scale this matrix, and we take

$$E_1 = D_1 = \text{diagonal}(2, 6, 6, 12, 18, 56) \tag{6.125}$$

Even $c_1 = cE_1$ can be scaled and we will use [1 5 0 0 0 0] to calculate the vector in formula 6.111. The calculation is complicated and we recommend using a computer program that provides functions to do linear algebra calculations. When this step is completed we have

$$p = (0.0276, 0, 6669, 0, 0401, -0.5930, -0.3799, 0.2384) \tag{6.126}$$

$$\mathbf{1}^T + p = (1.0276, 1.6669, 1.0401, 0.4070, 0, 6201, 1.2384) \tag{6.127}$$

Though formula 6.114 calls for division by $n+1$, this is not necessary since everything will be scaled at the end.

The calculation of x_1 in formula 6.107 yields

$$x_1 = (0.0198, 0.0965, 0.0602, 0.0471, 0.1076, 0.6688) \tag{6.128}$$

It can be checked directly that this is a feasible solution of the constraints in 6.119. Since this problem is scaled from the original problem, we have to multiply this solution by 50 to obtain a feasible solution to the problem given by 6.117 and 6.115. This gives

$$(0.9910, 4.8226, 3.0090, 2.3548, 5.3819, 33.4407) \tag{6.129}$$

as a feasible solution for the original problem.

We knew that the value of the optimal solution was 36. The feasible value obtained for the initial feasible solution is 18. The value of the objective variable for this new solution is 27.0859.

The development of ellipsoidal algorithm, the Karmarkar algorithm, and the growing field of interior point methods represents an interesting turning point in mathematics. It has been a tradition that linear methods were more successful than nonlinear methods. It is ordinarily not possible to obtain exact solutions for nonlinear problems except in concept. A common approach to a nonlinear problem is to find a linear problem that is close to the nonlinear problem in some sense and use the solution of the linear problem as an approximate solution for the nonlinear problem.

Here we see nonlinear methods used to solve linear problems, and very successfully. So far the methods are not considered to be elementary, but research will undoubtedly lead to improvements and a better understanding of the methods. The calculations involved are also not easy, but the number of iterations climb slowly with an increase in the size of the problem and this makes them practical for large problems.

Exercises

1. Verify that the product of C, formula 6.46, and R, formula 6.47, is equal to A, formula 6.35.

2. Determine the LU decomposition of

$$\begin{bmatrix} 1^* & 2 & 3 \\ -1 & 2^* & 1 \\ 3 & 1 & 2^* \end{bmatrix}$$

Use the entries that are starred as pivot entries in order from top down.

3. Determine the LU decomposition of

$$\begin{bmatrix} 1^* & 2 & 3 \\ -1 & 2 & 1^* \\ 3 & 1^* & 2 \end{bmatrix}$$

Use the entries that are starred as pivot entries in order from top down.

4. Determine the LU decomposition of

$$\begin{bmatrix} 1^* & 2 & 3 \\ -1 & 2 & 1^* \\ 3 & 1^* & 2 \end{bmatrix}$$

Use the entries that are starred as pivot entries in order from left to right.

5. Convert the canonical linear program

x_1	x_2	x_3	-1	
1	2	−3	4	$= -x_4$
2	3	1	7	$= -x_5$
−3	2	3	0	

into a standard linear program.

6. Convert the following representation of a linear program in standard form into the form required for the Karmarkar algorithm. Use $M = 50$ as an upper bound.

x_1	x_2	x_3	x_4	x_5	-1	
1	2	−3	1	0	4	$= 0$
2	3	1	0	1	7	$= 0$
−3	2	3	0	0	0	$= f$

Part II

Related Problems

Chapter 7

Matrix Games

7.1 Matching Games

We start by using some simple games as examples from which we can move to more general matrix games. Probably the most widely familiar game is matching pennies. Though rudimentary, this game shares several points of similarity with those we wish to discuss.

- It is a *two-person game*.
- Each player has a finite number of actions that he can take.
- Each player takes his action independently of the action taken by the other player.
- There is a payoff specified to each player for each pair of actions taken by the two players.
- What one player wins (or loses) the other player loses (or wins).

These features can be represented for matching pennies in the game matrix 7.1

	H	T
H	1	−1
T	−1	1

(7.1)

One of the players is the *matching player* and the other player is the *nonmatching player*. The matching player chooses a row of the game matrix and the nonmatching player chooses a column of the game matrix.

Generally, the player that chooses a row is called the *row player*, and the player that chooses a column is the *column player*. The entry in the row and column chosen is the payoff to the row player. By implication, the negative of that entry is the payoff to the column player.

There is a general tendency to regard tossing the penny in air as a part of the rules of the game. We wish to promote the view that each player can make his choice any way he wishes and that tossing the coin is just a way to make that choice. By using the penny a player is doing two things. First, using this device makes the choice of heads or tails equally likely. That is, he chooses heads with probability $1/2$ and he chooses tails with probability $1/2$. Second, by using the penny openly he is revealing that he is choosing heads and tails with equal probabilities.

Though each player has only two actions that he can take, choosing either heads or tails, he is free to select the device or rule by which he makes that choice. He can choose heads with probability p and he can choose tails with probability q, where p and q are nonnegative and $p+q=1$. This gives him an infinite number of choices.

Before looking at any other possibilities, let us examine the effect of using a coin to make the choice. Suppose the matching player chooses heads and tails with equal probability. That is, $p = q = 1/2$. If the nonmatching player chooses heads, the expected value of the payoff to the matching player is $1/2 - 1/2 = 0$. If the nonmatching player chooses tails, the expected value of the payoff to the matching player is still 0. With this choice of probabilities, at least the expected value of the payoff to the matching player is not negative.

Suppose the nonmatching player chooses heads and tails with equal probability. If the matching player chooses heads, the expected value of the payoff to the matching player is 0. If the matching player chooses tails, the expected value of the payoff to the matching player is still 0. By this means the nonmatching player can be assured that the expected value of the payoff to the matching player will not be positive.

Thus, the matching player can assure himself that the expected value of his payoff will not be negative. The nonmatching player can assure himself that the expected value of the payoff to the matching player will not be positive. Since the interests of the matching and the nonmatching player are opposed, this is the best that each can do.

To sharpen the ideas involved in this example, let us change the values of the payoffs in the game matrix. Consider a matching game with the matrix 7.2.

	bk 2	rd 3
bk 1	1	−3
rd 4	−2	4

(7.2)

This game can be described in terms that may appear more realistic. Give each of the players two cards. Give the matching player a black 1 and a red 4. Give the nonmatching player a black 2 and a red 3. Each player chooses a card from his pair and they are compared. If they match in color, the matching player wins the amount shown on his card. If they do not match in color, the nonmatching player wins the amount shown on his card.

The cards given to each player add up to 5. Is the game fair to both players? Would you rather play the role of the matching player or the role of the nonmatching player? Before reading further, try seriously to answer these two questions.

Suppose the nonmatching player chooses the black 2 with probability 7/10 and the red 3 with probability 3/10. The expected value of the payoff to the matching player is then −2/10 regardless of what choices he makes.

This looks disadvantageous to the matching player. Assuming that he does not have the option of refusing to play at all, what is the best he can do? Suppose he chooses the black 1 with probability 6/10 and the red 4 with probability 4/10. Then, regardless of the choices made by the nonmatching player, the matching player's expected payoff is −2/10.

If the game is played repeatedly, the matching player can assure himself that he will lose, on the average, no more than 2/10 point per play. The nonmatching player can assure himself that the matching player will lose, on the average, at least 2/10 point per play. Most people would regard this game as unfair to the matching player. But that raises the question as to what would make a game a "fair" game.

A game is said to be *fair* if the expected value of the payoff to both players is zero. The unfair matching game described could be made fair by requiring the nonmatching player to pay the matching player 2/10 point per play of the game. Thus, by requiring each player to pay the other player an amount equal to his expected value, the game can be made fair. (In this example, the nonmatching player's expected value is 2/10 point, that is, he pays 2/10 point.) We say the expected value of the payoff function to each player is the *value of the game* to that player.

As a further example, consider the game with game matrix 7.3.

(7.3)

	bk 1	rd 1	rd 2
bk 1	1	−1	−2
rd 1	−1	1	1
bk 2	2	−1	−2

The matching player is given three cards: a black 1, a red 1, and a black 2. The nonmatching player is given three cards: a black 1, a red 1, and a red 2. Each player chooses a card and they are compared. If the colors match, the matching player receives the value of the card he shows. If the colors do not match, the nonmatching player receives the value of the card he shows.

Let us ask ourselves the same two questions. Is this game fair? If the game is not fair, would you rather play the role of the matching player or the role of the nonmatching player? Superficially, the game seems to favor the nonmatching player because of the larger payoff on the nonmatching 2's. So, before answering these two questions for this game, let us alter the game slightly by making a match on the 2's a tie. That is, consider the game matrix 7.4.

(7.4)

	bk 1	rd 1	rd 2
bk 1	1	−1	−2
rd 1	−1	1	1
bk 2	2	−1	0

With a little examination a few things become evident. For the matching player, the black 2 is at least as favorable as the black 1 for every choice of his opponent. Assuming that the matching player will not show the black 1, the nonmatching player sees that the red 1 is at least as favorable as the red 2. On this basis we examine the possibility that the matching player will never show the black 1 and the nonmatching player will never show the red 2.

Suppose the matching player chooses the red 1 with probability 3/5 and the black 2 with probability 2/5, and suppose the nonmatching player chooses the black 1 with probability 2/5 and the red 1 with probability 3/5. The results of these choices are shown in 7.5.

(7.5)

	2/5	3/5	0	
	bk 1	rd 1	rd 2	
0 bk 1	1	−1	−2	$= -1/5$
3/5 rd 1	−1	1	1	$= 1/5$
2/5 bk 2	2	−1	0	$= 1/5$
	$= 1/5$	$= 1/5$	$= 3/5$	

The probabilities of making the various choices are shown on the left and top margins. On the right margin are shown the expected values of the payoff for each choice of the matching player if the nonmatching player makes his choices with the indicated probabilities. On the bottom margin are shown the expected values of the payoff for each choice of the nonmatching player if the matching player makes his choices with the indicated probabilities.

If the choices are made with the probabilities indicated, the matching player is assured an expected value of at least $1/5$ against any choice made by the nonmatching player, and the nonmatching player is assured that the expected value will not be greater than $1/5$ against any choice made by the matching player. We regard $1/5$ as the value of this game to the matching player (and $-1/5$ as the value to the nonmatching player).

This certainly favors the matching player, so our attempt to make the game fair by calling a nonmatch on 2's a tie has failed. Let us re-examine the game with game matrix 7.3. From the point of view of the matching player this still leaves the black 2 at least as favorable as the black 1 against all choices of the nonmatching player. However, for the nonmatching player this makes the red 2 at least as favorable as the red 1 in all cases.

(7.6)

	1/2	0	1/2	
	bk 1	rd 1	rd 2	
0 bk 1	1	−1	−2	$= -1/2$
2/3 rd 1	−1	1	1	$= 0$
1/3 bk 2	2	−1	−2	$= 0$
	$= 0$	$= 1/3$	$= 0$	

For the probabilities shown on the margins of 7.6, the matching player can assure himself of a nonnegative expected value for the payoff and the nonmatching player can assure himself that the matching player will do no better than that. Thus, the value of this game to both players is zero, and we regard this as a fair game.

7.2 Optimal Strategies

We shall generalize the examples of matching games given in the preceding section to matrix games, define optimal strategies for playing matrix games and, in the following section, show how to determine optimal strategies.

Consider the game matrix 7.7.

$$\begin{array}{|ccc|} a_{11} & \cdots & a_{1n} \\ \vdots & & \vdots \\ a_{m1} & \cdots & a_{mn} \end{array} \tag{7.7}$$

One player selects a row and, independently, the other player selects a column. If the row player (RP) selects row i and the column player (CP) selects column j, then a_{ij} is the payoff to RP. We agree that by implication CP receives the payoff $-a_{ij}$.

Only a finite number of actions is available to either player. The row player selects a row and the column player selects a column. However, each player can use a probability distribution as a rule for making his choice. Let RP select row i with probability p_i. The set of probabilities $p = (p_1, p_2, \ldots, p_m)$ is called a *mixed strategy* for RP. The probabilities p_i must be nonnegative and sum to 1. If the strategy has one $p_i = 1$, it is called a *pure strategy*. Similarly, let CP select column j with probability q_j. The set of probabilities $q = (q_1, q_2, \ldots, q_n)$ is a mixed strategy for CP.

If RP uses the mixed strategy p then

$$\begin{aligned} e_1 &= p_1 a_{11} + \cdots + p_m a_{m1} \\ \vdots & \quad\ \vdots \qquad\quad \vdots \\ e_n &= p_1 a_{1n} + \cdots + p_m a_{mn} \end{aligned} \tag{7.8}$$

give the expected values of the payoff to CP against each column. If CP uses the mixed strategy q then

$$\begin{aligned} f_1 &= a_{11} q_1 + \cdots + a_{1n} q_n \\ \vdots & \quad\ \vdots \qquad\quad \vdots \\ f_m &= a_{m1} q_1 + \cdots + a_{mn} q_n \end{aligned} \tag{7.9}$$

give the expected values of the payoff to RP against each row.

To simplify notation let us write 7.8 in the form

$$e = pA \tag{7.10}$$

and 7.9 in the form

$$f = Aq \tag{7.11}$$

Then

$$eq = (pA)q = pAq = p(Aq) = pf \tag{7.12}$$

The tableau 7.13 shows 7.8 and 7.9 in a format similar to that used in the preceding section for the matching games.

$$\begin{array}{c|ccc|l} & q_1 & \cdots & q_n & \\ \hline p_1 & a_{11} & \cdots & a_{1n} & = f_1 \\ \vdots & \vdots & & \vdots & \vdots \\ p_m & a_{m1} & \cdots & a_{mn} & = f_m \\ \hline & = e_1 & \cdots & = e_n & \end{array} \tag{7.13}$$

The row player is concerned with the minimum value of e_j. That is the amount of which he is assured regardless of the action chosen by the column player. For a given mixed strategy p that minimum establishes a "floor" under the expected gains for RP. Similarly, the column player is concerned with the maximum value of f_i. That is the most that the row player can get for any action available to him. That maximum establishes a "ceiling" over the expected gains for RP and, therefore, over the expected losses for CP.

If u is the minimum of the e_j, then for any mixed strategy q we have

$$u = u(q_1 + q_2 + \cdots + q_n) \leq e_1q_1 + e_2q_2 + \cdots + e_nq_n = eq \tag{7.14}$$

This is because the q_i sum to 1 and are nonnegative. Similarly, if v is the maximum value of the f_i, then for any mixed strategy p we have

$$v = (p_1 + p_2 + \cdots + p_m)v \geq p_1f_1 + p_2f_2 + \cdots + p_mf_m = pf \tag{7.15}$$

Therefore, by 7.12,

$$u \leq eq = pAq = pf \leq v \tag{7.16}$$

The goal of RP is to establish a floor under his expectations that is as high as possible, and the goal of CP is to establish a ceiling over these expectations that is as low as possible. The inequality 7.16 says that the floor for RP established by selecting a mixed strategy p is not more than the ceiling established by the mixed strategy q selected by CP.

An *optimal strategy* for RP is a mixed strategy that maximizes RP's floor, and an optimal strategy for CP is a mixed strategy that minimizes the ceiling. In the spirit of the other problems considered in this book, 7.16 establishes a sufficient condition for optimality.

Theorem 7.51 *Mixed strategies p for RP and q for CP are optimal if they make $u = v$.* ◫

The sufficient condition stated in theorem 7.51 can be restated in terms of a tableau.

Theorem 7.52 *Mixed strategies p and q are optimal if there is a value v such that*

$$\begin{array}{c|ccc|c} & q_1 & \cdots & q_n & \\ \hline p_1 & a_{11} & \cdots & a_{1n} & \le v \\ \vdots & \vdots & & \vdots & \vdots \\ p_m & a_{m1} & \cdots & a_{mn} & \le v \\ \hline & \ge v & \cdots & \ge v & \end{array} \tag{7.17}$$

Proof. $v \le \min e_j = u$. By 7.16, $u \le u$. $\square$

In 1928 John von Neumann published a theorem that became the "main" theorem for his theory of games. This theorem asserts that for all matrix games the maximum ceiling is equal to the minimum floor.

Theorem 7.53 *(Minimax Theorem) For all matrix games*

$$\max_p \min_j e_j = \min_q \max_i f_i \tag{7.18}$$

We shall prove this theorem in the following section. When George B. Dantzig first described a linear programming problem to John von Neumann in 1947, von Neumann immediately saw that the problem of finding an optimal strategy for a player of a matrix game was a linear programming problem. But since there are two players in a matrix game, von Neumann suggested that linear programming problems came in pairs. This conjecture was proved by David Gale, Harold W. Kuhn and Albert W. Tucker in 1948.

Matrix games and linear programming were intertwined in the early years of the development of linear programming. Von Neumann regarded matrix games as the more fundamental and he considered a linear program as solved if it could be reduced to a game. We now regard linear programming as more fundamental and in the following section we shall provide methods for solving matrix games by reducing them to linear programming problems. Actually, the minimax theorem and the existence-duality theorem of linear programming are closely related, as we shall show in the following two sections.

The term "matrix game" (introduced by H. W. Kuhn) is an informal equivalent of the more technical term "zero-sum two-person game in normal form," used by von Neumann. Matrix games are a small part of the theory of games, but they constitute the cornerstone of the theory. Multi-move games, like chess and checkers, can, in principle, be reduced to matrix games by regarding a single sequence of choices for all conceivable alternatives as a single action in a matrix game. Games like tic-tac-toe can be analyzed by this method but, unfortunately, the game matrix for chess is too large to be realistically considered. Von Neumann estimated that the game matrix for chess would have more entries than there are particles in the universe!

Chance events, like rolling dice or dealing cards, can be woven into matrix games. The payoff values are then expected values rather than specific payments. Matrix games are basic to an understanding of games involving more than two players. These are many-person games that permit side payments and agreements between the players so that coalitions can be formed.

To every coalition **S** there is a complementary coalition $\bar{\mathbf{S}}$. If we regard the latter as merely aiming to thwart the former, we obtain a zero-sum two-party game involving these two coalitions. The minimax value of this game becomes the "characteristic value" of the coalition **S**. That is why von Neumann named the minimax theorem the Main Theorem of his theory of games.

We cannot pursue these general extensions of matrix games in this book. We mention them to emphasize that there is much more to game theory than is suggested by the very elementary examples given in this chapter. Even so, the ideas included here are fundamental to the theory.

7.3 Matrix Games and Linear Programs

In this section we shall describe two ways by which a matrix game can be converted into a linear program. Either conversion will accomplish two things. First, it will provide an effective, algorithmic method for computing optimal strategies for both players. Second, it will provide a means of proving the minimax theorem by reducing the minimax theorem to the existence-duality theorem of linear programming.

Consider the tableau 7.13 introduced for matrix games in the preceding section. Since u is the minimum of the e_j and v is the maximum of the f_i, tableau 7.13 can be converted into a slightly more familiar looking tableau 7.21 by introducing nonnegative slack variables

$$r_j = e_j - u \tag{7.19}$$

and

$$s_i = v - f_i \tag{7.20}$$

$$
\begin{array}{c|ccc|c|l}
 & q_1 & \cdots & q_n & -v & \\
\hline
p_1 & a_{11} & \cdots & a_{1n} & 1 & = -s_1 \\
\vdots & \vdots & & \vdots & \vdots & \vdots \\
p_m & a_{m1} & \cdots & a_{mn} & 1 & = -s_m \\
\hline
-u & 1 & \cdots & 1 & 0 & = 1 \\
\hline
 & = r_1 & \cdots & = r_n & = 1 &
\end{array}
\tag{7.21}
$$

Notice that the conditions that the p_i and q_j sum to 1 are also included in tableau 7.21.

The goal of RP is to maximize u, the floor under his expectations. The goal of CP is to minimize v, the ceiling over his losses. Tableau 7.21 is almost the tableau of a linear program, except that the objective variables seem to be in the wrong places. This can be remedied by introducing two new objective variables, $f = -v$ and $g = -u$. In this way we obtain tableau 7.22.

$$
\begin{array}{c|ccc|c|c|l}
 & q_1 & \cdots & q_n & -v & -1 & \\
\hline
p_1 & a_{11} & \cdots & a_{1n} & 1 & 0 & = -s_1 \\
\vdots & \vdots & & \vdots & \vdots & \vdots & \vdots \\
p_m & a_{m1} & \cdots & a_{mn} & 1 & 0 & = -s_m \\
\hline
-u & 1 & \cdots & 1 & 0 & 1 & = 0 \\
\hline
-1 & 0 & \cdots & 0 & 1 & 0 & = f \\
\hline
 & = r_1 & \cdots & = r_n & = 0 & = g &
\end{array}
\tag{7.22}
$$

In tableau 7.22 we are to maximize f (minimize v) and minimize g (maximize u). In this context we regard u and v as free variables. There are two artificial variables, indicated by the 0's, dual to the free variables $-u$ and $-v$. All other variables on the margins of tableau 7.22 are canonical variables. Thus, tableau 7.22 represents a pair of general dual linear programs.

As described in Chapter 5, one way to reduce general dual linear programs to canonical dual linear programs is to pivot the free variables to the positions of basic variables and the artificial variables to the positions of nonbasic variables. In tableau 7.22 this is easily achieved by two pivot exchanges, one in the u-row and one in the v-column.

Let us be specific about the calculations involved. Suppose the pivot entries are chosen in the q_k-column and in the p_h-row, as shown in tableau 7.23. We also show a typical p_i-row and q_j-column.

(7.23)

	$\cdots$	q_k	$\cdots$	q_j	$\cdots$	$-v$	-1	
		$\vdots$		$\vdots$		$\vdots$	$\vdots$	
p_h	$\cdots$	a_{hk}	$\cdots$	a_{hj}	$\cdots$	1^*	0	$= -s_h$
		$\vdots$		$\vdots$		$\vdots$	$\vdots$	
p_i	$\cdots$	a_{ik}	$\cdots$	a_{ij}	$\cdots$	1	0	$= -s_i$
		$\vdots$		$\vdots$		$\vdots$	$\vdots$	
$-u$	$\cdots$	1^*	$\cdots$	1	$\cdots$	0	1	$= 0$
-1	$\cdots$	0	$\cdots$	0	$\cdots$	1	0	$= f$
	$\cdots$	$= r_k$	$\cdots$	$= r_j$	$\cdots$	$= 0$	$= g$	

After the first pivot exchange in the q_k-column we have

(7.24)

	$\cdots$	0	$\cdots$	q_j	$\cdots$	$-v$	-1	
		$\vdots$		$\vdots$		$\vdots$	$\vdots$	
p_h	$\cdots$	$-a_{hk}$	$\cdots$	$a_{hj} - a_{hk}$	$\cdots$	1^*	$-a_{hk}$	$= -s_h$
		$\vdots$		$\vdots$		$\vdots$	$\vdots$	
p_i	$\cdots$	$-a_{ik}$	$\cdots$	$a_{ij} - a_{ik}$	$\cdots$	1	$-a_{ik}$	$= -s_i$
		$\vdots$		$\vdots$		$\vdots$	$\vdots$	
r_k	$\cdots$	1	$\cdots$	1	$\cdots$	0	1	$= -q_k$
-1	$\cdots$	0	$\cdots$	0	$\cdots$	1	0	$= f$
	$\cdots$	$= -u$	$\cdots$	$= r_j$	$\cdots$	$= 0$	$= g$	

The column headed by the zero in 7.24 does not contribute to the row system of equations and can be deleted. That column could be saved if it were important to know the value of u. However, the value of u is available since $u = -g$. Thus, we drop the column headed by zero, pivot in the p_h-row, and obtain tableau 7.25.

(7.25)

	q_j	s_h	-1	
0	$a_{hj} - a_{hk}$	1	$-a_{hk}$	$= v$
p_i	$a_{ij} - a_{ik} - a_{hj} + a_{hk}$	-1	$a_{hk} - a_{ik}$	$= -s_i$
r_k	1	0	1	$= -q_k$
-1	$a_{hk} - a_{hj}$	-1	a_{hk}	$= f$
	$= r_j$	$= p_h$	$= g$	

Again, the row headed by zero will be dropped.

After the row headed by zero is deleted, the b-column contains a 1 and entries of the form $a_{hk} - a_{ik}$. Thus, we can obtain a nonnegative b-column by choosing as pivot entry a_{hk} the maximum entry in the q_k-column. This shows that the max program in tableau 7.22 is feasible.

The c-row contains a -1 and entries of the form $a_{hk} - a_{hj}$. Thus, we could obtain a nonpositive c-row by choosing as pivot entry a_{hk} the minimum entry in the p_h-row. This shows that the min program in tableau 7.22 is feasible.

The conclusions of the preceding two paragraphs are valid for every game matrix. That is, for every game matrix 7.13 both of the general dual linear programs in the corresponding tableau 7.22 are feasible. By the existence-duality theorem of linear programming, $\max f = \min g$. In turn, this shows that $\min v = \max u$. This proves the minimax theorem.

Returning our attention to tableau 7.25, note that the entry in the d-corner is $a_{hk} = -f$. We can make the objective variable f large by making a_{hk} small. This reasoning leads to the following rule for the first two pivot exchanges.

Find the maximum entry in each column of the game matrix. Choose the column in which this maximum is minimal. This determines the pivot column, column h. The pivot row, row k, is the row in which this maximum occurs.

The first two pivot exchanges in tableau 7.22 are on 1's and are easy to perform. But it is even easier to obtain tableau 7.25 (with the row headed by zero deleted) directly from the game matrix 7.13. Between tableau 7.13 and tableau 7.22 we inserted a row and a column. Between tableau 7.22 and tableau 7.25 we deleted a row and a column. We can consider the new row and column as replacements for the row and column deleted. The new row consists of 1's and the new column consists of −1's, except for the 0 where they intersect.

The other entries can be computed directly from the expressions shown in tableau 7.25.

Replace a_{ij} by $a_{ij} - a_{ik} - a_{hj} + a_{hk}$, for $i \neq h$, $j \neq k$.

Set $b_i = a_{hk} - a_{ik}$, for $i \neq h$.

Set $c + j = a_{hk} - a_{hj}$, for $j \neq k$.

Set $d = a_{hk}$.

Let us see how these ideas work in an example. Consider the game matrix 7.26.

2	−3	4
−3	1*	−6
0	−1	1

(7.26)

The column maxima are 2, 1, 4, and the minimum maximum occurs in column 2. Thus, the pivot column and pivot row are those containing the entry with the asterisk in 7.26. The linear programming tableau 7.27 can be obtained directly by the rules described above.

(7.27)

	q_1	s_2	q_3	-1	
p_1	9	−1	14	4	$= -s_1$
r_2	1	0	1	1	$= -q_2$
p_3	5*	−1	9	2	$= -s_3$
-1	4	−1	7	1	$= f$
	$= r_1$	$= p_2$	$= r_3$	$= g$	

Tableau 7.27 represents a pair of dual linear programs. We apply the simplex algorithm and perform a pivot exchange on the entry indicated with the asterisk.

(7.28)

	s_3	s_2	q_3	-1	
p_1	−9/5	4/5	−11/5	2/5	$= -s_1$
r_2	−1/5	1/5	−4/5	3/5	$= -q_2$
r_1	1/5	−1/	9/5	2/5	$= -q_1$
-1	−4/5	−1/5	−1/5	−3/5	$= f$
	$= p_3$	$= p_2$	$= r_3$	$= g$	

From tableau 7.28 we obtain the optimal strategy $p = (0, 1/5, 4/5)$ for the row player and the optimal strategy $q = (2/5, 3/5, 0)$ for the column player. The value of the game is the common value of $u = v = -f = -3/5$.

The values obtained for the optimal strategies should be checked in the original game matrix.

$$
\begin{array}{c|ccc|l}
 & 2/5 & 3/5 & 0 & \\
\hline
0 & 2 & -3 & 4 & = -1 \\
1/5 & -3 & 1 & -6 & = -3/5 \\
4/5 & 0 & -1 & 1 & = -3/5 \\
\hline
 & = -3/5 & = -3/5 & = -2/5 &
\end{array}
\tag{7.29}
$$

On the bottom margin we see that the floor is $-3/5$, and on the right margin we see that the ceiling is also $-3/5$. Even without a general justification for the method, this shows that the two strategies are optimal.

There is something else that can be read from the tableau yielding the optimal solutions. Note from tableau 7.22 that s_i is the variable dual to p_i and r_j is the variable dual to q_j. These are slack variables. Specifically, s_i is positive for an optimal strategy for the column player if the payoff for that choice is unfavorable for the row player. In that case every optimal strategy for the row player will use that choice with probability zero. Thus, in tableau 7.28, since s_1 is positive, p_1 must be zero, and since r_3 is positive q_3 must be zero.

If h and k are selected by the rule given above, after the initial pair of pivot exchanges the b-column will be nonnegative. It might happen that the c-row is also nonpositive. This will occur if the pivot entry is both the greatest entry in its column and the least entry in its row. An entry with that property is called a *saddle point.* In this case we have optimality immediately and $p_h = 1$, $p_i = 0$ for $i \neq h$, and $q_k = 1$, $q_j = 0$ for $j \neq k$ are the optimal strategies. For example,

$$
\begin{array}{c|cccc|l}
 & 0 & 0 & 1 & 0 & \\
\hline
0 & 2 & 2 & 3 & 9 & = 3 \\
1 & 6 & 5 & 4^* & 6 & = 4 \\
0 & 7 & 6 & 2 & 1 & = 2 \\
\hline
 & = 6 & = 5 & = 4 & = 6 &
\end{array}
\tag{7.30}
$$

The entry 4 is a saddle point. In this case each player has an optimal strategy that chooses one action with probability 1. Such a strategy is called a *pure strategy.*

A second method for reducing a matrix game to a linear program is available if the maximum floor u and the minimum ceiling v (which we now know to be equal) are both positive. This can be guaranteed if, for example, the entries in the game matrix are all positive. In fact, it is sufficient for the game matrix to have at least one row with positive entries. Every matrix game is equivalent to a matrix game with a positive game matrix, in the following sense. If we add a constant to every entry in a game matrix we obtain another game matrix in which the relative advantages of the choices remain the same. It is the same as if the game were played first and then a bonus equal to the added constant were awarded later.

Return to the tableau 7.21; divide the entries on the top and right margins by v, and divide the entries on the left and bottom margins by u. We obtain tableau 7.31.

$$
\begin{array}{c|cccc|c|l}
 & q_1/v & \cdots & q_n/v & & -1 & \\
\hline
p_1/u & a_{11} & \cdots & a_{1n} & & 1 & = -s_1/v \\
\vdots & \vdots & & \vdots & & \vdots & \vdots \\
p_m/u & a_{m1} & \cdots & a_{mn} & & 1 & = -s_m/v \\
\hline
-1 & 1 & \cdots & 1 & & 0 & = 1/v \\
\hline
 & = r_1/u & \cdots & = r_n/u & & = 1/u &
\end{array}
\tag{7.31}
$$

The systems of row equations in tableaux 7.21 and 7.31 are equivalent, and the systems of column equations in tableaux 7.21 and 7.31 are equivalent. In tableau 7.31, the problem of maximizing $1/v$ is equivalent to minimizing v, and the problem of minimizing $1/u$ is equivalent to maximizing u.

After the solutions for the linear programs in tableau 7.31 are obtained by the simplex algorithm, the solutions must be rescaled by the factor $u = v$ to obtain the probabilities of the optimal strategies.

A choice between these two methods is largely a matter of personal preference. The amount of work required seems to be about the same for either method.

7.4 Linear Programs and Symmetric Games

The minimax theorem of game theory and the existence-duality theorem of linear programming are equivalent. By this we mean that either theorem can be proved from the other by an argument that is relatively direct. In the preceding section we showed that the minimax theorem follows from the existence-duality theorem.

Historically, the minimax theorem was the earlier theorem, proved in von Neumann's 1928 paper. Over the years several proofs, all nonelementary, of the minimax theorem were given by use of topological arguments. In the book *Theory of Games and Economic Behavior* by von Neumann and Morgenstern (1944), a proof was given based on a separating hyperplane theorem, also a nonelementary theorem. These proofs were nonconstructive. That is, they proved that optimal strategies existed for both players, but they did not suggest a computational method for finding these optimal strategies.

On a theoretical level, a number of properties of optimal strategies were identified. Before linear programming was available as a computational method, these properties could be used to determine optimal strategies, but the methods employed were more ingenious than systematic. Nevertheless, in the early days of the development of linear programming much more was known about games than about linear programming, and there was considerable interest in reducing linear programs to matrix games. That is what we are going to do here.

Actually, we are going to reduce linear programs to "symmetric games." In all matrix games the amount paid by the row player to the column player is the negative of the amount paid by the column player. Thus, if we interchange the row player and the column player, we would interchange rows and columns and change the signs of the entries. This amounts to taking the negative transpose of the game matrix. So if a game matrix is skew-symmetric it does not matter whether one plays the role of the row player or the role of the column player. Thus, a game with a skew-symmetric game matrix is a *symmetric game*. In a symmetric game both players have the same optimal strategies and the value of the game is 0.

Let

(7.32)

<table>
<tr><td></td><td>x</td><td>-1</td><td></td></tr>
<tr><td>v</td><td>A</td><td>b</td><td>$= -y$</td></tr>
<tr><td>-1</td><td>c</td><td>0</td><td>$= f$</td></tr>
<tr><td></td><td>$= u$</td><td>$= g$</td><td></td></tr>
</table>

be the tableau for a pair of dual canonical linear programming problems. Optimal x and v are nonnegative solutions of

$$Ax - b \leq 0 \tag{7.33}$$

$$vA - c \geq 0 \tag{7.34}$$

such that

$$vb = cx \tag{7.35}$$

But 7.33 and 7.34 imply

$$vb \geq vAx \geq cx \tag{7.36}$$

for $x \geq 0$ and $v \geq 0$. So 7.35 can be replaced by

$$vb \leq cx \tag{7.37}$$

Conversely, any nonnegative solutions of 7.33, 7.34, and 7.37 provide optimal x and v for the dual linear programs in 7.32.

Theorem 7.54 *The canonical dual linear programs in 7.32 have optimal solutions determined by x and v if, and only if, x and v are nonnegative solutions of the linear inequalities 7.33, 7.34, and 7.37.* □

Now rewrite 7.33, 7.34, 7.37 as

$$Ax - b \leq 0, -A^T v^T + c^T \leq 0, b^T v^T - cx \leq 0 \tag{7.38}$$

and again (by negative transposition) as

$$-x^T A^T + b^T \geq 0, vA - c \geq 0, -vb + x^T c^T \geq 0 \tag{7.39}$$

We combine these in tableau 7.40, with 7.38 as the row inequalities and 7.39 as the column inequalities.

(7.40)

	v^T	x	1	
v	0	A	$-b$	≤ 0
x^T	$-A^T$	0	c^T	≤ 0
1	b^T	$-c$	0	≤ 0
	≥ 0	≥ 0	≥ 0	

The thing that is striking about tableau 7.40 is that the matrix is skew-symmetric. This suggests constructing a symmetric game with the skew-symmetric matrix in 7.40 as the game matrix. Define $t > 0$ by

$$t(v_1 + \cdots + v_m + x_1 + \cdots + x_n + 1) = 1 \tag{7.41}$$

and then set $p = tv$, $q = tx$. This gives $p \geq 0$, $q \geq 0$ and

$$p_1 + \cdots + p_m + q_1 + \cdots + q_n + t = 1 \tag{7.42}$$

Now, multiplying by t on the left and top margins of 7.40, we obtain 7.43—the game matrix for a symmetric game.

	p^T	q	t	
p	0	A	$-b$	≤ 0
q^T	$-A^T$	0	c^T	≤ 0
t	b^T	$-c$	0	≤ 0
	≥ 0	≥ 0	≥ 0	

(7.43)

If the tableau 7.32 has optimal solutions for both programs, then the matrix game defined by 7.43 has optimal strategies for both players with $t > 0$. There may be many optimal strategies, but if there is at least one optimal strategy with $t > 0$ for the symmetric game attached to 7.43, then the steps from 7.32 to 7.43 can be retraced to provide optimal solutions for the dual linear programs in 7.32.

Theorem 7.55 *The canonical dual linear programs in 7.32 have optimal solutions determined by x and v if, and only if, the symmetric game in 7.43 has optimal strategies determined by $p = tv$, $q = tx$, and $t > 0$.* □

Exercises

Solve the following matrix games. That is, determine optimal strategies for each player and the value of the game. Where there is more than one optimal strategy, find all the extreme strategies—those that are not linear combinations of other optimal strategies.

1.

	C_1	C_2	C_3	C_4
R_1	2	3	−1	2
R_2	1	0	3	2
R_3	−1	2	1	−1

2.

	C_1	C_2	C_3	C_4
R_1	4	6	−2	4
R_2	2	0	6	4
R_3	−2	4	2	−2

3.

	C_1	C_2	C_3	C_4
R_1	4	5	1	4
R_2	3	2	5	4
R_3	1	4	3	1

4.

	C_1	C_2	C_3	C_4	C_5
R_1	-1	0	-1	1	-1
R_2	2	-1	-1	0	1
R_3	-1	-1	1	0	1
R_4	0	2	-1	-2	0
R_5	0	1	1	0	-1

5. **Scissors–Paper–Stone.** This is a traditional game. Two players simultaneously name one of the three objects. If both name the same object, the game is a draw. Otherwise, Scissors cuts Paper, Paper wraps Stone, and Stone breaks Scissors. The player with the superior choice wins one point from the other player. Determine the optimal strategies and the value of the game.

6. Two players simultaneously announce a nonnegative integer. If both players choose the same integer it is a draw and the payoff is zero. Otherwise, the player that chooses the smaller integer wins one point from the other player unless it is smaller by exactly one. In that case he loses two points. Find optimal strategies for both players and the value of the game.

7. **Morra.** Morra is a traditional game. Two players clench their fists. At a given signal each shows his hand with a number of fingers extended and simultaneously announces a guess of the number of fingers extended by the other player. The game is usually played with the number of fingers a player can extend limited to 1, 2, or 3. If only one player guesses correctly, he wins a number of points equal to the total number of fingers extended. Find optimal strategies for each player and the value of the game.

8. A classical book on game theory is *The Compleat Strategyst,* written by J. D. Williams. The narratives for the games are a delight of whimsy. It was written (1954) before it was generally known how to solve games with linear programming and the methods he gives for solving games are pretty much ad hoc.

The Bass and the Professor. The hero of this story is a bright young centrarchoid (Micropterus), unimaginatively described as a "perchlike fish much esteemed for food." The villain is Angler Kleene. These are familiarly known as the Bass and the Professor. Together with certain insects and water, they constitute a natural-habitat group.

The insects—horntails, dragonflies, and bumblebees—are also much esteemed as food. They are not equally common on the surface of the pool; if the Professor adds one (complete with hook) to any species and the Bass feeds on that species, the lethalities are as $2 : 6 : 30$, which means there are 5 times as many dragonflies as bumblebees and 3 times as many horntails as dragonflies. How should the Bass feed, and the Professor angle?

(Part of the inside humor in Williams's book is that it helps to know who his friends were. The Professor is Stephen Kleene, a prominent logician.)

9. **The Detective and the Mobster**. A detective is assigned the mission of tailing a mobster that a tip says is meeting in a certain building with three of his henchmen. The detective has never seen the man he is after. If the detective tails the right person, he accomplishes his goal. If he tails the wrong person, the mobster gets away and the mission is a failure.

The mobster has reason to believe that he is being tracked. His strategy is to have his cohorts leave the building one-at-a-time. He will choose the time to leave himself. The detective's option is to choose to tail the first, second, third, or fourth person who leaves the building.

If the detective picks a person to follow at random, his chances of success are one-fourth. While the detective is waiting he tries to think of some way to improve his chances. He then remembers that the mobster is known to be the tallest person at the meeting. That means he will not follow anyone who is shorter than anyone who has already left. He decides that he will pick a time to follow someone by a random device, but he follows someone at that time or later only if he is taller than everyone who has already left. What are the optimal strategies for the detective and for the mobster? How likely is the detective to get his man?

(Morton D. Davis, *Game Theory, a Nontechnical Introduction*, p36)

10. Formulate the **Maintenance Manager's Problem** of Chapter 1 as a "game against nature." That is, the maintenance manager considers that there are unknown malevolent forces against him and his objective is to minimize his loses. This is the way one used to describe such problems. Currently, one wonders whether the situation is not man as the malevolent force against nature.

11. **Simple Poker**. Under certain conditions poker can be cast as a matrix game. Analysis of poker in these terms is very complex and only a few poker games have been analyzed, including blackjack and faro. There are $\binom{52}{5} = 2,598,960$ different five-card poker hands. The usual colorful description of these hands serves only to give them an ordering. Thus, one could look on the deal as giving each player an integer between 1 and 2,598,960, and then the betting takes place. The game matrix for a simple

poker game would have a row and column for each integer in that range, and even more if there were draws and subsequent bets!

We are going to look at a very simple abstraction of poker to make an interesting point. The deck has only four cards, two high cards and two low cards. Each player is dealt one card. Wagers are made simultaneously. Each player can bet 8 units or 12 units. If they bet the same amount, the cards are compared. If the cards are different, the higher card wins. If they are the same it is a draw. If they bet different amounts, the player who bet the smaller amount can either fold, in which case he loses without comparing the cards, or call by raising the amount of his bet to 12.

Determining the game matrix is not very interesting and it has nothing to do with the subject of this book, so we will give it to you. Each player has three choices for each card he will receive, nine choices in all. To allow compact notation we will denote the choices for each card by an integer: 1 means bet low and fold if the other player bets high, 2 means bet low and call if the other player bets high, 3 means bet high. Then (i, j) means use choice i for the low card and use choice j for the high card. In these terms the game matrix is

	$(1,1)$	$(1,2)$	$(1,3)$	$(2,1)$	$(2,2)$	$(2,3)$	$(3,1)$	$(3,2)$	$(3,3)$
$(1,1)$	0	0	−2	0	0	−2	−6	−6	−8
$(1,2)$	0	0	0	0	0	0	−1	−1	−1
$(1,3)$	2	0	0	3	1	1	1	−1	−1
$(2,1)$	0	0	−3	0	0	−3	−4	−4	−7
$(2,2)$	0	0	−1	0	0	−1	1	1	1
$(2,3)$	2	0	−1	3	1	0	3	1	0
$(3,1)$	6	1	−1	4	−1	−3	0	−5	−7
$(3,2)$	6	1	1	4	−1	−1	5	0	0
$(3,3)$	8	1	1	7	0	0	7	0	0

Show that $(0, 0, 0, 0, 0, 0.5, 0, 0, 0.5)$ is an optimal strategy for both players.

Questions

Q1. A matrix game always has optimal strategies for both players.

Q2. A matrix game with only integer entries always has optimal solutions that are rational numbers.

Q3. A matrix game can be fair without being symmetric.

Q4. Every canonical slack variable in the constructed linear program for a matrix game corresponds to a pure strategy for the opposing player.

Q5. If a slack variable in an optimal strategy for the constructed linear program is positive, the choice corresponding to the dual variable in the opponent's list of pure strategies will not be played with a positive probability in any optimal strategy.

Q6. If an optimal strategy for one player does not use all pure strategies with positive probabilities, then the optimal strategy for the opposing player is not unique.

Q7. If the linear program corresponding to a matrix game has only one optimal basic solution, the corresponding player for the matrix game has a unique optimal strategy.

Q8. A player of a matrix game must always use an optimal strategy to assure himself that he will obtain the optimal expected value of the game.

Q9. If one player uses an optimal strategy and the other player knows that he is using an optimal strategy, the other player does not have to use an optimal strategy.

Q10. If one player uses an optimal strategy, it does not matter what strategy the other player uses.

Chapter 8

Assignment and Matching Problems

8.1 The Assignment Problem

Consider a situation where we have a number of jobs to fill and an equal number of candidates for those positions. Suppose each candidate has been tested or otherwise rated for each available position. The problem is to assign the candidates to the positions in some optimal way. If the ratings are, for example, the times it takes each candidate to perform each task, we might choose as a measure of the proposed assignment pattern the sum of the ratings (times) of the candidates in the positions to which they are assigned. An assignment would be optimal if the sum of the assigned ratings were minimal.

There is no intrinsic reason why the assignment problem should be formulated to minimize the sum of the ratings. If a high rating indicates a high skill level, for example, we might seek to maximize the sum of the ratings. However, we shall take the minimization formulation as the standard form of the assignment problem. A further motivation for choosing this formulation is that, if we interpret the ratings as costs, there is an interesting connection between the transportation problem (to be discussed in the following chapter) and the assignment problem.

Suppose we are given a cost matrix in the form

(8.1)

12	17	8	16	20
9	5	12	6	15
15	16	15	14	17
22	24	17	28	26
15	10	12	12	15

Here, the number c_{ij} in row i and column j is the cost of assigning candidate i to task j. Our goal is to select one number from each row and each column so that the sum is as small as possible.

Since we must pick exactly one number from each row, we can add or subtract a constant from all numbers in a row without changing the relative advantage of any choice. We can simplify the problem by subtracting the minimum number in each row from each number in that row. In 8.2 we write the minimum for each row on the left margin and show the remainder within the matrix.

(8.2)

8	4	9	0	8	12
5	4	0	7	1	10
14	1	2	1	0	3
17	5	7	0	11	9
10	5	0	2	2	5

Since we must pick exactly one number from each column, we can add or subtract a constant from all numbers in a column without changing the relative advantage of any choice. In 8.3 we write the minimum for each column on the top margin and show the remainder within the matrix.

(8.3)

	1	0	0	0	3
8	3	9	0	8	9
5	3	0	7	1	7
14	0	2	1	0	0
17	4	7	0	11	6
10	4	0	2	2	2

A set of entries from a cost matrix is said to be *independent* if no two are in the same row or column. Our problem is to choose five independent entries from the cost matrix 8.1 for which the sum is minimal. For each choice of five independent entries from 8.1, the sum of these five entries differs from the sum of the corresponding entries from 8.3 by the sum of all entries on the margins of 8.3. Thus, if we choose five independent entries from 8.3 with a minimal sum, the optimal assignment in the cost matrix 8.1 will be obtained by making the choices in the same positions. The minimum sum in 8.1 would be larger than the minimum sum in 8.3 by the sum of the entries on the margins. Since all entries in 8.3 are nonnegative, if we could select five independent zeros we would obtain a minimal sum.

In general, a matrix obtained from the original cost matrix by subtracting constants from all entries in a row or column is called a *reduced cost matrix.* By subtracting these common costs from the rows and columns in different ways we can obtain different reduced cost matrices. For example, if we had subtracted the minimum entry in each column from the entries in the column before subtracting the minimum from each row, we would have obtained another reduced cost matrix. If we can obtain a reduced cost matrix with five (n, in the general $n \times n$ assignment problem) independent zeros, these independent zeros yield an optimal choice for the reduced cost matrix and an optimal choice for the original cost matrix.

An independent set of zeros with a largest number of elements is called a *largest independent set* of zeros. In the reduced cost matrix 8.4 we show a set of three independent zeros, each marked with an asterisk. Since there is no independent set of zeros with four or five elements this is a largest independent set.

(8.4)

	1	0	0	0	3
8	3	9	0*	8	9
5	3	0*	7	1	7
14	0*	2	1	0	0
17	4	7	0	11	6
10	4	0	2	2	2

In the matrix 8.3 it is not difficult to see by inspection that there is no independent set of zeros with more than three elements. But in a much larger matrix, how can we be sure that a set of zeros we claim is largest actually is largest? We need some criterion or test that is decisive for any reduced cost matrix. Our immediate goal is to describe such a criterion.

A *cover* for the set of zeros in a cost matrix is a set of vertical and horizontal lines that covers every zero in the matrix. In the cost matrix 8.5 we show a set of three lines that covers the zeros in that matrix. A cover always exists since we could either draw a line through every row or through every column. On the other hand the number of lines in a cover cannot be less than the largest number of independent zeros since each line can cover at most one zero in an independent set of zeros.

(8.5)

	1	0	0	0	3
8	3	9	0*	8	8
5	3	0*	7	1	7
14	0*	2	1	0	0
17	4	7	0	11	6
10	4	0	2	2	2

Since no two of the starred zeros are in the same line, no line can cover two starred zeros. The cover shown in 8.5 contains three lines. Since there is a set of three independent zeros shown, three is the maximum number of independent zeros and three is the minimum number of lines in a cover. It turns out to be true in general that the number of zeros in a largest independent set of zeros is equal to the number of lines in a smallest cover.

This theorem is known as König's theorem. In a following section we shall prove this theorem and describe an algorithm for finding a largest independent set of zeros and a smallest cover. For small-scale assignment problems, up to 10×10 or so, it is relatively easy to find a largest independent set of zeros and a smallest cover by inspection. If the number of zeros is small there are not many possibilities to consider, and if the number of zeros is large it is likely that we can find n independent zeros. In this section we assume we can find a cover for which the number of lines is equal to the number of zeros in a largest independent set of zeros.

Let us designate an entry in row i and column j of the original cost matrix 8.1 by c_{ij}. Designate the number on the left margin of row i in 8.5 by u_i and the number on the top margin of column j by v_j. Let w_{ij} denote the corresponding entry within the reduced cost matrix 8.5. Then

$$w_{ij} = c_{ij} - (u_i + v_j) \tag{8.6}$$

Suppose we subtract a constant e from every number on the left margin for a row that is covered in 8.5, and add e to every number on the top margin for a column that is not covered. The new numbers on the margins are shown in 8.7.

(8.7)

	$1+e$	0	0	$0+e$	$3+e$
8	3	9	0*	8	9
5	3	0*	7	1	7
$14-e$	0*	2	1	0	0
17	4	7	0	11	6
10	4	0	2	2	2

Let us compute the new $w'_{ij} = c_{ij} - (u'_i + v'_j)$, using the new numbers u'_i and v'_j on the margins. If an entry is not covered, it is

$$\begin{aligned} w'_{ij} = c_{ij} - (u_i + v_j + e) &= c_{ij} - (u_i + v_j) - e \\ &= w_{ij} - e \end{aligned} \tag{8.8}$$

If an entry is covered once, it is either

$$w'_{ij} = c_{ij} - (u_i + v_j) = w_{ij} \tag{8.9}$$

or

$$w'_{ij} = c_{ij} - (u_i - e + v_j + e) = w_{ij} \tag{8.10}$$

If an entry is covered twice, it is

$$w'_{ij} = c_{ij} - (u_i - e + v_j) = w_{ij} + e \tag{8.11}$$

In summary, we decrease every noncovered entry by e, increase every twice-covered entry by e, and leave every once-covered entry unchanged. We wish to obtain a new w-matrix (reduced cost matrix) with nonnegative entries and with at least one new zero that is located at a noncovered entry. Since all noncovered entries are decreased by e, we choose the smallest noncovered entry for the value of e. The smallest noncovered entry is positive since all zeros are covered. In 8.5 this smallest entry is 1. Using these new values for the u'_i and v'_j on the margins we obtain 8.12.

(8.12)

	2	0	0	1	4
8	2	9	0*	7	8
5	2	0	7	0*	6
13	0*	3	2	0	0
17	3	7	0	10	5
10	3	0*	2	1	1

In 8.12 we have identified four independent zeros and we show four lines that cover all zeros. All the noncovered entries are positive and the minimum noncovered entry is 2. Again, we subtract 2 from every left margin entry corresponding to a covered row, add 2 to every top margin entry corresponding to a noncovered column, subtract 2 from every noncovered entry, and add 2 to every twice-covered entry. In this way we obtain table 8.13.

(8.13)

	4	2	0	3	6
8	0*	7	0	5	6
3	2	0	9	0*	6
11	0	3	4	0	0*
17	1	5	0*	8	3
8	3	0*	4	1	1

In 8.13 there is an independent set of zeros with five elements, one from each row and one from each column. They represent an optimal assignment in the reduced cost matrix 8.13. We mark the corresponding positions of the original cost matrix with asterisks to indicate the assignment. We obtain

(8.14)

12*	17	8	16	20
9	5	12	6*	15
15	16	15	14	17*
22	24	17*	28	26
15	10*	12	12	15

For the assignment shown in 8.14, no entry chosen is the smallest in both its row and its column. If each entry chosen were the smallest in its row and column it would be easy to conclude that we had made the optimal choice. Nothing so easy or obvious will show that the assignment indicated in 8.14 is a minimum. It is clearly desirable to have a general and effective test for optimality. In the next section we shall describe such a criterion and an algorithm, the Hungarian algorithm, for obtaining an optimal assignment.

8.2 Kuhn's Hungarian Algorithm

In general, consider a cost matrix of the form

$$\begin{array}{|cccc|} c_{11} & c_{12} & \cdots & c_{1n} \\ c_{21} & c_{22} & \cdots & c_{2n} \\ \vdots & \vdots & & \vdots \\ c_{n1} & c_{n2} & \cdots & c_{nn} \end{array} \qquad (8.15)$$

The assignment problem is to select one c_{ij} from each row and each column for which the sum is least. For notational convenience we introduce an $n \times n$ permutation matrix X with entries $x_{ij} = 0$ or 1, arranged so that each row and column of X contains a single 1. For example, the permutation matrix 8.16 has a 1 in each position corresponding to a starred entry in the cost matrix 8.13.

1	0	0	0	0
0	0	0	1	0
0	0	0	0	1
0	0	1	0	0
0	1	0	0	0

(8.16)

Clearly, there is a one-to-one correspondence between the possible assignments in the $n \times n$ cost matrix 8.15 and the possible $n \times n$ permutation matrices. Furthermore, the sum of the entries selected by the assignment can be written in the form

$$\sum_{(i,j)} c_{ij} x_{ij} \tag{8.17}$$

where the notation in 8.17 means that the sum is taken over all possible combinations of values of i and j.

The assignment problem can now be cast in the following form.

Assignment Problem: Given any $n \times n$ cost matrix $C = [c_{ij}]$, find an $n \times n$ permutation matrix $X = [x_{ij}]$ that minimizes

$$\sum_{(i,j)} c_{ij} x_{ij} \tag{8.18}$$

Now introduce a variable u_i for each row i and a variable v_j for each column j, and define

$$w_{ij} = c_{ij} - (u_i + v_j) \tag{8.19}$$

for each i, j. The u_i and v_j are free, but we require $w_{ij} \geq 0$.

Then, for any permutation matrix X we have

$$\begin{aligned} \sum_{(i,j)} c_{ij} x_{ij} &= \sum_{(i,j)} u_i x_{ij} + \sum_{(i,j)} v_j x_{ij} + \sum_{(i,j)} w_{ij} x_{ij} \\ &= \sum_{i=1} u_i + \sum_{j=1} v_j + \sum_{(i,j)} w_{ij} x_{ij} \end{aligned} \tag{8.20}$$

The second equality in 8.20 follows from the observation that

$$\sum_{j=1}^{n} x_{ij} = 1 \tag{8.21}$$

and

$$\sum_{i=1}^{n} x_{ij} = 1 \tag{8.22}$$

are true for every permutation matrix.

Now, since the x_{ij} and w_{ij} are nonnegative, we have

$$\sum_{(i,j)} c_{ij}x_{ij} \geq \sum_{i=1}^{n} u_i + \sum_{j=1}^{n} v_j \tag{8.23}$$

We are now in a position to formulate the dual of the assignment problem.

Dual Problem: Find u_i and v_j that maximize

$$\sum_{i=1}^{n} u_i + \sum_{j=1}^{n} v_j \tag{8.24}$$

subject to the condition

$$u_i + v_j \leq c_{ij} \tag{8.25}$$

for all i, j.

The inequality in 8.23 tells us that feasible values of the objective variable for the dual problem are always less than or equal to feasible values of the objective variable for the assignment problem. Thus, if we can find feasible values for these two objective variables that are equal, each will be optimal for its problem. From 8.20 we see that these two objective variables are equal if, and only if,

$$\sum_{(i,j)} w_{ij}x_{ij} = 0 \tag{8.26}$$

Since w_{ij} and x_{ij} are both nonnegative, this sum will be zero if and only if either w_{ij} or x_{ij} is zero for each i and j.

It is convenient to represent the variables of the dual problem in a reduced cost matrix of the form

(8.27)

	v_1	v_2	$\cdots$	v_n
u_1	w_{11}	w_{12}	$\cdots$	w_{1n}
u_2	w_{21}	w_{22}	$\cdots$	w_{2n}
$\vdots$	$\vdots$	$\vdots$		$\vdots$
u_n	w_{n1}	w_{n2}	$\cdots$	w_{nn}

Though we did not use the present line of argument to obtain tableau 8.3, the entries in that tableau are related by the inequality in 8.25 just as those in 8.27 are.

The u_i, v_j, and w_{ij} are related by equation 8.19 and the w_{ij} are nonnegative. Thus, the variables u_i and v_j shown in 8.27 represent a feasible solution for the dual problem. If we can find a reduced cost matrix with n independent zeros, there is a permutation matrix X with 1's in the positions of these zeros. In this case the w_{ij} and x_{ij} satisfy equation 8.26. That is, we would have optimal solutions for the assignment problem and its dual problem.

Notice that the sum of the margin entries in tableau 8.13 is equal to the sum of the starred entries in tableau 8.14. The u_i, v_j and w_{ij} in tableau 8.13 and the c_{ij} in tableau 8.14 are related by equation 8.19. That is, the u_i, v_j, and c_{ij} are related by inequality 8.25. This is sufficient to show that the assignment in tableau 8.14 is optimal.

Harold W. Kuhn devised an algorithm to obtain a reduced cost matrix with n independent zeros. His algorithm is based on work by the Hungarian mathematicians König and Egerváry. In honor of them Kuhn called his algorithm the Hungarian Method or the Hungarian Algorithm.

Kuhn's Hungarian Algorithm

1. START. We start with any set of u_i, v_j, and $w_{ij} \geq 0$ satisfying 8.19. A good start is obtained by setting

$$\begin{aligned} u_i &= \min_j c_{ij} \\ v_j &= \min_i (c_{ij} - u_i) \\ w_{ij} &= c_{ij} - (u_i + v_j) \end{aligned} \tag{8.28}$$

   Note: This is the way we started the example discussed in Section 8.1.

2. Find a maximal independent set of zeros and a minimal cover of the zeros in the w-matrix. For relatively small assignment problems this can be done by inspection. In the next section we shall describe an algorithm, a subalgorithm of the Hungarian algorithm, for finding a maximal independent set of zeros and a minimal cover. Let k be the number of zeros in a maximal independent set of zeros. If $k = n$, STOP: We have optimality. If $k < n$, go to step 3.

3. Let e be the smallest entry of the w-matrix not covered by a line. Change u_i to $u_i - e$ for i in the cover. Change v_j to $v_j + e$ for j not in

the cover. Recompute the w-matrix by $w_{ij} = c_{ij} - (u_i + v_j)$. Return to step 1.

Suppose the cover obtained in step 2 has $k < n$ lines, r horizontal lines and $k - r$ vertical lines. Then, in step 3, r of the margin entries will be decreased and $n - k + r$ of the margin entries will be increased. Thus, the sum of the margin entries will be increased by $e(n - k)$. If, for example, each entry in the original cost matrix is an integer, then every increase in the sum of the margin entries is an integer. Since this sum is bounded in 8.24, only a finite number of integral increases is possible. Thus, the Hungarian algorithm must terminate in a finite number of iterations if we start with an integral initial cost matrix. Actually, the Hungarian algorithm will terminate in a finite of iterations even if the initial cost matrix is not integral, but we shall not prove that here. At termination we have n independent zeros, an optimal assignment, and an optimal solution for the dual problem.

8.3 König's Theorem

In this section we shall prove König's theorem and describe an algorithm that serves as a subalgorithm in the Hungarian algorithm. This algorithm will find a largest set of independent zeros and a smallest cover for the zeros in a w-matrix of reduced costs.

Theorem 8.56 *(König's Theorem) For every rectangular array with zeros in some of the cells, the number of zeros in a largest independent set of zeros is equal to the number of lines in a smallest cover of the zeros.*

Since a largest independent set of zeros and a smallest cover for the zeros depends only on the zeros, we shall simplify the appearance of the reduced cost matrix by suppressing the nonzero entries. For any independent set of zeros and any cover, no two zeros in the independent set can be covered by the same line. Thus, the number of lines in a cover is always at least as large as the number of zeros in an independent set. To prove the theorem it is sufficient to show that we can always find a cover and an independent set of zeros with the same number of elements. The rest of this section is devoted to describing an algorithm, the König algorithm, which will produce both a smallest cover and a largest independent set of zeros. A proof of König's theorem will be deferred until our description of the algorithm is complete.

The cost matrix for the assignment problem is square, but König's theorem and the König algorithm do not require that the matrix containing the

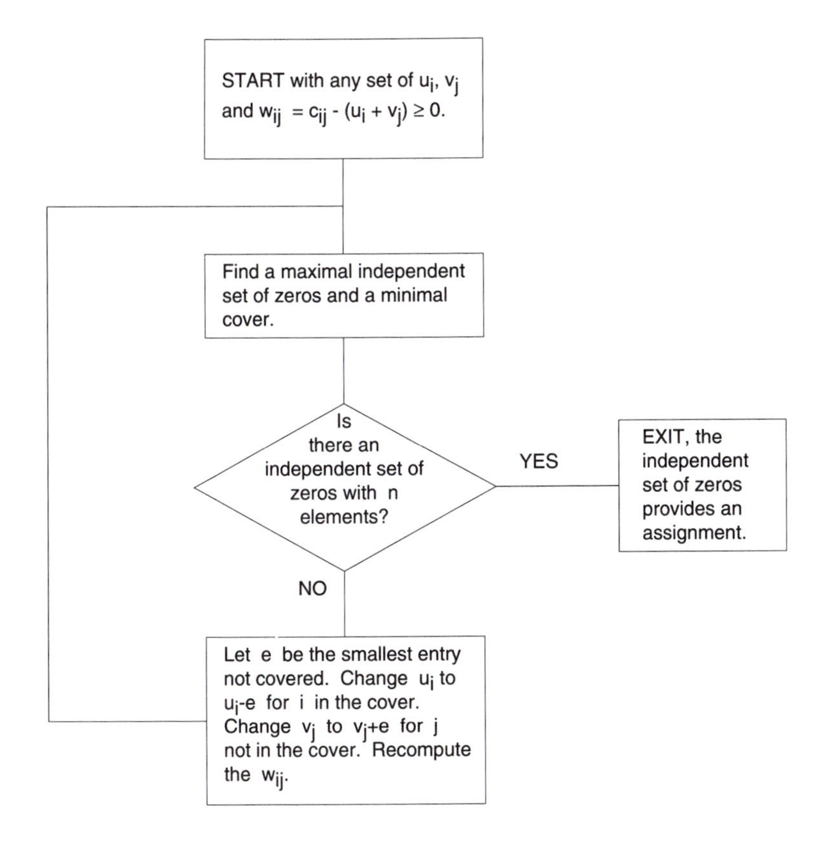

Figure 8.1: Flow diagram for the Hungarian algorithm.

zeros be square. As a first step in describing the König algorithm, consider the array of zeros in the 8×9 matrix 8.29.

$$\left[\begin{array}{ccccccccc}
 & 0 & & 0^* & & & & & \\
0^* & & & & & 0 & & 0 & \\
 & & 0^* & & 0 & & 0 & & \\
 & 0^* & & & 0 & & & & \\
0 & & & 0 & & & & & 0^* \\
 & & 0 & & 0^* & & & & \\
 & & & & 0 & & 0^* & & \\
 & & & & & & 0 & &
\end{array}\right] \tag{8.29}$$

The zeros designated with asterisks in 8.29 form an independent set. There are seven starred zeros in this independent set. Draw a horizontal line through each starred zero. Since the zeros are independent, each line will cover only one starred zero. That is, the number of lines equals the number of starred zeros. Label the uncovered rows 0.

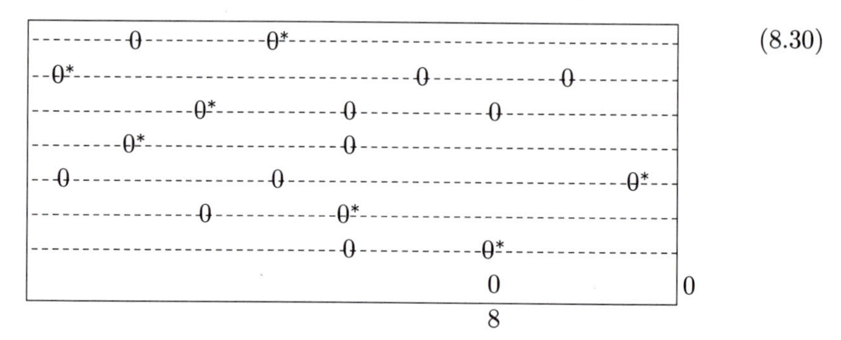

(8.30)

Starting at the top, look for an uncovered row with an uncovered zero in it. In 8.30 we find such a zero in the eighth row. Label the column containing the zero with the index of the row; 8 in this case. This labeling is shown in 8.30.

Then, look for a starred zero in the column just labeled. If there is such a starred zero, erase the horizontal line through the starred zero, draw a vertical line through the starred zero, and label the row just uncovered with the column index of the starred zero. This step is shown in 8.31.

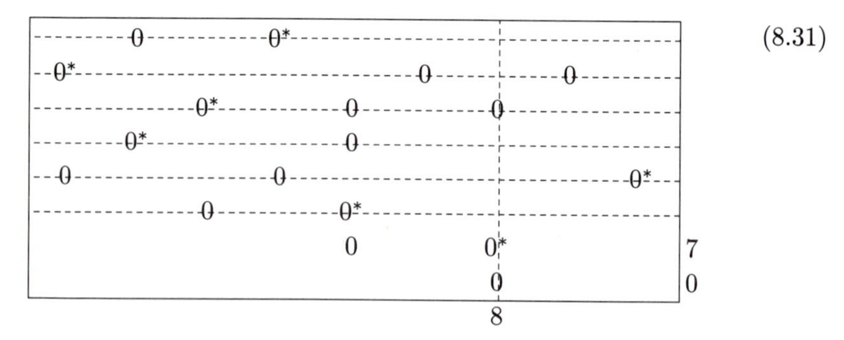

(8.31)

Again, we look for an uncovered zero, and we find one in the seventh row. We label the column with the index of the row and look for a starred zero in that column. We find one, and we again erase the horizontal line through that starred zero and draw a vertical line through it.

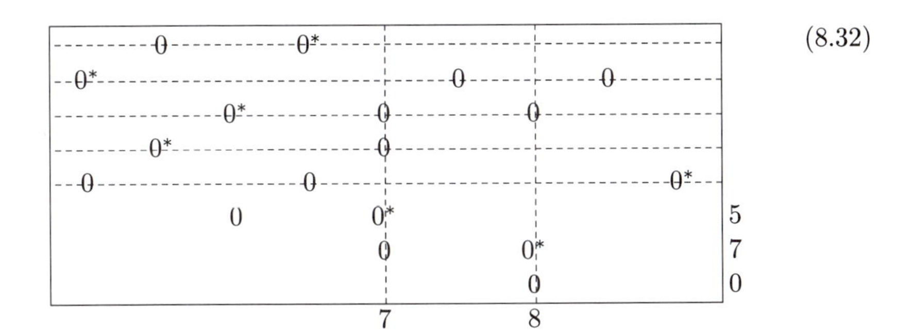

(8.32)

Again, we look for an uncovered zero. We find one in the sixth row. When we find the starred zero in that column, enter the new labels, and switch the covering line, we obtain 8.33. Here, there are no more uncovered zeros and we stop. We started with a number of lines equal to the number of starred zeros. Each step preserved the number of lines. Thus, we end with a cover for all the zeros with the number of lines in the cover equal to the number of starred zeros.

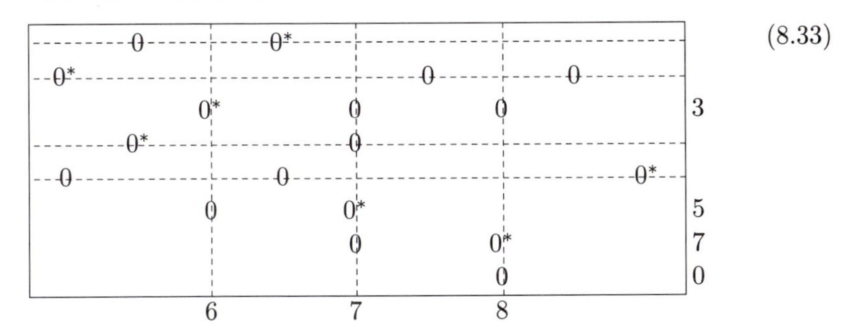

(8.33)

In 8.29 we started with a largest independent set of zeros, and we must now examine what to do if we start with an independent set of zeros that is not largest. For this purpose consider the array in 8.34.

	0*		0					
0*					0		0	
		0*		0		0		
	0			0*				
0			0*					0
		0		0				
				0		0*		
						0		

(8.34)

The pattern of zeros in 8.34 is the same as the one in 8.29. We know that there is a set of seven independent zeros, but the starred set in 8.34 has only six zeros. However, this set of zeros cannot be expanded to a larger independent set. Every zero in 8.34 is in a row or a column of a starred zero. An independent set of zeros that cannot be expanded to a larger independent set is said to be *complete.* A complete independent set of zeros is easy to obtain by starring any zero that is in neither a row or a column of a starred zero until no further zero can be starred. The König algorithm does not depend on starting with a complete set of starred zeros. However, such a set is so easy to obtain that we assume that we start with a complete set of starred zeros.

We start just as we did with matrix 8.29. The first labels obtained are shown in 8.35.

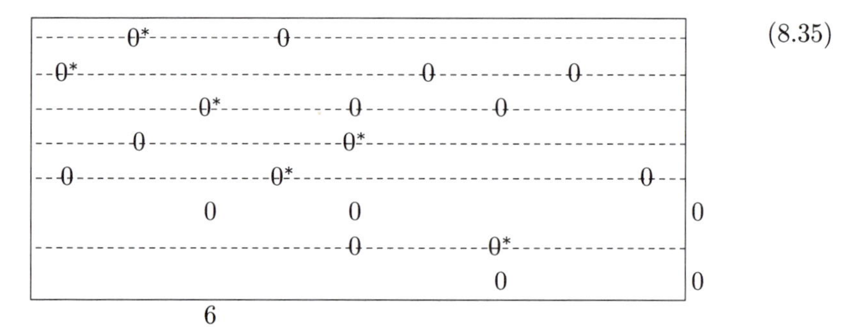

(8.35)

We look for a starred zero in the labeled column, label the row and switch the cover line through the starred zero. We obtain 8.36.

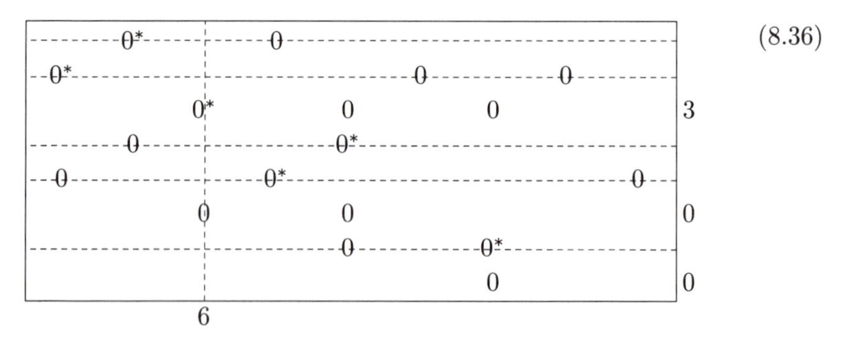

(8.36)

We continue to search for uncovered zeros, and we find one in the third row. Label and switch the cover line again to obtain 8.37. Continue in this fashion until we obtain the matrix in 8.41.

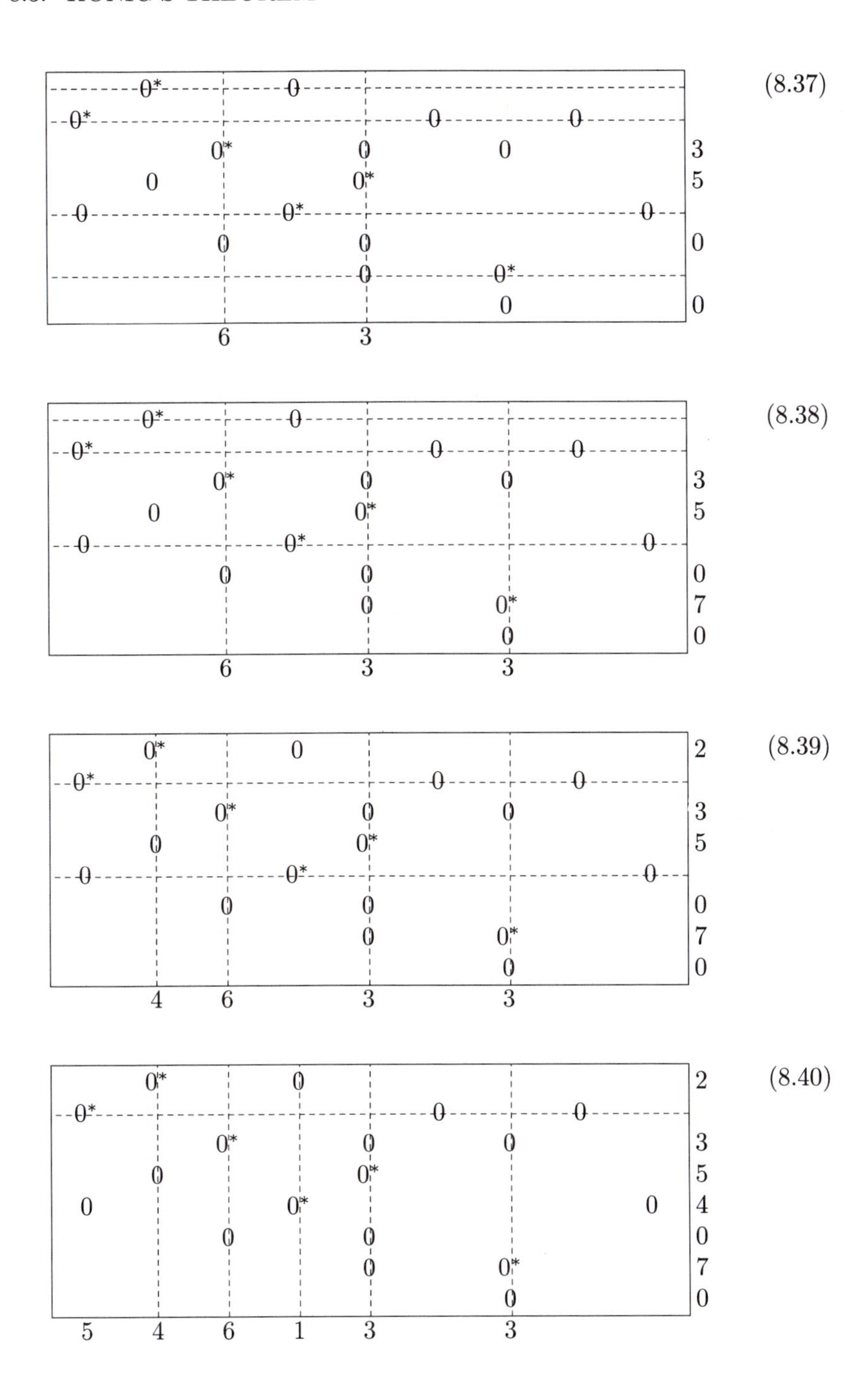
(8.37)
(8.38)
(8.39)
(8.40)

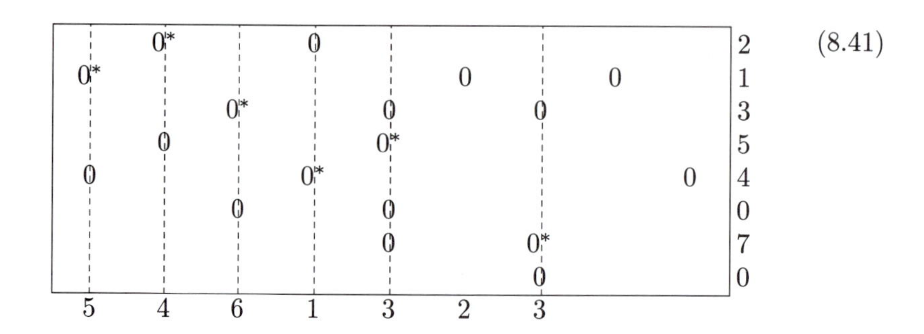

(8.41)

Finally, in 8.41 we label a column containing an uncovered zero but containing no starred zero. Then the labels identify a path starting with this uncovered zero. The column label 2 identifies the row containing this zero. The row label 1 identifies the column containing the starred zero in that row. The 5 in that column identifies the row containing the next zero, etc. The 0 label identifies the end of the path. The path described in this way is shown in 8.42.

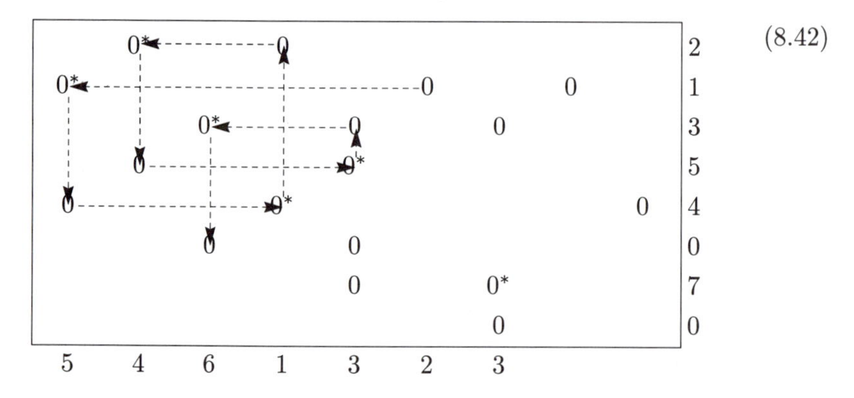

(8.42)

The phenomenon described above is called *breakthrough*. The path obtained has starred and unstarred zeros alternating along the path, and the path begins and ends with unstarred zeros. If we erase the asterisk on each starred zero and star each unstarred zero we increase the number of starred zeros by one. It is easily seen that the set of starred zeros obtained in this way is also independent and complete.

We now erase the labels on the margins and the cover lines and start over. Each run at this process will either result in a larger set of independent zeros or it will produce a cover with one line for each starred zero. This rationale provides the basis for the König algorithm.

The König Algorithm

1. START. We start with a complete set of starred zeros. Such a start is always easy to produce. For example, in the first row star the first zero in the row. In each subsequent row star the first zero not in a column of a previously starred zero. (The starred zeros in 8.35 were produced this way.) Cover each starred zero with a horizontal line. Label all uncovered rows with a 0.

2. Look for a row with an uncovered zero. If there is no uncovered zero, STOP. We have a cover for which the number of lines in the cover is equal to the number of independent starred zeros. If there is an uncovered zero, label the column containing the zero with the index of the row. Go to step 3.

3. In the column labeled in step 1, look for a starred zero. If there is no starred zero in that column, we have *breakthrough.* Go to step 4.

   If there is a starred zero in this column, it is covered with a horizontal line. Label the row containing the starred zero with the index of the column, erase the horizontal line through the starred zero and draw a vertical line through the starred zero. Go to step 2.

4. Breakthrough gives an alternating path along which the number of 0*'s can be increased by one. The path is traced by going to the row in each column identified by the column label, and going to the column in each row identified by the row label. When the label 0 is reached, we are at the end of the path.

   The path begins and ends with an unstarred zero. That is, the number of unstarred zeros is one more than the number of starred zeros. Star each unstarred zero and remove the star from each previously starred zero. This will produce an independent set of starred zeros with one more element. Erase all labels and cover lines and go to step 1.

Proof of König's Theorem: Each iteration of the algorithm will either result in an increase in the number of independent zeros or produce both a smallest cover and a largest independent set of zeros. Thus, the algorithm must terminate in a finite number of steps. This proves König's theorem. ◻

Once König's theorem has been proven and the algorithm validated it is not necessary to draw the lines in the cover. This makes it easier to implement the algorithm by hand and in a computer program. Start again with a copy of 8.34.

(8.43)

	0*		0						
0*					0		0		
		0*		0		0			3
	0			0*					5
0			0*					0	
		0		0					0
				0		0*			7
						0			0
		6		6	6	8			

The result of the first cycle through the algorithm is shown in 8.43. A subtle vagueness in this implementation of the algorithm is that neither the column chosen nor the label for that column is unique. This allows column 5 to be labeled with either a 3 or a 6. Also, column 7 could be labeled before column 3 is labeled. The row label produced in step 3 is always unique. Where there are alternatives, any choice will lead to eventual termination, but it is difficult to anticipate which choice will provide a shorter alternating path. The choices made here produce a shorter path.

When the labeling is carried through to termination, we obtain 8.44. It terminates with breakthrough when the last column is labeled and contains no starred zero. Notice that the alternating path produced here is shorter than the alternating path produced in 8.41. The labeling produced in 8.41 was obtained by choosing the left-most column when there was more than one choice for a column, and the top-most zero in a column when there was more than one choice for a row. Notice also that dogged use of the left–right, top–down priority rule would extend the alternating path through column 1, even though breakthrough was already reached.

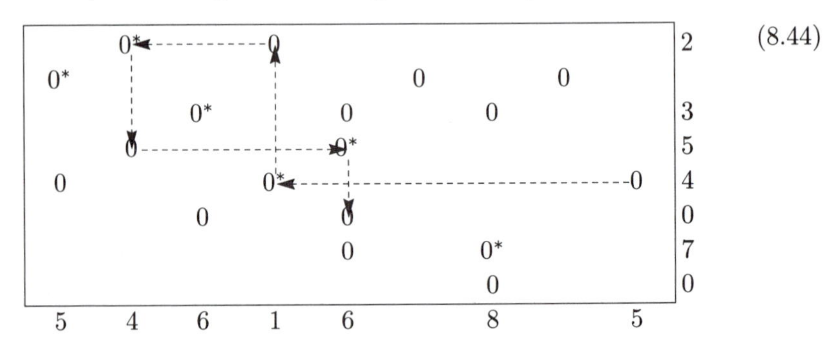

(8.44)

8.4 Matching and Hall's Theorem

There is an interesting and important interpretation of König's theorem in terms of graph theory. Consider the finite graph illustrated in Figure 8.2. The nodes of this graph are divided into two classes, indicated by displaying the nodes in two parallel rows. The arcs of this graph connect only a node in one class with a node in the other class. Such a graph is called a *bipartite graph*.

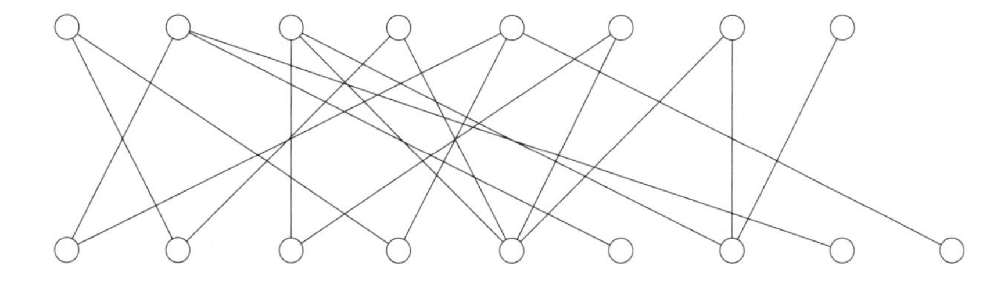

Figure 8.2: A bipartite graph.

Figure 8.2 is a graphical interpretation of the array of zeros in 8.34. Each row of 8.34 corresponds to a node in the upper row of Figure 8.2, and each column of 8.34 corresponds to a node in the lower row of Figure 8.2. Each zero in 8.34 can be considered as connecting the row and column in which it occurs. The arcs shown in Figure 8.2 correspond to the zeros in 8.34.

It is more in keeping with the traditions of graph theory to represent a connection between two nodes by a 1 rather than by a 0. An *adjacency matrix* for a bipartite graph is a matrix of 0's and 1's in which each row corresponds to a node in one of the two groups of nodes and each column corresponds a node in the other group. A 1 at the intersection of a row and column indicates that the nodes corresponding to the row and column are connected. A 0 indicates that they are not connected. The adjacency matrix for Figure 8.2 is shown in 8.45. The pattern of 1's and the stars in 8.45 are the same as the pattern of 0's and stars in 8.34.

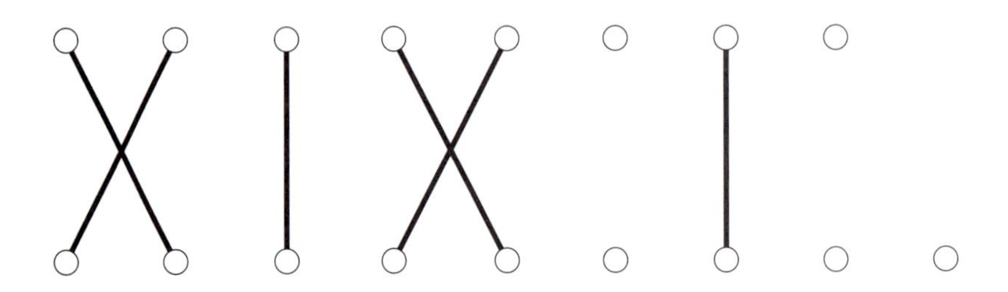

Figure 8.3: A matching.

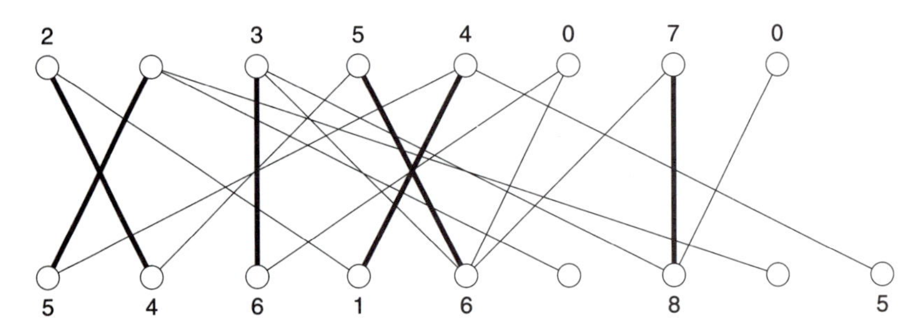

Figure 8.4: A matching covering a bipartite graph.

$$
\begin{array}{|ccccccccc|}
\hline
0 & 1^* & 0 & 1 & 0 & 0 & 0 & 0 & 0 \\
1^* & 0 & 0 & 0 & 0 & 1 & 0 & 1 & 0 \\
0 & 0 & 1^* & 0 & 1 & 0 & 1 & 0 & 0 \\
0 & 1 & 0 & 0 & 1^* & 0 & 0 & 0 & 0 \\
1 & 0 & 0 & 1^* & 0 & 0 & 0 & 0 & 1 \\
0 & 0 & 1 & 0 & 1 & 0 & 0 & 0 & 0 \\
0 & 0 & 0 & 0 & 1 & 0 & 1^* & 0 & 0 \\
0 & 0 & 0 & 0 & 0 & 0 & 1 & 0 & 0 \\
\hline
\end{array}
\qquad (8.45)
$$

A *matching* in a bipartite graph, as in Figure 8.2, is a set of arcs that connect disjoint pairs of nodes. That is, no two arcs in the matching share a common node. There is a one-to-one correspondence between matchings in the graph in Figure 8.2 and independent set of ones in 8.45. The matching corresponding to the starred ones in 8.45 is shown in Figure 8.3.

In Figure 8.4 we show the matching of Figure 8.3 with heavy lines and the other arcs from Figure 8.2 with light lines. Matrix 8.46 is the adjacency matrix for the bipartite graph for which the König algorithm has

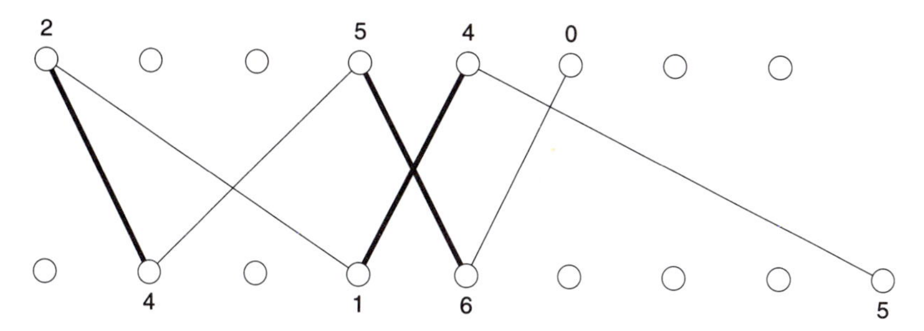

Figure 8.5: An alternating path.

been carried out. The alternating path determined there is illustrated in Figure 8.5.

$$
\begin{array}{|ccccccccc|c}
\hline
0 & 1^* & 0 & 1 & 0 & 0 & 0 & 0 & 0 & 2 \\
1^* & 0 & 0 & 0 & 0 & 1 & 0 & 1 & 0 & \\
0 & 0 & 1^* & 0 & 1 & 0 & 1 & 0 & 0 & 3 \\
0 & 1 & 0 & 0 & 1^* & 0 & 0 & 0 & 0 & 5 \\
1 & 0 & 0 & 1^* & 0 & 0 & 0 & 0 & 1 & 4 \\
0 & 0 & 1 & 0 & 1 & 0 & 0 & 0 & 0 & 0 \\
0 & 0 & 0 & 0 & 1 & 0 & 1^* & 0 & 0 & 7 \\
0 & 0 & 0 & 0 & 0 & 0 & 1 & 0 & 0 & 0 \\
\hline
\multicolumn{1}{c}{5} & 4 & 6 & 1 & 6 & & 8 & & \multicolumn{1}{c}{5} &
\end{array}
\tag{8.46}
$$

The node corresponding to the last column is the last node in the bottom row. If we trace a path from this node back through nodes according to the marks, we will obtain the path shown in Figure 8.5.

The path shown in Figure 8.5 is called an alternating path since the arcs are alternately in the matching and not in the matching. Since the first and last arcs in the path are not in the matching, there is one more nonmatching arc than matching arc. By changing matching arcs to nonmatching arcs, and changing nonmatching arcs to matching arcs, we can increase the number of arcs in the matching by one. This corresponds exactly to the role of the path shown in the array 8.46.

We have interpreted the 1's in a zero–one matrix as arcs in a bipartite graph, starred 1's as arcs in a matching, and the labeling process of the König algorithm in terms of these pairings. A largest set of independent 1's must correspond to a largest matching. The König algorithm then produces a largest matching. How do we interpret the smallest cover produced by the König algorithm?

Mark the nodes of the graph that correspond to the rows and columns of the array of 1's that are covered in a smallest cover. Then every arc has one or the other (or both) of its nodes marked since every 1 is covered. Since each line in the covering covers only one starred 1, every arc in the matching is marked at one end only. Thus, this gives the minimum number of nodes that can be marked in such a way that every arc is marked at one end or the other.

In 8.47 the rows and columns corresponding to marked nodes are marked with 1's. They are the rows and columns in the cover obtained in 8.32.

(8.47)

			1		1		1		
1	0	1	0	1*	0	0	0	0	0
1	1*	0	0	0	0	1	0	1	0
	0	0	1*	0	1	0	1	0	0
1	0	1*	0	0	1	0	0	0	0
1	1	0	0	1	0	0	0	0	1*
	0	0	1	0	1*	0	0	0	0
	0	0	0	0	1	0	1*	0	0
	0	0	0	0	0	0	1	0	0

In Figure 8.6 we show the result of changing the assignment as indicated by the matching augmenting path shown in Figure 8.5. We have also darkened the nodes corresponding to the cover shown in 8.47. Note that every arc is incident at one end or the other with a darkened node. This is equivalent to the fact that we have a cover for all the arcs. Also, every arc in the matching is incident with one and only one darkened node. The fact that the arcs in the matching are not adjacent is equivalent to the 1's being independent in 8.47. And, finally, the number of lines in the cover is equal to the number of independent arcs in the matching. That is, this is a largest matching.

It should be clear that one cannot expect a largest independent set of 1's to be unique, nor can one expect a smallest cover to be unique. One cannot even expect that the cover that goes with a largest independent set of 1's or the largest independent set of 1's that goes with a smallest cover to be unique. However, there is a kind of stable relationship between a given largest independent set of 1's and a given smallest cover.

For example, consider the independent set of 1's and the cover given in 8.47. Suppose the columns represent jobs and the rows represent candidates for those positions. A 1 indicates that the candidate is qualified for the position and a 0 indicates that he is not qualified for the position. To select a largest independent set of 1's, then, is to assign a largest number

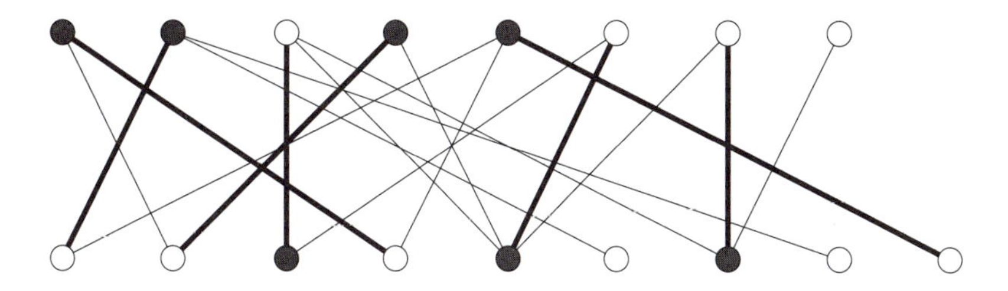

Figure 8.6: A matching as a cover.

of candidates to jobs for which they are qualified. The asterisks in 8.47 represent a particular assignment of candidates to jobs.

The candidates in marked rows hold special positions. They cannot have the jobs to which they are assigned reassigned to someone else without causing a further reassignment of that person. In contrast, the candidate represented by row 7 is assigned a job that could be reassigned to the candidate represented by row 8.

The jobs corresponding to marked columns also hold special positions. The persons assigned to those jobs cannot be reassigned to other jobs without displacing the person assigned to that job. Thus, while other assignments are possible, a new optimal assignment cannot be obtained by a single reassignment involving any of the marked rows or marked columns.

There are other related questions that can be asked about matchings. Is it possible to give every candidate a job for which he is qualified? Is it possible to fill every position with a qualified candidate? These two questions are symmetric and we will consider only the first question.

There is a circumstance in which it is obvious that not everyone can be assigned a job: if there are fewer jobs than there are candidates. Also, it will be impossible if there is any subset made up of candidates that are qualified for fewer collective jobs than the number of people in the subset.

Clearly, an assignment cannot be made unless every subset of the set of candidates is qualified, collectively, for at least as many jobs as there are candidates in the subset. It is interesting that this necessary condition is also sufficient.

Theorem 8.57 *Consider an $m \times n$ array of 0's and 1's. If every set of rows has 1's in at least as many columns as there are rows in the set, then there is a set of independent 1's with a 1 in every row.*

Proof. Find a largest independent set of 1's, using the König algorithm if necessary. Suppose there is no set of independent 1's with a 1 in every row. There is a smallest cover that has fewer than m lines (where m is the number of rows). If r rows are covered, fewer than $m-r$ columns are covered. So the $m-r$ uncovered rows have their 1's in fewer than $m-r$ columns. This contradicts the hypothesis of the theorem. Thus, the König algorithm can terminate only with a starred 1 in every row. ◻

A matching between a collection of sets and the members of those sets provides an interesting form of Theorem 8.57. If the sets in the collection correspond to the rows in Theorem 8.57 and the elements correspond to the columns, then a matching that satisfies Theorem 8.57 is called a *system of distinct representatives* (SDR). In Congress, for example, there are many committees. Each committee contains several members and many members serve on several committees. Is it possible to make up a new committee with exactly one representative from each of the existing committees?

We denote the set of committees by X, and the set of members of Congress by Y. An arc is drawn between each committee and each member serving on that committee. A matching involving every committee would produce a set of distinct representatives for these committees in the new committee. Theorem 8.57 says that this is possible if and only if every subset of committees contains at least as many members as the number of committees in the subset. This version of Theorem 8.57 is a theorem due to Philip Hall.

Theorem 8.58 *(Hall's Theorem) A collection of sets has an SDR (system of distinct representatives) if, and only if, each subcollection has a combined membership at least as large as the number of sets in the subcollection.* ◻

8.5 Assignments and Linear Programs

The assignment problem is not a linear program because the condition that X be a permutation matrix is not a linear constraint. There is, however, a closely related linear program that is equivalent to the assignment problem. Consider the following

LP-Minimum Assignment Problem: Given any $n \times n$ cost matrix $C = [c_{ij}]$, find nonnegative x_{ij} that minimize

$$g = \sum_{(i,j)} c_{ij} x_{ij} \tag{8.48}$$

subject to the constraints

$$\sum_{j=1}^{n} x_{ij} - 1 = y_i \tag{8.49}$$

$$\sum_{i=1}^{n} x_{ij} - 1 = z_j \tag{8.50}$$

where y_i and z_j are artificial.

Permutation matrices satisfy the feasibility constraints 8.49 and 8.50 with nonnegative x_{ij}. That is, this linear program is feasible. Since the x_{ij} are bounded, the values of g are bounded. Thus, the LP-assignment problem has an optimal solution.

The only feasible solutions for the feasibility constraints that are integral are permutation matrices. We will show in the following chapter that all basic solutions for the feasibility constraints are integral. Since the LP-assignment problem has an optimal solution, it has a basic optimal solution. That is, the LP-assignment problem has an optimal solution that is a permutation matrix.

The feasible set for the assignment problem is a subset of the feasible set for the LP-assignment problem. Thus, the minimum value of g for the assignment problem is not smaller than the minimum value of g for the LP-assignment problem. Since a basic optimal solution for the LP-assignment problem is a solution for the assignment problem, their optimal values are equal. Furthermore, any basic optimal solution for the LP-assignment problem is an optimal solution for the assignment problem, and any optimal solution for the assignment problem is also an optimal solution for the LP-assignment problem. For this reason we will cease making any distinction between these two formulations of the problem.

The advantage of formulating the assignment problem as a linear program is that it allows us to formulate the dual problem as a linear program. It will also allow us to pose a number of interesting related problems.

Consider the tableau 8.51 as a suggestive representation of a full tableau. We will find such suggestive representations useful for the assignment problem and a variety of related problems that will be considered in the following chapters.

(8.51)

	u_i	v_j	-1	
x_{ij}	1	1	c_{ij}	$= -w_{ij}$
-1	1	1	0	$= f$
	$= y_i$	$= z_j$	$= g$	

Such a suggestive representation provides a convenient way to formulate the dual problem and the dual feasibility specifications. It is easily seen that the dual of the minimum assignment problem is to maximize

$$f = \sum_{i=1}^{n} u_i + \sum_{j=1}^{n} v_j \tag{8.52}$$

subject to the conditions

$$w_{ij} = c_{ij} - u_i - v_j \tag{8.53}$$

where the u_i and v_j are free and w_{ij} are nonnegative.

It is easily seen that this is the dual problem discussed in Section 8.2. If the c_{ij} are integers, the Hungarian algorithm provides only integral values for the u_i and v_j. However, this was never suggested as a condition on the u_i and v_j and we can regard the dual of the linear programming formulation of the assignment problem as the dual of the assignment problem. Furthermore, the Hungarian algorithm solves both the assignment problem and its dual.

We also wish to state the assignment problem in a maximization form.

The LP-Maximum Assignment Problem: Given any $n \times n$ rating matrix $r = [r_{ij}]$, find nonnegative x_{ij} that maximize

$$f = \sum_{(i,j)} r_{ij} x_{ij} \tag{8.54}$$

subject to the constraints

$$\sum_{j=1}^{n} x_{ij} - 1 = -y_i \tag{8.55}$$

$$\sum_{i=1}^{n} x_{ij} - 1 = -z_j \tag{8.56}$$

where the y_i and z_j are artificial.

Tableau 8.57 is a simple but suggestive representation of this problem.

(8.57)

	x_{ij}	-1	
u_i	1	1	$= -y_i$
v_j	1	1	$= -z_j$
-1	r_{ij}	0	$= f$
	$= w_{ij}$	$= g$	

The row equations represent the maximum assignment problem and the column equations represent the dual minimum problem, which is to minimize

$$g = \sum_{i=1}^{n} u_i + \sum_{j=1}^{n} v_j \tag{8.58}$$

subject to

$$w_{ij} = u_i + v_j - r_{ij} \tag{8.59}$$

where the u_i and v_j are free and the w_{ij} nonnegative.

Suppose we are asked to maximize the sum of the ratings in the rating matrix 8.60.

1*	2	6
3	4	9*
5	8*	7

(8.60)

The assignment that maximizes the sum of the ratings is starred in 8.60. The use of the term "rating" instead of "cost" is intended to imply that we are considering the maximum assignment problem. Maximizing the sum is equivalent to minimizing the sum of the negatives. Thus, the maximum assignment problem in 8.60 is equivalent to the minimum assignment problem in 8.61.

−1*	−2	−6
−3	−4	−9*
−5	−8*	−7

(8.61)

We suggest using the Hungarian algorithm to solve this equivalent problem and then selecting the corresponding entries in the rating matrix 8.60. If you choose to work the maximum assignment problem directly with the rating matrix, because of constraint 8.59 it is necessary to use reduced rating matrices with nonpositive entries. Thus, the first step starting with 8.60 is to subtract the maximum in each row.

6	−5	−4	0
9	−6	−5	0
8	−3	0	−1

(8.62)

Then we subtract the maximum in each column to obtain

(8.63)

	−3	0	0
6	−2	−4	0
9	−3	−5	0
8	0	0	−1

From this point on the algorithm proceeds in the same way in which the algorithm for the minimum assignment problem does, except that we look for the maximum uncovered entry, the −2 in transportation table 8.63.

We have introduced the maximum assignment problem not for computational purposes, but because we wish to describe several variations of the assignment problem and state their dual programs. For example, if we require u_i and v_j to be nonnegative, then y_i and z_j would be nonnegative in the dual program. That is, the assignment problem would be to maximize the sum of the ratings by selecting at most one entry from each row and column.

As an illustration of what is involved, consider the 2×2 array 8.64.

(8.64)

4	1
1	−1

If the y_i and z_j are artificial, the assignment must select exactly one entry from each row and each column. The optimal assignment and its optimal dual are shown in 8.65. The value of the optimal assignment is 3.

(8.65)

	3	0
1	4*	1
−1	1	−1*

If the y_i and z_j are nonnegative, the assignment must select at most one entry from each row and each column. By making a subassignment, the value of the assignment is increased to 4. Notice that requiring the u_i and v_j to be nonnegative results in a larger minimum value for the dual program.

(8.66)

	3	0
1	4*	1
0	1	−1

In tableau 8.66 note that the constraints 8.55 and 8.56 are satisfied since the subassignment selects at most one entry from each row and column. We can also check that 8.59 is satisfied with nonnegative u_i and v_j. That is, both solutions shown in 8.66 are feasible. A variable y_i in 8.55 or z_j in 8.56 will be positive for any row or column that does not contain an entry in the assignment. Thus, complementarity requires that a variable on the margin should be zero for any row or column that does not contain an entry in the assignment. Since this is true, the solutions shown are complementary and optimal.

We do not want to make a catalog of all possible variations of assignment type problems. We will illustrate only a few possibilities and leave the reader to see how the duality equation can be used to formulate the correct dual problems and the correct sufficient conditions for optimality. In this spirit, let us return to the minimum assignment problem and describe one variation of that problem.

In the constraints 8.49 and 8.50 take y_i and z_j to be nonnegative. This amounts to selecting at least one entry from each row and column. Again, the min program is feasible since the permutation matrices satisfy the column equations.

This problem has a new twist. If there is a negative entry in the cost matrix C, it is possible to have the value of the min program unbounded from below. If there is a zero entry in C, it is possible to have arbitrarily large values for some of the x_{ij}. The basic solutions will still assign only integral values to the variables of the column equations. We must impose conditions to assure that this does not happen. Let us assume that the c_{ij} are positive. Then, since C is positive, no x_{ij} would be greater than 1 in any optimal solution.

To illustrate these ideas, consider the cost matrix

7	1	8
6	2	9
3	5	4

(8.67)

The Hungarian algorithm provides the minimum assignment, and the solution for the dual shown in 8.68. The common optimal value for the assignment problem and its dual is 11.

	4	0	5
1	7	1*	8
2	6*	2	9
−1	3	5	4*

(8.68)

The solution for the minimum assignment with "at least one entry from each row and column" is shown in 8.69. Since the y_i and z_j are nonnegative, the u_i and v_j are also nonnegative. Note that the solution of the min program shown on the margins of 8.68 is infeasible for this problem. Also, where more than one entry is chosen in a row or column, the corresponding y_i or z_j is positive. Then complementarity requires the dual u_i or v_j must be 0. The dual solutions shown in 8.69 are feasible. Complementarity is checked by noting that wherever a margin variable is positive, there is just one entry selected in the corresponding row or column. That is sufficient to make them optimal. The common optimal value for their objective variables is 10.

(8.69)

	3	0	4
1	7	1*	8
2	6	2*	9
0	3*	5	4*

Another variation of these problems is illustrated in tableau 8.70. We will be concerned with this type of problem extensively in the following chapter.

(8.70)

	u_i	v_j	-1	
x_{ij}	-1	1	c_{ij}	$= -w_{ij}$
-1	-1	1	0	$= f$
	$= y_i$	$= z_j$	$= g$	

All variables are canonical. The column constraints mean that we are to select at most one entry from each row and at least one entry from each column. For an $n \times n$ cost matrix this means we select at most n entries from the rows and at least n entries from the columns. Thus, exactly n entries are selected, forcing the y_i and z_j to be zero in a feasible solution.

It is possible to formulate a large number of similar assignment type problems for non square cost matrices. One could require exactly one from every row and at most one from every column; at least one from every row and exactly one from every column; etc. These are formulated by giving appropriate feasibility specifications for the y_i and z_j, and determining the appropriate feasibility specifications for the dual variables.

8.6 Egerváry's Theorem

Theorem 8.59 *(Egerváry's Theorem) If $R = [r_{ij}]$ is a given $n \times n$ matrix with nonnegative integers as elements, and if u_i and v_j are nonnegative integers such that*

$$u_i + v_j \geq r_{ij} \tag{8.71}$$

then

$$\min \sum_{k=1}^{n} (u_k + v_k) = \max \sum_{(i,j)} r_{ij} x_{ij} \tag{8.72}$$

where X is an $n \times n$ permutation matrix.

The maximum problem, the right side of 8.72, resembles the maximum assignment problem. The only difference is that Egerváry requires that R be nonnegative. The curious thing is that the minimum problem, the left side of 8.72, is not the dual of the assignment problem in the sense in which we have defined the dual. Recall from the previous section that the variables u_i and v_j in the dual problem are free, while Egerváry requires them to be nonnegative. It is the nonnegativity of R that makes this possible.

Egerváry proved his theorem by mathematical induction, a nonconstructive method of proof. The Hungarian algorithm, applied to the negative of 8.73, will determine an optimal solution for the maximum assignment problem and its dual. This is a constructive proof.

To see what then must be done, let us finish the 3×3 maximum assignment problem we started in the previous section, which we copy here as 8.73.

(8.73)

1	2	6
3	4	9
5	8	7

We apply the Hungarian algorithm to the negative of this table, and then take the negative again when the Hungarian algorithm terminates. Omitting the details, we obtain

(8.74)

	−5	−2	0
6	0*	−1	0
9	−1	−3	0*
10	0	0*	−3

We can see that the maximum assignment and its dual are optimized, and the common value is 18. While the assignment problem also solves

Egerváry's maximum problem, the solution for the dual problem does not solve Egerváry's minimum problem. It fails to be nonnegative.

Actually, the assignment problem is degenerate and the dual problem always has multiple solutions even when the assignment is unique. We can add 5 to each entry on the top margin and subtract 5 from each entry on the left margin. This will give a new solution for the dual problem that satisfies Egerváry's requirements.

Proof. To prove Egerváry's theorem we need to show that what was done in the example can be done in general. It is easily seen that the Hungarian algorithm will produce u_i and v_j satisfying 8.71 and that their sum is the value of the maximum assignment. We must make use of the assumption that the r_{ij} are nonnegative.

We can add a constant to all the entries on the top margin and subtract the same constant from the entries on the left margin. Since there are as many u_i on the left margin as there are v_j on the top margin, this will preserve both the sum of the margin entries and condition 8.71. In this way we can make the minimum entry on one margin, say the top margin, equal to zero. Then the entries on the top margin are nonnegative. Also, by condition 8.71, each entry on the left margin must be at least as large as the r_{ij} in the column of that zero. □

König's theorem and Egerváry's theorem were published in papers appearing consecutively in 1931. König's theorem was originally stated and proved in terms of graph theory. Egerváry restated König's theorem in a form that would generalize to his theorem. This restated form is very close to the form given in the preceding section. Consider a maximum assignment problem in which the values of the ratings are only 0's and 1's. This can be interpreted as a situation in which a candidate for a position is qualified if his rating is 1 and he is unqualified if his rating is 0. This is the kind of problem we considered in the preceding section.

In this case the maximum value of the assignment is just the maximum number of people that can be assigned to positions for which they are qualified, and this is the maximum number of independent 1's in the matrix. By König's theorem, this is also the minimum number of lines required to cover the 1's in the matrix. In this context, consider the example shown in 8.75.

(8.75)

	0	0	1	1
1	1*	1	1	0
0	0	0	1*	0
0	0	0	1	1*
0	0	0	0	1

The rows and columns that are used for the cover can be assigned the value of 1 on the margin and the other rows and columns assigned the value of 0. This will give values to the u_i and v_j that will satisfy 8.71. This clearly gives the optimal value to the objective function of the dual problem.

8.7 Von Neumann's Hide-and-Seek Game

In 1951 von Neumann described a reduction of the maximum assignment problem to a "hide-and-seek" game. Although König and Egerváry had published relevant papers 20 years earlier, von Neumann was not aware of their work. Kuhn's Hungarian algorithm had not yet been devised. Even today, knowing the work of König, Egerváry, and Kuhn, we find von Neumann's game to be a strikingly imaginative equivalent of the assignment problem.

As we remarked earlier, von Neumann regarded matrix games as more fundamental and he considered a linear program solved when he reduced it to a matrix game. It is much easier to solve an assignment problem using the Hungarian algorithm than it is to find the optimal strategies for the corresponding hide-and-seek game. Since the hide-and-seek game is interesting in its own right, there is more to be gained by reducing the hide-and-seek game to an assignment problem than there is in reducing the assignment problem to a game. The Hungarian algorithm provides a convenient and effective method for finding optimal strategies for the hide-and-seek game.

Consider a square array of positive numbers, as in 8.76.

13	24	7	10
9	13	5	6
12	25	8	9
7	9	4	5

(8.76)

One player hides at an entry of the array. The other player seeks the hider by choosing a row or a column. If the row or column chosen contains the entry where the hider is hiding, the seeker has "found" him. If the hider is found he pays the seeker an amount equal to the entry where he is hiding. Otherwise, there is no payment.

What are the optimal strategies for the two players? What is the value of the game?

The hide-and-seek game can be formulated as a matrix game. For an $n \times n$ array $A = [a_{ij}]$, the hider has n^2 actions and the seeker has $2n$ actions. We take the seeker as the row player of the matrix game. While either player could be designated as the row player, this choice leads to a game matrix with nonnegative entries. Thus, the game matrix for this game has $2n$ rows and n^2 columns. We are not going to write down the game matrix. All we need to know is the content of the minimax theorem, that there are optimal strategies for both players and that there is a value for the game.

Let p'_i be the probability with which the seeker chooses row i of the array A, and let p''_j be the probability with which he chooses column j of A. If s is the floor under the seeker's expectations then for every choice of the hider, that is for every pair (i, j), we have

$$(p'_i + p''_j)a_{ij} \geq s \tag{8.77}$$

Let q_{ij} be the probability with which the hider chooses entry a_{ij} as the place to hide. If t is the ceiling over the seeker's expectations, then for every i we have

$$\sum_{j=1}^{n} a_{ij}q_{ij} \leq t \tag{8.78}$$

and for every j we have

$$\sum_{i=1}^{n} a_{ij}q_{ij} \leq t \tag{8.79}$$

Since all a_{ij} are positive, we can assume that the value of the game is positive, and that the floor s and the ceiling t are positive. Let

$$r_{ij} = 1/a_{ij} \tag{8.80}$$
$$u_i = p'_i/s \tag{8.81}$$
$$v_j = p''_j/s \tag{8.82}$$

Then 8.77 takes the form

$$u_i + v_j \geq r_{ij} \tag{8.83}$$

Further, let

$$x_{ij} = a_{ij}q_{ij}/t \tag{8.84}$$

Then 8.78 takes the form

$$\sum_{j=1}^{n} x_{ij} = \sum_{j=1}^{n} a_{ij}q_{ij}/t \leq 1 \tag{8.85}$$

and 8.79 takes the form

$$\sum_{i=1}^{n} x_{ij} = \sum_{i=1}^{n} a_{ij}q_{ij}/t \leq 1 \tag{8.86}$$

The seeker wants to maximize the floor s under his expectation. That is, he wishes to minimize

$$\sum_{i=1}^{n} u_i + \sum_{j=1}^{n} v_j = (\sum_{i=1}^{n} p'_j + \sum_{j=1}^{n} p''_j)/s = 1/s \tag{8.87}$$

The hider wants to minimize the ceiling t over the seeker's expectation. That is, he wishes to maximize

$$\sum_{(i,j)} r_{ij} x_{ij} = \sum_{(i,j)} r_{ij} a_{ij}/t = 1/t \tag{8.88}$$

The seeker wants to minimize $1/s$ for nonnegative u_i and v_j subject to the constraints 8.83. This is clearly Egerváry's minimization problem.

The hider wants to maximize $1/t$ for nonnegative x_{ij} subject to the constraints 8.85 and 8.86. The difference between the hider's problem and Egerváry's maximization problem is in the way the constraints are given.

Through relation 8.80 we obtain a rating matrix $R = [r_{ij}]$. For our example 8.76 the rating matrix is shown in 8.89.

(8.89)

1/13	1/24	1/7	1/10
1/9	1/13	1/5	1/6
1/12	1/25	1/8	1/9
1/7	1/9	1/4	1/5

We can use the Hungarian algorithm to solve this maximum assignment problem, as described in the preceding section. We obtain

(8.90)

		(2/24)	(1/24)	(4/24)	(3/24)
		1/24	0	3/24	2/24
(0)	1/24	1/13	1/24*	1/7	1/10
(1/24)	2/24	1/9	1/13	1/5	1/6*
(0)	1/24	1/12*	1/25	1/8	1/9
(2/24)	3/24	1/7	1/9	1/4*	1/5

In 8.90 the maximum assignment is indicated with asterisks. Two solutions (one in parentheses) for the minimization problem are shown on the margins. It can be seen that the value of the maximum assignment is $13/24$. The common optimal values for the floor and ceiling are the reciprocals of the maximum assignment, as shown in 8.87 and 8.88. The value of the hide-and-seek game is then $24/13$.

The optimal strategies for the hider and the seeker can now be obtained using 8.80, 8.81, 8.82, and 8.84. These optimal strategies are shown in 8.91. Within the box we show the probabilities that the hider hides at the indicated position. Blanks are zeros. On the margins we show two optimal strategies for the seeker, each derived from the corresponding margin entries in 8.90.

(8.91)

		(2/13)	(1/13)	(4/13)	(3/13)
		1/13	0	3/13	2/13
(0)	1/13		1/13		
(1/13)	2/13				4/13
(0)	1/13	2/13			
(2/13)	3/13			6/13	

Notice that at each position chosen by the hider with positive probability, the expected value is 24/13, the value of the game. Thus, the hider assures himself that his expected payout is not more than 24/13. For the seeker, the probability that he finds the hider at a particular entry is the sum of the probabilities for the row and column containing that entry. For every entry the expected value is at least 24/13. Thus, the seeker is assured of an expected payoff of at least 24/13 for any action of the hider.

Before the simplex algorithm became available to solve matrix games, solving a matrix game was an ad hoc affair. That is, each game required some ingenuity. The Hungarian algorithm is even faster than the simplex algorithm wherever it can be used. It is often worth the effort to recast the problem so that it can be interpreted as an assignment problem to take advantage of the computational efficiency of the Hungarian algorithm, even if recasting the problem requires a high level of ingenuity itself.

8.8 Doubly Stochastic Matrices

A *stochastic matrix* is a matrix with nonnegative entries in which the sum of the entries in each row is 1. The word "stochastic" is simply a technical word meaning "probabilistic." Stochastic matrices arise in a part of the theory of probability that deals with finite state systems that change from one state to another in a random way at discrete time intervals. Such problems occur in the study of subjects as diverse as gambling, physics, genetics, economics, and psychology.

Suppose the system has n possible states and that the probability of making a transition from state i to state j is p_{ij}. Then $P = [p_{ij}]$ is the stochastic matrix that is considered. Each p_{ij} is nonnegative because the numbers are probabilities, and the sum in each row is 1 because one of the possible states must occur. The questions that are asked include the following: On the average, how frequently will the system be in state i? If the system starts in state i, on the average, how long will it take to get to state j? In this section we will confine our attention to some properties of stochastic matrices and we will not attempt to deal with any problems in the applications of stochastic matrices.

A doubly stochastic matrix is a stochastic matrix in which the sum of the entries in each column is also 1. The interesting thing in the context of this chapter is that for a given n the set of all doubly stochastic matrices is precisely the feasible set for the linear programming formulation of an assignment problem with n assignments.

The linear constraints for an assignment problem are of the form

$$\sum_{i=1}^{n} x_{ij} = 1 \tag{8.92}$$

$$\sum_{j=1}^{n} x_{ij} = 1 \tag{8.93}$$

where $x_{ij} \geq 0$. There are $2n$ equations in n^2 variables.

The permutation matrices are a special kind of doubly stochastic matrix. The permutation matrices also satisfy the linear constraints 8.92 and 8.93. For a permutation matrix, only n of the n^2 variables are nonzero, and n^2-n of the variables are zero. If P is a permutation matrix, it is the only solution of the constraints with those variables equal to zero. By Theorem 2.2 of Chapter 2, this means that a permutation matrix is a basic feasible solution. Furthermore, by Theorem 3.21 of Chapter 3, the permutation matrices are also the extreme points of the set of doubly stochastic matrices.

We wish to show that the permutation matrices play an even more interesting role in the set of doubly stochastic matrices.

Theorem 8.60 *Let W be any $n \times n$ doubly stochastic matrix. Then there exist permutation matrices $P_1, P_2, \ldots, P_r$ and nonnegative scalars $a_1, a_2, \ldots, a_r$ such that*

$$W = a_1 P_1 + a_2 P_2 + \cdots + a_r P_r \tag{8.94}$$

where

$$a_1 + a_2 + \cdots + a_r = 1 \tag{8.95}$$

Proof. We will consider a sequence of $n \times n$ matrices with nonnegative entries for which the row sums and column sums are constant. The doubly stochastic matrices are merely special cases in which the row and column sums are equal to 1. Let h denote the number that is the sum of the entries in each row or column. The sum of all the entries in such a matrix in nh.

Let $W_0 = W$. Use the König algorithm to find a minimal cover for the positive entries in W_0. If this cover contains k lines, the sum of all the covered entries will be not more than kh since the sum in each covered row or column is h ($h = 1$ in this particular case). If $k < n$, this sum is too small (the sum of all positive entries must be nh). Thus, the minimal cover must contain n lines. Then there must be n independent positive entries in W_0. There may be several such independent sets of positive entries. Choose any such set. Let a_1 be the smallest positive entry in this set.

Let P_1 be a permutation matrix with a 1 in each location of an entry in the independent set of positive entries. Then

$$W_1 = W_0 - a_1 P_1 \tag{8.96}$$

is a matrix with nonnegative entries and constant row and column sums. Notice that W_1 has at least one fewer positive entry than W_0.

If W_1 is not the zero matrix, repeat this construction to obtain

$$W_2 = W_1 - a_2 P_2 \tag{8.97}$$

where P_2 is a permutation matrix and a_2 is positive, and W_2 has at least one fewer positive entry than W_1.

Repeat this process as often as possible. Since each iteration reduces the number of positive entries, it must eventually terminate with some $W_r = 0$. Then

$$\begin{aligned} W &= W_0 = a_1 P_1 + W_1 \\ &= a_1 P_1 + a_2 P_2 + W_2 \\ &\vdots \\ &= a_1 P_1 + a_2 P_2 + \cdots + a_r P_r + W_r \\ &= a_1 P_1 + a_2 P_2 + \cdots + a_r P_r \end{aligned} \tag{8.98}$$

Finally, sum all the entries in the matrices on both sides of equation 8.98. We obtain

$$n = na_1 + na_2 + \cdots + na_r \tag{8.99}$$

This gives

$$\sum_{i=1}^{n} a_i = 1 \tag{8.100}$$

The proof of Theorem 8.60 provides a practical way to compute the expression 8.94. Just carry out the steps described.

There are two different ways to interpret the results of Theorem 8.60. Since the a_i are nonnegative and sum to 1, we can regard them as probabilities. That is, a doubly stochastic matrix is a probability distribution on a set of permutation matrices.

The other way to interpret Theorem 8.60 is much more geometric. In Section 3.6 we discussed the interpretation of a set of feasible solutions as a convex set with the basic feasible solutions as extreme points. There, we showed that if A and B are two points, the line through A and B can be represented in the form

$$W = (1-t)A + tB \tag{8.101}$$

If we set $s = 1 - t$, then 8.101 can be written in the form

$$W = sA + tB \tag{8.102}$$

where

$$s + t = 1 \tag{8.103}$$

If we further restrict s and t to be nonnegative, then W is in the line segment between A and B. We can also interpret 8.102 and 8.103 in physical terms. If masses are concentrated at points A and B with s the fraction at A and t the fraction at B, then W is at the center of gravity.

Generally, if P_1, P_2, ... , P_r is a set of points with a_1, a_2, ... , a_r representing the fractions concentrated at each of these points, where the a_i are nonnegative and sum to 1, then

$$W = a_1P_1 + a_2P_2 + \cdots + a_rP_r \tag{8.104}$$

is at the center of gravity.

In this interpretation, the point W is interior to the smallest convex set containing $P_1, P_2, \ldots, P_r$. However, the P_i are not necessarily the extreme points of that convex set.

A statement similar to Theorem 8.60 is true for the feasible set for a canonical linear program and the set of basic feasible solutions for the linear program. It is necessary to assume that the feasible set is bounded. Otherwise, there are not enough extreme points to enclose the feasible set. Expressions similar to 8.94 and 8.95 can be obtained, but the calculation is more difficult. We must find a nonnegative solution for the problem in tableau 8.105.

$$\begin{array}{cccc|c|l} a_1 & a_2 & \cdots & a_r & -1 & \\ \hline p_{11} & p_{12} & \cdots & p_{1r} & w_1 & = 0 \\ p_{21} & p_{22} & \cdots & p_{2r} & w_2 & = 0 \\ \vdots & \vdots & & \vdots & \vdots & \vdots \\ p_{n1} & p_{n2} & \cdots & p_{nr} & w_n & = 0 \\ 1 & 1 & \cdots & 1 & 1 & = 0 \\ \hline \end{array} \tag{8.105}$$

Here, the columns contain the coordinates of the P_j. The feasibility of this problem is equivalent to Farkas's lemma, which is Theorem 5.40 of Chapter 5. It must not be possible to find a hyperplane that separates the point W from the P_j.

Exercises

Solve the following three minimum assignment problems. There is no harm in your skipping some steps and guessing either the selection or the cover, but you should prove that your solution is correct by solving the dual problem.

1.

	A	B	C	D	E
a	3	14	18	2	5
b	14	9	19	2	14
c	14	10	4	17	12
d	7	2	9	9	19
e	18	11	16	2	10

2.

	A	B	C	D	E	F	G
a	3	14	18	14	2	5	12
b	14	9	19	3	2	14	9
c	14	10	4	13	17	12	8
d	7	2	9	3	9	19	18
e	10	14	19	5	8	7	20
f	18	11	16	8	2	10	15
g	19	5	5	16	13	15	4

3.

	A	B	C	D	E	F	G	H	I
a	3	4	14	17	18	14	2	5	12
b	11	5	5	1	1	19	13	17	2
c	14	19	9	15	19	3	2	14	9
d	14	20	10	19	4	13	17	12	8
e	11	20	2	17	17	6	3	18	2
f	7	8	2	12	9	3	9	19	18
g	10	11	14	2	19	5	8	7	20
h	18	13	11	13	16	8	2	10	15
i	19	15	5	7	5	16	13	15	4

4. Solve the following maximum assignment problem.

	A	B	C	D	E
a	3	14	18	2	5
b	14	9	19	2	14
c	14	10	4	17	12
d	7	2	9	9	19
e	18	11	16	2	10

5. We have nine applicants for eight jobs. In the following table each row represents a job and each column represents an applicant. There is a 1 wherever an applicant is suitable for the job, and the applicant is unsuitable otherwise. Determine a maximal assignment of applicants to suitable jobs.

	A	B	C	D	E	F	G	H	I
a	1	1	1		1		1	1	
b	1		1		1				
c	1	1		1					1
d	1	1			1		1		
e	1	1	1						
f			1	1		1	1	1	
g	1				1				
h	1		1						

6. Interpret the results in Exercise 5 in terms of systems of distinct representatives.

7. In the following table a partial assignment of 1's, at most one from each column and at most one from each row, has been made. Show that it is a maximal such assignment or find a larger assignment.

	A	B	C	D	E	F	G	H
a		1*						1
b	1*			1			1	
c			1*		1		1	
d			1			1*		1
e		1			1*			
f				1*		1		1
g	1						1*	
h	1			1		1	1	

Some of the various combinations of max/min, at-least-one/at-most-one don't make satisfactory problems. Let us look at a few that do.

8. In the following table make a selection of entries, at least one from each row and at least one from each column, for which the sum is minimal. Prove your solution is correct by formulating the dual problem and solving it.

	A	B	C
a	7	1	3
b	6	2	9
c	3	5	4

9. In the following table find a selection of entries, at least one from each row and at least one from each column, for which the sum is least. Prove your solution is correct by formulating the dual problem and solving it.

	A	B	C	D	E
a	4	5	6	3	5
b	1	3	3	2	2
c	2	2	4	1	3

10. In the following table find a selection of entries, at least one from each row and at least one from each column, for which the sum is least. Prove your solution is correct by formulating the dual problem and solving it.

	A	B	C	D	E
a	4	6	5	2	4
b	3	5	4	1	4
c	1	4	2	1	1
d	5	6	6	4	5

11. In the following table find a selection of entries, at most one from each row and at most one from each column, for which the sum is greatest. Prove your solution is correct by formulating the dual problem and solving it.

	A	*B*	*C*	*D*	*E*
a	4	6	5	2	4
b	3	5	4	1	4
c	1	4	2	1	1
d	5	6	6	4	5

12. We have four jobs to do and we have available six teams to do the jobs. Each team can do any of the jobs, but only one of them, in the time available. They have different ratings for each job, and a higher rating means better performance. Determine an optimal assignment.

	A	*B*	*C*	*D*	*E*	*F*
a	18	8	7	12	7	13
b	20	14	16	20	15	20
c	16	11	10	16	10	15
d	15	9	7	14	10	14

13. The chairman of the invitation committee for a post-season football bowl game has the following problem. His committee will invite one team from each of two football conferences. At this point in the season, they have determined that the pairings in the following diagram are possibilities.

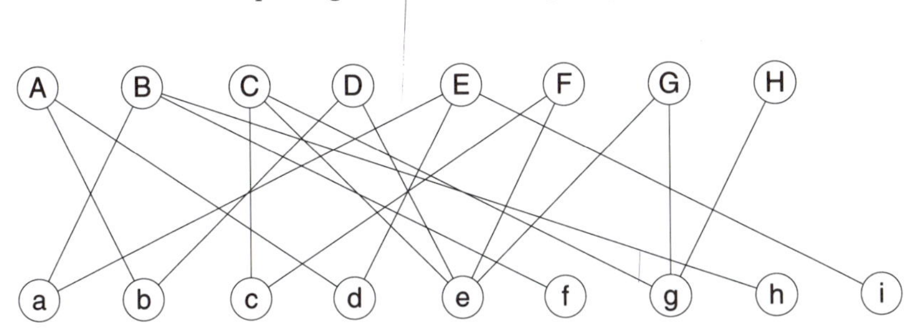

Each line connects a team in one conference with a team in the other conference that might be invited to play. The chairman wants to send scouts to see games played this Saturday. He wants to have the scouts see at least one team of any possible pairing that might eventually be made. On the coming Saturday no team in one conference will play a team in the other conference. What is the minimum number of games that have to be visited? Which teams should be visited? How can this problem be solved? (It may save you work to note that the pairings are the same as those in Figure 8.2.)

14. Find nonnegative values of u_i and v_j for which

$$\sum_{i=1}^{n} u_i + \sum_{j=1}^{n} v_j$$

is minimal and for which $u_i + v_j \geq r_{ij} \geq 0$, where R is

2	3	4	6
1	3	3	4
4	5	7	7
6	6	7	8

Questions

Q1. For a minimum assignment problem, an optimal assignment necessarily uses the least costly choice.

Q2. For a minimum assignment problem, an optimal assignment will never use the most costly choice.

Q3. A minimum assignment problem always has at least one optimal solution

Q4. A maximum assignment problem always has at least one optimal solution.

Q5. In the course of using the Hungarian algorithm to solve a minimum assignment problem, the solution for the dual problem using the values on the margins of the reduced cost table is always feasible.

Q6. An assignment problem cannot be represented as a linear program and, therefore, cannot be solved as a linear program by the simplex algorithm.

Q7. An optimal solution for an assignment problem is always degenerate.

Chapter 9

The Transportation Problem

9.1 The Transportation Problem

Suppose we have several *origins*, locations at which quantities of some material are available, several *destinations*, locations at which quantities of the material are needed, and routes connecting the origins with the destinations. Each route has associated with it a *shipping rate*, the cost of shipping a unit quantity of the material. We wish to satisfy the requirements at the destinations by shipping along routes so as to minimize the total cost.

This kind of problem, known as the *transportation problem*, is typical of an important class of economic problems. Even before George B. Dantzig devised his simplex method, such problems had been investigated by L. V. Kantorovich, (1939), F. L. Hitchcock (1941), and T. C. Koopmans (1947).

We have already considered a transportation problem in Section 1.3. There we set up a tableau to represent the constraints and used the simplex algorithm to solve the problem. Note that the A-matrix in that example consists mostly of zeros with just a few nonzero entries in each row. It turns out that every tableau equivalent to the initial tableau for a transportation problem contains only zeros, 1's, and -1's in the A-matrix. This special structure of the transportation problem allows the use of a special streamlined algorithm. This algorithm, an adaptation of the simplex algorithm, was devised by G. B. Dantzig.

Let us start by considering again the transportation problem worked in Section 1.3. Figure 9.1 shows the three origins, the two destinations, and the six routes joining the origins and destinations.

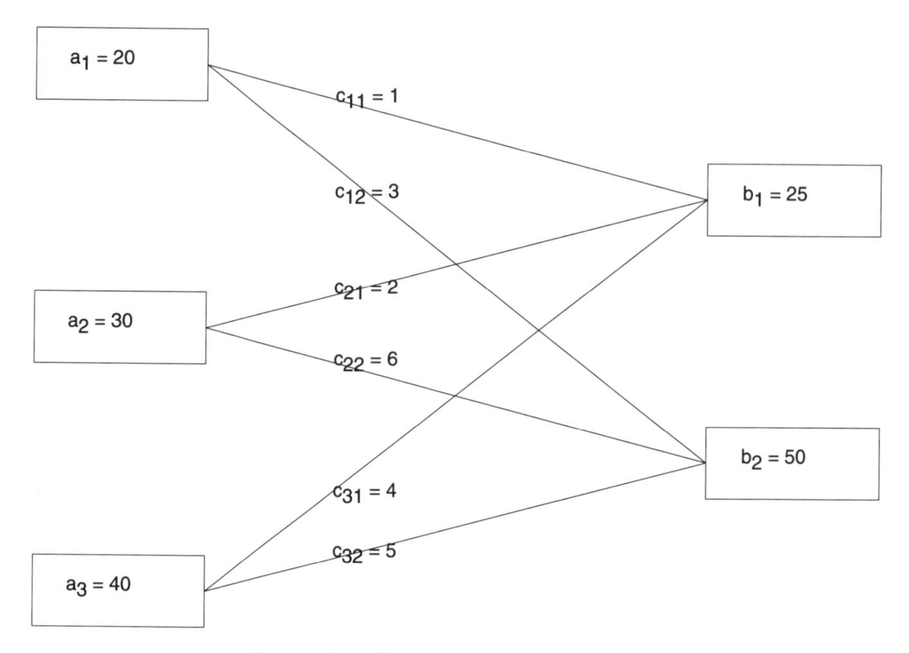

Figure 9.1: Transportation network.

The entries a_1, a_2, a_3 in the origin boxes are the supplies of the material available at these origins. The entries b_1, b_2 in the destination boxes are the demands at these destinations. The number c_{ij} associated with the route joining origin i and destination j is the cost of shipping a unit of the material along that route.

Let x_{ij} denote the quantity of the material shipped from origin i to destination j; y_i the unused supply at origin i; and z_j the excess over demand at destination j. Since we cannot ship more from an origin than is available nor undersupply a destination, the y_i and z_j are required to be nonnegative.

The variables satisfy equations that can be expressed compactly by table 9.1.

$$\begin{array}{c|c|cc|l}
 & 1 & 1 & 1 & \\
\hline
1 & t & -z_1 & -z_2 & = \sum_j b_j \\
\hline
1 & y_1 & x_{11} & x_{12} & = a_1 \\
1 & y_2 & x_{21} & x_{22} & = a_2 \\
1 & y_3 & x_{31} & x_{32} & = a_3 \\
\hline
 & \sum_i a_i & = b_1 & = b_2 &
\end{array} \tag{9.1}$$

The variable t is the total amount shipped,

$$t = x_{11} + x_{12} + x_{21} + x_{22} + x_{31} + x_{32} \tag{9.2}$$

The left column equation

$$t + y_1 + y_2 + y_3 = a_1 + a_2 + a_3 \tag{9.3}$$

says merely that the amount shipped plus the amount unused is equal to the total supply. A similar interpretation can be given to the top row equation

$$t - z_1 - z_2 = b_1 + b_2 \tag{9.4}$$

The other row equations have the form

$$x_{i1} + x_{i2} + y_i = a_i \tag{9.5}$$

or, the amount shipped out of origin i plus the amount unused at origin i is the supply available at origin i. Note that equation 9.3 is the sum of the three equations of the form of equations 9.5 and 9.2 and is, therefore, redundant. The other column equations have the form

$$x_{1j} + x_{2j} + x_{3j} - z_j = b_j \tag{9.6}$$

and equation 9.4 is the sum of these two equations and 9.2.

We wish to minimize the total cost given by

$$g = c_{11}x_{11} + c_{12}x_{12} + c_{21}x_{21} + c_{22}x_{22} + c_{31}x_{31} + c_{32}x_{32} \tag{9.7}$$

A transportation problem has associated with it in a natural way a network consisting of the origins, the destinations, and the routes connecting the origins and destinations. In the network in Figure 9.1, an arc connecting the origin i with the destination j can be associated with the variable x_{ij}. But this network does not provide arcs that correspond to the variables y_i, z_j, and t that also appear in the equations of the transportation problem

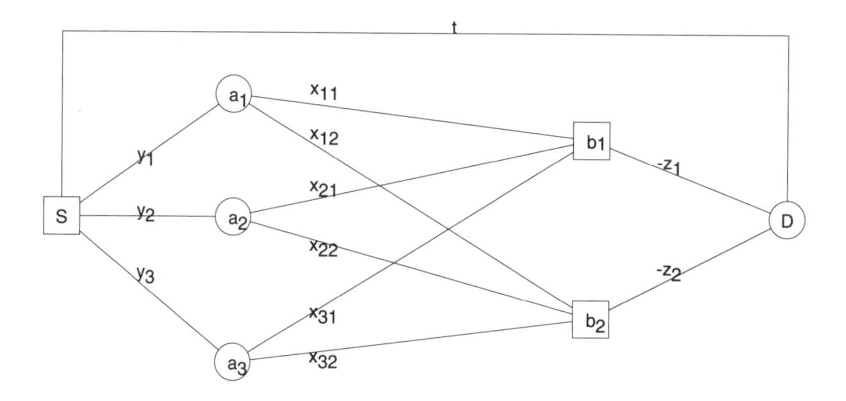

Figure 9.2: Bipartite transportation network.

Thus, we enlarge the network with two more nodes, a *source* S and a *sink*, D as shown in Figure 9.2.

In the network of Figure 9.2, each arc represents a variable in table 9.1 and each node represents an equation of that table. At each node representing an origin we show the supply a_i available there. The quantities on the arcs incident with each such node must sum to the supply available there. Also, the variables on the arcs incident with each destination must sum to the demand there. We can think of S as representing the total supply,

$$S = a_1 + a_2 + a_3 \tag{9.8}$$

so that equation 9.3 is represented by the source node. Similarly, let D represent the total demand,

$$D = b_1 + b_2 \tag{9.9}$$

so that equation 9.4 is represented by the sink node.

Each row or column of the transportation table 9.1 is thus represented by a node of the network. The arcs incident at a node correspond to the variables summed in the equation for that node. Two arcs are adjacent (have a node in common) if and only if they appear in a common equation and, therefore, if and only if their variables appear in the same row or column of the transportation table. Two nodes are adjacent (have an arc in common) if and only if their equations have a variable in common and, therefore, if and only if one represents a row and one represents a column of the transportation table.

In general, the transportation problem with m origins and n destinations has the following form:

Let a_i be the supply at origin i, let b_j be the demand at destination j, and let c_{ij} be the shipping rate from origin i to destination j. Let $x_{ij} \geq 0$ be the amount shipped from origin i to destination j. The objective is to minimize

$$g = \sum_{(i,j)} c_{ij} x_{ij} \tag{9.10}$$

subject to

$$\sum_{j=1}^{n} x_{ij} - a_i = -y_i, i = 1, \ldots, m \tag{9.11}$$

$$\sum_{i=1}^{n} x_{ij} - b_j = z_j, j = 1, \ldots, n \tag{9.12}$$

where all x_{ij}, y_i, z_j are nonnegative.

Cast in this form, it easy to see that the transportation problem is a linear program. There are mn decision variables, the x_{ij}, and there are $m+n$ linear constraints. Thus, if this were set up as a minimization problem in tableau form, the A-matrix would have mn rows and $m+n$ columns. A transportation table in the form of table 9.1 would have $m+1$ rows and $n+1$ columns. The advantage in using a transportation table to represent a transportation should not be underestimated. The tableau representation is much larger than the table representation.

Even more important is the fact that, with the table representation, the A-matrix never has to be computed. Furthermore, the arithmetic is easier. The steps that we will take in the Dantzig algorithm are fully equivalent to pivot exchanges in a corresponding tableau. However, the computation involved in each step amounts to computing the b-column and the c-row, without computing the A-matrix. This is an enormous saving in computation. Actually, the A-matrix is implicit in the transportation table, and we will show later how to obtain the A-matrix from the transportation table, if desired.

We develop a "streamlined" two-phase simplex method applied to the compact table 9.1. In the next section we will show how to do phase I. The Dantzig transportation algorithm omits so many computational steps, and so many more can be omitted, that some care must be used to ensure that we understand and preserve the structure of the problem. However, the calculations are so easy that we are going to describe the computational steps before we go into detail about the structure of the transportation problem.

Notice that the assignment problem can be cast as a special case of the transportation problem In this interpretation, each origin has a supply of one unit, each destination has a demand of one unit, and the number of origins and destinations is equal. An integral solution to the transportation problem of this type will be a solution to the assignment problem.

We are not going to pursue this connection since when solving the assignment problem, the Hungarian algorithm is easier to use than the algorithm for the transportation problem.

9.2 Phase One of Dantzig's Method

Let us proceed with the transportation problem used as an illustration in Section 9.1.

Initially, we take all $x_{ij} = 0$. The initial solution is most easily obtained by starting with the table 9.1. Suppress the x_{ij} and remember that the sum in each row is the supply and the sum in each column is the demand. This gives table 9.13.

0	25	50
20		
30		
40		

(9.13)

The variables corresponding to the blanks in this table, and in future tables, are assigned zeros. The remaining entries are determined, or computed, using equations 9.5, 9.6, and 9.3. The values shown in table 9.13 are the y_i, $-z_j$, and t for an initial shipping schedule in which nothing is shipped.

Because the z_j are negative, this solution is not feasible. The first step, phase I, is to find a feasible solution for the transportation equations. Is there a feasible solution?

The sum $S = \sum_{i=1}^{n} a_i$ is the *total supply* and the sum $D = \sum_{i-1}^{n} b_j$ is the *total demand.* If the total supply is smaller than the total demand, the demand cannot be met without exceeding the supply available. The problem is then infeasible and we shall not consider this possibility further. If the total supply is equal to the total demand we say the transportation problem is *balanced.* If the total supply is larger than the total demand we say the transportation problem is *unbalanced.*

The balanced case is usually regarded as the normalized form of the transportation problem. The unbalanced case is often reduced to the balanced case by creating another destination, called a *dump*, to which all surplus is sent. If there are positive costs associated with the surplus, this is necessary. If the costs associated with the surplus are zeros, the extra column is not needed since the entries in the left margin already represent unshipped surpluses. We will make no distinction between the balanced and the unbalanced case. If there is a positive cost associated with the dump, it is just another destination.

As long as all origins are connected to all destinations, the problem is feasible if the total supply is at least as large as the total demand. In Chapter 11 we will consider problems for which some routes are blocked. In that case, feasibility is not such a simple matter.

It is easy to find an initial feasible solution, and it is possible to find a large number of different initial feasible solutions. If the initial feasible solution is a reasonably good solution, the optimizing phase II will take fewer iterations. For this reason a lot of effort has been expended to find ways to obtain good initial feasible solutions. However, the Dantzig transportation algorithm is so efficient that the extra work expended in trying to find a good initial feasible solution is often more than the work saved in phase II.

> The technique we recommend for phase I is a rather arbitrary compromise between proceeding at random and expending a great deal of effort. We start by looking at each destination in turn and supply as much of its demand from one origin as we can as cheaply as we can. When we have scanned each destination once, we go through again and supply all the demand of each destination.

In this light, consider the two tables in 9.14. The right table shows the rates on the various routes. The table on the left shows our initial (infeasible) solution.

0^+	25^-	50
20^-	+	
30		
40		

	1	3
	2	6
	4	5

(9.14)

We look in the first column (the first destination) for the route with the lowest rate. It is the route from origin 1. We have entered a plus sign in that position of the left table to indicate that we are going to increase the amount shipped along that route. This will simultaneously increase the total amount shipped and decrease the surplus at the origin and the shortage at the destination by the same amounts. We have indicated those changes with the appropriate plus and minus signs.

We increase the amount shipped until it exhausts the available supply, or until it fulfills the demand at the destination, whichever is less. This leads to tables 9.15.

(9.15)

20^+	5	50^-
	20	
30		
40^-		+

	1	3
	2	6
	4	5

The minimum rate in the second column is in the first row, but that supply is already exhausted. The source with the lowest rate from which a supply is available is origin 3. The entries that are to be increased and decreased are indicated with plus and minus signs. The amount that will be shipped is 40. This leads to tables 9.16. This ends the first pass through the table.

(9.16)

60^+	5^-	10
	20	
30^-	+	
		40

	1	3
	2	6
	4	5

In the second pass through the table, we try to satisfy the demand at each destination before looking at the next destination. For this reason, the second pass will end with a feasible solution for the transportation equations. Again, we look in the first column for a route with the lowest rate from which a supply is available. If the demand is still unfulfilled, we look in the first column again, and again, until the demand is fulfilled at destination 1. When the second pass is finished, we obtain table 9.17.

(9.17)

75		
	20	
15	5	10
		40

While it is not very important how an initial feasible solution is obtained, it is important that it correspond to a basic feasible solution for the tableau representation. If we do not have a tableau, how do we know a solution is a basic solution?

In a transportation table, the blanks are the zeros "on purpose." The non blanks should correspond to computed variables.

In Chapter 2, Theorem 2.3, we showed that a solution that is the only solution with specified variables equal to zero is a basic solution. Thus, if the values for the non-blanks arc unique, the solution will be a basic solution. Suppose a pattern of positive variables appears in a transportation table as in table 9.18.

(9.18)

75		
	20	
10^+	5	15^-
5^-		35^+

Four values that form a rectangle are marked with + and − signs. If the values marked with + signs are increased and the values marked with − signs are decreased, all by the same amounts, the transportation equations will still be satisfied. That is, in this case, the solution with the specified blanks as zeros is not unique. This cannot be a basic solution.

Other, more complicated, patterns can be devised that will allow for multiple solutions. The algorithm must avoid these patterns if the solutions are to be basic solutions. Later, we will discuss this set of ideas completely. For the moment, we need only assure ourselves that with each step we preserve the number of non-blank variables, and we avoid creating a pattern that will allow for multiple solutions. For phase I, we must end up with a solution that satisfies these conditions.

For this reason a fine point must be resolved. In the example used, at each point where a blank was chosen the numbers on the margins were unequal. This meant that the position of the next blank was determined (at the smaller of the two margin entries). To illustrate what must be done if the two numbers on the margins are equal, consider table 9.19.

(9.19)

0^+	25^-	50
25^-	+	
30		
40		

We can increase x_{11} to 25, but that will reduce both margin entries to zero. If both were replaced by blanks, the number of computed variables would be too small. Therefore, we replace one variable by a blank and the other by a computed zero.

25		50
0	25	
30		
40		

(9.20)

Furthermore, whenever this situation occurs, we replace the entry on the top margin by a blank. Even though the statement of the problem only requires the z_j to be nonnegative, if the shipping rates are positive no destination will be oversupplied by a least-cost shipping schedule. That is, we expect the z_j to be zeros in the end.

Let us illustrate the process with an example that is rich enough to show the entire process. Consider the transportation table 9.21.

(0)	(11)	(9)	(6)	(12)
(6)	5	4	6	3
(10)	2	5	7	1
(8)	4	6	5	4
(12)	$1^{(11)}$	2	6	2
(9)	3	4	4	1

(9.21)

The entries involved in the transportation equations are enclosed in parentheses. The costs are not enclosed. The sum of the entries on the left margin is the total supply, 45 units. The sum of the entries on the top margin is the total demand, 38 units. This is enough to tell us that the problem is feasible.

In 9.21 the minimum cost route in the first column is indicated by a superscript, which represents the smaller of the supply and the demand for that route. Ship that amount (11) and reduce the supply and demand by that amount. The row and column sums must be preserved. We obtain table 9.22.

(9.22)

(11)		(9)	(6)	(12)
(6)	5	4	6	3
(10)	2	5	7	1
(8)	4	3	5	4
(1)	(11)	$2^{(1)}$	6	2
(9)	3	4	4	1

In table 9.22 we show the reduced supply and demand and the minimum cost in the second column. We then obtain table 9.23.

(9.23)

(12)		(8)	(6)	(12)
(6)	5	4	6	3
(10)	2	5	7	$1^{(10)}$
(8)	4	3	5	4
	(11)	(1)	6	2
(9)	3	4	$4^{(6)}$	1

In table 9.23 we show the choices in both the third and the fourth column. Table 9.24 shows the end of the first scan through all columns.

(9.24)

(28)		(8)		(2)
(6)	5	4	6	3
	2	5	7	(10)
(8)	4	3	5	4
	(11)	(1)	6	2
(3)	3	4	(6)	1

In the second scan through the columns, we fulfill each demand before going on to the next column. The minimal costs in the unfulfilled columns (the second and fourth columns) are indicated in table 9.25 with the superscripts (8) and (2). When the second scan is completed we have table 9.25.

(9.25)

(28)		(8)		(2)
(6)	5	4	6	3
	2	5	7	(10)
(8)	4	$3^{(8)}$	5	4
	(11)	(1)	6	2
(3)	3	4	(6)	$1^{(2)}$

The sequence of steps from table 9.21 to table 9.25 can be performed on one table without recopying. From table 9.25 we can obtain the initial feasible shipping schedule, shown in table 9.26.

(9.26)

(38)				
(6)				
				(10)
(0)		(8)		
	(11)	(1)		
(1)			(6)	(2)

Notice the (0) in the left margin. This is obtained at the step when the available supply and the available demand are equal. This will always occur in a balanced problem, but this example shows that it can occur in an unbalanced problem, too. We started with 10 computed variables, and we have 10 computed variables in table 9.26.

We have not shown that this procedure will lead to a basic feasible solution, but we can verify that this solution is basic by showing that it is the unique solution with these computed variables. Consider a transportation table with only the positions of the computed variables indicated.

(9.27)

( )				
( )				
				( )
( )		( )		
	( )	( )		
( )			( )	( )

The sums must remain constant in each row and each column from table to table. Thus, we can calculate the value of each variable where it is the only computed variable in its row or column. This gives us table 9.28.

(9.28)

(38)				
(6)				
				(10)
( )		( )		
	(11)	( )		
( )			(6)	( )

Next, we compute the values of those variables that are the only remaining uncomputed values in their row or column. We obtain table 9.29.

(9.29)

(38)				
(6)				
				(10)
( )		( )		
	(11)	(1)		
( )			(6)	(2)

Again, look for single remaining uncomputed values in each row and column. We obtain table 9.30.

(9.30)

(38)				
(6)				
				(10)
( )		(8)		
	(11)	(1)		
(1)			(6)	(2)

Finally, the last variable on the left margin can be calculated. It is the only uncomputed value in both its row and its column. It can be computed using either its row or its column. They must yield that same values since the system of equations is redundant and solvable. Since this solution is unique with the specified blanks, the solution is a basic solution.

9.3 The Dual of the Transportation Problem

Let us write the equations of the transportation problem considered in Section 9.1 in tableau form.

(9.31)

	u_1	u_2	u_3	v_1	v_2	0	-1	
x_{11}	-1			1		1	1	$= -w_{11}$
x_{12}	-1				1	1	3	$= -w_{12}$
x_{21}		-1		1		1	2	$= -w_{21}$
x_{22}		-1			1	1	6	$= -w_{21}$
x_{31}			-1	1		1	4	$= -w_{31}$
x_{32}			-1		1	1	5	$= -w_{32}$
-1	-20	-30	-40	25	50			$= f$
	$= y_1$	$= y_2$	$= y_3$	$= z_1$	$= z_2$	$= t$	$= g$	

The row equations are

$$- u_i + v_j + w_{ij} = c_{ij} \tag{9.32}$$

for all combinations of i, j with u_i, v_j, and w_{ij} nonnegative. The objective of the dual problem is to maximize

$$f = -a_1u_1 - a_2u_2 - a_3u_3 + b_1v_1 + b_2v_2 \tag{9.33}$$

The constraints 9.32 should be suggestive of the very similar constraints for the dual of the assignment problem. In a manner similar to the discussion in Section 8.5 we can suggest the transportation problem and its dual in a tableau that contains only the representative relations. Consider tableau 9.34.

(9.34)

	u_i	v_j	-1	
x_{ij}	-1	1	c_{ij}	$= -w_{ij}$
-1	$-a_i$	b_j	0	$= f$
	$= y_i$	$= z_j$	$= g$	

The dual of the transportation problem. Maximize

$$f = \sum_{j=1}^{n} b_j v_j - \sum_{i=1}^{n} a_i u_i \tag{9.35}$$

subject to the constraints

$$-u_i + v_j - c_{ij} = -w_{ij} \tag{9.36}$$

for nonnegative u_i, v_j and w_{ij}.

We see that the duality equation for the transportation problem and its dual is

$$\sum_{i=1}^{n} u_i y_i + \sum_{j=1}^{n} v_j z_j + \sum_{(i,j)} w_{ij} x_{ij} = g - f \tag{9.37}$$

At the start we have a table containing the data of the transportation problem. On the left in 9.38 we show a generic transportation data table, and on the right we show the data table for the example in Section 9.1.

(9.38)

	b_1	b_2
a_1	c_{11}	c_{12}
a_2	c_{21}	c_{22}
a_3	c_{31}	c_{32}

	25	50
20	1	3
30	2	6
40	4	5

Then, there are two tables, one containing the variables for the transportation problem and one containing the variables for the dual of the transportation problem. The tables for the example are shown in 9.39.

(9.39)

t	$-z_1$	$-z_2$
y_1	x_{11}	x_{12}
y_2	x_{21}	x_{22}
y_3	x_{31}	x_{32}

0	v_1	v_2
$-u_1$	w_{11}	w_{12}
$-u_2$	w_{21}	w_{22}
$-u_3$	w_{31}	w_{32}

Some terms are displayed with negative signs to make the arithmetic more convenient for hand calculation. The negative signs are not necessary for machine computation.

The left table in 9.39 represents the equations of the transportation problem. In each row the sum of the entries is the a_i of the corresponding row in table 9.38, and in each column the sum of the entries is the b_j of the corresponding column in table 9.38.

The right table in 9.39 represents the equations of the dual of the transportation problem. Each w_{ij} within the table is obtained by adding the two margin entries, $-u_i$ and v_j, and subtracting that sum from c_{ij}. This is equivalent to equation 9.36.

Each variable in the left table of 9.39 is in the same position as its dual variable in the right table. For convenience in discussing the two tables at the same time, we refer to the left table as the x-table and right table as the w-table. The x-table generates all the equations of the transportation problem except the objective function. Similarly, the w-table generates all equations of the dual program except the dual's objective function. Our immediate goal is, given any basic solution for the x-problem, to find a complementary basic solution for the w-problem.

In Section 9.2, at the end of phase I we found an initial basic feasible solution for the x-problem for a transportation problem. Let us repeat table 9.17 here for convenience. The right table is the data table for the example.

(75)		
	(20)	
(15)	(5)	(10)
		(40)

(0)	(25)	(50)
(20)	1	3
(30)	2	6
(40)	4	5

(9.40)

To keep the x-problem and the w-problem separate, we show the entries for the x-problem in parentheses. The basic variables for the x-problem appear in the left table of 9.40 in parentheses. Since the basic variables of one problem correspond to the nonbasic variables of the dual problem, wherever there is a parenthesis the corresponding variable in the dual problem is nonbasic and would be represented by a blank. This makes it possible to write the basic solution for the x-problem and the dual basic solution for the w-problem in the same table.

The equations 9.32 are the equations for the w-problem. Each equation has four terms, one of which is specified in the data table. Thus, whenever two of the variables are known, the third can easily be calculated. In the second row, we know that $u_2 = 0$, and $w_{21} = w_{22} = 0$, since they are nonbasic variables. This makes it possible to compute v_1 and v_2, which we show in table 9.41.

(75)	2	6
	(20)	
(15)	(5)	(10)
		(40)

(9.41)

Then w_{11} and w_{32} are zero because they are nonbasic, and u_1 and u_3 can be calculated. We obtain

(9.42)

(75)	2	6
−1	(20)	
(15)	(5)	(10)
−1		(40)

Once all the u_i and v_j have been determined, the remaining w_{ij} can be filled in. In this case we get

(9.43)

(75)	2	6
−1	(20)	−2
(15)	(5)	(10)
−1	3	(40)

The process by which we determined this solution to the w-problem shows that this solution is unique with the specified variables as zeros. Hence, this is a basic solution.

If the solution to the w-problem were feasible, we would have complementary basic feasible solutions and each would be optimal. Feasibility would require the entries on the left margin to be nonpositive (which they are) and the remaining entries to be nonnegative. The -2 within the table shows that the solution is not a feasible solution.

Let us illustrate these ideas in a more complex example. Consider the transportation table 9.44.

(9.44)

(0)	(6)	(5)	(5)	(8)	(11)
(9)	5	6	7	5	4
(17)	2	6	5	4	5
(8)	3	5	6	7	6
(6)	5	7	9	4	3

The entries within the table are the original rates, the c_{ij}. The entries in parentheses on the left margin are the supplies available at the origins (the a_i) and the entries in parentheses on the top margin are the demands at the destinations (the b_j). The network for this problem has four origins and five destinations.

In table 9.45 we show a set of basic variables of the x-problem by parentheses. We will show later how to be sure that the set of parentheses shown represents a set of basic variables.

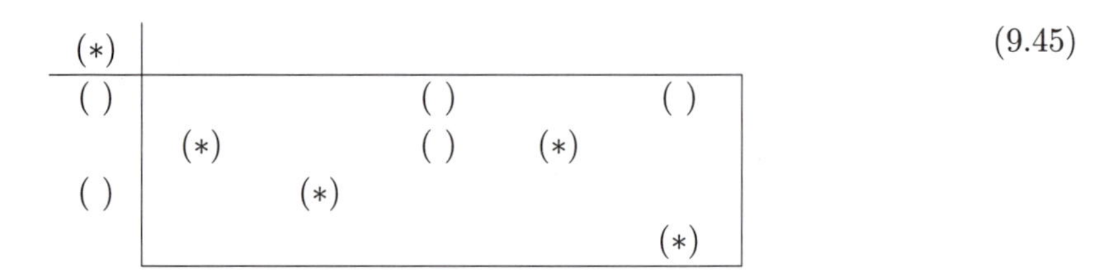

(*)					
( )			( )		( )
	(*)		( )	(*)	
( )		(*)			
					(*)

(9.45)

The basic variables that can be computed immediately are easily identified in table 9.45. They are represented by the rows and columns that contain only one parenthesis. In table 9.45 these entries are identified with asterisks. Each row and column represents an equation of the x-problem and the sum of the entries in each row (or column) must equal the supply (or demand). Thus, these single entries are easily computed. Their values are shown in table 9.46.

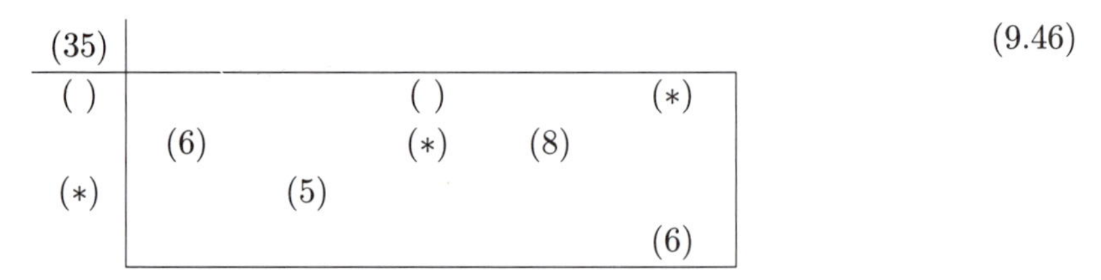

(35)					
( )			( )		(*)
	(6)		(*)	(8)	
(*)		(5)			
					(6)

(9.46)

The entries in parentheses that can be computed next are identified with asterisks in table 9.46. They are the entries that are the only uncomputed entries in their row or column. They are evaluated and shown in table 9.47.

(35)					
(*)			(*)		(5)
	(6)		(3)	(8)	
(3)		(5)			
					(6)

(9.47)

The rest of the entries in the basic solution can be evaluated one-by-one. The results are shown in table 9.48.

(35)		*	*		*
(2)			(2)		(5)
	(6)		(3)	(8)	
(3)		(5)			
					(6)

(9.48)

In table 9.48 the nonbasic variables u_1 and u_3 must be zero. The basic variables for the w-problem that can be calculated immediately are indicated with asterisks in table 9.48. In the u_1-row of table 9.48, w_{13} and w_{15} must be zero. This allows us to evaluate $v_3 = 7$ and $v_5 = 4$. In the u_3-row, w_{32} must be zero. Thus, we get $v_2 = 5$.

(9.49)

(35)		5	7		4
(2)			(2)		(5)
*	(6)		(3)	(8)	
(3)		(5)			
*					(6)

In table 9.49 we show the next few variables that can be computed. The results of this step are shown in table 9.50.

(9.50)

(35)	*	5	7	*	4
(2)			(2)		(5)
−2	(6)		(3)	(8)	
(3)		(5)			
−1					(6)

Then v_1 and v_4 can be calculated, as indicated in table 9.51.

(9.51)

(35)	4	5	7	6	4
(2)			(2)		(5)
−2	(6)		(3)	(8)	
(3)		(5)			
−1					(6)

The remaining values of the w_{ij} are determined by equation 9.32, and the results are shown in table 9.52.

(9.52)

(35)	4	5	7	6	4
(2)	1	1	(2)	−1	(5)
−2	(6)	3	(3)	(8)	3
(3)	−1	(5)	−1	1	2
−1	2	3	3	−1	(6)

In table 9.52 the basic solution for the x-problem is feasible. The basic solution for the w-problem is infeasible.

If the transportation problem had been set up as a linear program in tableau form, in a tableau equivalent to the initial tableau the c-row would give the basic solution for the x-problem (the negatives of the entries in parentheses) and the b-column would give the basic solution for w-problem (the remaining entries). Notice that these basic solutions can be obtained without computing, or even knowing, the corresponding A-matrix.

There are several important consequences of the observations of the previous paragraph. First is the obvious one that the computation is easier with transportation tables than with tableaux. Also, it is not necessary to go through a sequence of tableaux to arrive at a particular basic solution. It is merely necessary to know the set of basic variables. The basic solution is calculated directly from the entries in the initial transportation table. Finally, calculating the basic solution involves only additions and subtractions. If the entries in the initial tableau are integers, all subsequent calculations involve only integers.

9.4 Dantzig's Method

Just as with the simplex method, phase II begins when a feasible basic solution is obtained for the x-problem of the transportation problem. Let us consider the example first presented in Section 9.1. The information given in Figure 9.1 is shown in table 9.53.

(9.53)

	(25)	(50)
(20)	1	3
(30)	2	6
(40)	4	5

In Section 9.2 we obtained an initial feasible solution for the x-problem.

This solution is repeated in table 9.54, where the solution of the x-problem is identified by the use of parentheses.

(9.54)

(75)		
	(20)	
(15)	(5)	(10)
		(40)

We use the method described in the preceding section to find a complementary basic solution for the w-problem. We obtain

(9.55)

(75)	2	6	
-1	(20)	-2^{+}	
(15)	(5)	(10)	
-1	3	(40)	
			290

We have included the value of the objective variable, the shipping cost, in the lower right corner. This value can be obtained by evaluating the shipping cost objective variable

$$g = 1 \cdot 20 + 2 \cdot 5 + 6 \cdot 10 + 5 \cdot 40 = 290 \tag{9.56}$$

or by evaluating the objective variable of the dual problem

$$f = 20(-1) + 40(-1) + 25 \cdot 2 + 50 \cdot 6 = 290 \tag{9.57}$$

If the transportation problem were written in tableau form, the transportation equations would be the column equations, and the equations of the dual problem would be the row equations. The numbers in parentheses represent the values of the basic variables of the min program. Thus, they are the negatives of the corresponding c-row. The numbers not in parentheses represent the values of the basic variables of the max program, except for the entries in the left margin, which are the negatives of these values. They correspond to entries in the b-column. We can test for a basic feasible solution by noting the signs of the entries in the transportation table.

Infeasibility occurs when any entry within the table or on the top margin is negative, or when an entry on the left margin is positive. The solution for the x-problem in table 9.55 is feasible, but the solution for the w-problem is infeasible since $w_{12} = -2$.

We have marked a position where a basic variable of the dual program has an infeasible value. We intend to increase the value of the corresponding nonbasic variable of the transportation equations. Since the sum in each row and in each column must remain constant, if we increase the value of that variable, we must decrease the values of one or more of the other variables in its row and in its column. It turns out to be possible to change the values of exactly none or two variables in each row and column and preserve the validity of all equations.

In table 9.58 we show a pattern of increases and decreases that will preserve all of the equations.

(9.58)

(75)	2	6	
-1	$(20)^-$	-2^+	
(15)	$(5)^+$	$(10)^-$	
-1	3	(40)	
			290

We wish to obtain a new basic feasible solution for the transportation equations after this change. Since that requires six basic variables, we increase x_{12} enough to reduce one of the other variables to zero, but not enough to make any variable negative. It is easily seen that we can increase x_{12} by 10. We obtain table 9.59.

(9.59)

(75)		
	(10)	(10)
(15)	(15)	
		(40)

We claim that this solution for the transportation equations is a basic solution. We could show that by showing that this is the unique solution with the blanked variables set equal to zero. However, we prefer to establish this, and to fill in all other gaps in this discussion, in Sections 9.6 and 9.7, where we discuss the structure of network problems in general.

The solution for the w-solution can be obtained by the method described in the preceding section, and it is shown in table 9.60.

(9.60)

(75)	2	4	
-1	$(10)^-$	$(10)^+$	
$(15)^-$	$(15)^+$	2	
1^+	1	$(40)^-$	
			270

This passage from one pair of complementary basic solutions to another pair is equivalent to a pivot exchange, and that fact will be established in Section 9.7. The solution of the w-problem is basic (since it is uniquely obtained by the method of the previous section) but it is still not feasible because of the positive entry on the left margin. We enter a + sign there and repeat the process. Notice that the pattern of + and − signs forms a loop in the table. We obtain table 9.61.

(9.61)

(75)	2	5	
-2	1	(20)	
(5)	(25)	1	
(10)	2	(30)	
			270

The basic solutions given in table 9.61 are feasible and complementary. Hence, they are optimal.

Let e denote the amount redistributed around the loop in table 9.58. Then the net change in the shipping costs will be

$$\begin{aligned} &ec_{12} - ec_{22} + ec_{21} - ec_{11} \\ &\quad = e(-u_1 + v_2 + w_{12}) - e(-u_2 + v_2) + e(-u_2 + v_1) - e(-u_1 + v_1) \\ &\quad = ew_{12} \end{aligned} \tag{9.62}$$

If e denotes the amount redistributed around the loop in table 9.60, the net change in the shipping costs will be

$$\begin{aligned} &-ec_{11} + ec_{12} - ec_{32} + ec_{21} \\ &\quad = -e(-u_1 + v_1) + e(-u_1 + v_2) - e(-u_3 + v_2) + ev_1 \\ &\quad = eu_3 \end{aligned} \tag{9.63}$$

We see that for either type of loop, one starting at a negative w_{ij} or one starting at a negative u_i (remember that the entries on the left margin are the negatives of the u_i), the net change is the product of the infeasibility marked with a + sign and the minimum term in the loop marked with a − sign. This observation motivates the commonly used rule of marking the maximum infeasibility. This choice is likely to result in the largest change in the shipping costs. However, if the minimum entry in the loop marked with a − sign is zero, there will be no change in the shipping costs. This is the way degeneracy occurs in a transportation problem. This will be illustrated later in this section.

The Dantzig algorithm for the transportation problem consists of the following steps.

Dantzig's Transportation Algorithm

1. Find an initial basic feasible solution for the x-problem by any method; for example by using the phase I method described in Section 9.2. Mark the entries corresponding to the basic variables with parentheses.

2. Construct a complementary basic solution for the w-problem using the method described in Section 9.3.

3. If the solution for the w-problem is feasible STOP. We have obtained optimality.

4. If the solution is not feasible, select the entry with the greatest infeasibility. Construct the unique loop L with this entry marked with a + sign. All other entries in the loop are located at entries of the basic x-solution.

5. Mark the nonbasic position (for the x-problem) to be increased with a + sign, and all other entries on the loop L alternately with − signs and + signs. The smallest entry with a − sign is the amount to redistribute around the loop. Increase each entry with a + sign by that amount; decrease each entry with a − sign by that amount. The variable at the marked nonbasic position becomes a new basic variable. One basic entry that is reduced to zero is dropped from the basic solution. If two or more entries are reduced to zero by the redistribution, only one is dropped and the others are retained as zero basic entries. Return to step 2.

Let us tie things together by considering a fresh example. Let table 9.64 represent a new transportation problem. This example is designed to illustrate degeneracy in both the x-problem and the w-problem.

(0)	(15)	(15)	(35)	(30)
(15)	1	3	5	5
(10)	7	5	7	11
(50)	4	7	13	8
(30)	2	3	11	7

(9.64)

We shall use the phase I method described in Section 9.2. Scan the columns and in each column choose the least costly route over which a positive quantity can be shipped. In the very first step we face reducing both the supply and the demand to zero. Where there is a choice in blanking one margin entry and leaving the other as a computed zero, we enter zero on the left margin.

(95)				
(0)	(15)			
			(10)	
(10)			(25)	(15)
		(15)		(15)

(9.65)

Once a feasible solution for the x-problem is obtained we compute the complementary basic solution for the w-problem, as shown in table 9.66.

(95)	1	4	13	8	
$(0)^-$	(15)	-1	-8^+	-3	
-6	12	7	(10)	9	
$(10)^+$	3	3	$(25)^-$	(15)	
-1	2	(15)	-1	(15)	
					680

(9.66)

We find the largest infeasibility, which is the -8 in the first row. We mark it and the loop, and perform the pivot exchange to obtain table 9.67. Because of the zero value for a basic variable on the left margin, the basic solution is degenerate. Notice that there is no change in the value of the objective variable.

(9.67)

(95)	9	4	13	8	
−8	$(15)^-$	7	$(0)^+$	5	
−6	4	7	(10)	9	
(10)	−5	3	$(25)^-$	$(15)^+$	
−1	-6^+	(15)	−1	$(15)^-$	
					680

We have marked the maximum infeasibility and the loop in table 9.67. We realize that it will not always be easy to determine the loop. It is not too difficult in small problems, and we will describe an effective general method for finding the loop in Section 9.7. Here there are two choices for the minimum entry marked with a − sign. This is a symptom of degeneracy, which will show up in table 9.68. Notice that there is no improvement in the value of the objective variable between table 9.68 and table 9.69. Cycling in transportation problems does not seem to be much of a problem, but it is easy to give a rule that will prevent cycling, the Bland rule.

Refer to the tableau representation of the transportation problem in tableau 9.31. Marking the infeasibility with a + sign is equivalent to choosing the pivot row. Choosing the basic variable to blank is equivalent to choosing the pivot column. There is a natural and convenient ordering of the variables in a transportation table that can be used for the Bland rule. Each position in the table represents a variable and we can order the variables by their positions.

The Bland rule for transportation problems. Whenever there is more than one choice, either for marking an infeasibility or choosing the minimum x-entry in the loop with a − sign, choose the first available position with the following priority rule: first the left margin from top to bottom, then the positions within the table by rows from top to bottom and within rows from left to right, then the top margin from left to right.

We have not used the Bland rule for marking infeasibilities since cycling is not a problem as long as the objective variable is decreased. We did use the Bland rule in table 9.67 for choosing the variable to blank.

(9.68)

(95)	3	4	13	8	
−8	6	7	(15)	5	
−6	10	7	(10)	9	
(10)	1	3	$(10)^-$	$(30)^+$	
−1	(5)	(15)	-1^+	$(0)^-$	
					590

There is only one choice for the pivot exchange in table 9.68 and we obtain optimality in table 9.69.

(9.69)

(95)	4	5	13	8	
−8	5	6	(15)	5	
−6	9	6	(10)	9	
(10)	0^+	2	$(10)^-$	(30)	
−2	$(5)^-$	(15)	$(0)^+$	1	
					590

The zero value for w_{31} in table 9.69 means that the w-problem is degenerate. Degeneracy in the w-problem is not a complication until optimality is obtained because the marked infeasibilities are always nonzero. However, when optimality is obtained it often means there is more than one optimal solution. A different optimal shipping schedule is shown in table 9.70.

(9.70)

(95)	4	5	13	8	
−8	5	6	(15)	5	
−6	9	6	(10)	9	
(10)	(5)	2	(5)	(30)	
−2	0	(15)	(5)	1	
					590

9.5 Economic Interpretation of the Dual Program

It is possible to give an economic interpretation to the dual of the transportation problem that gives considerable insight into the transportation problem itself.

Suppose the business, whose transportation problem we have been considering, owns the material that is available at the origins and wishes to have the material made available at the destinations in the specified quantities. The business is not concerned with how the material is transported. Suppose there is an entrepreneur who has supplies of the material that the business wants transported. He proposes to buy the material from the business at origin i for a price u_i per unit and sell it (or another unit) back to the business at destination j at a price of v_j per unit.

To make his proposal reasonable and competitive, the entrepreneur must arrange his price schedule so that the difference between the price at destination j and the price at origin i does not exceed the shipping rate c_{ij}. Otherwise, the business will elect to ship the material rather than accept the entrepreneur's proposal. This condition can be expressed in the form of equation 9.71.

$$-u_i + v_j + w_{ij} = c_{ij} \tag{9.71}$$

with w_{ij} nonnegative.

The net income for the entrepreneur is

$$-\sum_{i=1}^{m} a_i u_i + \sum_{j=1}^{n} b_j v_j = f \tag{9.72}$$

The entrepreneur wants to maximize his profit, so his problem is the max program of tableau 9.34.

If the transportation problem is feasible and all rates are nonnegative, the transportation problem has an optimal solution, and the optimal solution for the min program and the optimal solution for the max program are equal. Thus, the entrepreneur's proposal would be reasonable. It would be a matter of financial indifference whether the business accepted or rejected the proposal.

The existence of such an entrepreneur is fanciful. A more reasonable interpretation is to consider the business and the entrepreneur to be identical. The prices in the dual problem, then, are *marginal prices* which represent the increases in cost or value incurred by moving the material from one location to another to meet demand. These costs are "values added" as a result of the interplay between supply, demand, and shipping costs.

To illustrate these ideas, consider the example we worked out in the preceding section. We repeat table 9.70 here as table 9.73.

(9.73)

(95)	4	5	13	8	
−8	5	6	(15)	5	
−6	9	6	(10)	9	
(10)	(10)	2	0	(30)	
−2	(5)	(15)	(10)	1	
					590

Except for the fact that there is a surplus at origin 3, we are less interested in the solution for the transportation problem than we are in the solution for the dual problem. In Figure 9.3 we show the arcs corresponding to the basic variables of the basic optimal solution of the transportation problem and the rates associated to those arcs.

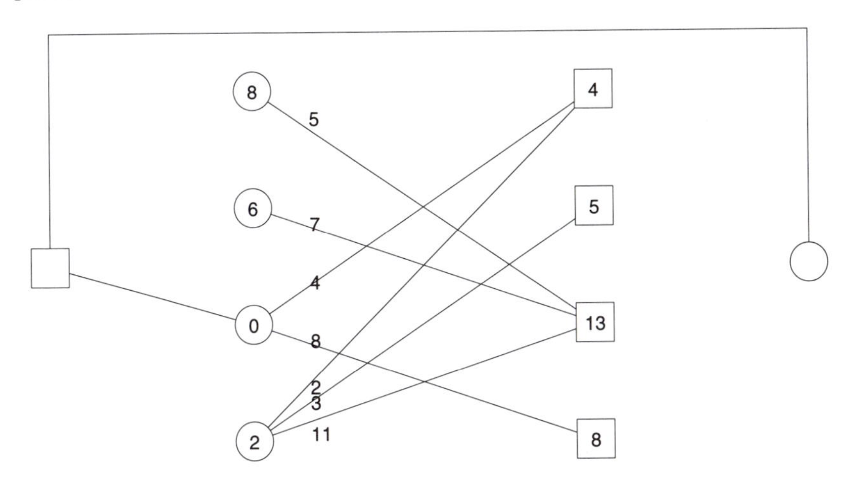

Figure 9.3: Optimal basic solution for a transportation problem.

The numbers at the nodes are the prices at the origins and destinations for the optimal solution of the dual problem. They were established by the optimal solution of the dual problem. For the routes actually used by the solution of the transportation equations, the costs over those routes are exactly equal to the prices differences between the ends of those routes. Of course, that is equivalent to the condition that the nonbasic variables $w_{ij} = 0$ for those routes. For the routes that are not used, those blanked in Figure 9.3, the costs are at least as large as the price differentials. That is equivalent to the feasibility condition that w_{ij} be nonnegative.

The prices are marginal prices, not absolute prices. At the origin where there is a surplus, the price is zero. There was a supply of 50 units of the material to be shipped at that origin. If there had been a little more or less available, but with a change insufficient to remove the surplus, the solutions to both problems would have been the same except for the surplus at that origin. This would not have changed the total shipping cost. That is, at an origin where there is a surplus the marginal price is zero.

Consider a destination where the demand is increased slightly. For example, suppose the demand at destination 2 is increased from 15 units to 16 units. Destination 2 receives all 15 units from origin 4. If its shipment is increased to 16, origin 4 will have to decrease the amount it ships to destination 1, and the amount shipped from origin 3 will have to be increased accordingly. Eventually, any increase at any destination will have to come out of the surplus. These changes in the amounts shipped will change the cost by $3 - 2 + 4 = 5$, the marginal price at destination 2.

Similarly, consider an origin where the supply is increased slightly. For example, suppose the supply at origin 1 is increased from 15 units to 16 units. Then, destination 3 can receive one more unit from origin 1, and will receive one less from origin 4, which will ship one less to destination 1, which will receive one less from origin 3, where a surplus occurs. These changes in the amounts shipped will change the cost by $5-11+2-4 = -8$, the negative of the marginal price at origin 1.

The dual problem is very informative to the business that requires the shipping. The prices in the solution to the dual problem indicate what the relation between the supply and the demand is costing the business, it and they can be interpreted in order to invoke changes that would reduce costs. Even if this material is used entirely within the business enterprise and need not otherwise have prices associated with, this analysis will tell the businessman what changing the demand or supply at any point will actually cost him or save him.

Each $w_{ij} = c_{ij} - (v_j - u_i)$ is a residual cost, the excess of the shipping rate over the price differential. If the residual cost of a route is positive for an optimal solution of the dual problem, the route cannot be used in an optimal solution of the transportation problem.

Because of the duality equation, the marginal prices will be zero at all origins where the supplies are not exhausted. If the transportation problem is unbalanced, an optimal solution will have at least one y_i nonzero. That is, at least one surplus will be positive. This will force the corresponding marginal price to be zero.

If the transportation problem is balanced, our convention of forcing blanks on the top margin will leave at least one zero basic variable on the left margin. This pegs at least one origin price at zero. In the balanced case, one could add a constant amount to all prices, at the origins and at the destinations. This would not affect the price differentials, and it would have no practical effect on the solution. We could, in the balanced case, regard the variables y_i and z_j as artificial variables and the variables u_i and v_j as free variables. If the transportation problem is worked in this way, we could, at the end, add a constant to all prices to make the origin prices nonnegative with at least one equal to zero.

The balanced case is usually regarded as the normalized form of the transportation problem, with u_i and v_j as free variables. The unbalanced case is then reduced to the balanced case by creating another destination, called a dump, to which all surplus is sent. However, this adds an extra column to the transportation table and offers no computational advantage. We recommend, therefore, that the balanced and unbalanced cases be treated alike, as in Sections 9.2 and 9.4.

9.6 Graphs and Trees

The transportation problem network is a special case of a geometric figure called a graph. In the 1940s and 1950s, graph theory was a fascinating topic that seemed to have few practical applications. Now it is considered fundamental in a variety of subjects. Of particular importance today is the usefulness of graph theory in computer applications.

A *graph* is a finite set of nodes, some pairs of which are connected by arcs. Graph theory lacks a standard terminology for its fundamental objects, and almost every term we will use has several synonyms. In particular, nodes are also called points and vertices, and arcs are also called edges. Our choices for the terms used are arbitrary and we do not claim that they are more desirable than other choices. It is the ideas that are important and not the terminology.

We cannot get involved in all the generalities of graph theory, so we will make a few restrictions appropriate to our needs. Sometimes arcs are permitted that begin and end at the same node: We will assume that we have no such arcs. An arc is said to be *oriented* if one end is designated as the tail end and the other is designated as the head end. In this case a pair of nodes can be connected by an arc with one of two different orientations, or they can be connected by two arcs—one in each orientation. An arc is unoriented if neither end is distinguished. Sometimes multiple arcs connecting a pair of nodes are permitted: We will assume that there is no duplication of connections. In the context of this book we consider only the following possibilities: A pair of nodes is not connected, the pair is connected by a single arc (unoriented or oriented in one of two possible orientations), or the pair is connected by two arcs (one in each orientation).

These restrictions allow us to identify an arc by specifying the nodes that it connects. If i and j are two nodes, (i, j) is the arc connecting them. If the arc is considered to be oriented, i is the tail and j is the head. In this case (i, j) and (j, i) are two different arcs. If the arc is unoriented, then (i, j) and (j, i) are the same arc.

If the arcs in a graph are oriented, the graph is called a *directed graph* or *digraph*. Most of the graphs we consider are directed graphs. For example, in the graph of a transportation network, the connection between an origin and a destination is oriented, with the tail at the origin and the head at the destination.

We must now introduce a lot of the terminology of graph theory. If this were a book on graph theory these terms could have been introduced more gradually. Fortunately, the ideas involved are really quite simple.

An arc and a node are *incident* if the node is at one end of the arc. Two arcs are *adjacent* if they are incident with a common node. Two nodes are *adjacent* if they are incident with a common arc. A *path* is a sequence of adjacent arcs. Even if the arcs are oriented, a path does not require that the arcs in the path be oriented in any particular way. The path is a *loop* or a *cycle* or a *circuit* if the first and last arcs are adjacent.

A path can equally well be defined in terms of nodes. That is, a path is a sequence of adjacent nodes. A path is a loop or a cycle or a circuit if the first and last nodes are identical. A path is simple if it does not cross itself. That is, a *simple path* is a path in which all the nodes are distinct or, in the case of a loop, all except the first and last nodes are distinct. Two nodes in a graph are connected if there is a path within the graph in which these two nodes are the first and last nodes. A graph is *connected* if every pair of nodes in the graph is connected.

A *graph* G formally consists of two sets: One is a set N of nodes and the other is a set A of arcs. These two sets are related by the fact that every arc in the arc set is incident with two nodes in the node set. A *subgraph* G' of G is a graph whose node set N' is a subset of N and whose arc set A' is a subset of A. Since G' is a graph, each arc in A' must be incident with two nodes in N'.

Let us illustrate these ideas in Figure 9.4. The graph in Figure 9.5 is a subgraph of the graph in Figure 9.4.

The connection between network problems and graph theory is this. The transportation problem is a prototype network problem. The nodes and arcs of the network form a graph. Associated with each arc (i, j) is a variable x_{ij}. In a transportation problem x_{ij} is the amount shipped from node i (an origin) to node j (a destination). Associated with each node is a linear equation. In a network problem this equation involves with nonzero coefficients only the variables of the arcs incident with the node. The usual

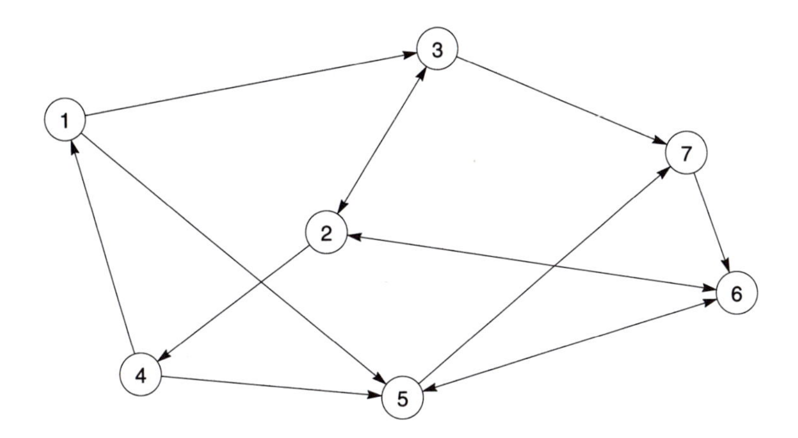

Figure 9.4: A generic graph.

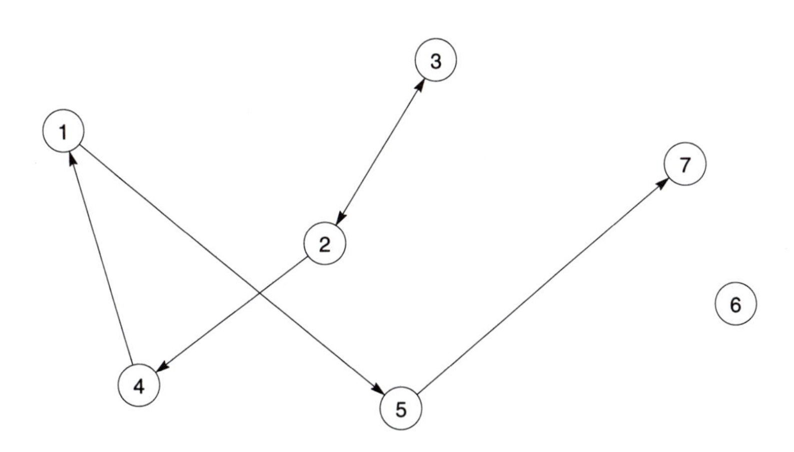

Figure 9.5: A subgraph.

equation in a network problem is a conservation equation: The sum of all flows into the node minus the sum of all flows out of the node is a constant. In this interpretation, the only nonzero coefficients are $+1$ and -1. The coefficient is $+1$ for the variable associated with an arc whose head is at the node, and the coefficient is -1 for the variable associated with an arc whose tail is at the node.

An arc and a node are incident if the variable appears in the equation with a nonzero coefficient. Two arcs are adjacent if their variables appear in the same equation. Since each arc is incident with only two nodes, each variable appears in only two equations with a nonzero coefficient. Since there is at most one arc connecting a pair of nodes, no two variables appear in the same pair of equations.

The purpose of this section is to associate the basic solutions of the system of linear network equations with special subgraphs of the graph associated with the system. Now let us discuss these special subgraphs. A *tree* is a connected graph with no loops. The fundamental properties of trees are embodied in the following theorem.

Theorem 9.61 *Let T be a graph with M nodes. (i) If T is connected and has no loops, it has $M-1$ arcs. (ii) If T is connected with $M-1$ arcs, it has no loops. (iii) If T has $M-1$ arcs and no loops, it is connected.*

Proof. We can easily verify the three assertions of this theorem for a graph with one node and for a graph with two nodes. Assume the theorem is true for any graph with less than M nodes.

Let T be connected with no loops. Remove any arc. The result is two connected graphs, each with no loops. If one contains r nodes the other contains $M-r$ nodes, both numbers less than M. One component has $r-1$ arcs and the other has $M-r-1$ arcs. The original graph then has $(r-1)+(M-r-1)+1 = M-1$ arcs. This proves assertion (i).

Suppose T is connected and has $M-1$ arcs. The number of arcs of the network can be computed by counting arc-ends and dividing by 2. Since T is connected, every node is incident with at least one arc. If every node were incident with at least two arcs, the number of arcs would be at least $2M/2 = M$. Thus, T has at least one node incident with exactly one arc. Such a node is a terminal node L, also called a *leaf*. If L and its incident arc are removed, we obtain a graph with $M-1$ nodes and $M-2$ arcs, which is connected. By the induction assumption, this reduced graph has no loops. Thus, T also has no loops. This proves (ii).

Finally, suppose T has $M-1$ arcs and no loops. Each connected component has no loops. If a connected component has r nodes, it has $r-1$ arcs, by (i) of this theorem. If there were more than one component, there would be fewer than $M-1$ arcs. This proves (iii). ◻

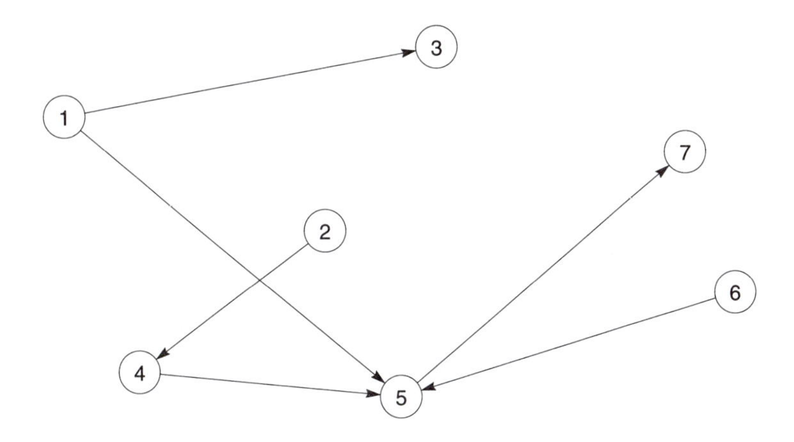

Figure 9.6: A spanning tree.

We have shown that for a graph with M nodes, any two of the following three properties imply the third: (i) The graph is connected; (ii) the graph has no loops; (iii) the graph has $M - 1$ arcs.

Theorem 9.62 *Every tree has at least one terminal node, a node incident with only one arc.*

Proof. Start with any node and trace a path, starting in any direction. Whenever we reach a node not previously touched, if it is not a terminal node we can depart from that node on an arc different from the arc on which we arrived. We cannot arrive at a node that we previously visited since that would complete a loop. But since there are only a finite number of nodes we must eventually arrive at a node we cannot leave, a terminal node. ◻

If G is a graph, a *spanning subtree* of G is a subgraph that is a tree and contains all the nodes of G. Since a spanning tree is connected, the graph G must be connected. It turns out that any connected graph has at least one spanning subtree. We will not take time to prove this since we have other objectives, though the proof is quite easy. Our goal is to show that there is a one-to-one correspondence between basic solutions and spanning subtrees in the network.

Figure 9.6 is a tree, and it is a spanning tree in the graph of Figure 9.4.

9.7 Structure of Network Problems

Let us turn our attention to the system of linear equations associated with a network.

1. We assume that each variable appears in exactly two equations with nonzero coefficients and no two variables appear in the same two equations. With this assumption, an arc associated with a variable connects two nodes, which are identified with the equations, and each pair of nodes is connected by at most one arc.

2. If there are M equations we assume the rank of the system is $M - 1$ and that the system is consistent. That is, the system of equations has at least one solution and, if the system of equations is represented in tableau form, exactly $M - 1$ independent pivot exchanges can be performed.

3. Since our network is to have spanning trees, we have to impose a condition that will guarantee that the network is connected. Hence, assume that the system of equations can not be broken into two disjoint systems with disjoint sets of variables.

At the moment we are not concerned with feasibility, only with solvability. Thus, we will take all variables to be free. In all the other sections in this chapter we write the network equations in the columns of the tableau. Here we are not interested in either the min program or the max program, only in the system of equations. Since most of us are more comfortable when the equations are written in rows and the variables are in the columns, we will use tableaux in which the network equations appear in the rows.

Consider the tableau 9.74 representing this system of linear equations.

(9.74)

<table>
<tr><td></td><td>arc variables</td><td>-1</td><td></td></tr>
<tr><td rowspan="3">M linear node equations</td><td rowspan="3"></td><td rowspan="3"></td><td>$= 0$</td></tr>
<tr><td>$\vdots$</td></tr>
<tr><td>$= 0$</td></tr>
</table>

Let us introduce the variables of the dual system of equations for this tableau. In this way we obtain tableau 9.75.

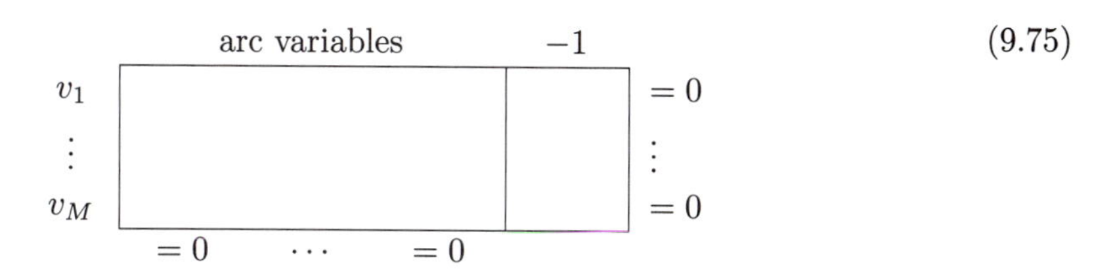

(9.75)

Perform any $M-1$ independent pivot exchanges. For the sake of simplicity in representing the tableau, let us assume the pivot entries for the pivot exchanges occur in the first $M-1$ rows and the first $M-1$ columns. Then the resulting tableau has the appearance of tableau 9.76.

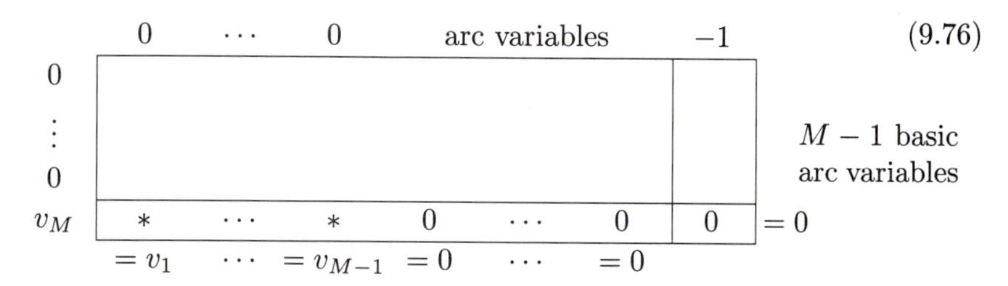

(9.76)

The entries shown as zeros in the last row that are within the A-matrix must be zero or else there would be more than $M-1$ independent pivot exchanges. The entry in the last row of the b-column must be zero in order for the system of equations to have a solution.

The column system of equations has a nonzero solution since we can take v_M to be nonzero. We claim that all $v_1, \ldots, v_M$ are nonzero. Suppose that r, where $0 < r < M$, of the $v_1, \ldots, v_M$ are nonzero. Rearrange the rows of tableau 9.75 so that these nonzero dual variables are in the first r rows. Then, rearrange the columns so that all variables with a nonzero coefficient in one of the first r rows are in the first block of columns.

(9.77)

	arc variables						−1	
*				0	⋯	0		= 0
*				⋮		⋮		⋮
*				0	⋯	0		= 0
0	0	⋯	0					= 0
0	⋮		⋮					⋮
0	0	⋯	0					= 0
	= 0		⋯			= 0		

The upper right block of the A-matrix must be all zeros since we rearranged the columns to bring that about. The lower left block of the A-matrix must be all zeros for the following reasons. Each of those columns contains a nonzero term in one of the first r rows. The column equations would not be satisfied unless at least two nonzero terms were in the first r rows. Since no column contains more than two nonzero terms, there are none in the lower left block. This violates condition 3 given above, since the system of equations would then break down into two disjoint systems.

Now things get much simpler. Multiply row equation i by v_i. For the new system of equations, $v_1 = \cdots = v_M = 1$ is a solution of the column equations. That is, the sum of the equations is zero. This implies that the two nonzero terms in each column are negatives of each other. Then, in each column we can change the scale of the variable so these coefficients become $+1$ and -1.

This means that in addition to the three conditions on the system of equations given above we can, without loss of generality, assume that each column of the A-matrix contains one $+1$, one -1, and all other entries are zeros.

Note that in tableau 9.76, the zeros on the top and right margins do not represent variables that correspond to arcs in the corresponding graph. For the following discussion, we consider tableaux obtained from tableau 9.74 by pivoting in which there are $M - 1$ basic variables corresponding to arcs in the graph of the system of equations.

Theorem 9.63 *For a system of linear equations satisfying the conditions 1, 2, and 3 given above, there is a one-to-one correspondence between basic solutions of the system of equations and spanning subtrees in the corresponding graph.*

Proof. For any tableau obtained from tableau 9.74 by pivoting as many variables as possible to the top margin, we have a basic solution with $M - 1$ basic variables. That is, every other basic solution will also have $M - 1$ basic variables corresponding to arcs. $M - 1$ is the correct number of arcs for a spanning tree in the graph, and it is the correct number of variables for the basic variables of a basic solution.

First, suppose we have a spanning tree in the graph. Set all variables not in the spanning tree to zero. We wish to show that there is just one solution with those variables set to zero. From Theorem 9.62, there is a terminal node in the spanning tree. In the equation corresponding to that node, there is just one variable corresponding to an arc in the spanning tree. That means that all other variables either do not appear in the equation or have values already assigned. This variable has a nonzero coefficient, and we can solve the linear equation for the value of that variable.

Now consider the graph that remains after removing that node and that arc. What remains is a tree, and it also has a terminal node. Again, the value of one more variable is uniquely determined.

Continue in this way until we arrive at a tree with one arc and two nodes. That is, we have one variable in two equations. It is at this point that the consistency of the system of equations plays a role. It is sufficient to solve either equation for the single variable. Since the solution with the prescribed variables set to zero exists and is unique, the solution is a basic solution and the variables corresponding to arcs in the spanning tree are basic variables.

Now, suppose we have a basic solution. The subgraph consisting of arcs corresponding to the basic variables has $M-1$ arcs. Suppose this subgraph contains a loop with r nodes and r arcs (there are possibly other arcs of the subgraph attached to these nodes). Let us reorder the equations and variables in tableau 9.74 so that these are the first r equations and the first r variables. Then the system of equations will have the appearance of tableau 9.78.

(9.78)

x_1	x_2	.	.	.	x_r	
1					-1	
-1	1					
.	.	.	.	.	.	
.	.	.	-1	1	.	
.	.	.	.	-1	1	

Since the variables are free, we can change the sign of a variable to change the sign of a coefficient. Since the constraints are equations, we can change the sign of all the coefficients in an equation. Thus, there is no loss of generality to assume the pattern of signs is that shown in tableau 9.78. For any solution of the system of equations, we can add a constant to all of $x_1, x_2, \cdots, x_r$ to obtain a different solution. Since this is a basic solution, the solution must be unique. Thus, the subgraph does not have a loop. It must be a subtree. □

Let us connect the ideas of this section with the transportation problem we have been considering. We will write down the network tableau for the example that we introduced in Section 9.1 and obtain the tableau 9.79 from it. For this example let us return to writing the transportation problem as a minimization problem.

(9.79)

x_{12}	-1			1				1
x_{22}	-1				1			3
x_{21}		-1		1				2
x_{22}		-1			1			6
x_{31}			-1	1				4
x_{32}			-1		1			5
y_1	-1^*					1		0
y_2		-1^*				1		0
y_3			-1^*			1		0
z_1				-1^*			1	0
z_2					-1^*		1	0
t						1^*	-1	0
-1	-20	-30	-40	25	50	90	-75	0
	$= 0$	$= 0$	$= 0$	$= 0$	$= 0$	$= 0$	$= 0$	$= g$

The network for this problem has seven nodes. Six independent pivot exchanges are indicated with asterisks. When these pivot exchanges are performed, we will obtain a tableau with the sixth and seventh columns the negatives of each other. That is, one is redundant. When that column and the last six rows, headed by zeros, are deleted we will obtain precisely tableau 9.31, because we will have the same margin variables for equivalent tableaux.

We have promised that we would show that even the A-matrix can be reconstructed from any spanning tree and that it is not necessary to update the A-matrix with every iteration. Suppose we have a spanning tree for the network. Write down a skeleton tableau with the variables of the spanning tree as basic variables on the bottom margin and the variables not in the spanning tree as nonbasic variables on the left margin. It is sufficient to show how any particular row can be computed, and in tableau 9.80 we show a typical row.

(9.80)

x_{ij}	0	1	-1	$*$	

The arc corresponding to x_{ij} is not in the spanning tree, but there is a loop including x_{ij} with its remaining arcs in the spanning tree. The variables corresponding to the arcs in the spanning tree are basic variables and are on the bottom margin. Traverse the loop in the direction indicated by the orientation of the arc (i, j). That is, start around the loop moving from the tail to the head of this arc. For any arc of the loop traversed from tail to head, enter a $+1$ in the column of the corresponding variable. For any arc of the loop traversed from head to tail, enter a -1 in the column of the corresponding variable. For those arcs that are not in the loop, enter a zero in the column of the corresponding variable.

If we repeat this procedure for every row of the tableau, we can fill in the A-matrix. Now we must show that this is correct. Notice that the basic solution for tableau 9.80 satisfies the equations in the initial network tableau in the form of 9.74 and 9.79 regardless of the entries in the A-matrix. We merely have to show that as we increase the nonbasic variable x_{ij} it remains a solution.

In the analogy between the system of equations 9.74 and the graph, each node corresponds to an equation and each arc corresponds to a variable. A loop consists of a sequence of adjacent arcs. That is, each equation in the loop contains two variables corresponding to arcs in the loop. If one arc has its tail at the node and the other arc has its head at the node, they appear in the equation with opposite signs. Increasing both variables by the same amount will preserve the solution. If both arcs have their heads at the node or both have their tails at the node, they appear in the equation with the same signs. Increasing one and decreasing the other by the same amount will preserve the solution. Because of the way we have assigned values to the entries in the x_{ij}, row of the A-matrix, as x_{ij} is increased, the variables for arcs not in the loop are unaffected, those in the loop oriented in the same direction as x_{ij} are increased by that amount, and those in the loop oriented in the opposite direction are decreased by that amount. Thus, the values obtained are also solutions of the network equations. Since the A-matrix is uniquely determined by the variables on the margins, the assignments we made are correct. That is, we can completely reconstruct the A-matrix from knowledge of the arcs in a spanning tree.

9.8 Pivoting in Graphs

We have seen that pivoting in a tableau amounts to going from one basic solution to another. Since there is a one-to-one correspondence between basic solutions in a linear system of network equations and spanning trees in a graph, the analogous operation in a graph amounts to going from one spanning tree to another, and we will call it *pivoting in graphs*.

In a typical network problem, we would have to show how to find a basic solution, then how to find a basic feasible solution, if there is one, then how to find an optimal basic solution, if there is one. We will describe these actions in terms of pivoting in a graph.

We are not going to discuss pivoting in graphs in its full generality. We will illustrate the essential ideas by examples in terms of a transportation network and the Dantzig algorithm.

Before going any further, let us verify that the system of equations for the transportation problem satisfies the three conditions we have used to define a network system of linear equations. The following general transportation equations are equivalent to 9.6, 9.5, 9.4, and 9.3, respectively.

$$\sum_{i=1}^{m} x_{ij} - z_j - b_j = 0, \quad j = 1, \ldots, n \tag{9.81}$$

$$-\sum_{j=1}^{n} x_{ij} - y_i + a_i = 0, \quad i = 1, \ldots, m \tag{9.82}$$

$$\sum_{j=1}^{n} z_j - t + D = 0 \tag{9.83}$$

$$\sum_{i=1}^{m} y_i + t - S = 0 \tag{9.84}$$

It is easily seen that each variable appears in exactly two equations, once with a $+1$ coefficient and once with a -1 coefficient, and that no two variables appear in the same pair of equations. Since $\sum_{i=1}^{m} a_i = S$ and $\sum_{j=1}^{n} b_j = D$, the sum of the equations is zero. This shows that $v_1 = \cdots = v_{m+n+2} = 1$ is a solution for the dual problem, and that the rank of the system is less than $m + n + 2 = M$.

In a general network problem, the hardest parts are to show that the system of equations has at least one solution and that it satisfies condition 3, the condition that makes the graph connected. In a general problem these might not be true, and we would have to devise methods for showing that the system of equations has a solution, or does not have a solution, and to show that the graph is connected, or to show that it is not connected. For the transportation problem, it is quite obvious that it is connected from looking at the graph of the transportation problem. For example, look at Figure 9.2.

In general, the easiest way to show that the graph is connected is to find a basic solution corresponding to a spanning tree. We prefer to use a tabular representation of the problem to do computational work, and we can use the tabular representation to find a spanning tree. In table 9.85 we duplicate the initial table for the transportation problem we used as an example in Section 9.2.

(0)	(25)	(50)
(20)	1	3
(30)	2	6
(40)	4	5

(9.85)

The entries in parentheses correspond to variables for the transportation equations while those not in parentheses are values of variables in the dual problem. The entries in parentheses correspond, in fact, to basic variables in a basic solution.

The entries on the top margin are not the values of the basic variables; they are the negatives of those values. The reason for this is that the equations 9.81 through 9.84 are rewritten in the following forms to facilitate hand calculations.

$$\sum_{i=1}^{m} x_{ij} + (-z_j) = b_j, \quad j = 1, \ldots, n \tag{9.86}$$

$$\sum_{j=1}^{n} x_{ij} + y_j = a_j, \quad i = 1, \ldots, m \tag{9.87}$$

$$\sum_{j=1}^{n} (-z_j) + t = D \tag{9.88}$$

$$\sum_{i=1}^{m} y_i + t = S \tag{9.89}$$

When the equations are written in this form, the sums in each row and each column of the table are constants.

The graph corresponding to the variables in parentheses in table 9.85 is shown in Figure 9.81.

The arcs that are blanked correspond to nonbasic variables, and their values are zero. The values of the basic variables are shown on the arcs in Figure 9.7. It is easily verified that the sum on the arcs around each node is equal to the value of the constant indicated at the node.

That the variables indicated with parentheses are basic variables is also easily verified directly in the table. Each variable on the left margin as well as each variable on the top margin, except for the top left corner, is the only variable in its row or column. Thus the value of the variable is uniquely determined. This is also seen in the graph of Figure 9.7 since the nodes corresponding to those variables are terminal nodes. Notice that the value

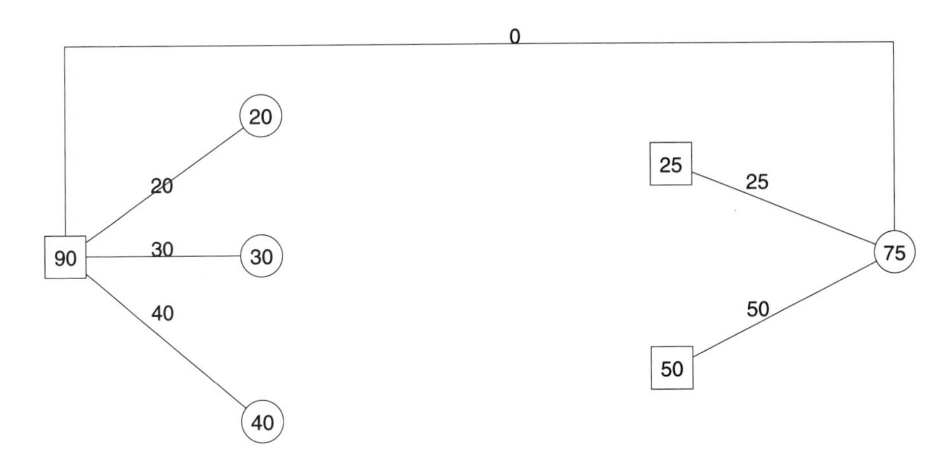

Figure 9.7: Initial spanning tree.

of t in the upper-left corner of 9.85 is determined by either the sum across the top margin or the sum down the left margin. This is where consistency and the rank equal to $M-1$ play a role. For the transportation problem we made sure of these properties when we expanded the network to include the source and the sink.

Once a basic solution is obtained, the next step is to obtain a basic feasible solution, or show that there is no feasible solution. The phase I method described in Section 9.2 is adequate for the transportation problem and we are not going to describe a phase I algorithm for general graphs here. We will merely describe the graphical interpretation that lies behind phase I pivoting.

Graphical pivoting consists of the following sequence of steps.

1. Start with a spanning subtree in the graph of the network. Choose any arc in the graph that is not in the spanning tree. This is equivalent to choosing a nonbasic variable when we pivot in a tableau.

2. This additional arc, together with arcs in the spanning subtree, forms a loop. This loop exists because the nodes connected by the additional arc are also connected within the tree. This loop is unique because the existence of two such loop would imply the existence of a loop within the spanning tree.

3. Remove any arc in the loop. Since the resulting subgraph will have $M-1$ arcs and connect all nodes, it will also be a spanning subtree.

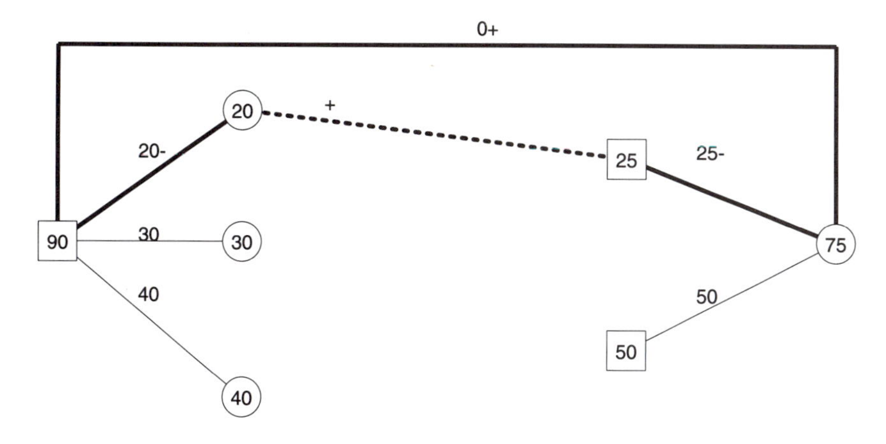

Figure 9.8: Loop created by out-of-tree arc.

If we add the arc corresponding to x_{11} to the tree in Figure 9.7 the loop created is shown in Figure 9.8.

Notice that the arcs of the loop shown in Figure 9.8 correspond to the positions marked with + and − signs in table 9.90.

$(0)^+$	$(25)^-$	(50)
$(20)^-$	$+$	
(30)		
(40)		

(9.90)

In general, we start with a transportation table in the form of table 9.90 in which nothing is shipped. There will be $m+n+1$ computed entries. At each step one assigned zero (a nonbasic variables) becomes a computed value (a basic variable) and one computed value becomes an assigned zero. This shows that at each step we have $m+n+1$ computed entries. This is the reason for the rule given in Section 9.2 for deleting only one basic variable when two or more become zero in the calculation. It is essential that we end phase I with a basic feasible solution, and this will be achieved if we preserve spanning trees from iteration to iteration.

Graphical pivoting in phase II is the same, except that the loops are more complex and more difficult to find, and we need a selection rule to determine which arc to introduce and which arc to delete.

To illustrate what is involved, let us look again at the example we consider in Section 9.4, where we discussed phase II. We repeat table 9.64 here for convenience.

(9.91)

(0)	(15)	(15)	(35)	(30)
(15)	1	3	5	5
(10)	7	5	7	11
(50)	4	7	13	8
(30)	2	3	11	7

Let us assume that we have completed phase I and that we have obtained table 9.92.

(9.92)

(95)				
(0)	(15)			
			(10)	
(10)			(25)	(15)
		(15)		(15)

The entries shown in table 9.92 are the values of the basic variables for the transportation equations. As in the discussion of phase II in Section 9.4, the first thing to do is to obtain the complementary basic solution for the dual problem.

We have mentioned several times that an advantage of the Dantzig algorithm is that we do not have to update the A-matrix with each pivot exchange. A further advantage of the graphical representation is that we do not have to update either basic solution as long as we maintain a sequence of spanning trees in the graph. We do not have to obtain the basic solution for the dual problem until phase I is completed. Now we will show how to obtain the basic solution for the dual problem.

A typical equation for the dual problem is in the form of equation 9.93.

$$v_j - u_i + w_{ij} = c_{ij} \tag{9.93}$$

If the arc (i, j) is in the spanning tree, x_{ij} is a basic variable and its dual variable w_{ij} must be nonbasic. That is, $w_{ij} = 0$. For arcs in the spanning tree, knowing either u_i or v_j allows us to compute the other. We can think of u_i as being associated with a node at one end of the arc (i, j), and v_j as being associated with the other node. Knowing the value associated with a node at one end of an arc in the spanning tree allows us to compute the value associated with the node at the other end.

A rooted tree is just a tree in which a particular node is designated as the root. When there is a root, it is possible to talk of moving from one node to another away from the root or towards the root. Though we are accustomed to visualizing our trees with their roots firmly planted in the ground, computer scientists prefer to have their rooted trees with their roots at the top. Thus, "down" the tree is away from the root and "up" the tree is towards the root. At any given node there may be several arcs leading away from the root but, other than at the root, there is always only one arc leading towards the root.

For the transportation problem, we take the root to be the source node. Notice that we compute the basic solution for the transportation equations by starting with the variables associated with terminal nodes, the leaves, and proceeding up the tree, towards the root. We will evaluate the basic solution for the dual problem by starting at the root. We assign an arbitrary value, zero, to the variable associated with the root, and compute the values associated with the other nodes as we move away from the root.

(9.94)

(95)	1	4	13	8
(0)	(15)			
−6			(10)	
(10)			(25)	(15)
−1		(15)		(15)

Table 9.94 shows the result of computing the u_i and v_j. The next step is to compute the w_{ij} using equation 9.93. When all these values are filled in we have table 9.95.

(9.95)

(95)	1	4	13	8
$(0)^-$	(15)	−1	-8^+	−3
−6	12	7	(10)	9
$(10)^+$	3	3	$(25)^-$	(15)
−1	2	(15)	−1	(15)

Feasibility for the dual problem requires that the entries on the left margin, which are $-u_i$, be nonpositive and all other entries be nonnegative. The −8 in the first row is the greatest infeasible value, and we elect to introduce the nonbasic variable in that position in the pivot exchange. As we can see from the economic interpretation of the dual problem, a negative sign for one of the w_{ij} indicates a cheap route, an arc for which the shipping costs are less than the price differential.

The one part of the Dantzig algorithm for which we have yet to provide an effective procedure is that of finding the loop that is completed when a new variable is introduced.

If the spanning tree for a basic solution is drawn as a graph on paper and another arc is introduced, it is easy enough to find the unique path within the tree that forms a loop with the introduced arc. We wish to be able to find this loop in a transportation table without using a drawing of the tree. Consider the transportation table 9.96. We are not interested in the values of the basic variables, only their positions in the table and in the spanning tree. We show these positions in table 9.96 with parentheses.

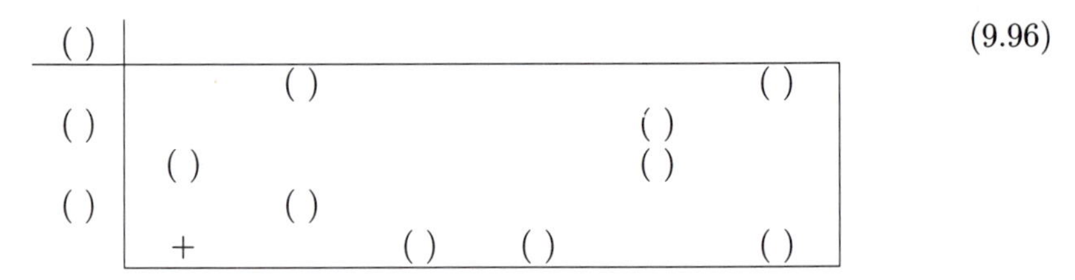

(9.96)

The + sign in table 9.96 represents an arc not in the spanning tree corresponding to the basic solution. How do we find the loop containing that arc?

Each row or column in a transportation table corresponds to a node of the transportation network. Thus, labeling the row and columns in table 9.96 can be regarded as equivalent to labeling the nodes of the spanning tree. Start in table 9.96 by labeling the root (the left margin) with 0, and the rows with parentheses on the left margin with 1. These are the nodes adjacent to the root. Then label with 2 those columns with parentheses in the rows labeled 1. Then label with 3 those unlabeled rows with parentheses in the columns labeled 2. Continue in this fashion. When the row and column containing the + sign are both labeled we have obtained a path within the tree connecting both ends of the introduced arc. When this is achieved we have table 9.97.

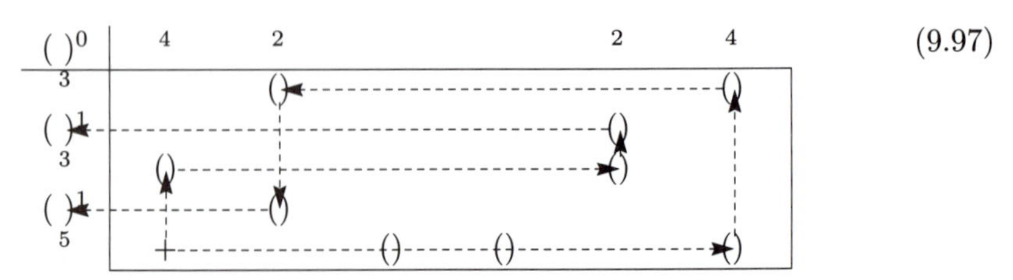

(9.97)

Notice that each labeled row has parentheses in exactly one column with the next lower label, and each labeled column has parentheses in exactly one row with the next lower label. Start with the plus sign and trace paths in both directions through rows and columns with lower labels until these paths meet in a common row or common column. The node represented by that common row or column is the node where these two paths join, and the loop is completed. For our example, the two paths back to the common node are indicated in table 9.97.

Exercises

If a random number generator is used to construct transportation problems, they all seem to be quite similar in structure, offering few unusual situations. These exercises have been constructed to confront you with situations where there are degenerate solutions, long loops, and multiple solutions. There are usually many choices available at each step. We suggest that you make your choices according to the rules given in the text. It is not that these rules are necessarily better than other rules. If you make other choices you may find the problem easier to solve, but you may avoid the situations that we have constructed for you to experience.

1. Find an initial basic feasible solution for the flow equations of the transportation problem with the following data table.

(0)	(4)	(4)	(5)	(2)
(5)	5	9	9	8
(4)	5	2	3	5
(2)	6	2	4	4
(5)	5	4	4	7

2. Complete the process of solving the transportation problem given in Exercise 1.

3. Find an initial basic feasible solution for the flow equations of the transportation problem with the following data table.

(0)	(7)	(3)	(6)	(4)	(5)	(7)
(8)	4	6	5	7	4	6
(9)	5	3	6	4	3	4
(4)	3	5	5	2	4	6
(5)	4	4	1	5	3	5
(8)	5	5	5	6	2	6

4. Complete the process of solving the transportation problem given in Exercise 3.

5. Solve the transportation problem represented by the following transportation table.

(0)	(4)	(5)	(5)	(7)
(8)	3	4	5	4
(6)	6	5	5	6
(10	5	7	8	4

6. Solve the transportation problem represented by the following transportation table.

(0)	(4)	(5)	(5)	(7)
(3)	3	6	5	2
(4)	6	5	5	6
(5)	5	7	8	4
(9)	3	6	4	5
(4)	4	7	6	4

7. Solve the transportation problem represented by the following transportation table.

(0)	(4)	(2)	(8)	(3)	(10)
(5)	5	7	3	8	9
(3)	3	4	5	7	6
(4)	4	6	4	10	8
(6)	6	5	5	5	10
(5)	5	5	4	6	8
(6)	8	7	8	7	6

8. Solve the transportation problem represented by the following transportation table.

(0)	(1)	(1)	(1)	(1)
(1)	5	9	9	8
(1)	5	2	3	5
(1)	6	2	4	4
(1)	5	4	4	7

9. Consider the following transportation table.

(0)	(4)	(5)	(5)	(7)
(3)	3	6	5	2
(4)	6	5	5	6
(5)	5	7	8	4
(9)	3	6	4	5
(4)	4	7	6	4

In the following table, the empty parentheses indicate the location of the basic variables for the transportation equations. Determine the complementary basic solutions for both the primal and the dual problem corresponding to that pattern of basic variables.

( )				
	( )	( )		
( )				( )
	( )		( )	
		( )		
	( )			( )

10. Reconstruct the initial transportation table from the equivalent table given below.

(12)	4	5	6
−1	(4)	(2)	2
(1)	−1	(1)	(3)
−2	3	2	(2)

11. Construct the Tucker tableau representing the table in Exercise 10 considered as a linear program.

12. Perform one iteration of the Dantzig algorithm by increasing the flow at the -1 (w_{21})in the table of Exercise 10. Then, perform the corresponding exchange on the tableau representing the problem. For this purpose, use the tableau given as the answer to Exercise 11.

13. In the following transportation table, both solutions are optimal. Determine if these optimal solutions are unique or whether there are multiple optimal solutions for either program.

(17)	5	4	4	7
(1)	(4)	5	5	1
−2	2	(4)	1	(1)
−3	4	1	3	(2)
(1)	0	0	(5)	(1)

14. In the following transportation table, both solutions are optimal. Determine if these optimal solutions are unique or whether there are multiple optimal solutions for either program.

(15)	5	4	4	7
(1)	(4)	5	5	1
−2	2	(4)	1	(0)
−3	4	1	3	(2)
(1)	1	1	(5)	(0)

15. Consider the transportation table given below. The table contains optimal solutions for both problems of the dual pair. Find other optimal solutions, if there are any. If there are no others, prove that there are no other optimal solutions.

(15)	5	4	4	7
(1)	(4)	5	5	1
−2	2	(4)	1	(0)
−3	4	1	3	(2)
(0)	0	0	(5)	(0)

16. Find an assignment problem equivalent to the following transportation problem.

(0)	(2)	(1)	(3)
(3)	2	4	3
(2)	3	2	5
(1)	4	3	4

17. Find a transportation problem equivalent to the assignment problem with the following cost table.

2	4	1	3
5	3	5	4
2	4	3	5
4	5	4	3

18. In most newsstands one can pick up magazines containing logic puzzles. Typically, they describe the activities of several people in the form of clues. The solver is asked to match the first name with the last name with the activity. The problem is a three-dimensional (or more) assignment problem.

There is no objective variable to minimize. The clues are designed to have a unique feasible solution for an assignment problem. The clues usually work by excluding certain assignments.

Suppose you are the designer of such a puzzle. For simplicity, consider a two-dimensional puzzle. What is the smallest number of assignments that your clues can prohibit and yet provide a unique solution?

19. A transportation problem for which the total supply is at least as large as the total demand is feasible. However, if some routes are not available, the problem might be infeasible. Such problems can be attacked by assigning very high rates to the forbidden routes and solving the resulting transportation problem. If the optimal solution does not use one of the expensive routes, you have solved the original problem. If the optimal solution requires the use of one of the expensive routes, the problem is infeasible. Use this approach to solve the following problem. Forbidden routes are indicated by X's.

(0)	(4)	(5)	(5)	(7)
(3)	2	3	6	7
(4)	4	4	3	5
(7)	6	2	5	X
(10)	5	X	X	X

20. Solve the following problem with forbidden routes.

(0)	(4)	(5)	(5)	(7)
(3)	2	3	6	7
(6)	4	4	3	5
(8)	6	2	5	X
(10)	5	X	X	X

Questions

Q1. For a transportation problem, an optimal solution always exists.

Q2. For a transportation problem, an optimal solution will always use the least costly route.

Q3. For a transportation problem, an optimal solution will never use the most costly route.

Q4. Phase II of the Dantzig algorithm for the transportation problem always provides a feasible solution for the transportation equations, even when an optimal solution has not yet been achieved.

Q5. A transportation problem can be cast as a linear program and can, therefore, be solved by the simplex algorithm.

Q6. A transportation problem can be solved with no arithmetic steps that require multiplication or division.

Q7. An optimal solution for a transportation problem does not use more than $m + n$ of the routes, where m is the number of sources and n is the number of destinations.

Q8. For a graph (network) with n nodes, a spanning subgraph with $n - 2$ arcs can sometimes be found.

Chapter 10

Network Flow Problems

10.1 Network Flow Problems

As a model for the kind of problem we wish to focus on in this chapter, consider an oil pipeline network. Oil is fed into the network at various points, which we call *sources*, and it is removed from the network at other points, which we call *sinks*. There are other nodes where the various segments of the pipeline network have junctions. Each segment of the pipeline has a limit on the quantity of oil that can be pumped through it in a given amount of time. We wish to determine the capacity of the network. We regard the supply and demand as sufficiently large that they are not limiting considerations. What is the maximum rate at which oil can be moved from the sources to the sinks and, to achieve this maximum, how much should be transported along each segment?

It is sufficient for our purpose to consider relatively simple networks. In particular, our examples will have only one source and one sink. In the example in Figure 10.1 the node labeled A is the source and the node labeled F is the sink.

The numbers written near the arcs are the capacities of those arcs. The number on an arc near a node represents the capacity out of the node. Where no capacity is specified it is implied that the capacity is zero. In this example the capacities are not the same in both directions along an arc.

Let x_{ij} represent the amount to be sent along the arc from node i to node j. Let c_{ij} represent the capacity of the arc from node i to node j. To be a *feasible flow* the set of x_{ij} must satisfy the following conditions.

1. The x_{ij} are nonnegative.

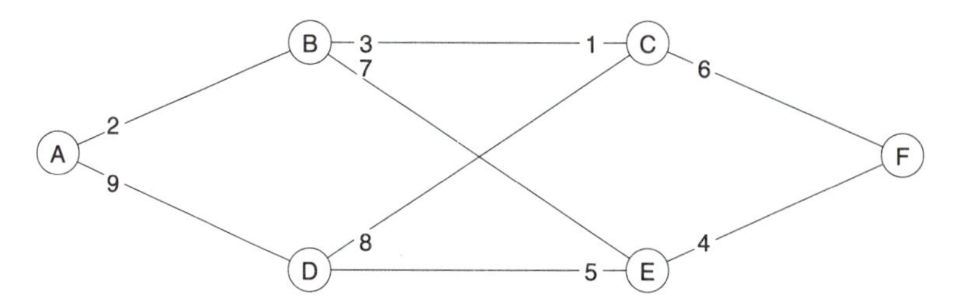

Figure 10.1: Flow network.

2. $x_{ij} \leq c_{ij}$

3. At each node, other than a source or a sink, the amount sent into a node must equal the amount sent out.

If we have a network with several sources or several sinks, we can introduce an ultimate source and an ultimate sink, just as we did for the transportation problem. This would introduce additional arcs and their capacities would be unlimited, but otherwise the resulting network would resemble the network in Figure 10.1. We shall consider the network flow problem only in this standardized form, with one source and one sink.

We also introduce an additional arc in the network connecting the sink back to the source. This closes the circuit. Let f denote the flow in this arc, from the sink to the source. Introducing this arc and its associated variable serves two purposes. First, it is the variable that must be maximized. Second, it allows us to impose the conservation condition (number 3 above, that the total flow into a node must equal the total flow out of the node) uniformly for all nodes, including the source and the sink.

The maximum flow problem is to find the maximum value for the flow f subject to the conditions

$$x_{ij} \geq 0 \tag{10.1}$$

$$x_{ij} \leq c_{ij} \tag{10.2}$$

$$\sum_i x_{ik} = \sum_j x_{kj} \tag{10.3}$$

for each node k, excluding the source and the sink, and

$$\sum_i x_{in} = \sum_j x_{0j} \tag{10.4}$$

where 0 is the index of the source and n is the index of the sink.

$$f = \sum_i x_{in} \tag{10.5}$$

is the variable to be maximized.

This formulation is enough to establish that the maximum flow problem is a linear programming problem. It is clearly feasible since a zero flow in every arc is feasible. The flow is bounded since it is limited by the sum of the capacities out of the source. Thus, the maximum flow problem has an optimal solution.

Figure 10.2 is the new network for the problem given in Figure 10.1.

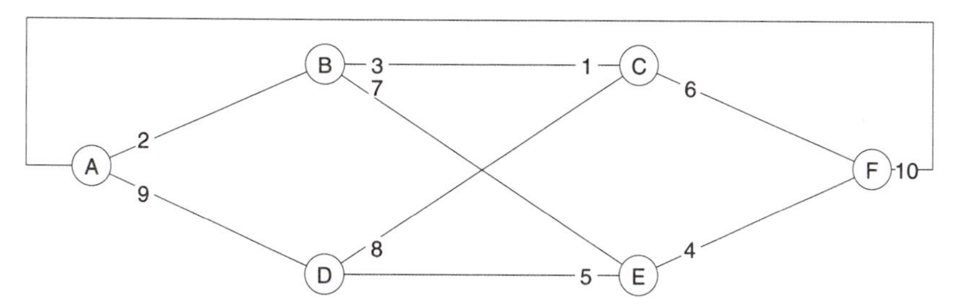

Figure 10.2: Closed network flow network.

In principle, the flow f is not limited by the capacity of its arc, but it certainly cannot exceed either the total capacity out of the source or the total capacity into the sink. Since it is convenient to have a finite capacity when using the algorithm to be described, we assign a capacity of 10 (the minimum of the total capacity out of the source and the total capacity into the sink) for the arc from the sink to the source.

L. R. Ford, Jr., and D. R. Fulkerson devised an algorithm to solve the maximum network flow problem. In principle, the algorithm is quite straightforward.

1. Find a path of arcs leading from the source to the sink for which every arc in the path has a positive capacity.

2. Assuming that no other arcs are used, the arc in the path with the smallest capacity determines the quantity that can be sent along that path. Assign that flow to all arcs in the path.

3. The difference between the capacity of each arc and the amount sent along the arc is the *residual capacity* of that arc. Use these residual capacities as the capacities of a new network. With these capacities, return to step 1.

4. When no further paths from the source to the sink with positive residual capacities can be found, STOP.

This description is just an outline of the Ford–Fulkerson algorithm. We must be careful about a number of details, but they are relatively easy to handle. For example, an arc with a capacity of 10 in each direction, and a flow of 10 in each direction, is not saturated. The net flow is zero. Counterflows must be adjusted to yield a net flow. That is, we will normalize the flows so that the flow in one direction or the other, or both, is zero.

As a model for a more detailed description of the algorithm, let us carry out this procedure for the network in Figure 10.2 and see what is involved. An obvious path in the network is shown in Figure 10.3.

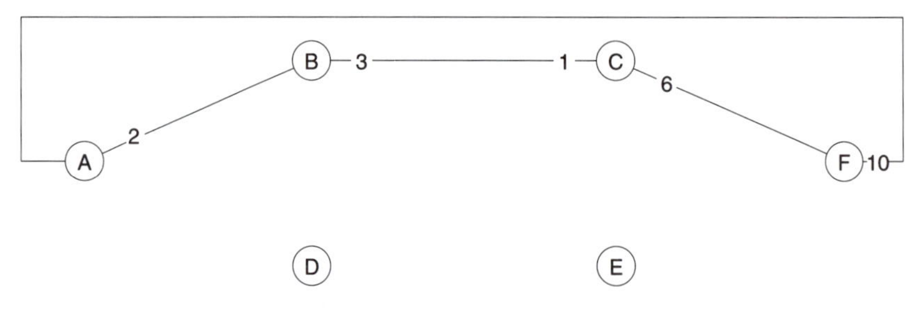

Figure 10.3: Flow-increasing loop.

The maximum flow that can be sent along the path in Figure 10.3 is the minimum capacity in the path, 2 units. We subtract this flow from the capacity of each arc to obtain the residual capacity, and we add this amount to the capacity in the opposite direction.

The reason for adding to the capacity in the opposite direction is that this is a way of balancing opposite flows to obtain a net flow. In the example mentioned above, when we have a capacity of 10 in each direction in an arc and a flow of 10 in one direction, adding 10 to the capacity in the opposite direction provides a capacity of 10 to offset the first flow before using the capacity in the opposite direction. Furthermore, as the counterflow is increased, increasing the capacity in the first direction is equivalent to reducing the flow in the first direction by the same amount.

In this way we obtain the network in Figure 10.4 with the new residual capacities.

A path of positive residual capacity is shown in Figure 10.5.

The maximum line of flow in this path is 4. When the residual capacities in Figure 10.4 are recomputed we have Figure 10.6.

The two lines of flow obtained so far both go through the arc CF and, together, they fully utilize (saturate) the capacity of that arc.

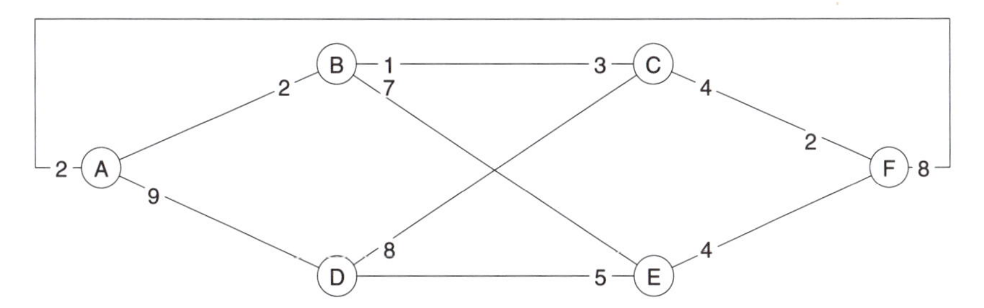

Figure 10.4: Residual capacities.

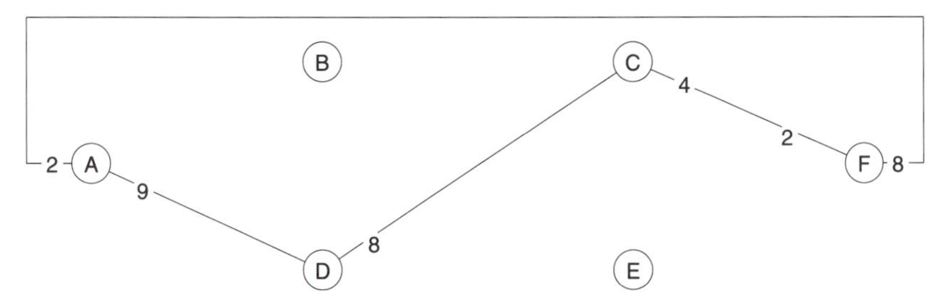

Figure 10.5: Flow-increasing loop.

There is a path of positive flow in Figure 10.6 that is not very obvious. It is shown in Figure 10.7.

The maximum line of flow in this path is 3. Here is an opportunity to illustrate the reason given earlier for adding a flow in one direction to the capacity in the opposite direction. The original capacity of the arc CB is 1. We cannot send 3 units of flow through this arc in that direction. However, the line of flow in Figure 10.3 sent 2 units of flow through the path $ABCFA$. Now we can divert this flow to the path $ABEFA$. The line of flow we have just generated in Figure 10.7 is used in the following way. Two units are diverted to the arc CF to make up for the loss from

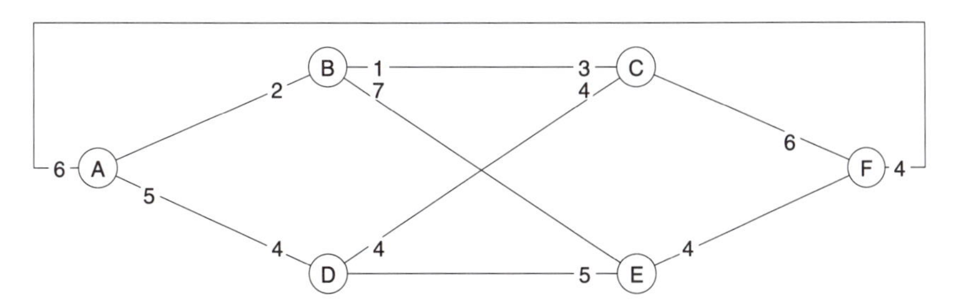

Figure 10.6: New residual capacities.

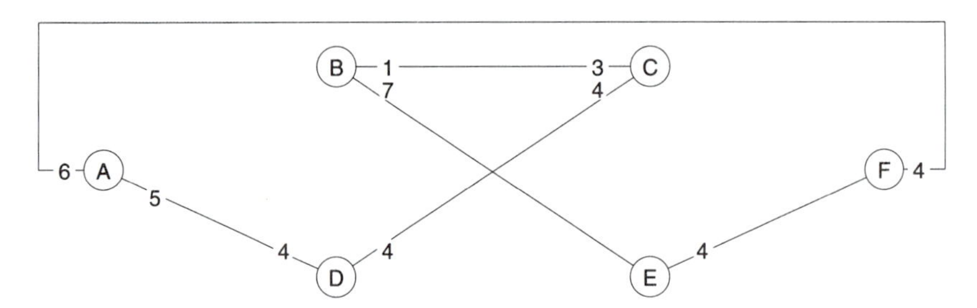

Figure 10.7: Last flow-increasing loop.

diverting the flow in the first path. One unit is sent to B. There it joins with the 2 units from the first path to make up the 3 units in this new line of flow.

Actually, it is not necessary to go through this sort of readjustment of the flows at this point. The adjustments in the capacities are sufficient to account for everything. We proceed from Figure 10.7 just as we have with the previous steps to obtain the residual capacities in Figure 10.8.

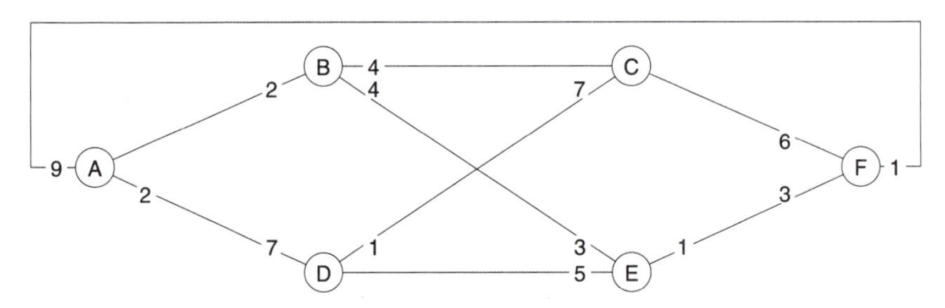

Figure 10.8: Final residual capacities.

Perhaps we can see by exhausting all possible cases that it is not possible to find an assignment of flows that will increase the amount sent from the source to the sink. But this doesn't reach the potential of 10 units of flow suggested in Figure 10.2. Is this really the best we can do? The answer is "yes," and this can be seen in Figure 10.9.

In Figure 10.9, several arcs have been cut. The nodes are divided into two groups, one connected to the source and the other connected to the sink. The only capacities shown in Figure 10.9 are those for the cut arcs leading from the source side to the sink side. We have a "wall" dividing the source from the sink, and the cut arcs are the only channels leading across this barrier. No matter how the flows are circulated among the nodes on each side of the barrier separately, the most that can cross this barrier is 9 units. This is an upper bound for the flow that could be sent from the

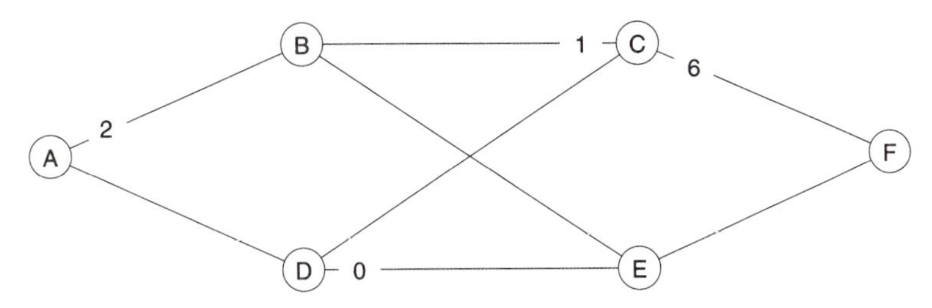

Figure 10.9: Minimum cut.

source to the sink. Since the flow we obtained actually sends this much, that is the maximum.

10.2 The Ford–Fulkerson Algorithm

A path of arcs leading from the source to the sink for which every arc in the path has a positive residual capacity is called a *flow-augmenting path*. We need an effective description of how such paths can be found. We also need a notation more convenient than a drawing of the network that must be drawn and redrawn, or on which erasures are required.

The process by which a flow-augmenting path can be obtained is called the *labeling process* by Ford and Fulkerson.

The Ford–Fulkerson Algorithm

1. Label the source with a 0.

2. Label with a 1 every node that can be reached from the source by an arc with a positive residual capacity.

3. Label with a 2 every node not already labeled that can be reached from a node labeled 1 by an arc with a positive residual capacity.

4. Continue in this fashion. That is, at a stage where the highest label is i and no unlabeled node can be reached from a node labeled $i-1$ by an arc with positive residual capacity, label with $i+1$ every node not already labeled that can be reached from a node labeled i by an arc with a positive residual capacity.

5. If the sink receives a label, trace a path from the sink back through nodes with descending labels until the source is reached. The path traced will be flow-augmenting.

6. If the labeling process stops without a label for the sink, no flow-augmenting path from the source to the sink exists. The Ford–Fulkerson algorithm STOPS.

7. If a flow-augmenting path is obtained, recompute the residual capacities. Since each arc in the path has a positive capacity, the minimum capacity is positive. Subtract this minimum from each capacity in the direction from the source to the sink for each arc in the flow-augmenting path, and add this minimum to each capacity in the direction from the sink to the source for each arc in the flow-augmenting path. Return to step 1.

In the following section we will show that this algorithm actually works. That is, we will show that it terminates in a finite number of iterations and that an optimal solution is obtained. In this section our objective is to recast the notation in a form that is more convenient for larger problems.

We shall use a matrix to represent the connections and the capacities, and their residual capacities. To illustrate the notation we will rework the problem discussed in the previous section. The matrix 10.6 represents the network in Figure 10.2.

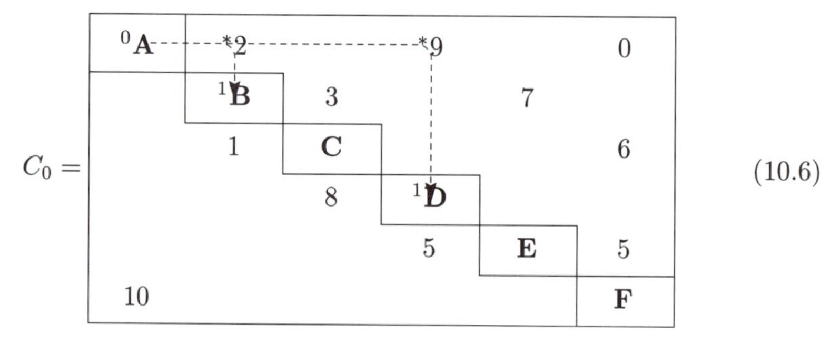

In 10.6 there is a row and a column for each node in the network in Figure 10.2. The name of each node is entered on the main diagonal. The numbers entered in the array are the capacities of the arcs. In each row, the entries are the capacities out of the node corresponding to the row. In each column, the entries are the capacities into the node corresponding to the column. The zero in the upper right corner represents the variable to be maximized. All other zeros are shown as blanks.

We have started the labeling process by writing a 0 as a prefix on the A in the main diagonal. The positive capacities in this row represent the positive capacities out of the source, and they are prefixed with asterisks. The bent arrows show arcs with positive capacities out of the source, and they identify the nodes labeled 1. The next step in the labeling process is shown in 10.7.

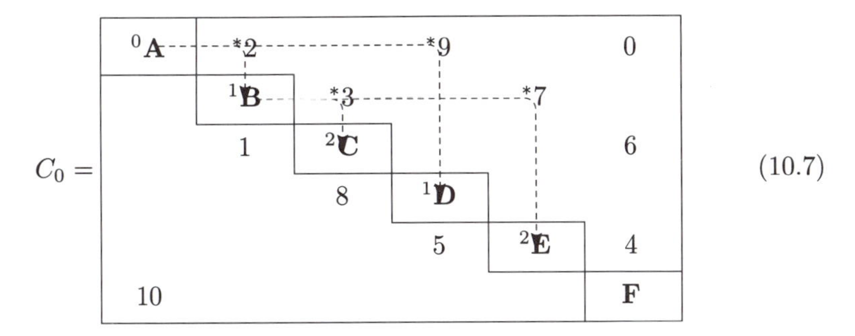

Notice that the node C could also have received its label of 2 through the arc DC with capacity 8. As far as the effectiveness of the algorithm is concerned it does not matter how the node receives its label. However, to make all choices unambiguous we shall use the following rules. (These rules also make it unnecessary to use the bent arrows.)

First, determine some strict ordering of the nodes of the network. We will use the alphabetical ordering.

1. Give the source the label 0. Label each positive entry in the row of the source (except the entry representing the return from the source to the sink, the upper right entry in 10.6) with a star. Then label the letters in the starred columns with a 1.

2. For $h = 1, 2$, etc., iterate the following procedure until further labeling is not possible or until the sink is labeled. For each row labeled h, the rows considered in the given order, star each positive entry that is not in the column of a previously labeled letter. Label with an $h + 1$ the letters in these newly starred columns.

3. If the labeling procedure described in step 2 terminates without labeling the sink, STOP. A flow-augmenting path from the source to the sink does not exist.

4. If the sink is labeled, a flow-augmenting path can be obtained by tracing the path from the sink back to the source through starred entries and nodes with decreasing labels. The flow can be increased by the minimum capacity in this path. Recompute the residual capacities (we will show how to do this in the example). Remove the labels and stars and return to step 2.

In applying the rule in step 2 to the matrix 10.7, in the C row the 6 would be starred but the 1 would not. The F in the column containing the 6 would be labeled 3. The labeling would stop here because the sink would be labeled, but to illustrate the rule in more detail let us continue. Our attention would then pass to the E row. Neither positive entry would be starred. In particular, the F was labeled when the C row was considered and we interpret the rule to say that this is a previous labeling.

When this iteration of the labeling rule is carried out we obtain matrix 10.8.

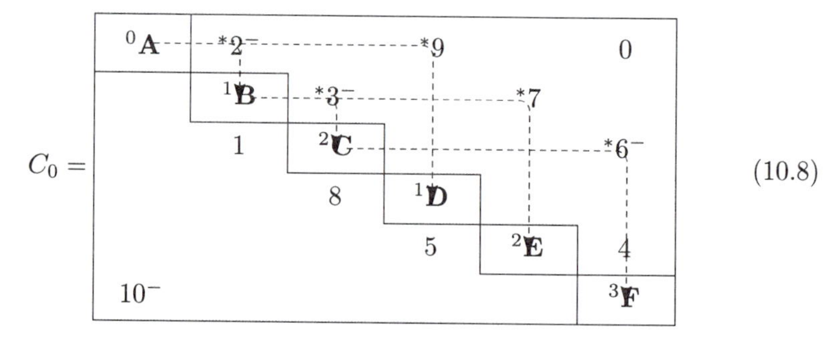

As we trace the path from the sink back to the source we place a minus sign as a postfix on each starred entry as we pass it. These are the residual capacities in the path and they will be reduced when the flow is increased. The amount of the reduction is the smallest capacity marked with a minus sign. Notice that we always mark the capacity in the lower left corner since it represents the return arc from the sink to the source and is always in the loop.

We are now in a position to illustrate how the new residual capacities are computed. For each position that is marked with a minus sign, we mark the position that is symmetric with respect to the main diagonal with a positive sign. When this is completed we obtain 10.9.

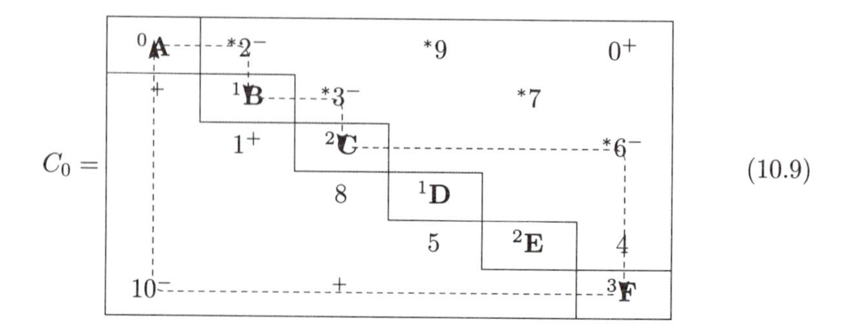

The entries marked with a minus sign indicate the arcs in the line of flow. These capacities will reduced by the minimum capacity in this path. The positions marked with plus signs are the capacities in the opposite directions. These will be increased by the same amount. This corresponds to the step of adding this amount to the capacity at the other end of the arc as described when we worked this example graphically in Section 10.1.

When these changes are made we obtain the new residual capacity matrix 10.10.

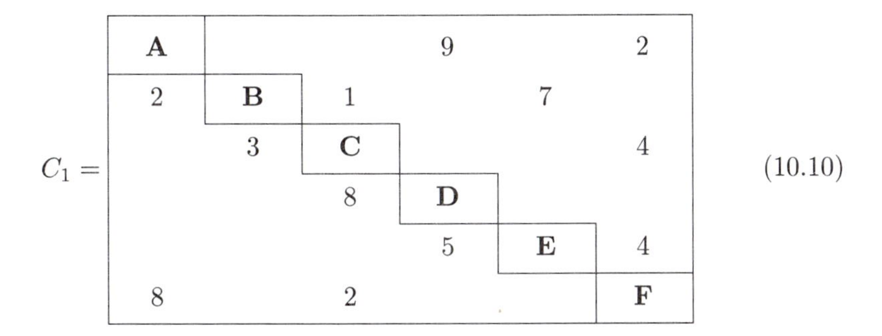

$C_1 =$ (10.10)

A			9		2
2	**B**	1		7	
	3	**C**			4
		8	**D**		
			5	**E**	4
8		2			**F**

We must repeat this process until the algorithm terminates. We star the arcs, label the nodes, and (assuming the sink gets labeled) retrace the flow-augmenting path and affix the minus and plus signs. When this is completed for the residual capacity matrix 10.10 we obtain

$C_1 =$ (10.11)

^{0}**A**			$^{*}9^{-}$		2^{+}
2	^{3}**B**	1		7	
	$^{*}3$	^{2}**C**	$+$		$^{*}4^{-}$
$+$		$^{*}8^{-}$	^{1}**D**		
			5	**E**	4
8^{-}		2^{+}			^{3}**F**

The flow-augmenting path identified in 10.11 is $ADCFA$. The minimum capacity in this path is 4, and the flow in the network can be increased by that amount. When this is completed and the new residual capacities are computed we obtain 10.12.

$C_2 =$ (10.12)

A			5		6
2	**B**	1		7	
	3	**C**	4		
4		4	**D**		
			5	**E**	4
4		6			**F**

When another iteration with the starring and labeling is completed we have 10.13.

$C_2 =$ (10.13)

^{0}**A**			$^*5^-$		6^+
2	^{3}**B**	1^+		$^*7^-$	
	3^-	^{2}**C**	4^+		
4^+		4^-	^{1}**D**		
	+		5	^{4}**E**	$^*4^-$
4^-		6		+	^{5}**F**

When the residual capacities are recomputed we obtain 10.14.

$C_0 =$ (10.14)

^{0}**A**			^{*2}		9
2	**B**	4		4	
		^{2}**C**	7		
7		^{*1}	^{1}**D**		
	3		5	**E**	1
1		6		3	**F**

The new starring and labeling is also shown in 10.14. This time the labeling does not reach the sink. The algorithm terminates.

Several things remain to be resolved. We must show the algorithm actually works—that it terminates in a finite number of iterations and that it yields an optimal solution. Assuming that is true, it is easy to see the value of the maximal flow. It is the 9 in the upper right corner. However, we have made no attempt to show the values of the flows that must be used in the individual arcs. We will do this in the following section.

If we transfer the labeling from table 10.14 to the initial table, we get table 10.15. We show an asterisk wherever there is an arc from a labeled node to an unlabeled node. These represent the cuts and you can see that they correspond to the cuts in Figure 10.9.

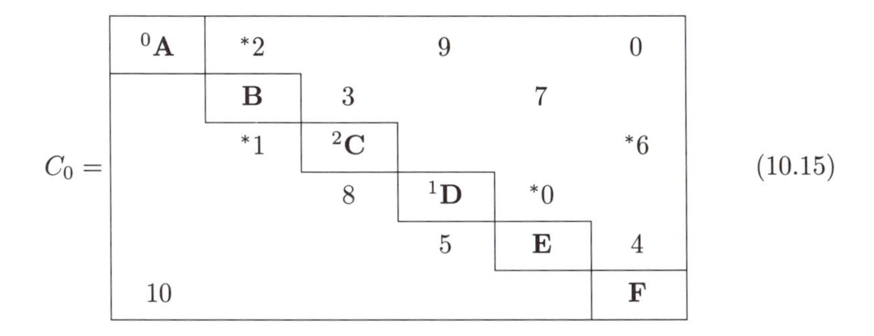

10.3 The Max-Flow, Min-Cut Theorem

We shall attack the proof of the termination of the Ford–Fulkerson algorithm by assembling a number of relations that persist among the entries in the sequence of residual capacity matrices produced by the algorithm. The optimality of the result is expressed in the max-flow, min-cut theorem.

First, we shall show how to calculate the flows in each arc for any of the residual capacity matrices. We start with the capacity matrix C_0. Let X_1 be the matrix of the flows x_{ij} obtained in the first line of flow. The next residual capacity matrix C_1 is obtained by subtracting the x_{ij} from each c_{ij} in the line of flow and adding the same quantity to the position that is symmetric with respect to main diagonal. In matrix form, this means

$$C_1 = C_0 - X_1 + X_1^T \tag{10.16}$$

where X_1^T is the transpose of X_1. The entries of C_1 are of the form

$$c'_{ij} = c_{ij} - x_{ij} + x_{ji} \tag{10.17}$$

Symmetric-pair-sums remain constant. By this we mean that the sum of each pair of symmetrically situated entries in the residual capacity matrices remains constant from iteration to iteration.

From 10.17 we see that

$$\begin{aligned} c'_{ij} + c'_{ji} &= (c_{ij} - x_{ij} + x_{ji}) + (c_{ji} - x_{ji} + x_{ij}) \\ &= c_{ij} + c_{ji} \end{aligned} \tag{10.18}$$

The same reasoning shows that this sum remains constant through each succeeding iteration.

Therefore, for each residual capacity matrix C_h the difference $C_0 - C_h$ must be a matrix in which the symmetric-pair-sums are all zero. Let X be the matrix consisting of all the positive entries that appear in $C_0 - C_h$, and for which the remaining entries are zeros. Then

$$C_0 - C_h = X - X^T \tag{10.19}$$

Since the entries in C_0 and C_h are all nonnegative, and X^T is zero wherever X has a positive entry,

$$0 \leq x_{ij} \leq c_{ij} \tag{10.20}$$

That is, for every residual capacity matrix, the corresponding X represents a feasible flow.

Line-sums remain constant. By this we mean that the sum of the entries in each row and the sum of the entries in each column of the residual capacity matrices remain constant from iteration to iteration.

Whenever an entry in the row of a node (a capacity out of the node) in a capacity matrix is marked with a minus sign, an entry in the column of that node (a capacity into the node) is also marked with a minus sign. Thus, when the plus signs are entered both the row and the column of that node will also contain a plus sign. Since each row and column either contains no mark or contains both a minus sign and a plus sign, the sum of the entries in each row and and in each column is unchanged.

Since the row and column sums in C_0 and C_h are equal, the row and column sums in $X - X^T$ must all be zero. Therefore, each row sum in X must be equal to the column sum for the same node. This means

$$\sum_i x_{ik} = \sum_j x_{kj} \tag{10.21}$$

for each k.

Conditions 10.20 and 10.21 mean that **the x_{ij} in the flow matrix X form a feasible flow**.

In our example, let us see how we obtain the feasible flow for the residual capacity matrix C_3. We have

$C_0 - C_3 =$ (10.22)

A	2		7		−9
−2	**B**	−1		3	
	1	**C**	−7		6
−7		7	**D**		
	−3			**E**	3
9		−6		−3	**F**

The flow matrix X is easily obtained from the positive entries in $C_0 - C_3$. We have

$X =$ (10.23)

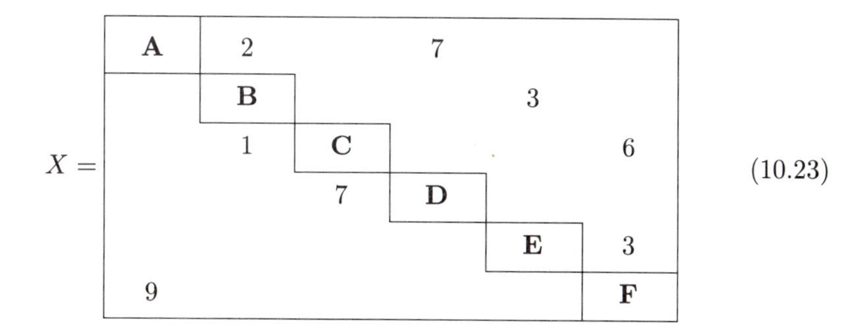

A	2		7		
	B			3	
	1	**C**			6
		7	**D**		
				E	3
9					**F**

Notice in table 10.22 that the sum of the entries in each row and in each column is zero. This is the way the rule "what goes in equals what goes out" is realized in the table. In table 10.23, for each node the sum of the entries in its row is equal to the sum of the entries in its column. In this way the same rule is realized in that table.

Quarter-sums remain constant. First we must explain that we mean by a quarter-sum. Partition the nodes of the network into two disjoint subsets in any way whatever. In the capacity matrix C_h draw a horizontal line through each row corresponding to a node from one subset, and a vertical line through each column corresponding to a node from the other subset. The matrices 10.23 and 10.24 show two different ways this might be done for the original capacity matrix C_0. In 10.24 the partition is represented by (ABC/DEF) and in 10.25 it is represented by (ACD/BEF).

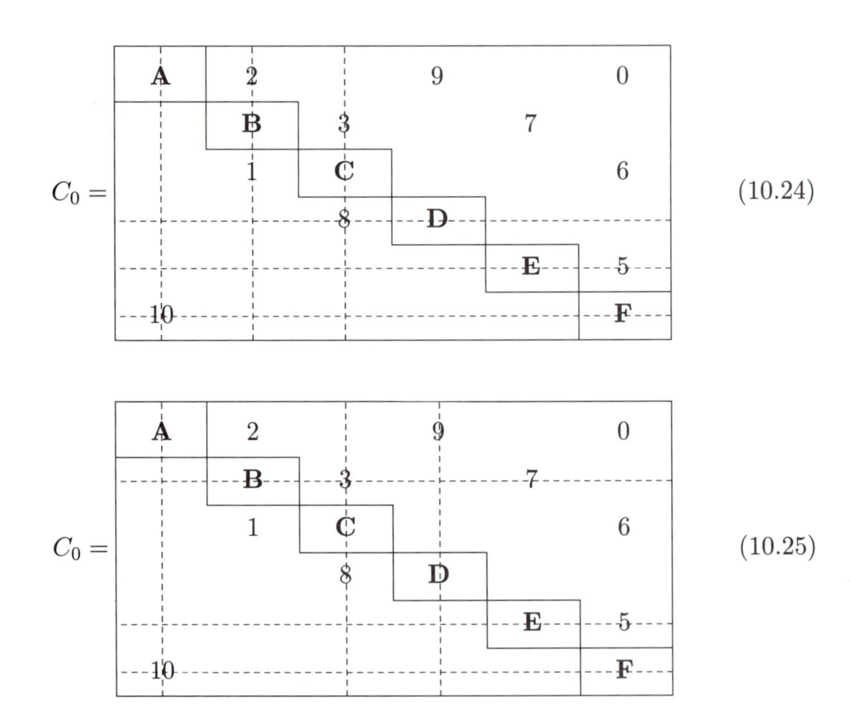

The entries in C_h are partitioned into four subsets: those that are covered once by a horizontal line, those that are covered once by a vertical line, those that are covered by both a horizontal line and a vertical line, and those that are not covered. This defines the four quarters. The sum of all the entries in any one of the quarters is called a *quarter-sum*.

The two quarters that are covered once are called the *main-diagonal quarters*, and the other two quarters are the *off-diagonal quarters*. The terminology is suggested by the configuration in 10.24 where the nodes in each set are contiguous. The upper left and the lower right quarters are the main-diagonal quarters, and the upper right and the lower left are the off-diagonal quarters.

Each main-diagonal quarter contains either both members of a symmetric pair or neither member of a symmetric pair. Since symmetric-pair-sums remain constant, the two main-diagonal quarter-sums remain constant. The two off-diagonal quarter-sums remain constant because the line-sums remain constant.

Though the invariance of the quarter-sums is true for any partition of the nodes of the network, we are primarily interested in partitions for which one subset contains the source and the other contains the sink. This is the case for the two partitions illustrated in 10.24 and 10.25. For a partition of this kind, the subset containing the source is called the *source set*, and the set containing the sink is called the *sink set*. A partition of the nodes into a source set and a sink set is called a *cut*.

If the nodes of the source set correspond to covered columns and the nodes of the sink set correspond to covered rows, the noncovered quarter represents arcs leading from the source set to the sink set. This quarter is called the *cut quarter*. The quarter-sum for the cut quarter is called the *value of the cut*.

Since the cut-quarter quarter-sum remains invariant, the value of the cut can be computed from any one of the residual capacity matrices produced by the algorithm. From 10.24 we see that the cut (BCA/DEF) has the value 22, and from 10.25 we see that the cut (ACD/BEF) has the value 9.

The entry in the upper right corner represents the return capacity from the source to the sink. It is the sum of the flow in each line of flow and is, therefore, the total flow. It is the variable to be maximized. This entry is in the source row and the sink column. Thus, it is in the cut quarter for every cut. Since the entries in each C_h are nonnegative, the value of the flow is always bounded by the value of the cut, for every possible cut. That is,

$$\max\ (\text{flow}) \leq \min\ (\text{cut}) \tag{10.26}$$

Suppose the Ford–Fulkerson algorithm terminates with an attempt to label the nodes which fails to label the sink. In our example, this happens in C_3. The nodes are then partitioned into two subsets, those that are labeled and those that are not labeled. The source is labeled and the sink is not. That is, this partition is a cut. The lines in C_3 corresponding to this cut are shown in 10.27.

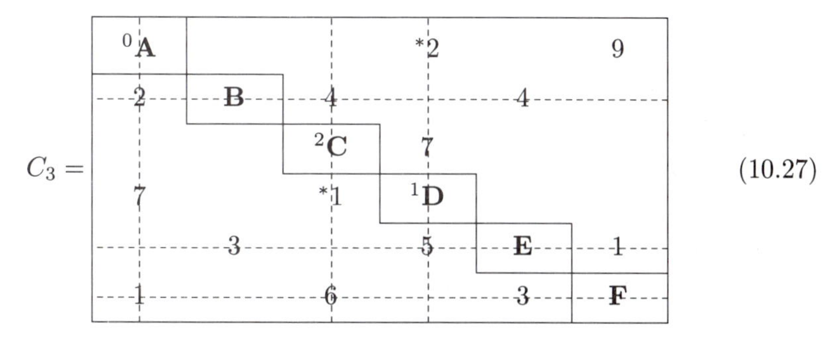

The only nonzero entry in the cut quarter is the objective variable. This will always be the case when the algorithm terminates, for otherwise we could label at least one more node. When this happens, the value of the flow and the value of the cut are equal. Because of the inequality 10.26 this common value is the maximum flow and the minimum cut.

Thus, if the algorithm terminates we have

$$\max\ (\text{flow}) = \min\ (\text{cut}) \tag{10.28}$$

The only thing that remains to be done is to show that termination always occurs.

Each time a flow-augmenting path is found, the total flow through the network is increased. The arithmetic that is performed to compute the residual capacities for the next iteration involves only addition and subtraction. If the initial capacities are integers or rational numbers the only numbers that will be produced from addition and subtraction will be integers (if the initial capacities are integers) or rational numbers. In the latter case, the rational numbers will not have denominators larger than the least-common-denominator of the rational numbers in the initial data.

This means that the amount of increase of the flow in each iteration is a positive amount bounded away from zero. Since the total flow is bounded by any cut, there can be only a finite number of such increases. This proves that the algorithm must terminate in a finite number of iterations.

Theorem 10.64 *(The Max-Flow, Min-Cut Theorem) For any finite flow network, the maximum flow is equal to the minimum cut.* ◻

Exercises

1. Figure 10.10 shows a network in which there are two sources, A and B, and two sinks, D and E. Construct a network flow problem with one source and one sink which can be used to solve this network flow problem. The new problem will not be equivalent, but all elements of the original problem can be obtained from it.

2. Write down the network flow table that represents the network in Figure 10.11.

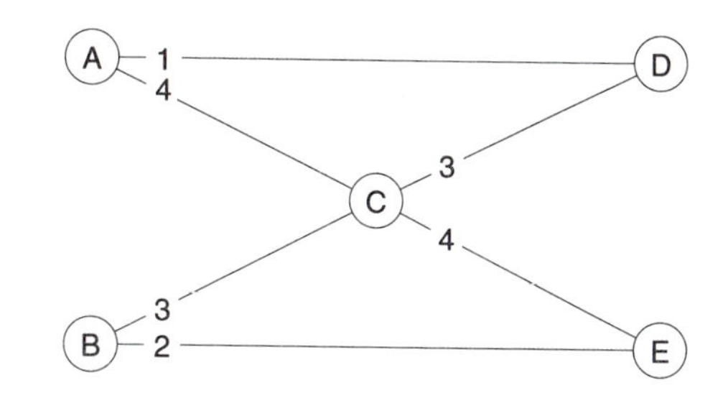

Figure 10.10: Figure for Exercise 1.

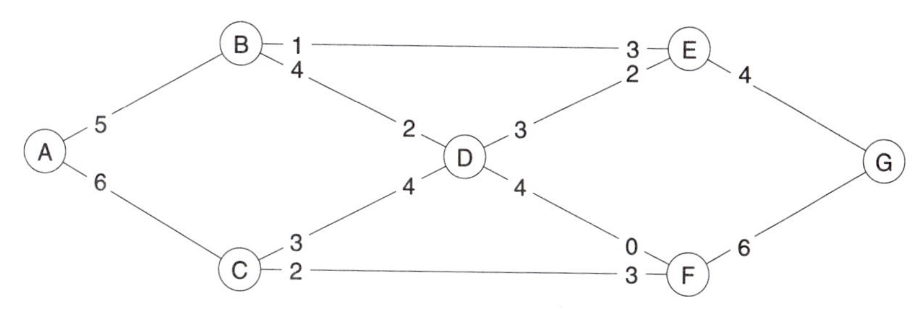

Figure 10.11: Figure for Exercise 2.

3. Draw a network represented by the following network table.

A	2	4	3			
	B		3	7		
		C	4		2	
	5	3	**D**	1	1	5
	2		6	**E**		2
		2	6		**F**	2
9						**G**

4. Find an upper bound for the maximum flow from the source to the sink in the network in Exercise 2. Node A is the source and node G is the sink. We are not asking for the optimal upper bound. Any upper bound is acceptable, but we expect an upper bound that is easy to obtain.

5. Find an upper bound for the maximum flow from the source A to the sink G in the network represented by the following table. Any upper bound is acceptable.

A	2		4			
	B			7		
		C		2	1	
	5	3	**D**	1	2	
	2		6	**E**		5
		2	6		**F**	2
6						**G**

6. Find the maximum flow attainable from node A to node F in the following network.

A	6	1	1		
6	**B**		1	3	1
1		**C**	3	1	
1	1	3	**D**		6
	3	1		**E**	1
8	1		6	1	**F**

7. Find the maximum flow attainable from node A to node G in the following network.

A	4	2	8			
4	**B**		8	3		
2		**C**	2	6	1	
8	8	2	**D**	1	4	1
	3	6	1	**E**		11
		1	4		**F**	2
14			1	11	2	**G**

8. Find the maximum flow attainable from node A to node G in the following network.

A	7	6	2			
	B		1	5		
		C	3		2	
	2	4	**D**	4	4	5
	3		2	**E**		2
		5	6		**F**	5
12						**G**

9. Find the maximum flow attainable from node A to node G in the following network.

A	4	2	7			
4	**B**		8	3		
2		**C**	2	6	1	
	8	2	**D**	1	4	1
	3	6	1	**E**		11
		1	4		**F**	2
13			1	11	2	**G**

10. Find the maximum flow attainable from node A to node H in the following network.

A	8	3					
8	**B**		2	5			
3		**C**	1	4			
	2	1	**D**		2	7	
	5	4		**E**	7	2	
			2	7	**F**		2
			7	2		**G**	9
11					2	9	**H**

11. Consider the following network table. Suppose it is asserted that the min cut involves those and only those arcs with capacity 1. Find the cut with this property. Show that this produces a cut and that the cut is a minimum cut.

A		5		2		3		4	
	B		3		5		4		2
		C		5		2	1	2	
	2		**D**		3		3		5
	1	3		**E**		5		3	
	3		2		**F**		5		3
		2	1	2		**G**		5	
	4		5		3		**H**		2
		2		3	1	2		**I**	
12									**J**

Questions

Q1. A network flow problem for a connected network with finite capacities in each arc always has an optimal solution.

Q2. A feasible solution for the min program (the dual of the max-flow problem) of a network flow problem for a connected network can always be found by cutting the network in any way into two connected pieces, with one piece containing the source and one piece containing the sink, and adding all capacities of arcs that cross from the piece containing the source to the piece containing the sink.

Q3. A feasible solution for the min program of a network flow problem can always be found from a network flow table by drawing a vertical or horizontal line (but not both) through each diagonal cell, where a vertical line is drawn through the cell of the source and a horizontal line is drawn through the cell of the sink, and adding up all uncovered flow capacities.

Chapter 11

The Transshipment Problem

11.1 The Transshipment Problem

In this section we describe a transshipment problem and show how it can be reduced to a transportation problem. In that form it can either be solved as a transportation problem or it can be handled with an algorithm we will discuss later in this chapter. Consider the network shown in Figure 11.1.

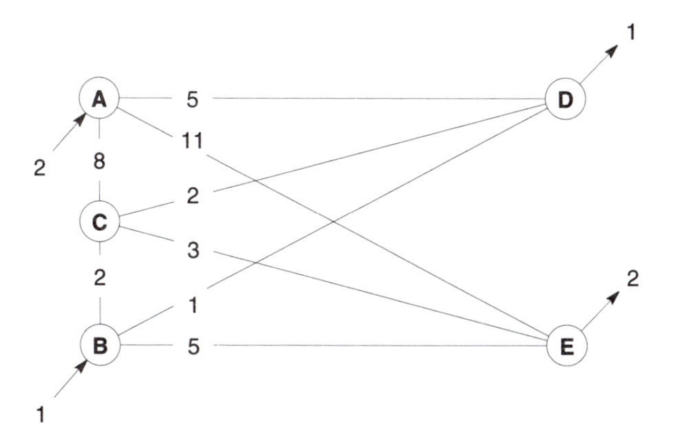

Figure 11.1: A simple transshipment network.

Nodes A and B are origins, with a supply of 2 at A and a supply of 1 at B. Nodes D and E are destinations, with a demand of 1 at D and a demand of 2 at E. The numbers written near or on the arcs are the unit costs for shipment along those routes. In this example, the costs are the same in both directions. If the unit cost were different in the two directions, the unit cost would be written near the node from which the shipment can be made. If no cost is shown, no shipment is permitted in that direction. We have chosen to make the costs the same in both directions to simplify the figure. The material that must be distributed from the origins to the destinations can be transshipped through any of the nodes, including an origin or a destination. The transshipment problem for the network in Figure 11.1 is to determine a shipping schedule for supplying the demands from the origins at minimum cost.

We index the nodes 1, 2, ..., n in any order, and denote the arc from node i to node j by (i, j). Then c_{ij} is the cost for shipping a unit of the material from node i to j along the arc (i, j), and x_{ij} is the quantity to be shipped from node i to node j along the arc (i, j). The total shipping cost is then

$$g = \sum_{(i,j)} c_{ij} x_{ij} \tag{11.1}$$

This is the quantity to be minimized.

It is easily seen that the route A–D–C–E is cheaper than the direct route from A to E. This cheaper, longer route by transshipment through intermediate nodes, one of which is itself a destination, is an allowable shipment route, hence the name transshipment problem.

We shall show first that a transshipment problem can be cast in the form in which the constraint equations and objective function resemble those of a transportation problem. At node i let a_i denote the supply available, and let b_i denote the demand. If both of these are positive, some of the supply would be used locally and would not contribute to the shipping costs. By reducing both the supply and the demand by the same amount, we could reduce at least one of these numbers to zero without affecting the problem. Using the same reasoning, we can increase the supply and demand by the same amount at any node without affecting the problem, and we will take advantage of this possibility to construct the related transportation problem.

Let t_k denote the amount that is transshipped through node k. Then, for each node k we have the following equations.

$$\sum_j x_{kj} = a_k - y_k + t_k \tag{11.2}$$

$$\sum_i x_{ik} = b_k + z_k + t_k \tag{11.3}$$

In equation 11.2 the sum is taken over the allowable arcs out of node k.

It says that the total amount shipped out of node k must equal the supply minus the amount left behind plus the amount shipped through. In equation 11.3 the sum is taken over the allowable arcs into node k. It says that the total amount shipped into node k must equal the demand plus the oversupply plus the amount shipped through. We require $y_k \geq 0$ and $z_k \geq 0$.

It is convenient to set

$$t_k = M_k - x_{kk} \tag{11.4}$$

where M_k are taken large enough to ensure that all x_{kk} are positive. Then the sums 11.2 and 11.3 can be written in the forms

$$\sum_j x_{kj} = (a_k + M_k) - y_k \tag{11.5}$$

$$\sum_i x_{ik} = (b_k + M_k) + z_k \tag{11.6}$$

The sums 11.2 and 11.3 do not include the terms x_{kk}, whereas the sums 11.5 and 11.6 do.

Notice that if M_k is small and t_k is large, then x_{kk} can be negative. The purpose of this discussion is to show that we should take the x_{kk} to be free variables. When that is done we can take $M_k = 0$ and drop them from further consideration. We then have the normalized transshipment problem.

Normalized transshipment problem. Minimize

$$g = \sum_{(i,j)} c_{ij} x_{ij} \tag{11.7}$$

subject to the constraints

$$\sum_j x_{kj} = a_k - y_k \tag{11.8}$$

$$\sum_i x_{ik} = b_k + z_k \tag{11.9}$$

where x_{ij}, y_i, and z_j are canonical, except that x_{kk} is free, and where c_{ij} is the cost of shipment along the arc (i, j). Since the original objective function 11.1 did not include terms in x_{kk} we agree to take $c_{kk} = 0$ so that 11.7 and 11.1 will always have the same values. Furthermore, either a_i or b_i is zero and the other is nonnegative.

In this form these equations resemble supply and demand equations for a transportation problem. The differences are that here each node appears both as an origin and as a destination and the equations for the transshipment problem include x_{kk} terms.

Dual to the transshipment problem. Maximize

$$f = \sum_j b_j v_j - \sum_i a_i u_i \tag{11.10}$$

subject to

$$-\, u_i + u_j - c_{ij} = -w_{ij} \tag{11.11}$$

where u_i, v_j, and w_{ij} are canonical, except for w_{kk}, which is artificial.

Since $-u_i + v_j - c_{ij} = -w_{ij}$ and $c_{kk} = 0$ and w_{kk} is artificial, we have $u_k = v_k$. Thus, the objective function is equivalent to

$$f = \sum_k (b_k - a_k) u_k \tag{11.12}$$

Notice in 11.12 that adding or subtracting the same quantity from both the supply and the demand at any node will not change the value of f. This reinforces the earlier observation. Let us write down the transportation table for the transshipment problem shown in Figure 11.1.

(11.13)

(20)				(1)	(2)
(2)	(0)		8	5	11^+
(1)		(0)	2	1^+	5
(0)	8	2	(0)	2	3
	5	1	2	(0)	
	11	5	3		(0)

We have arranged things so that the first two columns represent nodes that are origins, the last two columns represent nodes that are destinations, and the third column represents an intermediate (a node that is neither an origin nor a destination, but through which shipments can be made). This introduces a little order to make it easier to interpret what is going on, but we can usually not expect to have things arranged so nicely.

Where both a supply and a demand are zero, we show the supply on the left margin. The zero shipment for the intermediate node in 11.13 is required to preserve the tree structure of the basic solution.

Notice several differences between this table and the transportation tables we encountered earlier. The blank spaces in this table indicate the absences of arcs connecting the corresponding nodes. In other words, these are forbidden or blocked paths. For this reason, we show the zero-cost routes explicitly instead of using omission to indicate zeros. The other important difference is that row k and column k both refer to the same node.

One difficulty with using the Dantzig transportation algorithm for a problem like 11.13 is that a large number of gaps in the table might make it difficult or impossible to find a feasible shipping schedule. This would be particularly true if there were a large number of one-way routes. The system of equations 11.5 and 11.6 might be infeasible. There is an obvious way to handle this difficulty. Insert all the forbidden routes at such a high cost that an optimal solution should avoid them. If an optimal solution requires the use of one of these high-cost routes, the original problem does not have a feasible solution. If an optimal solution does not require the use of one of these routes, it is also an optimal solution for the original problem.

In practice it is not necessary to open all blocked routes. We can use any method to obtain an initial feasible solution. If a feasible solution is not obvious, open blocked routes as needed and assign high costs to them as they are introduced, and leave the other blocked routes closed. We could fill in the blanks, which represent blocked arcs, with a symbol to represent a high-cost route and proceed to find a feasible solution as if it were a transportation problem. However, in this example a feasible solution can be found easily and directly.

If we use the Dantzig transportation algorithm, the first step is to find an initial feasible solution for the transportation equations. To this end increase the shipments where the + signs are shown in 11.13. This yields table 11.14.

(11.14)

(23)					
(0)	(0)		?	?	(2)
(0)		(0)	?	(1)	?
(0)	?	?	(0)	?	?
	?	?	?	(0)	
	?	?	?		(0)

In 11.14 we have deleted the costs and replaced them by question marks since two of them are covered by the new shipments. Starting with table 11.14, the next step is to compute the residual costs using a method that is essentially the same as that used for the transportation problem. The only

difference is that we do not compute residual costs where the routes are closed. That is, we compute prices on the margins and the residual costs where the question marks are shown.

The first transportation table obtained in this way is shown in 11.15.

(11.15)

(23)	0	0	0	1	11
(0)	(4)		8	4	(2)
(0)		(4)	2	(1)	−6
(0)	8	2	(4)	1	−9
−1	6	2	3	(4)	
−11	22	16	13		(4)

The reader may wish to continue with the Dantzig algorithm to solve this problem. In the last section of this chapter we will present an alternate algorithm. For now we will just show the final table that is obtained with either algorithm. The reader can verify that the solutions to both dual problems are feasible and complementary. In particular, we wish to call your attention to the negative entries in the main diagonal. These are the negatives of the quantities transshipped through those nodes. The common solution is $f = g = 18$.

(11.16)

(3)	0	5	7	5	9
(0)	(0)		1	(2)	2
−5		(0)	(1)	1	1
−7	15	4	(−2)	4	(2)
−5	10	1	(1)	(−1)	
−9	20	9	4		(0)

11.2 Shortest Path Problems

There are several interesting and important problems that can be cast as transshipment problems. The shortest path problem is one, and it also has a special algorithm that is short and simple. Not only can this algorithm be used for the shortest path problem, but it is useful as a subalgorithm in another algorithm which we will discuss in this chapter.

Consider the network shown in Figure 11.2.

The numbers associated with the arcs are the distances between the

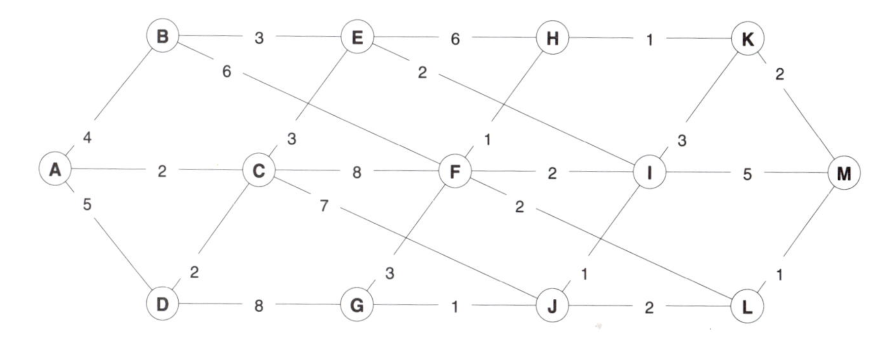

Figure 11.2: Shortest path network.

nodes incident with each arc. If the nodes represent cities and the arcs represent roads between the cities, the distances could be measured in road miles or in driving time. In neither case would it be necessary that the triangle inequality for distance be satisfied. Nor would it be necessary for the driving times in both directions to be equal. Some of the routes could also be one-way. The triangle inequality is not satisfied for the distances shown in Figure 11.2. In this example the distances in both directions are equal. It would clutter the drawing and add little to the example to make the distances in opposite directions different.

The problem is to find the shortest distance between two given nodes, and the path that gives this shortest distance. In the network of Figure 11.2 we might ask, for example, to find the shortest distance between node A and node M. If a graph were constructed at random with a lot of one-way streets, it might not be possible to get from any given point to another given point. That is, the problem might be infeasible.

This shortest path problem can be reduced to a transshipment problem in the following way. Interpret the distances between nodes as costs and find the optimal way to ship a unit of a material between a start (node A) and a destination (node M). The least expensive route for the transshipment problem is the shortest path.

The algorithm we will introduce can provide the shortest path from a given start node to all nodes that can be reached almost as easily as the shortest path to a particular node. For this reason, we choose to formulate the shortest path problem in the following form.

> **Shortest path problem.** For a given node as the start, find the shortest path from the start to each of the other nodes in the network that can be reached from the start.

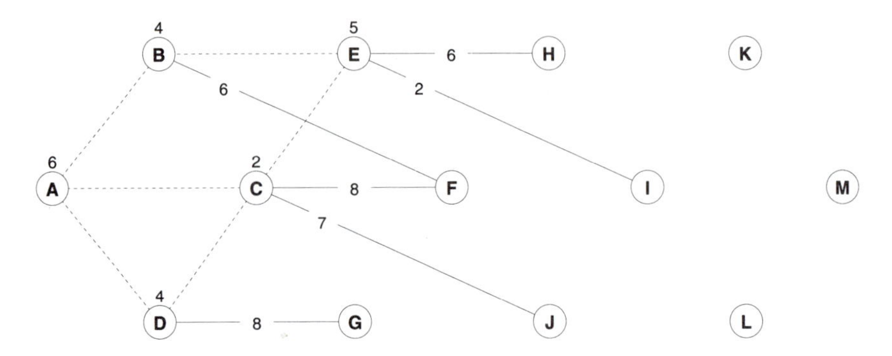

Figure 11.3: Partial labeling of a shortest path network.

This more general shortest path problem can also be reduced to a transshipment problem. Consider each node, other than the origin (start), to be a destination with a demand of 1. At the origin provide a supply sufficient to meet all the demands. If this problem is feasible, there is a path from the origin to each node. The price at each node is the distance from the origin, and the solution itself gives the shortest path to each node.

We are going to label every node of the network that can be reached from the start (node A) with a number that is the minimum distance from the start to that node. That will clearly solve the problem posed. We start by labeling the start 0 (it is zero distance from itself). The only subtlety, and taking care of it is the heart of the algorithm, is how to choose which node to label next when a portion of the nodes is already labeled correctly.

Suppose we are at a stage at which some of the nodes are labeled and some are not. We examine all unlabeled nodes that are connected by one arc to a labeled node. To see what is involved in the next step, consider the partial labeling of the network shown in Figure 11.3.

To simplify the figure, the arcs connecting unlabeled nodes are suppressed, the arcs connecting labeled nodes are shown as dotted lines and their lengths are suppressed, and the arcs connecting a labeled and an unlabeled node are shown as continuous lines and their lengths are shown. All labels affixed up to this point are shown.

Consider node F. The path from node A to node F through labeled node B has length 10, obtained by adding the length of the arc BF to the label on node B. The length of the path to node F through labeled node C is also 10. Thus, the shortest path to node F through labeled nodes is 10. There might, however, be a shorter path through some nodes that are as yet unlabeled.

In a similar way we can see that the shortest path from A to node G

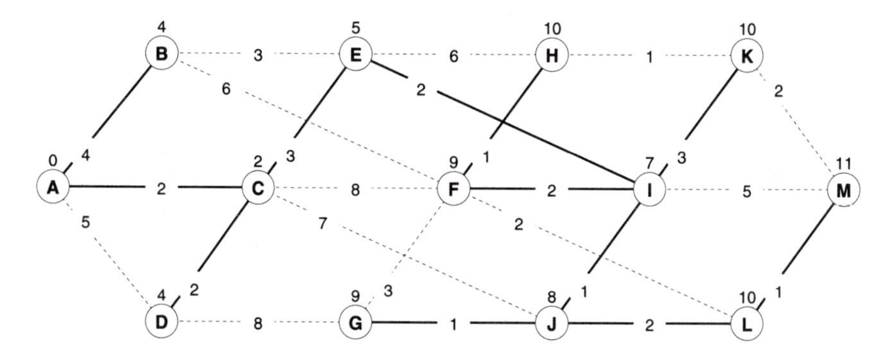

Figure 11.4: Shortest path tree in a network.

through labeled nodes is 12, the shortest path to node H through labeled nodes is 11, the shortest path to node I through labeled nodes is 7, and the shortest path to node J through labeled nodes is 9. From this we can conclude that the shortest path to node I by any path is 7. Any path to node I that includes unlabeled nodes must be at least as long as the paths to nodes F, G, H, and J since such a path would have to enter the unlabeled area through on of these nodes.

We label node I with a 7 and repeat this analysis. There are two ways this process can terminate. If not all nodes are labeled and we cannot label any more nodes, the network is not connected. We will have labeled every node reachable by a path from the start. If we label all nodes we have solved the problem posed. The complete labeling for the example is shown in Figure 11.4.

For each node in Figure 11.4 a shortest path from the start can be found by starting from that node and working back to the start. For example, consider node K. It is labeled with a 10, so its distance from the start is 10. Find a node adjacent to node K whose label is equal to 10 minus the length of the connecting arc. Node I has this property. There is a path of length 7 from the start to node I, and this path extended directly to node K will produce a shortest path to node K.

For each node in Figure 11.4 a shortest path can be found by tracing the arcs indicated with heavy lines back to the start. Notice that this network of heavy lines is a spanning tree in the network. A shortest path will not contain a loop. For a given node there might be more than one shortest path back to the start, but by making choices a spanning tree of shortest paths can always be obtained. Start with any node and trace a shortest path back to the start. For any node not on this path, trace a path back until it joins the paths already constructed. Continue until all nodes are joined to the tree.

To make this algorithm useful for large networks it is desirable to formulate the algorithm so that it can be directly applied to a transportation table without drawing the actual network. In table 11.17 we show the distances in Figure 11.2 as costs in a transportation table. The supplies (12 units) at the start and the demands (1 unit at each node other than the start) at the other nodes are not shown since they are not needed.

(11.17)

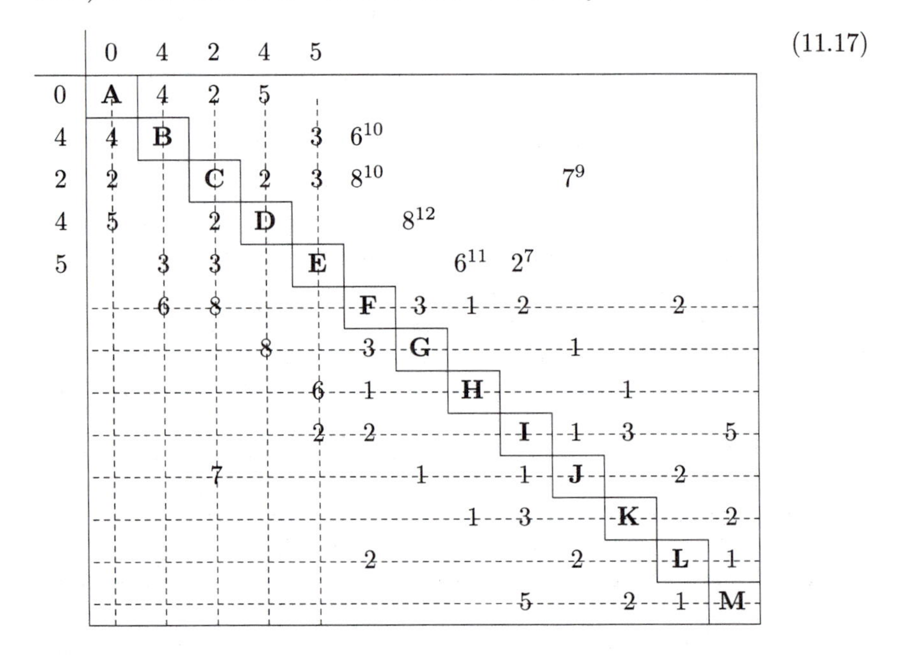

	0	4	2	4	5								
0	**A**	4	2	5									
4	4	**B**			3	6^{10}							
2	2		**C**	2	3	8^{10}				7^{9}			
4	5		2	**D**			8^{12}						
5		3	3		**E**			6^{11}	2^{7}				
		6	8			**F**	3	1	2			2	
				8		3	**G**			1			
					6	1		**H**			1		
					2	2			**I**	1	3		5
			7				1		1	**J**		2	
								1	3		**K**		2
						2				2		**L**	1
									5		2	1	**M**

The labels on the top margin of table 11.17 are the same as those appearing in Figure 11.3. The labels on the left margin are the same as those on the top margin since each node is represented by a row and a column. Dotted horizontal lines are drawn through the rows of the unlabeled nodes and vertical lines are drawn through the columns of the labeled nodes. Ordinarily, we cannot expect the labeled nodes to be contiguous in the table. We have arranged things that way here and drawn the lines only to assist in the description.

The uncovered zone in the upper right corner contains all the arcs that connect labeled nodes with unlabeled nodes. For each entry in this zone, the distance back to the start through a labeled node is the sum of that entry and the label on the left margin. These are shown as superscripts on those entries. The minimum such distance is the 7 above the 2 (just as it is in Figure 11.4) in the column for node I. This results in the label 7 for that column on the top margin, and this would be copied to the corresponding row on the left margin.

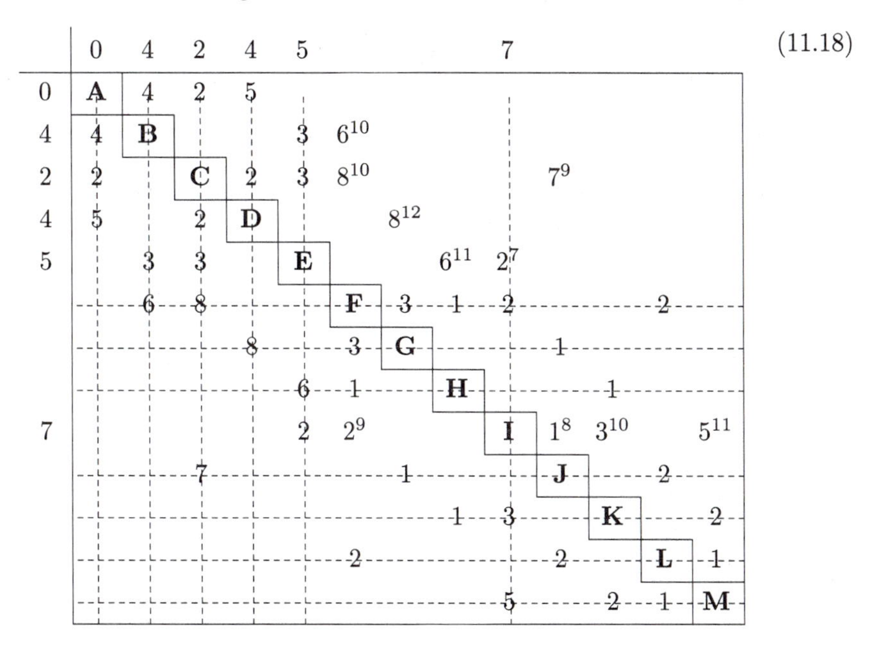

(11.18)

The next step would be to supply the superscripts in unlabeled columns in the newly labeled row. This would yield a label of 8 on the column for node J. Notice that once an arc is labeled its label will not be changed by subsequent labeling.

The algorithm described here is due to E. Dijkstra (1959). It is simple and fast. Only one pass through the table is required since once a label is assigned to a node it does not have to be recomputed. It terminates with all nodes labeled or with the knowledge that it is impossible to label all nodes. In the former case every node can be reached from the start, and in the latter case it is not possible to reach all nodes from the start. Let us state the algorithm formally.

Dijkstra's Shortest Path Algorithm

1. Construct a transportation table in the form of 11.17 with c_{ij} the distance from node i to node j. Label the column and row of the start with 0.

2. Write a superscript on each entry in a labeled row and an unlabeled column which is the sum of that entry and the label on the left margin.

3. Find the minimum such superscript and label the column with that superscript. Attach the same label to the row of that node.

4. If all columns are labeled, STOP—all nodes are labeled. If not all columns are labeled but there is no entry in a labeled row and an unlabeled column, STOP—the graph is not connected. If there is an entry in a labeled row and an unlabeled column, return to step 2.

In our example, all nodes will be labeled. The result is shown in table 11.19. All superscripts that would have been produced are also shown.

(11.19)

	0	4	2	4	5	9	9	10	7	8	10	10	11
0	**A**	4	2	5									
4	4	**B**			3	6^{10}							
2	2		**C**	2	3	8^{10}				7^{9}			
4	5		2	**D**			8^{12}						
5		3	3		**E**			6^{11}	2^{7}				
9		6	8			**F**	3^{12}	1^{10}	2			2^{11}	
9				8		3	**G**			1^{11}			
10					6	1		**H**			1^{11}		
7					2	2^{9}			**I**	1^{8}	3^{10}		5^{12}
8			7				1^{9}		1	**J**		2^{10}	
10								1	3		**K**		2^{12}
10						2				2		**L**	1^{11}
11									5		2	1	**M**

Although we have worked this problem directly, it is still a transshipment problem and the equations of the transshipment problem can be exploited in this table. The dual problem satisfies the equations

$$-u_i + v_j - c_{ij} = -w_{ij} \tag{11.20}$$

Since each node is represented by a row and column with the same index, $u_i = v_i$. That is, equation 11.20 can be written as

$$v_i - v_j + c_{ij} = w_{ij} \tag{11.21}$$

The v_i have been computed, and they are written on the top and left margins of table 11.19. We can then compute the w_{ij}. The result of this computation is shown in table 11.22.

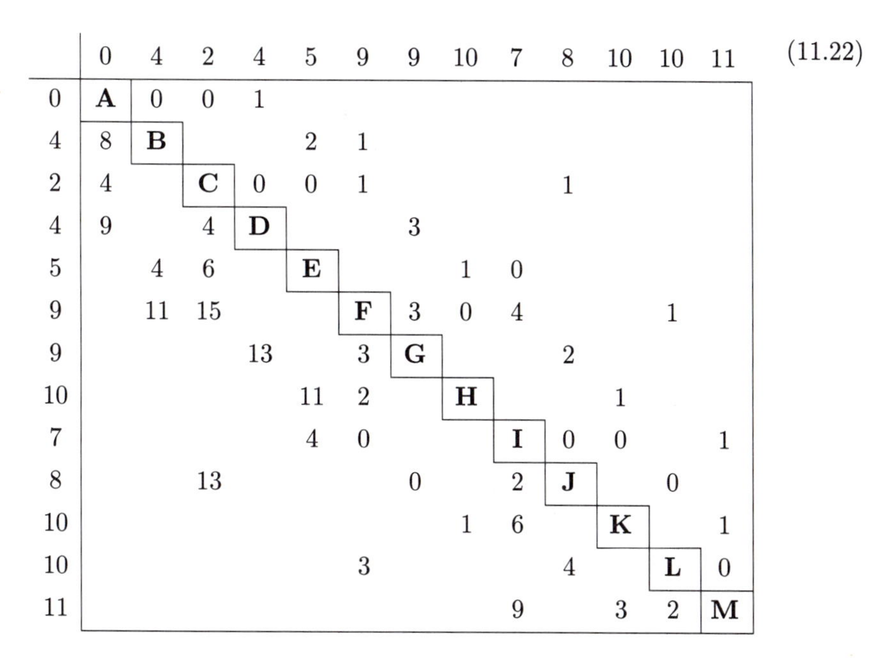

	0	4	2	4	5	9	9	10	7	8	10	10	11
0	**A**	0	0	1									
4	8	**B**			2	1							
2	4		**C**	0	0	1				1			
4	9		4	**D**			3						
5		4	6		**E**			1	0				
9		11	15			**F**	3	0	4			1	
9				13		3	**G**			2			
10					11	2		**H**			1		
7					4	0			**I**	0	0		1
8			13				0		2	**J**		0	
10								1	6		**K**		1
10						3				4		**L**	0
11									9		3	2	**M**

(11.22)

Wherever there is a zero $w_{ij} = 0$ in the table, equation 11.21 reduces to

$$v_j = v_i + c_{ij} \tag{11.23}$$

This means that wherever $w_{ij} = 0$, the corresponding arc connects two nodes whose distances from the start differ by exactly the length of the arc. These are the arcs that are on potential shortest paths. This suggests a simple way to find a tree of shortest paths.

1. Start with any node other than the start. (We will start with M.)
2. Move up or down in the column of that node to find a zero. There will be a zero in every column (except the start) if all nodes are labeled since the zero will occur at the position of the minimum entry that determined the label.
3. Move horizontally to the entry on the main diagonal. This identifies the next node down the tree.
4. Look for a zero in the column of that node and continue until you get back to the start.
5. Starting with any node that is not on this path, go through the same process and trace the path back until it joins a node already in the tree.

6. If the tree is not complete, return to step 5, else STOP.

Table 11.24 shows the tree that results from the tracing described above. The tree obtained is identical to that shown in Figure 11.4.

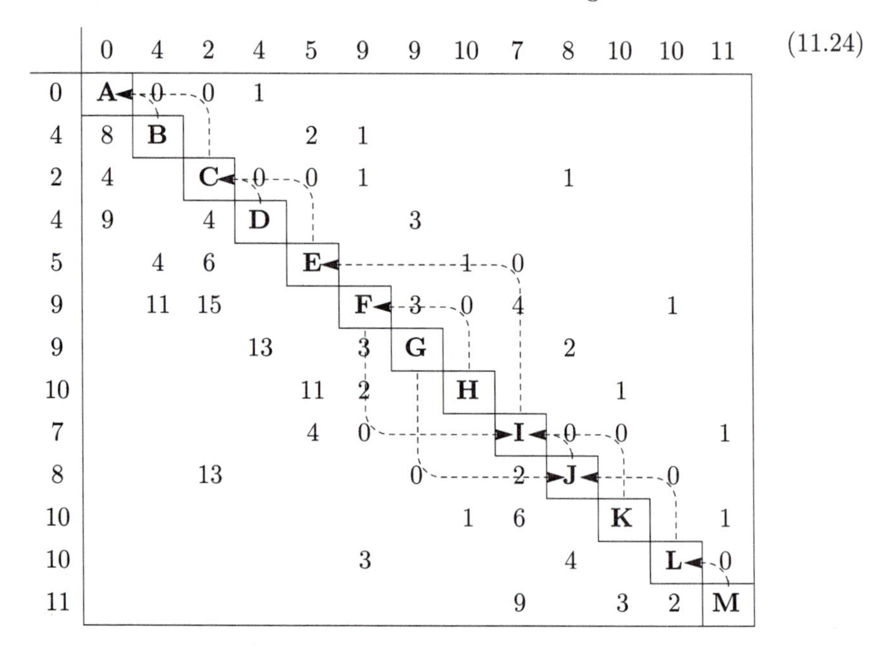

	0	4	2	4	5	9	9	10	7	8	10	10	11
0	**A**	0	0	1									
4	8	**B**			2	1							
2	4		**C**	0	0	1				1			
4	9		4	**D**			3						
5		4	6		**E**			1	0				
9		11	15			**F**	3	0	4			1	
9				13		3	**G**			2			
10					11	2		**H**			1		
7					4	0			**I**	0	0		1
8			13				0		2	**J**		0	
10								1	6		**K**		1
10						3				4		**L**	0
11									9		3	2	**M**

(11.24)

11.3 A Transshipment Algorithm

There are several algorithms available for transshipment problems, among them the Dantzig transportation problem algorithm as described in Section 11.1. The algorithm we are going to describe in this section is based on Kuhn's Hungarian algorithm. The primary difference is that the König subalgorithm is replaced by an adaptation of the shortest path algorithm. This algorithm also shows influences from the Dantzig transportation algorithm and the Ford–Fulkerson maximum flow algorithm.

The u_k are prices at the nodes. We raise prices by minimal amounts at the nodes until it becomes profitable to ship from a source to a destination at a cost just equal to the price differential. A path for a least-cost shipment is sought using the shortest path algorithm. If such a path cannot be found that reaches a destination, the labeling produces a cut in the network. The cheapest arc from a labeled node to an unlabeled node identifies the amount by which we must raise all prices at the unlabeled nodes to extend the path.

We start with a formulation of the transshipment problem as given in Section 11.1. Initialize the transshipment table as described in Section 11.1. However, the algorithm we will describe will not present solutions for the pair of dual problems that are complementary before termination. For this reason, we use two rows on the top margin, one for the variables of the primal problem and one for the variables of the dual problem. Each node is represented by both a row and a column, which intersect on the main diagonal. Initially, the basic variables for the transshipment equations are the $x_{kk} = 0$ (on the main diagonal), the $y_i = a_i$ (on the left margin) for the sources, the $y_i = 0$ (on the left margin) for the intermediate nodes, the $z_j = b_j$ (on the top margin) for the sinks, and $t = 0$ for the return from the universal sink to the universal source.

Let us take the problem described in Section 11.1 and set up the initial table to represent the problem in these terms. In table 11.25 we show the initial values of the basic variables for both the primal problem (in parentheses) and the dual problem. Since we use blanks to denote paths that are forbidden, we show the zeros explicitly. Paired dual variables occupy the same positions, except for the variables of the primal problem and the variables of the dual problem on the top margin.

(11.25)

(0)	0	0	0	(1)	(2)
(2)	(0)		8	5	11
(1)		(0)	2	1	5
(0)	8	2	(0)	2	2
0	5	1	2	(0)	
0	11	5	2		(0)

An Algorithm for the Transshipment Problem

1. Mark each row with 1 that represents a basic variable (a variable represented by a number in parentheses) for the primal problem (the transshipment equations) on the left margin. Set $k = 1$.

2. Mark each column with $k+1$ that has a zero for a variable of the dual problem in a row marked k. Note that basic variables for the primal problem (represented by numbers in parentheses) mask zeros for the dual problem.

3. If a new column is marked in step 2 go to step 4, else go to step 7.

4. If a column is marked that has a basic variable on the top margin go to step 9, else go to step 5.

5. Mark each row with $k+2$ that has a basic variable for the primal column in a column marked $k+1$.

6. If a new row is marked in step 4 go to step 2, else go to step 7.

7. We reach a point where no column with a basic variable on the top margin has been marked and it is impossible to mark a new row . Draw a line through each marked column and each unmarked row.

8. There will be no uncovered zeros for the dual problem (else another column could be marked). If there are no uncovered entries for the dual problem STOP—the problem is infeasible—else go to step 9.

   Table 11.26 shows the situation in our example when we complete step 8. The minimum noncovered entry is the 1 in column 4.

(11.26)

	0	0	0	0	0	
(0)				(1)	(2)	
(2)	(0)		8	5	11	1
(1)		(0)	2	1*	5	1
(0)	8	2	(0)	2	2	1
0	5	1	2	(0)		
0	11	5	2		(0)	
	2	2	2			

9. Star the minimum uncovered entry for the dual problem. Subtract this number from all entries on the left margin for covered rows, and add this number to all entries on the top margin for uncovered columns. Subtract this number from all uncovered entries and add it to all twice-covered entries within the table. Erase all lines and marks and return to step 1.

   Table 11.26 shows the situation when we arrive at step 9 for the first time. When we go back to step 1 and mark the rows and columns, we arrive at step 10 with the marks shown in table 11.27

(11.27)

	0	0	0	1	1	
$(0)^+$				$(1)^-$	(2)	
(2)	(0)		8	4	10	1
$(1)^-$		(0)	2	0^+	4	1
(0)	8	2	(0)	1	1	1
−1	6	2	3	(0)		
−1	12	6	3		(0)	
	2	2	2	2		

10. A basic variable for the primal problem on the top margin has been marked. This is called *breakthrough*. Trace a path of zeros for the dual problem back through rows and columns with descending marks until a basic variable on the left margin is reached. Starting with the basic variable on the top margin, mark these entries alternately with − and + signs. Determine the minimum basic variable for the primal problem that has been marked with a − sign. (It might be zero.) Increase all variables for the primal problem marked with a + sign by this amount, and decrease all variables marked with a − sign by this amount. Make one (and only one) of these variables nonbasic by erasing the parentheses on one of the zeros. If there is more than one choice, give highest priority to a basic variable on the top margin, then to a basic variable on the left margin, then to the choice in the left-most column of the top-most row. Erase all marks and + and − signs.

    Table 11.28 shows the recalculation of the primal solution indicated in step 10. The marking ends without breakthrough. The minimum noncovered entry is the 1 in column 5.

(11.28)

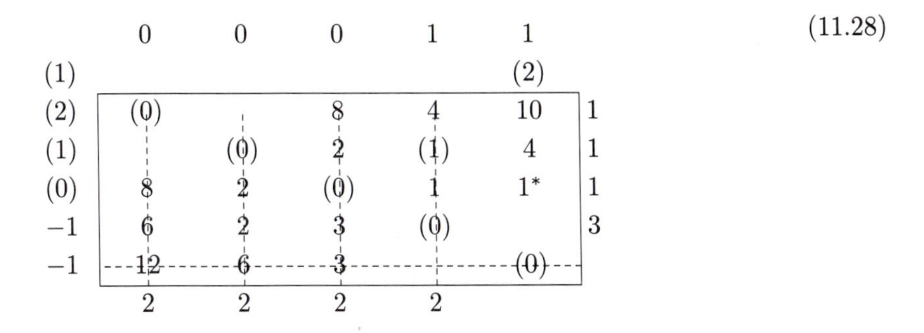

	0	0	0	1	1	
(1)					(2)	
(2)	(0)		8	4	10	1
(1)		(0)	2	(1)	4	1
(0)	8	2	(0)	1	1*	1
−1	6	2	3	(0)		3
−1	12	6	3		(0)	
	2	2	2	2		

11. If a basic variable for the primal problem remains on the top margin, return to step 1. (The priorities for making zero variables nonbasic will assure that a basic variable on the top margin is nonzero.) If no basic variable for the primal problem remains on the top margin, STOP—optimal solutions to both problems have been obtained.

In table 11.28 a basic variable remains on the top margin and we return to step 1. The following tables reiterate the steps described above until we obtain table 11.36, in which no basic variable remains on the top margin.

(11.29)

	0	0	0	1	2	
$(1)^+$					$(2)^-$	
(2)	(0)		8	4	9	1
(0)		(0)	2	(1)	3	1
$(0)^-$	8	2	(0)	1	0^+	1
−1	6	2	3	(0)		
−2	13	7	4		(0)	
	2	2	2	2	2	

In 11.29 the minimum entry marked with a − sign is zero. We have breakthrough but there is no change in the values of variables. There is a change in which variables are basic. This changes the marking, which is shown in 11.30.

(11.30)

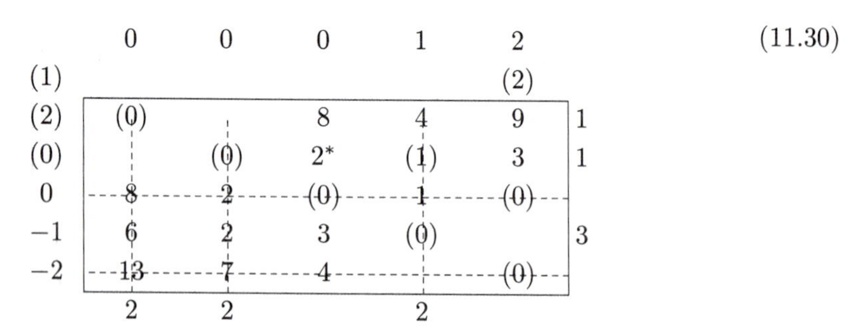

	0	0	0	1	2	
(1)					(2)	
(2)	(0)		8	4	9	1
(0)		(0)	2*	(1)	3	1
0	8	2	(0)	1	(0)	
−1	6	2	3	(0)		3
−2	13	7	4		(0)	
	2	2		2		

The minimum noncovered entry is the 2 in column 3. This produces a recalculation of the values of the dual variables.

(11.31)

	0	0	2	1	4	
$(1)^+$					$(2)^-$	
(2)	(0)		6	4	7	1
$(0)^-$		(0)	0^+	(1)	1	1
−2	10	4	$(0)^-$	3	$(0)^+$	3
−1	6	2	1	(0)		
−4	15	9	4		(0)	
	2	2	2		4	

This time the marking ends with a breakthrough. Again, the minimum entry with a – sign is zero and we have only a change in which variables are basic.

(11.32)

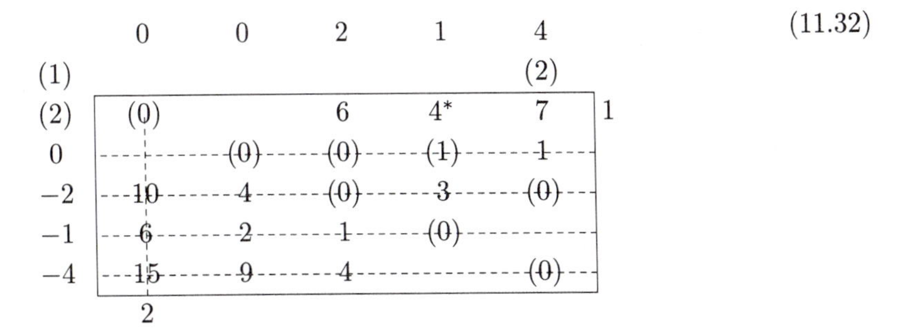

	0	0	2	1	4	
(1)					(2)	
(2)	(0)		6	4*	7	1
0		(0)	(0)	(1)	1	
−2	10	4	(0)	3	(0)	
−1	6	2	1	(0)		
−4	15	9	4		(0)	
	2					

The new marking does not produce breakthrough. The minimum non-covered entry is the 4 in column 4.

(11.33)

	0	4	6	5	8	
$(1)^+$					$(2)^-$	
$(2)^-$	(0)		2	0^+	3	1
−4		(0)	$(0)^+$	$(1)^-$	1	3
−6	14	4	$(0)^-$	3	$(0)^+$	5
−5	10	2	1	(0)		
−8	19	9	4		(0)	
	2	4	4	2	6	

The new marking produces breakthrough in 11.34. This time there is a – sign on one of the zeros in the main diagonal. These are free variables, so this redistribution can make this variable negative.

(11.34)

	0	4	6	5	8	
(2)					(1)	
(1)	(0)		2	(1)	3	1
−4		(0)	(1)	0	1	
−6	14	4	(−1)	3	(1)	
−5	10	2	1*	(0)		3
−8	19	9	4		(0)	
	2			2		

There is no breakthrough this time. The minimum noncovered entry is the 1 in column 4. We recalculate the values of the dual variables.

(11.35)

	0	5	7	5	9	
$(2)^+$					$(1)^-$	
$(1)^-$	(0)		1	$(1)^+$	2	1
−5		(0)	(1)	1	1	5
−7	15	4	$(-1)^-$	4	$(1)^+$	5
−5	10	1	0^+	$(0)^-$		3
−9	20	9	4		(0)	
	2	6	4	2	6	

This is the final breakthrough.

(11.36)

(3)	0	5	7	5	9
(0)	(0)		1	(2)	2
−5		(0)	(1)	1	1
−7	15	4	(−2)	4	(2)
−5	10	1	(1)	(−1)	
−9	20	9	4		(0)

The solutions for the two problems are feasible and complementary. Hence, both are optimal. The value of the objective variable for each problem can easily be computed. For the transshipment problem, multiply the amounts shipped along each arc (the numbers in parentheses in table 11.36) by the cost for that arc as shown in the original transshipment table 11.25. We obtain 18 as the shipping cost.

For the dual pricing problem, multiply each demand (in parentheses on the top margin of the original table 11.25) by the price (the number on the top margin of the final table 11.36) and multiply each supply (in parentheses on the left margin of the original table 11.25) by the price (the negative of the number on the left margin of the final table 11.36). Subtract the total costs at the sources from the total income at the destinations. The income is also 18.

Just as it does for the network flow problems of Chapter 10, the marking process identifies paths from the various sources. In this case the paths are paths along which the residual costs are zero. If no sink is marked, the lines drawn to form the cover provide a cut that divides the nodes into two disjoint sets, a source set and a sink set. The entries that are not covered represent arcs from the source set to the sink set. By raising the prices for the nodes in the sink set by an amount equal to the minimum residual cost on one of these connecting arcs, at least one arc is given a zero residual cost, which extends a path to a node in the sink set in the next marking.

If one of the sinks is marked, the marking identifies a path of zero residual costs from some source to that sink. Finding the path is only a matter of tracing the markings back to the source. The alternate + and − sign marking increases the flow from a source to a sink in the same way that this works for the network flow problem. The variables marked with − signs are the capacities.

If a marking produces a cut with no arcs joining the source set and the sink set, then there is no path from any of the remaining sources to any of the remaining sinks with a positive residual demand. In this case the transshipment problem is infeasible.

Each recomputation of the prices for the dual problem increases a price on at least one node. Each path from a source to a sink with zero residual costs is a path for which the price difference between the sink and the source is equal to the sum of the shipping costs along the path. Therefore, the greatest price differential can never be more than the sum of the shipping costs for all of the arcs. This is a very crude upper bound, but it shows that there is an upper bound for the price increases. The recomputation of the prices can be performed only finitely many times.

Also, each increase of a flow from a source to a sink reduces the remaining demand. Hence, if the supplies and demands are rational numbers, there can only be a finite number of such increases. The one way that the algorithm could fail to terminate in a finite number of steps is to have a cycle of iterations in which the flow is not increased. The example illustrated several instances where the path from a source to a sink does not produce an increase in the flow. The priority rule for selecting the variable that becomes nonbasic is a Bland selection rule. Thus, cycles cannot occur and the algorithm always terminates in a finite number of iterations.

Exercises

1. Construct the transportation table that represents the following transshipment network.

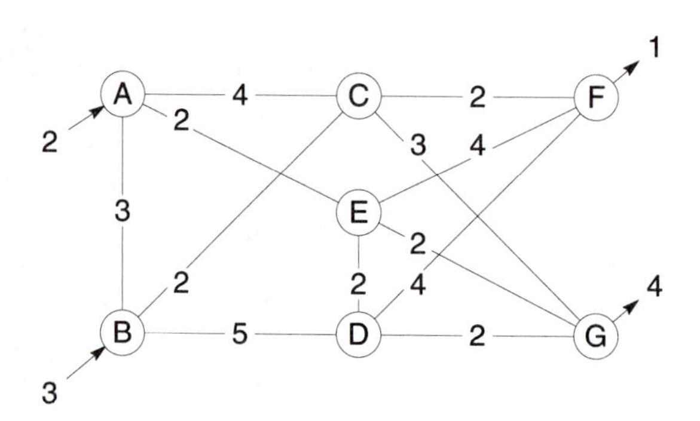

Transshipment network for Exercise 1

2. Find the shortest path from node A to each node in the network represented by the following table.

A	16	4						
16	**B**		14			12	5	
		C		14	8		6	
	14		**D**	15		5		
		14	15	**E**		14	6	26
		7			**F**		2	
	12		5	14		**G**		14
	5	6		6	2		**H**	32
				26		14	32	**I**

3. Find the shortest path from node A to each node in the network represented by the following table.

A	3	4						
3	**B**	2	7	6				
4	2	**C**	5	4				
	7	5	**D**	2	4	4		
	6	4	2	**E**	6	4		
			4	6	**F**	2	2	3
			4	4	2	**G**	4	5
					2	4	**H**	2
					3	5	2	**I**

4. Solve the transshipment problem represented by the following table. Simple though the structure of this problem is, it requires a reversal of a shipment made at an early step in the algorithm. It as an adequate test of your understanding of the algorithm.

(0)			(1)	(2)
(2)	(0)		2	
(1)		(0)	1	2
	2	1	(0)	
		2		(0)

5. Find the maximum flow from the sources to the sinks in the transshipment problem represented by the following table.

(0)						(1)	(4)
(3)	(0)	3	4		2		
(3)	3	(0)	2	5			
(0)	4	2	(0)			2	3
(0)		5		(0)	2	4	2
(0)	2			2	(0)		5
			2	4		(0)	
			3	2	5		(0)

6. Find an optimal solution for the transshipment problem given in Exercise 1.

7. **The Caterer Problem.** Suppose that a caterer must supply napkins over a period of several days, with a given number of napkins required

on each day. He can buy new napkins in an unlimited supply for $2 each, or he can have used napkins laundered for reuse. There is a two-day service available that charges $0.25 each or he can launder them himself in one day. Because of his labor costs, it will cost him $0.75 per napkin. Suppose he must supply napkins on four successive days, with 100, 60, 60, and 90 as the numbers required each day. How should he supply himself with napkins at least cost?

This is a classic transshipment problem. It was originally posed by W. Jacobs (1954), but with different data. He had in mind an aircraft maintenance shop that needed to overhaul aircraft engines. The shop manager knew when the planes would come in for overhaul. He could buy new engines at a high cost, he could rebuild the engines rapidly in the shop at a lower cost, or he could send them out for rebuilding at a still lower cost.

(W. Jacobs, "The caterer problem," *Naval Research Quarterly 1*: 1954, p154.

Chapter 12

Nonlinear Programs

12.1 Karush–Kuhn–Tucker Theorem

So far in this book we have considered a variety of optimization problems, but they were all linear problems. Nonlinear optimization problems are also important. However, the theory of nonlinear programs is more complex and obtaining numerical solutions is more difficult.

In this section we consider nonlinear programs in their most general form. The fundamental theorem is the *Karush–Kuhn–Tucker theorem*, formerly known as the *Kuhn–Tucker theorem*. We will briefly give the reason for this change of name at the end of this section.

For all the problems we have considered so far we have obtained necessary and sufficient conditions for optimality. For the most general nonlinear programs we can obtain only necessary conditions for optimality. This is similar to the situation in elementary calculus, where for a function that has a derivative in an interval and a maximum at a point in the interval the derivative is necessarily zero at that point. The necessary condition we obtain for nonlinear programs is very useful even though it is not sufficient. After we establish the Karush–Kuhn–Tucker conditions for general nonlinear programs, the rest of this chapter will be devoted to special nonlinear programs where the Karush–Kuhn–Tucker conditions are also sufficient.

The most general way to formulate an optimization problem (for a single optimizing variable) subject to constraints is that of maximizing

$$f(x) = f(x_1, x_2, \ldots, x_n) \tag{12.1}$$

subject to

$$a_i(x) = a_i(x_1, x_2, \ldots, x_n) = -y_i \tag{12.2}$$

for $i = 1, 2, \ldots, m$, where the x's and y's must satisfy feasibility specifications. Here, $x = (x_1, x_2, \ldots, x_n)$ is a column vector. We would customarily require the x's, the independent variables, to be canonical or free and the y's, the dependent variables, to be canonical or artificial. It does not seem to be useful to allow artificial independent variables or free dependent variables.

As a notational simplification we will abbreviate 12.2 in the form $a(x) = -y$. Here, $y = (y_1, y_2, \ldots, y_m)$ is a column vector.

Any (x, y) for which $a(x) = -y$ is called a solution. A solution that satisfies the feasibility specifications is called a feasible solution. The feasible set is the set of all feasible solutions. The feasible set can, of course, be empty. For nonlinear programs we must distinguish between a local optimal solution and a global optimal solution. For a linear program, all local optimal solutions are also global optimal solutions. A feasible solution for which $f(x)$ achieves a local/global maximum is called a local/global optimal solution. We must keep in mind that a nonlinear objective function can be bounded, and therefore have a least upper bound, without taking on a maximum value.

It is useful to think of relation 12.2 as defining a surface (or hypersurface) when y_i is artificial, and one side of the surface (or hypersurface) when y_i is nonnegative. Without some kind of restriction on the kinds of functions that are permitted in 12.2 such an interpretation is inaccurate, but by the time we place restrictions (e.g., that the functions $a_i(x)$ be differentiable) that will make particular problems workable this interpretation will not be misleading.

Calculus derives its power in applications from the fact that differentiation replaces a global nonlinear relation by a local linear relation. We assume the functions in 12.1 and 12.2 are differentiable and formulate the corresponding linear relations. Let

$$a_{ij} = \frac{\partial a_i}{\partial x_j} \tag{12.3}$$

and

$$f_j = \frac{\partial f}{\partial x_j} \tag{12.4}$$

Then

$$a_{i1}\, dx_1 + a_{i2}\, dx_2 + \cdots + a_{in}\, dx_n = -dy_i \tag{12.5}$$

for $i = 1, 2, \ldots, m$, and

$$f_1\, dx_1 + f_2\, dx_2 + \cdots + f_n\, dx_n = df \tag{12.6}$$

Each equation in 12.5 will be abbreviated in the form

$$a_i'(x)\,dx = -dy_i \tag{12.7}$$

and each equation in 12.6 will be abbreviated in the form

$$f'(x)\,dx = df \tag{12.8}$$

We make a further simplification of notation and write

$$a'(x)\,dx = -dy \tag{12.9}$$

to represent all m of the equations in the form of 12.7. In equation 12.8 $f'(x)$ is the gradient and in equation 12.9 $a'(x)$ is the Jacobian.

We can embody the equations in 12.9 and 12.8 in tableau 12.10.

(12.10)

dx	-1	
$a'(x)$	0	$= -dy$
$f'(x)$	0	$= df$

The general idea is to consider the equations 12.9 as representing a tangent plane to the surface represented by the equations 12.2. For a fixed (x, y) the variables dx and dy represent small deviations or displacements from the point (x, y). In the interior of the feasible set the displacements can be in any direction. At a boundary point of the feasible set the displacements can only be towards the interior of the feasible set, or along the boundary. Such a description makes sense only if the boundary surface is well-behaved and conforms to our intuition.

If (x, y) is a feasible solution then (dx, dy) is a *feasible variation* from (x, y) if

1. (dx, dy) is a solution for the equations in tableau 12.10, that is, $a'(x)\,dx = -dy$, and

2. $(x + dx, y + dy)$ satisfies the feasibility specifications for the original nonlinear program.

Specifically, we define feasibility specifications for the variables in tableau 12.10 according to the following rules.

1. If a variable x_j is free then dx_j is also free.

2. If a variable y_i is artificial then dy_i is also artificial.

3. If a variable x_j (or y_i) is canonical and positive then dx_j (or dy_i) is free.

4. If a variable x_j (or y_i) is canonical and zero then dx_j (or dy_i) is canonical.

In tableau 12.11 we have added the variables of a dual linear program.

(12.11)

	dx	-1	
v	$a'(x)$	0	$= -dy$
-1	$f'(x)$	0	$= df$
	$= u$	$= 0$	

The feasibility specifications for the variables u and v are dual to the feasibility specifications for dx and dy. However, we wish to express these feasibility specifications in terms of the variables x and y. We have:

1. If a variable x_j is free then dx_j is also free; therefore u_j is artificial.

2. If a variable y_i is artificial then dy_i is also artificial; therefore v_i is free.

3. If a variable x_j (or y_i) is canonical and positive then dx_j (or dy_i) is free; therefore u_j (or v_i) is artificial.

4. If a variable x_j (or y_i) is canonical and zero then dx_j (or dy_i) is canonical; therefore u_j (or v_i) is canonical.

Notice that these conditions always give $u_j x_j = 0$ and $v_i y_i = 0$. This motivates the following definition.

Definition: The point (x, y) satisfies the *Karush–Kuhn–Tucker conditions* if

1. (x, y) is a feasible solution of the equations $a(x) = -y$, and

2. there exists a feasible solution (u, v) of the equations in 12.11 for which

$$ux + vy = 0 \tag{12.12}$$

Notice that the feasibility specifications in 2 depend on the feasible solution obtained in 1. The complementarity condition 12.12 is included in the feasibility specifications and need not be explicitly stated.

There is another way to state condition 2 which is equivalent and somewhat easier to write down, but it does require stating the complementarity condition. If x_j is free, then u_j is artificial. If x_j is artificial, then u_j is free. If x_j is canonical, there are two possibilities for the feasibility specifications for u_j. But in either case it is nonnegative. If $x_j > 0$ then $x_j u_j = 0$ because u_i must be artificial, and if $x_j = 0$ then $x_j u_j = 0$. That is, we can say that if x_j is canonical then we require that u_j be nonnegative and $x_j u_j = 0$. In fact, this is the familiar way the Karush–Kuhn–Tucker conditions are stated since they are usually stated only for canonical problems.

To find a point that satisfies the Karush–Kuhn–Tucker conditions we regard x, y, u, and v as variables. We try to solve the equations in tableau 12.11 in which x and y appear as variables in the entries in the tableau, simultaneously with equation 12.12, for which all variables satisfy the duality specifications given above. In particular, if y_1 is zero in a solution of $a(x) = -y$, then v_1 will be canonical when we solve $va'(x) - f'(x) = u$ and x will also have to satisfy the equation $a_1(x) = 0$.

The Karush–Kuhn–Tucker theorem asserts that, under appropriate conditions, if the function f has a maximum at the point (x, y) then there is a feasible (u, v) for which the Karush–Kuhn–Tucker conditions are satisfied. Thus, the Karush–Kuhn–Tucker conditions become necessary conditions for the existence of a maximum.

The "appropriate conditions" mentioned above are conditions under which the linearized maximum program in tableau 12.11 correctly represents the local behavior of the variables. We will discuss these conditions, known as the constraint qualification, in detail in the following section.

Theorem 12.65 *(Karush–Kuhn–Tucker Theorem) Suppose a nonlinear program satisfies the constraint qualification at a feasible solution (x, y) at which f takes on a local maximum. Then (x, y) satisfies the Karush–Kuhn–Tucker conditions. That is, there exists a feasible solution (u, v) for the equation*

$$va'(x) - f'(x) = u \tag{12.13}$$

for which

$$ux + vy = 0 \tag{12.14}$$

Proof. We do not have to prove equation 12.14 since that requirement is already embodied in the feasibility specifications.

To prove the theorem we must show that if there is a local maximum at the feasible point (x, y) then the objective variable df in tableau 12.11 must be bounded. This is not true in all generality; certain conditions must be satisfied. We will discuss these conditions in the following section. The maximum program for tableau 12.11 is always feasible since $dx = 0$ and

$dy = 0$ is a feasible solution. Since the objective function for tableau 12.11 is bounded by zero, the dual program is feasible. That is, there exists a feasible solution to equation 12.13. ◻

For an unconstrained maximum problem with free variables, the set of equations in 12.2 is empty. In this case the Karush–Kuhn–Tucker conditions reduce to the equations $f_j = 0$, for $j = 1, \ldots, n$. This case is familiar to students of elementary calculus, at least for problems in one and two variables. Otherwise, the classical constrained optimization problems involved equality constraints and free variables. In these cases the v's are free and the u's artificial. The v's are then known as *Lagrange multipliers.* This term has now been extended to describe the corresponding variables that appear in problems with inequality constraints.

Let us consider a rather simple maximization problem. We want to maximize

$$f = x_1 x_2 \tag{12.15}$$

on the square with corners at $(1,0)$, $(0,1)$, $(-1,0)$, and $(0,-1)$. The constraints that describe this region are

$$\begin{aligned} x_1 + x_2 - 1 &= -y_1 \\ -x_1 - x_2 - 1 &= -y_2 \\ x_1 - x_2 - 1 &= -y_3 \\ -x_1 + x_2 - 1 &= -y_4 \end{aligned} \tag{12.16}$$

where the y's are canonical. Tableau 12.17 is the linearized tableau for this problem.

(12.17)

	dx_1	dx_2	-1	
v_1	1	1	0	$= -dy_1$
v_2	-1	-1	0	$= -dy_2$
v_3	1	-1	0	$= -dy_3$
v_4	-1	1	0	$= -dy_4$
-1	x_2	x_1	0	$= df$
	$= u_1$	$= u_2$	$= g$	

In the original problem the x's are free and the y's are nonnegative. Thus, the dx's are free and the u's are artificial. To determine the feasibility specifications beyond this point requires separating nine cases: the interior of the square, the four edges, and the four corners.

In the interior the y's are positive. Thus the v's are artificial. This gives the solution $x_1 = x_2 = 0$. This point satisfies the Karush–Kuhn–Tucker conditions, but it is a saddle point.

In the interior of the edge where $y_1 = 0$ the other y's are positive. Deleting rows of artificial variables, tableau 12.17 reduces to

v_1	1	1	0
-1	x_2	x_1	0
	$= 0$	$= 0$	$= 0$

(12.18)

The condition $x_1 + x_2 - 1 = y_1 = 0$ must also be satisfied and v_1 must be nonnegative. It is readily verified that the solution $v_1 = x_1 = x_2 = 1/2$ is feasible. This point satisfies the Karush–Kuhn–Tucker conditions and it is a point at which a local maximum occurs.

In the interior of the edge where $y_3 = 0$ the other y's are positive. Deleting rows of artificial variables, tableau 12.17 reduces to

v_3	1	-1	0
-1	x_2	x_1	0
	$= 0$	$= 0$	$= 0$

(12.19)

The condition $x_1 - x_2 - 1 = -y_3 = 0$ must also be satisfied and v_3 must be nonnegative. The only solution is $v_3 = x_2 = -x_1 = -1/2$, which is not feasible. Since the Karush–Kuhn–Tucker conditions cannot be satisfied on this edge, there is no local maximum in the interior of the edge.

We leave the other two edges to the reader to analyze.

At the corner where $y_1 = y_3 = 0$ the other y's are positive. We also know that $x_1 = 1$ and $x_2 = 0$. The tableau in this case reduces to

v_1	1	1	0
v_3	1	-1	0
-1	0	1	0
	$= 0$	$= 0$	$= 0$

(12.20)

Both v_1 and v_3 must be nonnegative. There is no feasible solution for tableau 12.20. This point does not satisfy the Karush–Kuhn–Tucker conditions. We leave the remaining three corners to the reader to analyze.

To find the Karush–Kuhn–Tucker conditions for a minimization problem requires some adjustments. We will suggest two methods here and illustrate both in the following sections.

Suppose we wish to minimize $g = g(v)$ subject to constraints expressed in the nonlinear equations $a(v) - c = u$, where u and v are to satisfy given feasibility requirements.

We can maximize $-g = -g(v)$ to bring the problem into the form described in this section. If any of the u_j are canonical, it is best to express the constraints in the form $-a(v) + c = -u$ in order to bring the problem that contains the Karush–Kuhn–Tucker conditions into the form of tableau 12.11.

An alternative is to leave the problem in minimization form and use a tableau analogous to 12.11 in the form

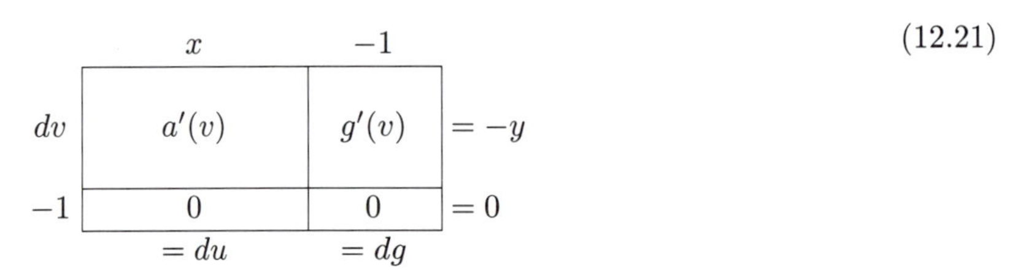

Here, x and y are required to satisfy feasibility specifications derived from u and v. If (u, v) is a feasible point at which a minimum for g occurs, then the Karush–Kuhn–Tucker theorem says there is a feasible solution to the row equations in tableau 12.21 in which $ux = 0$ and $vy = 0$. Any (u, v) for which such a solution exists satisfies the Karush–Kuhn–Tucker conditions.

Historical Note

In 1950 Harold W. Kuhn and Albert W. Tucker published a paper, "Nonlinear Programming," that contained the Kuhn–Tucker theorem. This was barely two years after Dantzig had published the simplex algorithm, and in the context of the interest that had been kindled in linear programming in general, the Kuhn–Tucker paper generated further interest in nonlinear programming.

In subsequent years the literature was searched for similar results that might have been published earlier. Several papers that contained similar results were found. All had been motivated by quite different sorts of problems, and all stated results that were somewhat different.

Almost 25 years later it was discovered that William Karush had written a master's thesis at the University of Chicago in 1939 that contained results fully equivalent to the Kuhn–Tucker theorem. Karush's thesis was written in the context of the calculus of variations and was regarded by him and his thesis advisor as a preliminary to a doctoral dissertation in that subject. For some reason he never published the results in his master's thesis.

This sort of thing has happened before, not only in mathematics but in virtually every scholarly discipline. It has often resulted in bitter and acrimonious cross and counter claims for priority. Even though Karush certainly recognized that ideas in his master's thesis had become an important part of nonlinear programming, he never raised his voice to make any claim. However, the existence of the master's thesis was rumored about and finally the thesis was dug out of the archives and examined carefully.

Harold W. Kuhn presented a talk on the history of nonlinear programming at a meeting of the Society for Industrial and Applied Mathematics in 1975 in which he described all the similar results that were known to exist and particularly emphasized the Karush thesis. The material in that talk is available in an issue of the *AMS-SIAM (American Mathematical Society, Society for Industrial and Applied Mathematics) Proceedings*, Volume IX, 1976. The title of that paper is "Nonlinear Programming: A Historical Perspective," pp 1–26. In effect, one can now consider Karush's thesis as published and available within Kuhn's paper.

Both Kuhn and Tucker have referred to the theorem as the "Karush–Kuhn–Tucker Theorem" in recent years. We are following that suggestion in this book. William Karush certainly deserves recognition for his achievement and his contribution should not be slighted simply because his work was done before its importance was evident.

12.2 The Constraint Qualification

When we differentiate a function $y = f(x)$ and obtain the relation $dy = f'(x)\,dx$, we obtain a linear relation between the "local" variables dx and dy at a point (x, y). Whether this linear relation has any significance relative to the original functional relation depends on how "well-behaved" the function is. In particular, differentiability alone means that the function is continuous. Additionally, if the derivative is continuous in a neighborhood of the point, the curve has a continuously turning tangent and the curve is said to be "smooth." In this case the tangent line approximates the curve in the neighborhood of the point.

For the nonlinear programs discussed in Section 12.1, the set of row equations $a'(x)\,dx = -dy$ is an analogous local linear approximation (generally a tangent plane or some higher-dimensional linear surface instead of a tangent line) of the nonlinear functions. The question that we raise in this section is, how "good" is this linear approximation? In the context of this chapter, "good" means that the Karush–Kuhn–Tucker theorem is valid. We are not looking to see how bad things can get and still support the Karush–Kuhn–Tucker theorem; we are looking for a reasonably general sufficient condition. The condition we are going to describe is the constraint qualification.

For a point in the interior of the feasible set a sufficiently small neighborhood of the point does not intersect any of the bounding surfaces (or hypersurfaces). That is, all variables are unconstrained in the neighborhood of an interior point. This is the situation in a classical optimization problem in several variables, and we have nothing new to say here. We are more interested in what can occur in the neighborhood of points on a boundary. Before trying to describe the situation in all generality, let us look at some special cases, including some situations in which things work out well and some in which there are problems.

Consider the problem of finding the maximum value of the function $f = x_1 + x_2$ inside the circle $x_1^2 + x_2^2 = 1$ and outside the circle $x_1^2 + (x_2 - 1)^2 = 1$ for which x_1 and x_2 are nonnegative. The feasible set for this problem is shown in Figure 12.1. It is not difficult to see that the maximum value of f occurs at the top cusp of the feasible set at the point $(\sqrt{3}/2, 1/2)$, labeled A.

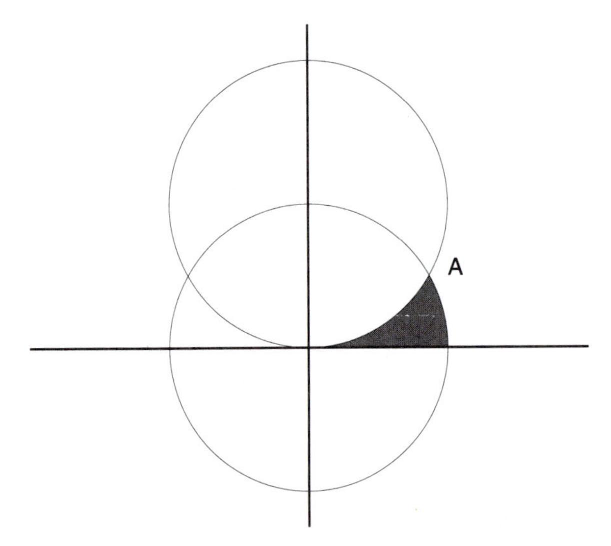

Figure 12.1: The feasible set is shaded.

The canonical constraints for this problem are

$$\begin{aligned} x_1^2 + x_2^2 - 1 &= -y_1 \\ -x_1^2 - (x_2 - 1)^2 + 1 &= -y_2 \end{aligned} \tag{12.22}$$

where all variables are nonnegative. The linearized problem is represented by the tableau 12.23.

(12.23)

	dx_1	dx_2	-1	
v_1	$2x_1$	$2x_2$	0	$= -dy_1$
v_2	$-2x_1$	$-2(x_2 - 1)$	0	$= -dy_2$
-1	1	1	0	$= df$
	$= u_1$	$= u_2$	$= g$	

At the point where we know the maximum occurs, this tableau becomes

(12.24)

	dx_1	dx_2	-1	
v_1	$\sqrt{3}$	1	0	$= -dy_1$
v_2	$-\sqrt{3}$	1	0	$= -dy_2$
-1	1	1	0	$= df$
	$= u_1$	$= u_2$	$= g$	

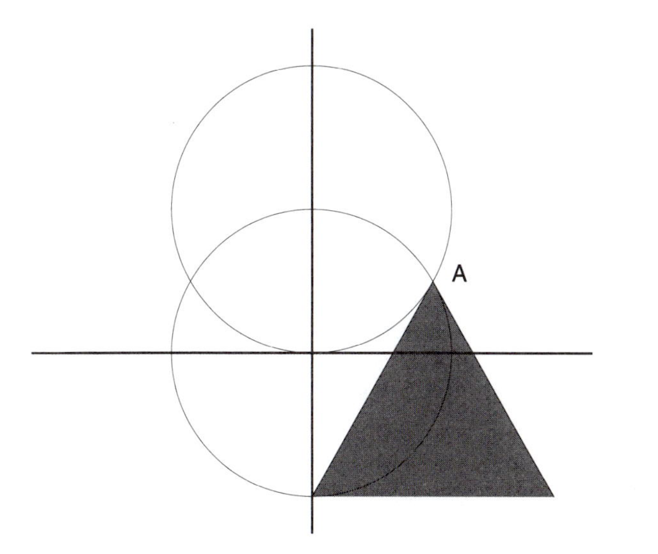

Figure 12.2: Well constrained feasible variations.

Notice that dx_1 and dx_2 are free and dy_1 and dy_2 are canonical. The feasible set for this linear program is the shaded region in Figure 12.2. The feasible region extends infinitely downward. The feasible region is a *cone.* That is, A is the vertex of the cone, and for any point in the feasible set, the entire half-line from the vertex through the point is also in the feasible set. This cone lies between the line where $dy_1 = 0$, which is tangent to the curve $y_1 = 0$, and the line where $dy_2 = 0$, which is tangent to the curve

$y_2 = 0$.

The important point here is the relation between the feasible set for the constraints 12.22 and the feasible set for the tableau 12.24 in the neighborhood of the point A. A curve in n dimensions is said to be smooth if there is a well defined tangent at each point and the direction of the tangent line varies continuously as a function of the position of the tangent point. Consider a smooth curve in Figure 12.1 drawn from the point A through points in the feasible set. For each such curve there will be a tangent line to the curve at the point A. Actually, we are interested in the half-line from the point A in the direction of the tangent line at that point and extending in the direction of the portion of the curve within the feasible set.

This set of tangent lines will also form a cone. We call the cone obtained from these tangent lines the *geometric cone* and the cone obtained from the linearized tableau 12.23 the *analytic cone.* In this example, the geometric cone and the analytic cone are the same. This is what we mean when we say that the linearized problem is a reasonable representation of the nonlinear problem in the neighborhood of the point. We will show later that the geometric cone is always a subset of the analytic cone.

Let us look at an example where the geometric cone and the analytic cone are not identical. The feasible set described by the canonical conditions 12.22 can also be described by the equations

$$\begin{aligned}(x_1^2 + x_2^2 - 1)(-x_1^2 - (x_2 - 1)^2 + 1) &= -y_1 \\ x_2 - 0.5 &= -y_2\end{aligned} \tag{12.25}$$

where all variables are nonnegative. The feasible set for this problem is also the region shown in Figure 12.1. The partial derivatives of the first expression in 12.25 are somewhat complicated, but every term contains one or the other of the terms in parentheses as a factor. That is, the partial derivatives vanish at the point A. The linearized problem in the neighborhood of the point A is

(12.26)

	dx_1	dx_2	-1	
v_1	0	0	0	$= -dy_1$
v_2	0	1	0	$= -dy_2$
-1	1	1	0	$= df$
	$= u_1$	$= u_2$	$= g$	

The feasible set for this problem is illustrated in Figure 12.3. It is a half-plane below the line $y_2 = 0.5$.

The nonlinear problem, whether the feasible set is defined by 12.22 or 12.25, has a maximum at the point A. In fact, any objective function for

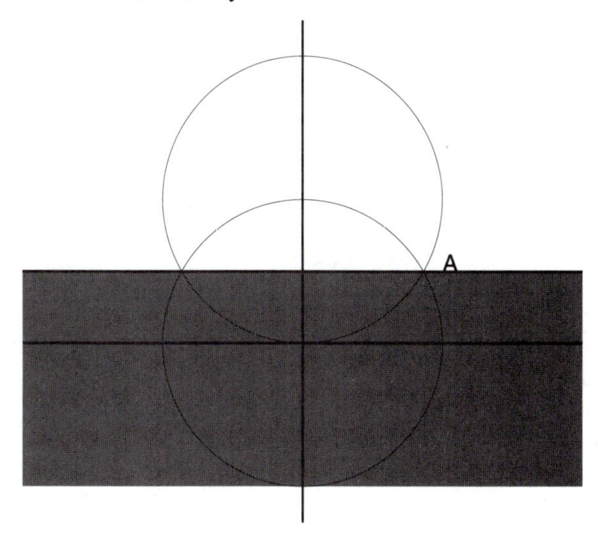

Figure 12.3: Ill constrained feasible variations.

which the gradient at A points to the exterior of the geometric cone will have a local maximum at that point.

The linearized problem represented by tableau 12.24 also has a maximum at A. Its objective variable has the same gradient at A and the same feasible cone at A. On the other hand the objective variable for the linearized problem represented by tableau 12.26 is unbounded. Thus, the constraints represented by the column equations of tableau 12.24 are feasible and the constraints represented by the column equations of tableau 12.26 are infeasible.

The linearized problem of tableau 12.26 does not correctly represent the nonlinear problem in the neighborhood of A. The rest of this section is devoted to describing conditions that will assure that the representation is satisfactory and that the Karush–Kuhn–Tucker theorem is true.

The point at which the conditions are to be imposed must be approachable within the feasible region from a range of directions equivalent to the range of allowable values for the variables in the maximum program of tableau 12.11. The following definition describes a condition that will ensure the correctness of this representation.

Consider a feasible point (x, y) and a curve in the feasible set with x at its endpoint. We assume the curve is parameterized by a parameter t so that $x(t)$ is a continuous function of t and $x(0) = x$. The assumption that $a_i(x)$ is differentiable at x is equivalent to the statement that the partial derivatives exists and

$$\frac{\partial a_i}{\partial x_1} dx_1 + \frac{\partial a_i}{\partial x_2} dx_2 + \cdots + \frac{\partial a_i}{\partial x_n} dx_n = -dy_i \tag{12.27}$$

That is, (dx, dy) satisfy the linearized constraints. It is easily seen that they also satisfy the feasibility specifications. Thus, the feasible geometric cone is a subset of the feasible analytic cone.

The constraint qualification is merely an assumption that the converse is true—that the analytic cone is a subset of the geometric cone. This amounts to assuming that every line from the vertex of the analytic cone within the analytic cone can be realized as the tangent line of a suitable curve approaching the vertex from within the feasible set (of the original nonlinear program).

The constraint qualification is often worded in a slightly different form, in terms of a sequence of points approaching x instead of as a curve with x at its endpoint.

Definition: A nonlinear program satisfies the *constraint qualification* at a feasible solution (x, y) if for every feasible (dx, dy) with $|dx| = 1$ for the linearized problem there exists a sequence of feasible solutions (x^k, y^k) such that

$$\lim_{x \to \infty} x^k = x \tag{12.28}$$

and

$$\lim_{k \to \infty} (x^k - x)/|x^k - x| = dx \tag{12.29}$$

Condition 12.28 says that x is accessible as a limit of feasible points. In condition 12.29, $|x^k - x|$ is the distance between the point x^k and the point x. Condition 12.29 says that the ratios dx_i/dx_j are the limits of the ratios $(x_i^k - x)/(x_j^k - x)$. That is, the direction represented by dx is a limit of directions from x to feasible points.

We are now in a position to prove the Karush–Kuhn–Tucker theorem. In view of the discussion of the Karush–Kuhn–Tucker theorem in the previous section, it is sufficient to prove the following lemma.

Lemma 12.66 *Suppose that (x, y) is a local maximum for the nonlinear program and that the constraint qualification is satisfied at that point. Then $df \geq 0$ for all feasible variations from (x, y).*

Proof. Since f is differentiable at (x, y), for any sequence of points $x^1, x^2, \ldots, x^k, \ldots,$ approaching x in the limit,

$$f(x^k) - f(x) = f'(x)(x^k - x) + d_k|x^k - x| \tag{12.30}$$

where

$$\lim_{k \to \infty} d_k = 0 \tag{12.31}$$

Since (x, y) is a local maximum, if we take the points x^k in the sequence to be feasible points,

$$f'(x)(x^k - x) + d_k|x^k - x| \leq 0 \tag{12.32}$$

for k large. Taking limits we have

$$df = f'(x)dx \leq 0 \tag{12.33}$$

□

The maximum program for tableau 12.11 is always feasible since $dx = 0$ and $dy = 0$ is a feasible solution. Since df is bounded, it has a maximum value. Therefore, the dual program is feasible. That is, the conclusion of the Karush–Kuhn–Tucker theorem follows.

12.3 Least Squares

An old and important nonlinear optimization problem arises in the method of least squares, devised by K. F. Gauss at about 1800 A.D. The problem to be solved is to find a "best" fit, or functional description, for a set of empirical data. For example, suppose at a sequence of times $t_1, t_2, \ldots, t_n$ the measurements $y_1, y_2, \ldots, y_n$ are made. Figure 12.4 shows a representative set of data plotted on a t, y coordinate system.

Eight points are shown in Figure 12.4, and it is possible to find a polynomial graph of degree seven, or less, that will pass through all these points. But that is usually not what the situation demands. There is usually some theoretical reason to believe that the data should obey a simpler relationship and that the scattering of the data is caused by unavoidable errors in the measurements. We might, in the example of Figure 12.4, expect y to be a linear or quadratic function of t. But since no such function can pass through all the points plotted, which function should we select, from among all those possible, to represent the data?

A family of possible functions defined by the model will usually be described by a set of parameters. For example, the set of quadratic polynomial functions of t,

$$y = at^2 + bt + c \tag{12.34}$$

has three parameters, the coefficients a, b, and c. For the method of least squares it does not matter whether the function is linear or not, but we do

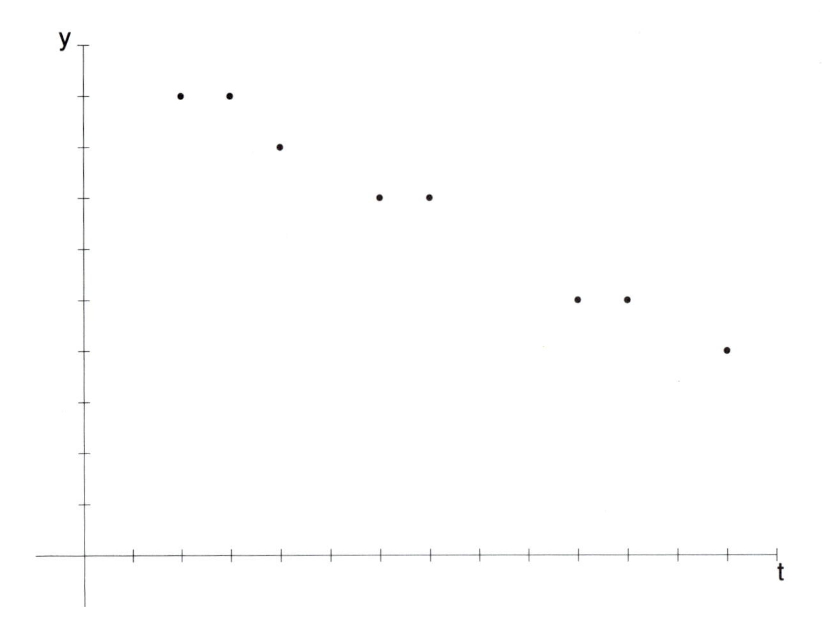

Figure 12.4: Data points for a least squares problem.

require that the parameters appear linearly in the functional form, as they do in 12.34.

For any particular function in the family, at time t_i, $at_i^2 + bt_i + c$ is the theoretical value of y. If e_i is the observed value then

$$r_i = e_i - (at_i^2 + bt_i + c) \tag{12.35}$$

is the deviation, or *residual*, between the observed and theoretical values.

The criterion used by the method of least squares is to choose the parameters so that the sum of the squares of the residuals is minimal. In most applications of least squares there are reasons inherent in the model to justify using this criterion but in the abstract, divorced from a rational model, it is impossible to give an argument to justify any particular criterion as yielding a best fit.

Consider the systems of linear equations represented by tableau 12.36, where D is a $k \times n$ matrix.

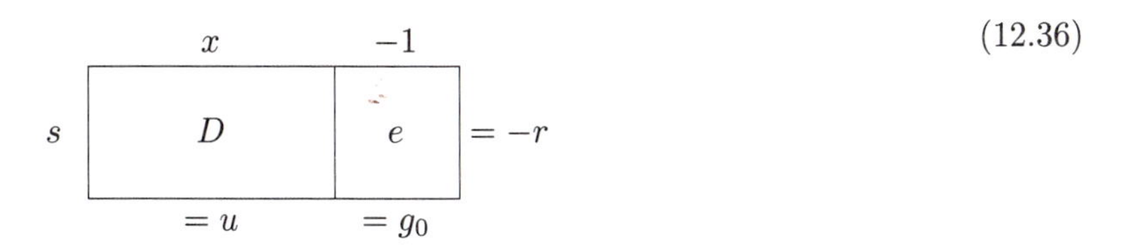

(12.36)

In the context of the discussion above, $x = (x_1, x_2, \ldots, x_n)$ are the parameters to be determined, $e = (e_1, e_2, \ldots, e_k)$ are the observed values, and $r = (r_1, r_2, \ldots, r_k)$ are the residuals. The *least squares problem* represented by tableau 12.36 is to find x that satisfies

$$Dx - e = -r \tag{12.37}$$

and minimizes $r_1^2 + \cdots + r_k^2$. But, for the purpose of this chapter, we maximize

$$f = -\frac{1}{2}r^T r = -\frac{1}{2}(r_1^2 + \cdots + r_k^2) \tag{12.38}$$

From the duality equation for tableau 12.36 we have

$$g_0 = ux + sr \tag{12.39}$$

Adding $\frac{1}{2}ss^T + \frac{1}{2}r^T r$ to both sides, we see that

$$(g_0 + \frac{1}{2}ss^T) + \frac{1}{2}r^T r = ux + \frac{1}{2}ss^T + sr + \frac{1}{2}r^T r = ux + \frac{1}{2}(s^T + r)^T(s^T + r) \tag{12.40}$$

On the right side $(s^T + r)^T(s^T + r)$ is nonnegative since it is a sum of squares. To force ux to be nonnegative we require that u and x satisfy dual feasibility specifications. Since x is free, this means we require $u = 0$. Thus,

$$-\frac{1}{2}r^T r \leq g_0 + \frac{1}{2}ss^T \tag{12.41}$$

for all feasible solutions for the row system and column system of equations in tableau 12.36.

In the spirit of all the problems considered previously in this book we have

$$\max(-\frac{1}{2}r^T r) \leq \min(g_0 + \frac{1}{2}ss^T) \tag{12.42}$$

The problem dual to the least squares problem is to find s to satisfy

$$sD = 0 \tag{12.43}$$

and minimize

$$g = g_0 + \frac{1}{2} ss^T \tag{12.44}$$

Notice that x, r, and s are free variables.

Since $u = 0$, a sufficient condition from 12.40 for equality to hold in 12.41 is

$$s^T + r = 0 \tag{12.45}$$

Substitution of 12.45 in 12.43 gives $(Dx - e)^T D = 0$, or

$$D^T Dx = D^T e \tag{12.46}$$

The linear equations in 12.46 are known as the *normal equations* for the least squares problem.

In a typical application of the method of least squares, the number of parameters (represented by x) is relatively small and the number of observations (represented by e) is relatively large. To require the residuals to be zero would result in the system of equations $Dx - e = 0$. This is a system with a large number of equations in a small number of unknowns, and it usually cannot be solved. The interesting and important fact is that the normal equations $D^T Dx - D^T e = 0$ can always be solved.

We can use Theorem 5.42 from Chapter 5 to prove that the normal equations can be solved. Consider the tableau 12.47

(12.47)

	x	-1	
v	$D^T D$	$D^T e$	$= -y$
	$= u$	$= g$	

where y and u are artificial. For any solution of the column equations, $vD^T D = 0$, we have $vD^T Dv^T = 0$, or $(vD^T)(vD^T)^T = 0$. Hence, $vD^T = 0$ for all feasible solutions and $g = vD^T e = 0$ for all feasible solutions. In 12.47 the column system is feasible (take $v = 0$) and the objective variable is bounded (always zero). Hence, the row system is feasible. That is, the normal equations can be solved.

Let x' be a solution of the normal equations. Set

$$r' = Dx' - e \tag{12.48}$$

$$s' = -r'^T \tag{12.49}$$

and

$$g'_0 = s'e \tag{12.50}$$

Then $D^T s'^T = -D^T(Dx' - e) = -D^T Dx' + D^T e = 0$. That is,

$$s'D = 0 \tag{12.51}$$

We have shown that (x', r') is a feasible solution for the row equations in tableau 12.36, and s' is a feasible solution for the column equations. Since 12.49 is the sufficient condition for optimality, both solutions are optimal and x' is a solution for the least squares problem.

The normal equations constitute a linear complementarity problem associated with the least squares problem. Later in this chapter we will discuss some techniques for solving some special linear complementarity problems.

Gauss devised the Gaussian elimination method for solving systems of linear equations to solve the normal equations. However, there is an interesting conceptual difference between the role of the normal equations here and their role in Gauss's theory. We have obtained the normal equations as sufficient conditions for optimality, while Gauss obtained them as necessary conditions for optimality.

To obtain the normal equations as necessary conditions, consider the objective function

$$\begin{aligned} f &= -\frac{1}{2}r^T r = -\frac{1}{2}\sum_{i=1}^{m} r_i^2 \\ &= -\frac{1}{2}\sum_{i=1}^{m}(\sum_{j=1}^{n} d_{ij}x_j - e_j)^2 \end{aligned} \tag{12.52}$$

Then

$$\begin{aligned} \frac{\partial f}{\partial x_k} &= -\frac{1}{2}\sum_{i=1}^{m} 2(\sum_{j=1}^{n} d_{ij}x_j - e_j)d_{ik} \\ &= -\sum_{j=1}^{n}(\sum_{i=1}^{m} d_{ik}d_{ij})x_j + \sum_{i=1}^{m} d_{ik}e_i \end{aligned} \tag{12.53}$$

or

$$\frac{\partial f}{\partial x} = -D^T Dx + D^T e \tag{12.54}$$

For the least squares problem, we have an objective function, 12.38, and no constraints. Thus, the tableau for the Karush–Kuhn–Tucker conditions takes the form

$$\begin{array}{c|c|l} & dx & \\ \hline -1 & \dfrac{\partial f}{\partial x} & = df \\ \hline & = 0 & \end{array} \tag{12.55}$$

That is, the Karush–Kuhn–Tucker conditions amount to setting 12.54 equal to zero, which is equivalent to the normal equations. In our discussion we established the normal equations as sufficient conditions for optimality. Now we see that the Karush–Kuhnl-Tucker conditions establish the normal equations as necessary conditions.

Let us work a simple problem with data used for Figure 12.4. The data are given in the following table.

t	y
2	9
3	9
4	8
6	7
7	7
10	5
11	5
13	4

Let us assume that we wish to find the equation of a straight line that best fits this set of data. That is, our model is linear functions of the form

$$y = at + b$$

The parameters of this family of functions are the coefficients a and b. These are the variables of the least squares problem. For this problem, the tableau corresponding to 12.36 is

a	b	-1	
2	1	9	$= -r_1$
3	1	9	$= -r_2$
4	1	8	$= -r_3$
6	1	7	$= -r_4$
7	1	7	$= -r_5$
10	1	5	$= -r_6$
11	1	5	$= -r_7$
13	1	4	$= -r_8$

The tableau representing the normal equations is

a	b	-1	
504	56	325	$= 0$
56	8	54	$= 0$

The system of equations is easily solved, yielding

$$a = -0.47$$

$$b = 10.06$$

Figure 12.5 shows the graph of this straight line against the background of the points of the data.

12.4 Least Distance

Consider tableau 12.56, where D is a $k \times n$ matrix.

$$\begin{array}{r|c|l} & x & \\ \hline s & D & = -r \\ \hline -1 & c & = f_0 \\ \hline & = u & \end{array} \qquad (12.56)$$

We seek to minimize

$$g = \frac{1}{2} s s^T \qquad (12.57)$$

for s satisfying

$$sD - c = u \qquad (12.58)$$

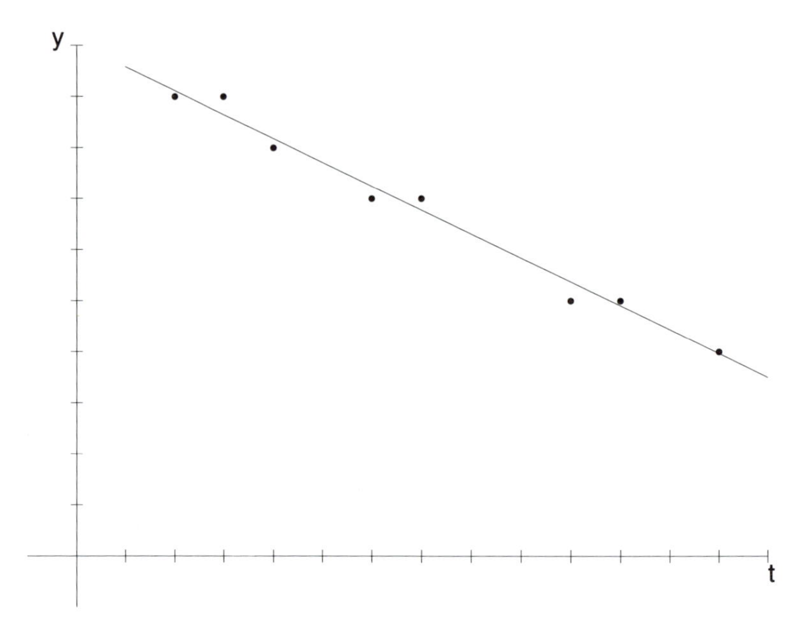

Figure 12.5: Linear approximation given by least squares.

We take s to be free, but u can be required to satisfy general feasibility specifications (canonical, artificial, or mixed).

If the feasible set is nonempty, the objective function 12.57 is half the square of the distance from a point in the feasible set to the origin. We seek the point closest to the origin and its minimum distance. Thus, we call this a *least distance problem.*

From the duality equation we have

$$sr + f_0 + ux = 0$$

or

$$-f_0 = sr + ux \tag{12.59}$$

Add $\frac{1}{2}ss^T + \frac{1}{2}r^Tr$ to both sides to "complete the square." Then

$$\frac{1}{2}ss^T - (f_0 - \frac{1}{2}r^Tr) = \frac{1}{2}ss^T + ux + sr + \frac{1}{2}r^Tr = ux + \frac{1}{2}(s^T + r)^T(s^T + r) \tag{12.60}$$

On the right side $(s^T + r)^T(s^T + r)$, as a sum of squares, is nonnegative. In order to have ux nonnegative, we require that u and x satisfy dual feasibility specifications (determined by those given initially for u). We take r to be free. Then

$$f_0 - \frac{1}{2}r^T r \leq \frac{1}{2}ss^T \tag{12.61}$$

for all pairs of feasible solutions.

If we let

$$f = f_0 - \frac{1}{2}r^T r \tag{12.62}$$

and

$$g = \frac{1}{2}ss^T \tag{12.63}$$

then we have dual programming problems.

I. Maximize f for feasible solutions of $Dx = -r$ and $cx = f_0$.

II. Minimize g for feasible solutions of $sD - c = u$.

In these problems, s and r are free, and u and x satisfy dual feasibility specifications.

Since $f \leq g$ under these conditions,

$$\max f \leq \min g \tag{12.64}$$

and equality can hold if and only if

$$ux = 0 \tag{12.65}$$

and

$$s^T + r = 0 \tag{12.66}$$

Thus, 12.65 and 12.66 are sufficient conditions for optimality.

As an example, consider the least distance problem associated with tableau 12.67.

(12.67)

	x_1	x_2	x_3	
s_1	-1	1	1	$= -r_1$
s_2	1	0	-1	$= -r_2$
-1	-2	-3	-4	$= f_0$
	$= u_1$	$= u_2$	$= u_3$	

Figure 12.6 shows an s_1, s_2 coordinate system with the lines $u_1 = 0$, $u_2 = 0$, and $u_3 = 0$.

For u_1, u_2, u_3 canonical, the feasible set is the shaded region including the origin. Thus, the feasible point closest to the origin is the origin itself, and the least distance is 0.

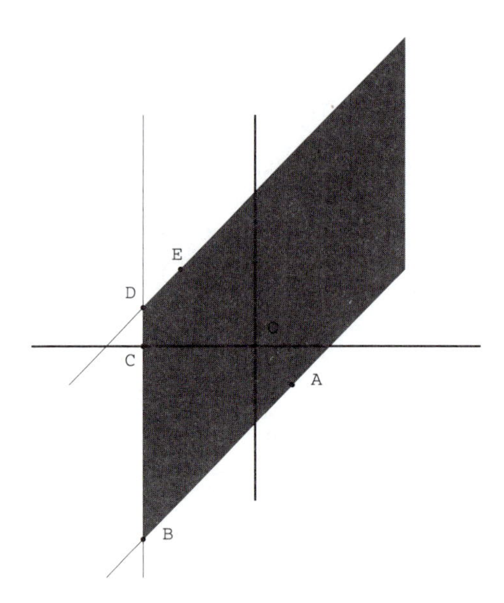

Figure 12.6: The feasible set for a least distance problem.

Other feasibility specifications are obtained by taking some of the u_i to be artificial. The feasible sets are then the various edges and vertices of the shaded region or, for u_1 and u_3 both artificial, the empty set. The several possibilities are shown in Table 12.1.

As we have for linear programs, we have used parentheses to denote variables in the dual problem, in this case the maximum problem. In Table 12.1, where the value of a variable is shown, its dual variable is taken to be zero. The points where the optimal values are taken on are easy to find in Figure 12.6. Where the feasible set is an edge the point nearest the origin is the foot of the perpendicular from the origin. Where the feasible set is a vertex, there is only one point in the set.

The sufficient conditions 12.65 and 12.66 can be formulated in a single problem. Use 12.66 to write $Dx = -r$ in the form

$$s = (Dx)^T \tag{12.68}$$

and substitute in 12.58 to obtain

$$x^T D^T D - c = u \tag{12.69}$$

Table 12.1: Solutions for the Least Distance Example.

	Feasibility						Solution					
	u_1	u_2	u_3	x_1	x_2	x_3	(x_1) u_1	(x_2) u_2	(x_3) u_3	$-r_1$ $= s_1$	$-r_2$ $= s_2$	f $= g$
$O:$	≥ 0	≥ 0	≥ 0	≥ 0	≥ 0	≥ 0	2	3	4	0	0	0
$A:$	$= 0$	≥ 0	≥ 0	free	≥ 0	≥ 0	(−1)	4	6	1	−1	1
$B:$	$= 0$	$= 0$	≥ 0	free	free	≥ 0	(−5)	(−8)	6	−3	−5	17
$C:$	≥ 0	$= 0$	≥ 0	≥ 0	free	≥ 0	5	(−3)	1	−3	0	4.5
$D:$	≥ 0	$= 0$	$= 0$	≥ 0	free	free	6	(−2)	(−1)	−3	1	5
$E:$	≥ 0	≥ 0	$= 0$	≥ 0	≥ 0	free	6	1	(−2)	−2	2	4

Then equation 12.69 can be represented in tableau 12.70.

$$\begin{array}{c|c|} \cline{2-2} x^T & D^T D \\ \cline{2-2} -1 & c \\ \cline{2-2} \multicolumn{1}{c}{} & \multicolumn{1}{c}{= u} \end{array} \tag{12.70}$$

We impose the conditions that u and x satisfy dual feasibility specifications and that they be complementary. That is, each $u_j x_j = 0$.

The problem in tableau 12.70 with these conditions is called a *linear complementarity problem* for the least distance problem. Suppose we can solve this problem. Then set $r = -Dx$ and $s = -r^T$ to obtain solutions for the least distance problem and its dual. When the feasible set for the least distance problem is nonempty the linear complementarity problem can, in fact, be solved.

Let us show that the linear complementarity problem described here is identical to the equations for the Karush–Kuhn–Tucker conditions. Our least distance problem is to find the minimum of

$$g = \frac{1}{2} s s^T \tag{12.71}$$

subject to the constraints

$$sD - c = u \tag{12.72}$$

where s is free and u satisfies given feasibility specifications. Since $dsD = du$ and $dg = ds\, s^T$, the appropriate tableau is

(12.73)

	x	-1	
ds	D	s^T	$= 0$
	$= du$	$= dg$	

with $ux = 0$ required.

The row equations in 12.73 take the form

$$Dx - s^T = 0$$
$$x^T D^T - s = 0 \tag{12.74}$$

and, multiplying by D on the right, we get

$$x^T D^T D - sD = x^T D^T D - c - u = 0 \tag{12.75}$$

With $ux = 0$, this is equivalent to the linear complementarity problem given in tableau 12.70.

We derived the linear complementarity problem for the least distance problem as a sufficient condition. The Karush–Kuhn–Tucker theorem shows that this is also a necessary condition.

12.5 Hybrid Problems

Consider the tableau 12.76, where D and A are $k \times n$ and $m \times n$ matrices.

(12.76)

	x	-1	
s	D	e	$= -r$
v	A	b	$= -y$
-1	c	0	$= f_0$
	$= u$	$= g_0$	

From the duality equation we have

$$g_0 - f_0 = ux + sr + vy \tag{12.77}$$

We complete the square by adding $\frac{1}{2}ss^T + \frac{1}{2}r^T r$.

$$(g_0 + \frac{1}{2}ss^T) - (f_0 - \frac{1}{2}r^T r) = ux + vy + \frac{1}{2}ss^T + sr + \frac{1}{2}r^T r$$

$$= ux + vy + \frac{1}{2}(s^T + r)^T(s^T + r) \quad (12.78)$$

On the right side $(s^T + r)^T(s^T + r)$ is a sum of squares. To make the other terms nonnegative we require that the pairs u, x and v, y satisfy dual feasibility specifications. The variables s and r are free. Then

$$f_0 - \frac{1}{2}r^T r \leq g_0 + \frac{1}{2}ss^T \quad (12.79)$$

for all pairs of feasible solutions. The dual pair of problems are,

I. Maximize

$$f = f_0 - \frac{1}{2}r^T r \quad (12.80)$$

subject to

$$Dx - e = -r \quad (12.81)$$

$$Ax - b = -y \quad (12.82)$$

and

$$cx = f_0 \quad (12.83)$$

II. Minimize

$$g = g_0 + \frac{1}{2}ss^T \quad (12.84)$$

subject to

$$sD + vA - c = u \quad (12.85)$$

and

$$se + vb = g_0 \quad (12.86)$$

The pairs u, x and v, y satisfy dual feasibility specifications. All other variables are free.

Note that if the rows containing the variables r and s are vacuous, the problems reduce to a pair of linear programs. On the other hand, if the rows containing the variables v, y, and f_0 are vacuous, the problems reduce to the least squares problem and its dual discussed in Section 12.3. Also, for v, y rows and g_0 column vacuous, the min program reduces to a least distance problem of Section 12.4. So we call the above *hybrid problems.*

The dual feasibility specifications and equation 12.78 imply

$$f \leq g \quad (12.87)$$

for all pairs of feasible solutions. Furthermore, we see that

$$s^T + r = 0 \tag{12.88}$$
$$ux = 0 \tag{12.89}$$

and

$$vy = 0 \tag{12.90}$$

are sufficient conditions for optimality for both problems.

Use condition 12.88 to replace s in 12.85 by the expression in 12.81. We obtain $(x^T D^T - e^T)D + vA - c = u$ which, by a slight rearrangement, becomes

$$x^T D^T D + vA - (e^T D + c) = u \tag{12.91}$$

Make a similar substitution for s in 12.86 to obtain

$$x^T D^T e + x^T c^T + vb = f_0 + g_0 + e^T e \tag{12.92}$$

With these factors in mind consider tableau 12.93.

$$\begin{array}{c|cc|c|l}
 & x & -v^T & -1 & \\
\hline
x^T & D^T D & -A^T & c^T + D^T e & = u^T \\
v & A & 0 & b & = -y \\
\hline
-1 & c + e^T D & -b^T & 0 & = f_0 + g_0 + e^T e \\
\hline
 & = u & = y^T & = f_0 + g_0 + e^T e &
\end{array} \tag{12.93}$$

Notice that the row system and the column system are identical. We can if we wish, therefore, ignore the variables associated with one of the systems.

Tableau 12.93 represents a linear complementarity problem. The variables x and u in dual pairs, and the variables y and v in dual pairs, must satisfy dual feasibility specifications. Thus, we seek feasible solutions to the equations

$$x^T D^T D + vA - c - e^T D = u \tag{12.94}$$

and

$$-x^T A^T + b^T = y^T \tag{12.95}$$

for which

$$ux = 0 \tag{12.96}$$

and

$$vy = 0 \tag{12.97}$$

Suppose the linear complementarity problem represented by tableau 12.93 has a solution. That is, we can find feasible solutions for the equations 12.94 through 12.97. Then define

$$\begin{aligned} r &= -Dx + e \\ s &= -r^T \\ g_0 &= se + vb \end{aligned}$$

and

$$f_0 = cx$$

Then

$$\begin{aligned} sD + vA - c &= -r^T D + va - c \\ &= (Dx - e)^T D + va - c \\ &= x^T D^T D + va - c - e^T D = u \end{aligned}$$

That is, we have feasible solutions for the dual hybrid problems. Conditions 12.96 and 12.97 imply optimality.

Consider the problem represented by tableau 12.98 as an example of a hybrid problem.

(12.98)

	x	-1	
s	2	-7	$= -r$
v	3	6	$= -y$
-1	4	0	$= f_0$
	$= u$	$= g_0$	

Since the D and A matrices are 1×1, this is as small an example as we can construct. There are four combinations of feasibility specifications that can be prescribed.

1. $x \geq 0,\ u \geq 0;\ y \geq 0,\ v \geq 0$

2. x free, $u = 0;\ y \geq 0,\ v \geq 0$

3. $x \geq 0,\ u \geq 0;\ y = 0,\ v$ free

4. x free, $u = 0;\ y = 0,\ v$ free

The tableau in the form of 12.93 representing the associated linear complementarity problem is

(12.99)

x	4	-3	-10
v	3	0	6
-1	-10	-6	0
	$= u$	$= y = f_0 + g_0 + e^T e$	

We shall postpone further consideration of this linear complementarity problem until Section 12.6. For the moment let us use geometry to solve the problems associated with tableau 12.98 and the four possible sets of feasibility specifications. The objective function 12.84 of the minimization problem II in this case is

$$g = g_0 + \frac{1}{2} s s^T = 6v - 7s + \frac{1}{2} s^2 \tag{12.100}$$

In the v, s plane, shown in Figure 12.7, the curves along which g has a constant value are parabolas. The minimum value of g occurs on the parabola farthest to the left that intersects the feasible set. The line shown in Figure 12.7 is the line $2s + 3v - 4 = u = 0$. For the feasibility conditions 1 above, the feasible set is the shaded area to the right of the s-axis and to the right of the line $u = 0$. The farthest left parabola touches this region at the point $(0, 7)$. The value of g at this point is -24.5.

For the feasibility specifications 2 above, the feasible set is the part of the line $u = 0$ to the right of the s-axis. This situation is illustrated in Figure 12.8. The farthest left parabola touches this region at the point $(0, 2)$. The value of g at this point is -12.

For the feasibility specifications 3 above, the feasible set is the area to the right of the line $u = 0$. For the feasibility specifications 4, the feasible set is the line $u = 0$. In both cases the farthest left parabola touches the feasible region at the point $(-6, 11)$. Both results are shown in Figure 12.9.

Again, let us show that the linear complementarity problem described above is equivalent to the Karush–Kuhn–Tucker conditions. Our problem is to maximize

$$f = cx - (Dx - e)^T (Dx - e) \tag{12.101}$$

subject to the constraints

$$Ax - b = -y \tag{12.102}$$

where x and y must satisfy specified feasibility conditions. Since

$$A\,dx = -\,dy \tag{12.103}$$

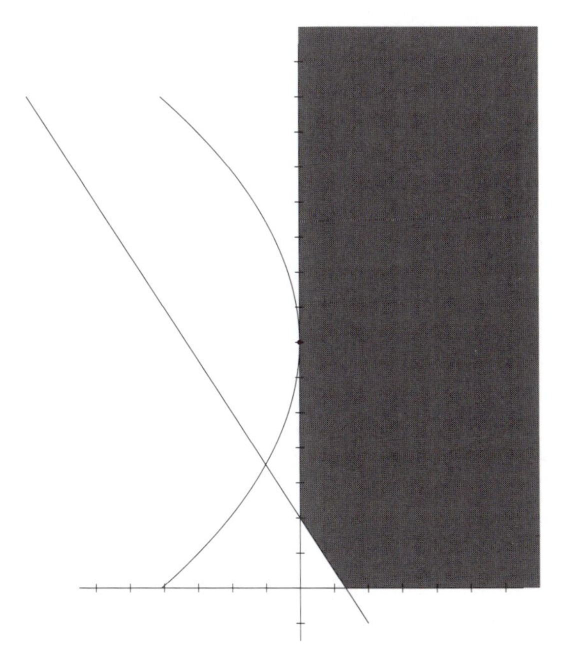

Figure 12.7: Feasibility Conditions 1.

and

$$df = c\,dx - (Dx - e)^T D\,dx = (x^T D^T D + e^T D + c)\,dx \tag{12.104}$$

the tableau that expresses the Karush–Kuhn–Tucker conditions takes the form

(12.105)

	dx	
v	A	$= -dy$
-1	$-x^T D^T D - e^T D + c$	$= df$
	$= u$	

where u and v must satisfy appropriate feasibility specifications and $ux = 0$ and $vy = 0$. These give conditions identical to the linear complementarity problem given above.

Our discussion established the complementarity problem as sufficient conditions for optimality. The Karush–Kuhn–Tucker conditions establish the same requirements as necessary conditions.

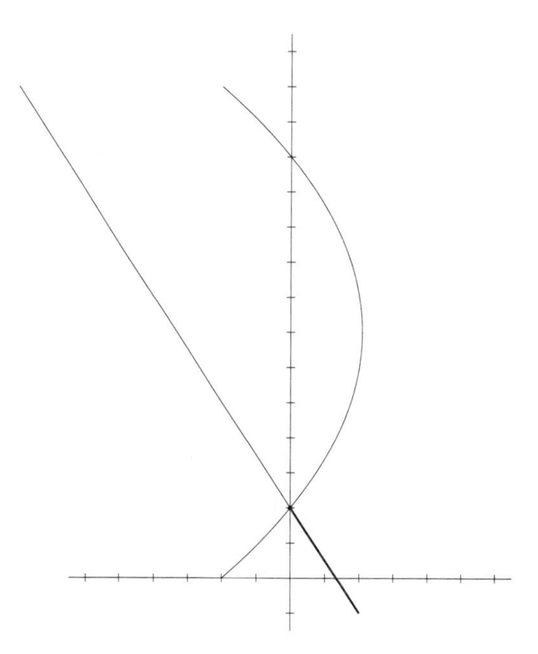

Figure 12.8: Feasibility Conditions 2.

12.6 Empirical Principal Pivoting

Consider a tableau in the form of tableau 12.106.

z	P
-1	q
	$= w$

(12.106)

We assume that the matrix P is square, $n \times n$, and that w and z satisfy dual feasibility specifications. Specifically, we assume that w_i and z_i are dual variables. The *linear complementarity problem* associated with tableau 12.106 is to find a feasible solution (w, z) for which

$$wz^T = 0 \tag{12.107}$$

The problem is *feasible* if the equations in tableau 12.106 have a feasible solution, whether or not that solution satisfies the complementarity conditions in 12.107. A solution that makes each term $w_i z_i = 0$ is said to be

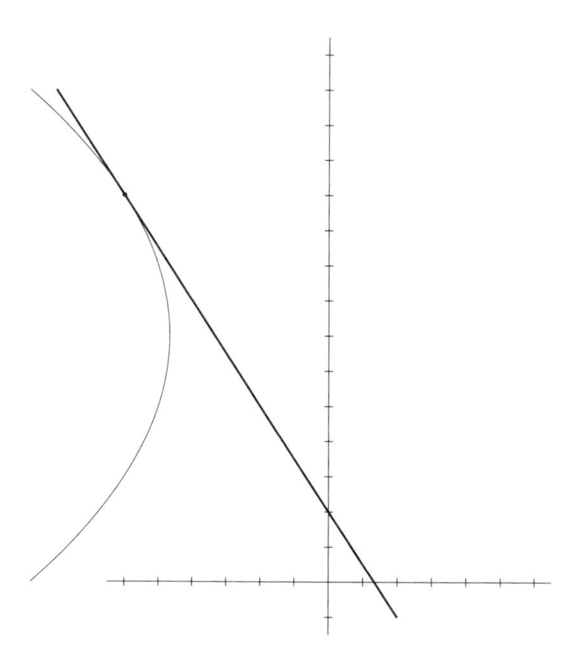

Figure 12.9: Feasibility Conditions 3 and 4.

complementary whether it is feasible or not. Our problem, then, is to find a complementary feasible solution.

A natural question is the following: Under what conditions does a feasible linear complementarity problem have a complementary feasible solution? It would be nice to have necessary and sufficient criteria formulated in simple terms that can easily be verified. It would also be nice to have a simple and efficient algorithm that would produce a solution when it is known to have one or, even better, an algorithm that would determine whether the problem has a solution and find one if one exists. Generally, the wider the class of problems for which a sufficient condition is known, the more difficult it is to establish the condition and the more difficult it is to provide an effective algorithm.

We will make no attempt to discuss linear complementarity problems in general. For the problems that arise from the programs of the previous and following sections we will discuss a method that will work for small problems.

It is easily seen from tableau 12.106 that a basic solution is complementary. We seek, therefore, to find equivalent tableaux for which the basic solutions are complementary and, among these, a tableau for which the basic solution is feasible. In order for a basic solution to be complementary it is sufficient that for every variable that is basic its dual variable is nonbasic, and conversely.

If we start with a tableau for which the basic variables and the nonbasic variables are dual and listed in the same order on the margins, this can be achieved by pivoting in such a way that we exchange a pair of dual variables with each pivot exchange. To do this we would select a pivot entry on the main diagonal, if a nonzero entry on the main diagonal is available. More generally, this can also be achieved by a block pivot exchange that interchanges a group of dual variables. A pivot exchange of this type is called a *principal pivot exchange.*

For each pair of dual variables there are two choices as to which is basic and which is nonbasic. Thus, there could be as many as 2^n different tableaux obtainable by principal pivot exchanges. For large n this is an unacceptably large number of tableaux to sort through to look for a solution. For small problems of the type offered in a textbook it is not unreasonable.

To solve the programs presented here we suggest that you pivot on the main diagonal of a column in which the entry in the q-row is positive or, if the entry on the main diagonal is not positive, that a principal pivot of order two be chosen where at least one of the two columns has a positive entry in the q-row.

In Section 12.4 we gave an example of a least distance problem represented by tableau 12.67, which we copy here as tableau 12.108.

(12.108)

	x_1	x_2	x_3	
s_1	-1	1	1	$= -r_1$
s_2	1	0	-1	$= -r_2$
-1	-2	-3	-4	$= f_0$
	$= u_1$	$= u_2$	$= u_3$	

The linear complementarity tableau corresponding to tableau 12.70 that is obtained from tableau 12.108 is the first tableau, labeled O, in Table 12.2. The other tableaux in table 12.2 are all those that can be obtained from tableau O by principal pivot exchanges. The superscripts on the main diagonals show the tableau that will be obtained by a principal pivot on that entry.

For the feasibility specifications $u_1 \geq 0$, $u_2 \geq 0$, $u_3 \geq 0$ the basic solution in tableau O is feasible. From tableau 12.108 we can calculate $s_1 = -r_1 = 0$ and $s_2 = -r_2 = 0$. These are the coordinates of the point identified by O in Table 12.1.

For the feasibility specifications $u_1 = 0$, $u_2 \geq 0$, $u_3 \geq 0$ the basic solution

Table 12.2: Set of Principally Equavalent Tableaux.

O

x_1	2^{A}	-1	-2
x_2	-1	1^{C}	1
x_3	-2	1	2^{E}
-1	-2	-3	-4
	$= u_1$	$= u_2$	$= u_3$

A

u_1	$1/2^{O}$	$-1/2$	-1
x_2	$1/2$	$1/2^{B}$	0
x_3	1	0	0
-1	1	-4	-6
	$= x_1$	$= u_2$	$= u_3$

B

u_1	1^{C}	1	-1
u_2	1	2^{A}	0
x_3	1	0	0
-1	5	8	-6
	$= x_1$	$= x_2$	$= u_3$

C

x_1	1^{B}	1	-1
u_2	-1	1^{O}	1
x_3	-1	-1	1^{D}
-1	-5	3	-1
	$= u_1$	$= x_2$	$= u_3$

D

x_1	0	0	1
u_2	0	2^{E}	-1
u_3	-1	-1	1^{C}
-1	-6	2	1
	$= u_1$	$= x_2$	$= x_3$

E

x_1	0	0	1
x_2	0	$1/2^{D}$	$-1/2$
u_3	-1	$1/2$	$1/2^{O}$
-1	-6	-1	2
	$= u_1$	$= u_2$	$= x_3$

in tableau A is feasible. From tableau 12.108 we can calculate $s_1 = 1$ and $s_2 = -1$. These are the coordinates of the point identified by A in Table 12.1.

We can continue in this way, associating each possible combination of feasibility specifications with one of the tableaux in Table 12.2 for which the basic solution satisfies the given feasibility specifications. Then it is a straightforward calculation to determine the remaining variables given in Table 12.1 and identify the corresponding point in Figure 12.6.

Now let us turn our attention to the hybrid problem of the previous section. Let us copy tableau 12.99 here for convenience.

(12.109)

x	4	−3	−10
v	3	0	6
−1	−10	−6	0
	$= u$	$= y$	$= f_0 + g_0 + e^T e$

If all variables in tableau 12.109 are canonical, the basic solution is feasible and complementary. Directly, we get $x = 0$, $y = 6$, $u = 10$, and $v = 0$. From tableau 12.98 we can calculate $r = -7$ and $s = 7$. From 12.80 we calculate $f = -24.5$. The solution obtained is the same as the one we found geometrically in Figure 12.7.

Tableau 12.109 is so small that only a small number of tableaux can be obtained by principal pivoting. If x is free and $y \geq 0$, the basic solution in tableau 12.110 is feasible and complementary.

(12.110)

u	1/4	−3/4	−5/2
v	−3/4	9/4	27/2
−1	5/2	−27/2	−25
	$= x$	$= y$	$= f_0 + g_0 + e^T e$

We can calculate $x = -5/2$, $y = 27/2$, $u = 0$, $v = 0$, $s = -r = 7$, and $f = -12$. The solution is the same as the one we found geometrically in Figure 12.8.

If $x \geq 0$ and y is artificial, the basic solution in tableau 12.111 is feasible and complementary.

(12.111)

u	0	1/3	2
y	−1/3	4/9	6
−1	−2	6	56
	$= x$	$= v$	$= f_0 + g_0 + e^T e$

We can calculate $x = 2$, $y = 0$, $u = 0$, $v = -6$, $s = -r = 11$, and $f = -105/2$. The solution is the same as the one we found geometrically in Figure 12.9. This solution is also obtained for x free and y artificial.

In the following sections we will discuss another important class of problems that can be reduced to linear complementarity problems and return to the question of solving these problems in Section 12.9.

12.7 Quadratic Programs

Given the n real variables $x = (x_1, x_2, \ldots, x_n)$, an expression of the form

$$Q(x) = \sum_{i=1}^{n} \sum_{j=1}^{n} x_i m_{ij} x_j \tag{12.112}$$

is called a *quadratic form*. It is generally more convenient to write 12.112 in the matrix form

$$Q(x) = x^T M x \tag{12.113}$$

where M is a square $n \times n$ matrix. We say that the matrix M *represents* the quadratic form $Q(x)$.

Except for the trivial cases where $n = 0$ or 1 the matrix M representing a quadratic form is not unique. For example, the quadratic form $Q(x) = x_1^2 + 4x_1x_2 + 3x_2^2$ is represented by the matrices

$$\begin{bmatrix} 1 & 2 \\ 2 & 3 \end{bmatrix}, \begin{bmatrix} 1 & 4 \\ 0 & 3 \end{bmatrix}, \begin{bmatrix} 1 & 6 \\ -2 & 3 \end{bmatrix} \tag{12.114}$$

among the many possibilities. The multiplicity of forms comes from representations of the mixed terms, like $4x_1x_2$, as $2x_1x_2 + 2x_2x_1$ or $6x_1x_2 - 2x_2x_1$, etc.

Since $x^T M x$ is a number (or a 1×1 matrix) $(x^T M x)^T = x^T M x = x^T M^T x$. Thus,

$$Q(x) = x^T M x = \frac{1}{2} x^T M x + \frac{1}{2} x^T M^T x = \frac{1}{2} x^T (M + M^T) x \tag{12.115}$$

Since $\frac{1}{2}(M + M^T)$ is symmetric, the quadratic form $Q(x)$ is representable by a symmetric matrix. Every quadratic form is representable by one and only one symmetric matrix and every symmetric matrix represents one and only one quadratic form. We shall consistently assume that the matrix representing a quadratic form is symmetric.

A quadratic form is said to be *positive definite* if $Q(x) > 0$ whenever x is not identically zero. It is said to be *positive semidefinite* if $Q(x) \geq 0$ for all x. For example, $Q_1(x) = x_1^2 + 4x_2^2$ is positive definite and $Q_2(x) = x_1^2 - 4x_1x_2 + 4x_2^2 = (x_1 - 2x_2)^2$ is positive semidefinite (but not positive definite since $Q_2(x) = 0$ for $x_1 = 2$, $x_2 = 1$).

There are obvious analogous definitions for *negative definite* and *negative semidefinite* quadratic forms. A quadratic form that is none of these four possibilities is said to be *indefinite*. That is, an indefinite quadratic form takes on both positive and negative values.

In the spirit of other problems considered in this chapter we are interested in problems of the following type:

Find the maximum value of the function

$$f = cx - \frac{1}{2}x^T Mx \tag{12.116}$$

where x must satisfy

$$Ax - b = -y \tag{12.117}$$

and (x, y) must satisfy some feasibility specifications.

In this case f, with the linear terms in cx, is a quadratic function rather than a quadratic form. The Karush–Kuhn–Tucker conditions are necessary conditions. However, the Karush–Kuhn–Tucker conditions do not imply that the function has a maximum (or a minimum) and a local maximum may not be a global maximum.

In this section we are interested in obtaining conditions that are both necessary and sufficient. This is not possible without some type of restriction on the class of functions to be considered. The condition we are going to impose here is that $Q(x) = x^T Mx$ shall be positive semidefinite.

The assumption that the matrix M is symmetric and positive semidefinite has several important consequences.

Any entry on the main diagonal of a positive semidefinite matrix must be nonnegative. For example, if all $x_i = 0$ except x_k then $x^T Mx = m_{kk}x_k^2$. Then the semidefinite property implies that $m_{kk} \geq 0$. With a similar argument we can show that every *principal submatrix* (a submatrix for which the rows and columns involve the same subset of indices in the same order) is also positive semidefinite.

Theorem 12.67 *Let M be symmetric and positive semidefinite. If an entry m_{ii} on the main diagonal is zero then every entry in the i-th row and i-th column is also zero.*

Proof. Let $m_{ii} = 0$. Take all $x_k = 0$ except $x_j = 1$ and x_i yet to be specified. Then $x^T Mx = m_{jj} + (m_{ij} + m_{ji})x_i$. If $m_{ij} + m_{ji}$ were not zero, by choosing the value of x_i appropriately we could make $x^T Mx$ negative, which would contradict the semidefinite property of M. Thus, $m_{ij} + m_{ji} = 0$. If M is symmetric, $m_{ij} = m_{ji} = 0$. ☐

Corollary 12.68 *If M is a nonzero, symmetric, positive semidefinite matrix, then at least one diagonal entry of M is positive.*

☐

Theorem 12.69 *Let M be a symmetric, positive semidefinite matrix. There exists a matrix D such that M can be represented in the form*

$$M = D^T D \tag{12.118}$$

Proof. Unless M is identically zero, there is a positive entry on the main diagonal. For notational convenience, assume it is m_{11} and subdivide M into submatrices as in 12.119.

$$\begin{bmatrix} m_{11} & M_{12} \\ M_{21} & M_{22} \end{bmatrix} \tag{12.119}$$

where $M_{21} = M_{12}^T$. Let $a_{11}^2 = m_{11}$, and set

$$D_1 = [\ a_{11} \quad M_{12}/a_{11}\ ] \tag{12.120}$$

It is easily verified that $D_1^T D_1$ is symmetric and with the same first row and first column as M. Hence,

$$M_1 = M - D_1^T D_1 \tag{12.121}$$

is a symmetric $n \times n$ matrix with the first row and first column all zeros.

Also, M_1 is positive semidefinite if M is. To see this note that $x^T M_1 x = x^T M x - x^T D_1^T D_1 x$ does not contain x_1 with a nonzero coefficient while $D_1 x$ does. By choosing the value of x_1 appropriately, we can make $D_1 x = 0$ without changing the value of $x^T M_1 x$. For this choice we have $x^T M_1 x = x^T M x$, which is nonnegative.

Since M_1 is symmetric and positive semidefinite, we can repeat the computation described above, ultimately obtaining

$$M_k = M - D_1^T D_1 - D_2^T D_2 - \cdots - D_k^T D_k = 0 \tag{12.122}$$

Corollary 12.68 assures us that we will not fail to find a positive entry on the main diagonal until we obtain a matrix identically zero.

Let $D_i = [\ d_{i1} \quad d_{i2} \quad \cdots \quad d_{in}\ ]$. Then form the $k \times n$ matrix $D = [d_{ij}]$. It is then a matter of direct calculation to verify that

$$\begin{aligned} M &= D_1^T D_1 + D_2^T D_2 + \cdots + D_k^T D_k \\ &= [d_{1i}d_{1j}] + [d_{2i}d_{2j}] + \cdots + [d_{ki}d_{kj}] \\ &= [d_{1i}d_{1j} + d_{2i}d_{2j} + \cdots + d_{ki}d_{kj}] = D^T D \end{aligned} \tag{12.123}$$

Conversely, if $M = D^T D$ then $x^T M x = x^T D^T D x = (Dx)^T (Dx) \geq 0$ for all x. That is, M is positive semidefinite. □

The factorization obtained in Theorem 12.69 is known as the *Cholesky factorization.* The method described in the proof is practical, though it is usually modified for numerical work. It is not necessary to know beforehand that the matrix M is positive semidefinite since the algorithm will also decide whether M is semidefinite. If M is positive semidefinite the algorithm will obtain the factorization; if the factorization can be obtained the matrix is positive semidefinite.

Let us illustrate the method with an example. Consider the matrix

$$\begin{bmatrix} 4 & 2 & 2 \\ 2 & 10 & -2 \\ 2 & -2 & 2 \end{bmatrix} \tag{12.124}$$

Following the procedure described in the theorem, we obtain at the first step

$$D_1 = \begin{bmatrix} 2 & 1 & 1 \end{bmatrix} \tag{12.125}$$

and

$$M_1 = \begin{bmatrix} 0 & 0 & 0 \\ 0 & 9 & -3 \\ 0 & -3 & 1 \end{bmatrix} \tag{12.126}$$

Notice here, and it is easy to prove in general, that M_1 can be obtained from M by performing a pivot exchange on 4 and setting the entries in the pivot row and pivot column equal to zero. This means that the successive matrices M_1, M_2, ... can be obtained by the familiar pivot operation, each time setting the pivot row and column to zero.

In the next iteration we have

$$D_2 = \begin{bmatrix} 0 & 3 & -1 \end{bmatrix} \tag{12.127}$$

This time we will get $M_2 = M_1 - D_2^T D_2 = 0$. Taking

$$D = \begin{bmatrix} 2 & 1 & 1 \\ 0 & 1 & -1 \end{bmatrix} \tag{12.128}$$

it is easy to verify that $M = D^T D$.

The minimum number of iterations required to obtain the factorization of M is the rank of M. While one is free to choose any positive entry in the main diagonal at each step, the number of iterations required will turn out to be the same for every sequence of choices. This number determines the minimum number of rows in the factor D. However, the rows of D can be permuted in any way whatever since this will not affect the inner products that must be computed to evaluate the $D_i^T D_i$.

It is convenient to arrange the work as shown in the following table.

(12.129)

4*	2	2			
2	10	−2			
2	−2	2			
			2	1	1
	9*	−3			
	−3	1			
			2	1	1
				3	−1
		0			

The blanks are planned zeros. Since the nonblank entries in the matrices in each level do not overlap the work can be carried out in one column of matrices. The zero shown is a computed zero. We start with M in the upper left box and a matrix of zeros in the upper right box. We pivot down the main diagonal of M. Each time we divide the pivot row by the square root of the diagonal entry and replace the row of zeros in the right box by this new row. After the pivot operation the pivot row and pivot column are replaced by zeros. When the left box is reduced to zeros, the right box will contain D. The rows of zeros in D, if any, can be deleted or retained. If at any time we obtain a negative entry in the main diagonal, or a zero entry in the main diagonal of a nonzero row, STOP. In this case the symmetric matrix we started with is not positive semidefinite.

We are now in a position to reduce a quadratic program with a positive semidefinite quadratic form to a hybrid program of the type considered in Section 12.5. Consider a tableau like tableau 12.76, except that it has a slightly different last column and last row.

(12.130)

	x	-1	
s	D	0	$= -r$
v	A	b	$= -y$
-1	c	0	$= f_0$
	$= u$	$= g_0$	

If we eliminate r by means of $Dx = -r$, the dual problems in Section 12.5 become

I. Maximize

$$f = cx - \frac{1}{2}x^T D^T Dx \tag{12.131}$$

subject to

$$Ax - b = -y \tag{12.132}$$

II. Minimize

$$g = vb + \frac{1}{2}ss^T \tag{12.133}$$

subject to

$$sD + vA - c = u \tag{12.134}$$

The pairs u, x and v, y satisfy dual feasibility specifications. All other variables are free.

If M is factored into the product $M = D^T D$, problem I above is precisely the quadratic programming problem we posed for ourselves at the beginning of this section. As in Section 12.5 this leads to a linear complementarity problem. Tableau 12.135 is the tableau for this problem.

(12.135)

	x	$-v^T$	-1	
x^T	M	$-A^T$	c^T	$= u^T$
v	A	0	b	$= -y$
-1	c	$-b^T$	0	$= f_0 + g_0$
	$= u$	$= y^T$	$= f_0 + g_0$	

Notice that we replace $D^T D$ by M in tableau 12.135. That is, we don't really have to factor M into $D^T D$ to write down the linear complementarity problem. In fact, as we will show below, the linear complementarity problem in tableau 12.135 can be obtained for quadratic programs that do not have a semidefinite quadratic form (and therefore cannot be factored). What, then, have we achieved by going through the factoring of M? Two things: One, we have established the linear complementarity problem associated with the quadratic programming problem independently of the Karush–Kuhn–Tucker conditions. The other, more important, is that we have established the linear complementarity problem as a sufficient condition for optimality. The Karush–Kuhn–Tucker conditions establish the linear complementarity problem as a necessary condition.

Finally, let us derive the linear complementarity problem of tableau 12.135 as necessary conditions from the Karush–Kuhn–Tucker conditions. Our problem is to maximize

$$f = cx - x^T Mx \tag{12.136}$$

subject to the constraints

$$Ax - b = -y \tag{12.137}$$

The variables x and y may be subject to general feasibility specifications.

The Karush–Kuhn–Tucker tableau for this problem is

(12.138)

<table>
<tr><td></td><td>dx</td><td></td></tr>
<tr><td>v</td><td>A</td><td>$= -dy$</td></tr>
<tr><td>-1</td><td>$c - x^T M$</td><td>$= df$</td></tr>
<tr><td></td><td>$= u$</td><td></td></tr>
</table>

This tableau, with the conditions in 12.137, contains the same equations as tableau 12.135. Thus, the complementarity conditions are equivalent to the Karush–Kuhn–Tucker conditions.

12.8 Semidefinite Quadratic Programs

To finish this chapter we are interested primarily in solving positive semidefinite quadratic programs. The tableaux we will deal with are like tableau 12.138 which we encountered in the previous section. We will first make some definitions and observations about tableaux, like 12.138, that we will encounter. Consider the tableau 12.139 for which M is square.

(12.139)

<table>
<tr><td>x^T</td><td>M</td></tr>
<tr><td>-1</td><td>c</td></tr>
<tr><td></td><td>$= u$</td></tr>
</table>

The problem of finding a feasible solution for the system of linear equations in tableau 12.139 for which $ux = 0$ is called a *linear complementarity problem.* The extra condition, $ux = 0$, is the *complementarity condition.*

The tableaux for the linear complementarity problems stemming from positive semidefinite quadratic programs have structures that permit designing a simple and efficient algorithm for their solutions. We want to describe that structure.

We are interested in problems in which the variables x_i and u_i are dual variables. That is, they satisfy dual feasibility specifications. A tableau is in *complementary form* if for each pair of dual variables, one is basic and the other is nonbasic. If the tableau is in complementary form, a basic solution automatically satisfies the complementarity condition, $ux = 0$. Thus, the problem of finding a feasible complementary solution is solved if we can find an equivalent tableau in complementary form for which the basic solution is feasible. We will use familiar pivoting techniques, but we will have new rules for selecting the pivot exchanges.

If all conceivable pivot exchanges were possible, an $n \times n$ tableau could have 2^n equivalent tableaux in complementary form. Usually, the number of possible equivalent complementary forms is much less, and, for small problems, empirical methods work quite well. However, the number of equivalent complementary forms is still large enough that an efficient algorithm to solve these problems is very desirable.

Usually, we will index the variables $x_1, x_2, \ldots, x_n$ and the dual variables $u_1, u_2, \ldots, u_n$ so that x_i and u_i are dual variables. If the tableau is in complementary form, we can also arrange the rows and columns so that for each i the row and column associated with x_i and u_i intersect on the main diagonal. Which of the two variables is associated with the row depends on which of the two is nonbasic. A tableau in this form is said to in *principal complementary form.*

Any sequence of pivot exchanges that transforms a tableau in complementary form into another tableau in complementary form is called a *principal pivot exchange.* If the entry on the main diagonal, where the row and column associated with x_i and u_i intersect, is nonzero, we can perform a principal pivot exchange exchanging one pair of dual variables. This is a simple principal pivot exchange. Generally, a block principal pivot exchange involves making k independent pivot exchanges to exchange the variables in k pairs of dual variables. If we start with a tableau in complementary form and hope to find a solution as a basic feasible solution for a tableau in complementary form, we can restrict our attention to principal pivot exchanges. If one tableau is obtained from another by a sequence of principal pivot exchanges we say one is a *principal pivot transform* of the other. Two tableaux are said to be *principally equivalent* if one is a principal pivot transform of the other.

A principal submatrix is a submatrix in which the columns are indexed by variables that are the duals of the variables that index the rows.

While it would not make any difference to a computing machine, it is much easier for a human to see what is going on if we maintain our tableaux in principal complementary form. In this case the entries for simple principal pivot exchanges are on the main diagonal. Even block pivots are easier to deal with because the pivot matrix is symmetrically situated with respect to the main diagonal. Furthermore, it is easier to refer to an "entry on the main diagonal," or a "principal submatrix," than it is to use complicated wording when the matrix is not in principal complementary form. But most important, the tableaux for positive semidefinite quadratic programs display a high degree of symmetry, or antisymmetry, if the tableaux are in principal complementary form.

Theorem 12.70 *Let M be a square matrix. M is positive semidefinite if and only if every principal submatrix of M is also positive semidefinite.*

Proof. Consider $Q(x) = x^T Mx$. We can drop from this expression all terms that would be multiplied by zero components of x. The terms that would be left have coefficients from a principal submatrix of M. $Q(x)$ is nonnegative for all x if and only if each of the quadratic forms generated by these principal submatrices in also nonnegative. That is, M is positive semidefinite if and only if every principal submatrix is positive semidefinite. □

Theorem 12.71 *Let M be a square matrix. M is positive definite if and only if every principal submatrix of M is also positive definite.*

Proof. With appropriate changes in wording, the proof given for Theorem 12.70 applies. □

A tableau, and its associated matrix, are *bisymmetric* if it is possible to rearrange the rows and columns so that the tableau is in principal complementary form and it has the appearance of tableau 12.140, where D and E are symmetric submatrices, x' and u' are dual pairs, and x'' and u'' are dual pairs.

(12.140)

x'	D	$-A^T$
x''	A	E
	$= u'$	$= u''$

In tableau 12.140, D and E are examples of principal submatrices.

In the previous section we encountered bisymmetric matrices obtained by performing principal pivot exchanges in a symmetric matrix. The definition of bisymmetry given here is slightly more general since it is possible to have a matrix that is explicitly bisymmetric by this definition but is not a principal pivot transform of a symmetric matrix. For example, if D and E in 12.140 are square matrices of zeros, the tableau is not principally equivalent to a symmetric matrix.

Theorem 12.72 *Let M be a square matrix in explicit bisymmetric form, like the matrix in tableau 12.140. M is positive semidefinite if and only if the principal symmetric submatrices D and E are positive semidefinite.*

Proof. The quadratic form for the matrix in this tableau is

$$\begin{aligned} Q(x) &= x'^T Dx' + x'^T(-A^T)x'' + x''^T Ax' + x''^T Ex'' \\ &= x'^T Dx' + x''^T Ex'' \end{aligned} \tag{12.141}$$

Thus, $Q(x)$ is nonnegative for all (x', x'') if and only if $x'^T Dx'$ is nonnegative for all x' and $x''^T Ex''$ is nonnegative for all x''. ◻

Theorem 12.73 *Let M be a square matrix in explicit bisymmetric form, like the matrix in tableau 12.140. M is positive definite if and only if the principal symmetric submatrices D and E are positive definite.*

Proof. Again, use the expression in 12.141 for the quadratic form. $Q(x)$ is positive for all nonzero (x', x'') if and only if $x'^T Dx'$ is positive for all nonzero x' and $x''^T Ex''$ is positive for all nonzero x''. ◻

Theorem 12.74 *If a tableau is bisymmetric, a simple principal pivot exchange or a principal pivot exchange of order two produces another bisymmetric tableau.*

Proof. A simple principal pivot exchange is based on a pivot entry in the main diagonal. If the diagonal entry is positive, the pivot exchange will not change the signs of the entries in the pivot row and it changes all the signs in the pivot column, except the pivot entry. If the diagonal entry is negative, the pivot exchange will change the signs of the entries in the pivot row and it does not change the signs in the pivot column, except for the pivot entry.

The results of the pivot exchange for the other entries can be calculated directly. Assume the pivot entry is in the main diagonal of the submatrix D. Then for the other entries in D we have

$$\begin{aligned} a_{ij} - \frac{a_{ir}a_{rj}}{a_{rr}} &= a_{ji} - \frac{a_{ri}a_{jr}}{a_{rr}} \\ &= a_{ji} - \frac{a_{jr}a_{ri}}{a_{rr}} \end{aligned} \tag{12.142}$$

For the entries in E we have

$$a_{ij} - \frac{a_{ir}a_{rj}}{a_{rr}} = a_{ji} - \frac{(-a_{ri})(-a_{jr})}{a_{rr}} = a_{ji} - \frac{a_{jr}a_{ri}}{a_{rr}} \tag{12.143}$$

For the entries in A we have

$$a_{ij} - \frac{a_{ir}a_{rj}}{a_{rr}} = -a_{ji} - \frac{(-a_{ri})(a_{jr})}{a_{rr}} = -(a_{ji} - \frac{a_{jr}a_{ri}}{a_{rr}}) \tag{12.144}$$

This shows that except for the pivot row and pivot column, the symmetry pattern throughout the tableau is preserved by the simple principal pivot exchange. The resulting tableau can be rearranged into bisymmetric form where one of the principal symmetric blocks is one row smaller and the other is one row larger.

A principal pivot exchange of order two is based on a 2×2 principal submatrix. This matrix must be invertible if it is the matrix of a legal block pivot exchange. That is, two independent pivot exchanges are possible within that submatrix. If either of the diagonal entries is nonzero, the first pivot exchange can be taken as a simple principal pivot exchange, and it will also be possible to take the second as a simple principal pivot exchange. If both diagonal entries are zero, we must make pivot exchanges on the two off-diagonal entries. Within the two pivot rows and pivot columns, these pivot exchanges act independently, and a calculation similar to that done above shows that the symmetry pattern for the rest of the tableau is preserved. Thus, the resulting tableau is also bisymmetric.

Consider principal pivot exchanges of order higher than two. If B is the principal submatrix on which the principal pivot exchange is based, it is invertible. We have seen that any invertible matrix can be inverted by a sequence of independent pivot exchanges. If B has a nonzero entry on the main diagonal, we can perform a simple principal pivot exchange on that entry, and the remaining independent pivot exchanges must be selected from the remaining rows and columns, which also form a principal submatrix. If B does not have a nonzero entry on the main diagonal, because B is bisymmetric we can select two symmetrically situated nonzero entries for a principal pivot exchange of order two. Then the remaining rows and columns form a principal submatrix from which the remaining independent pivot exchanges must be selected. We can continue in this way until the entire principal pivot exchange is effected. That is, every principal pivot exchange can be realized as a sequence of simple pivot exchanges and pivot exchanges of order two. Thus, if a matrix is bisymmetric, every matrix principally equivalent to it is also bisymmetric. □

Theorem 12.75 *If a matrix is positive semidefinite then every principal pivot transform of it is also positive semidefinite.*

Proof. Consider tableau 12.145 in which M is a square matrix.

x^T	M
	$= u$

(12.145)

Then the value of the quadratic form associated with M is

$$x^T M x = ux \tag{12.146}$$

A principal pivot exchange in tableau 12.145 interchanges pairs of dual variables and leaves the expression ux invariant. That is, principal pivot exchanges do not change the values of the associated quadratic form. ◻

Theorem 12.76 *If a matrix is positive definite then every principal pivot transform of it is also positive definite.*

Proof. The proof is identical to the proof of Theorem 12.75. Principal pivot exchanges preserve the product ux in tableau 12.145. ◻

Let us be rather precise about what these theorems do not say, as well as what they do say. Theorems 12.70, 12.71, 12.72, and 12.73 connect properties of the matrix M with properties of its principal submatrices, but they do not say anything about pivoting. Theorems 12.75 and 12.76 say something about preserving properties of the matrix under principal pivoting, but they say nothing about submatrices. In particular, it is not possible (from these theorems) to make inferences about the preservation of the properties of a principal submatrix under principal pivoting.

Suppose, for example, that D and E in tableau 12.140 are positive definite. Suppose, also, that D is the leading $k \times k$ principal submatrix. If M' is a principal pivot transform of M and the rows and columns are rearranged so that the leading principal submatrix of M' is indexed with the same variables as D, then that principal submatrix is also positive definite. But the inference does not follow from the assumption that D is positive definite. It follows from the fact that M and M' are positive definite. Also, in this context it does not matter whether D is symmetric.

A simple example should help illustrate the situation. Consider

x_1	1	-1
x_2	1	0
	$= u_1$	$= u_2$

This tableau is bisymmetric. The leading principal symmetric submatrix is 1×1 and positive definite. The trailing principal submatrix is 1×1 and positive semidefinite. The matrix M is positive semidefinite. A second order principal pivot exchange yields

u_1	0	1
u_2	-1	1
	$= x_1$	$= x_2$

Here, the leading 1×1 principal submatrix is not positive definite, and the trailing 1×1 principal matrix is positive definite.

A positive definite principal submatrix cannot be guaranteed to remain positive definite under principal pivoting unless all the pivot exchanges are on entries within that submatrix. If a matrix M is positive semidefinite, then for every matrix principally equivalent to M, all principal submatrices will be positive semidefinite.

Suppose the matrix M in tableau 12.139 is invertible, and consider the pair of tableaux 12.147.

(12.147)

x^T	M
-1	c
	$= u$

u	M^{-1}
-1	c'
	$= x^T$

Theorem 12.77 *If M is positive semidefinite then*

$$cc'^T \leq 0 \tag{12.148}$$

Proof. The proof of Theorem 12.77 is quite easy. Let $u = 0$ and calculate x. In the right tableau of 12.147, $x^T = -c'$. In the left tableau of 12.147, $x^T M - c = 0$, or $x^T M = c$. Then

$$0 \leq x^T M x = cx = -cc'^T \tag{12.149}$$

or $cc'^T \leq 0$. ∎

Since the matrix in the right tableau of 12.147 is the inverse of the matrix in the left tableau, the variables u and x are dual pairs in the same order. We say that c and c' display the same sign pattern if for each i, c_i and c'_i are either both nonpositive or both nonnegative. If c and c' display the same sign pattern and are nonzero for at least one dual pair, then $cc'^T > 0$. Theorem 12.77 says this cannot occur.

Theorem 12.78 *If a tableau with a square matrix M is nondegenerate and M is positive semidefinite, then no two tableaux principally equivalent to it can display the same sign pattern for those variables exchanged between the two tableaux.*

Proof. If two tableaux are principally equivalent and different, there is a nonempty set of dual pairs of variables that are exchanged between the two tableaux. Delete all rows and columns of pairs of dual variables that are not exchanged between the two tableaux. What remains are two tableaux that display the same sign pattern and for which the matrix of one is the inverse of the matrix for the other. The assumption of nondegeneracy assures that no entry in either c-row is zero. Theorem 12.77 says that they cannot display the same sign pattern. □

12.9 An Algorithm for Quadratic Programs

We are now ready to describe methods for solving positive semidefinite quadratic programs. Consider a quadratic program and its linear complementarity problem represented by tableau 12.135, which we repeat here for convenience as tableau 12.150.

(12.150)

	x	$-v^T$	-1	
x^T	M	$-A^T$	c^T	$= u^T$
v	A	0	b	$= -y$
-1	c	$-b^T$	0	$= f_0 + g_0$
	$= u$	$= y^T$	$= f_0 + g_0$	

We explicitly assume that M is positive semidefinite. Since the square submatrix of zeros is also positive semidefinite, the entire matrix of tableau 12.150 is positive semidefinite. We assume that x and u are dual variables, and y and v are dual variables. In this section we assume all variables are canonical. We will discuss the situation for noncanonical feasibility specifications in the following section.

Phase 1: Pivot in the A matrix using the feasibility algorithm. For each pivot in the A matrix, perform a symmetric pivot in the $-A^T$ matrix. If this process shows the problem is infeasible, STOP.

When we work examples we will not bother to perform the phase 1 step. We do it here to establish the simplifying condition that we are applying the algorithm to a problem that is known to be feasible.

We use the *Bland rule* to select a pivot column. We select the column with a positive entry in the basement row corresponding to a basic variable with the least index. Depending on the nonzero entries in this column we perform one of two types of principal pivot exchanges.

1. If the diagonal entry in the selected column is positive, perform a simple principal pivot exchange on the diagonal entry.

2. If the diagonal entry is zero, and if the nonzero entry in the selected column with the least index has an index less than the index of the column, perform a second-order principal pivot exchange for which one entry is that nonzero entry.

We refer to these rules as *rule 1* and *rule 2*.

The important thing about these two rules is that a decision to perform one of these pivot exchanges chooses a pivot column with least index and a positive entry in the c-row.

Suppose that these pivoting rules produce a cycle. Since the rules are specific, the cycle would be repeated indefinitely. Let k be the largest index of a dual pair that is exchanged by a principal pivot exchange in this cycle. Then there is one tableau in which x_k is basic and x_k and u_k are exchanged, and there is one tableau in which u_k is basic and x_k and u_k are exchanged.

The sequence of pivot exchanges between the two tableaux forms a principal pivot exchange. Delete all rows and columns in which the same variable of the dual pair is basic in both tableaux. In particular, this will delete all columns and rows with an index higher than k. What is left in each tableau is a square matrix and the matrices in the two tableaux are positive semidefinite and inverses of each other. They are, in fact, the pivot matrices of inverse principal block pivot exchanges. Then the resulting pair of tableaux must look like this:

$$
\begin{array}{r|c|}
\cline{2-2}
x^T & M \\
\cline{2-2}
-1 & c \\
\cline{2-2}
\multicolumn{1}{r}{} & \multicolumn{1}{c}{= u}
\end{array}
\qquad\qquad
\begin{array}{r|c|}
\cline{2-2}
u & M^{-1} \\
\cline{2-2}
-1 & c' \\
\cline{2-2}
\multicolumn{1}{r}{} & \multicolumn{1}{c}{= x^T}
\end{array}
\tag{12.151}
$$

where c and c' are nonpositive except for the entry in column k, which is positive. This would produce two tableaux, one the inverse of the other, in which the c-rows display the same sign and are nonzero. This contradicts Theorem 12.78.

Since the sequence of pivot exchanges cannot continue indefinitely, the selection rules must eventually fail. There are two ways the selection rules can fail. One is when there is no positive entry in the c-row. In that case we have a basic feasible solution and a feasible complementary solution. In the other, it is not possible to find a nonzero entry in the selected pivot column with an index smaller than the index of the column. If the matrix of tableau 12.150 is positive definite, selection rule 1 suffices to find an optimal solution. Every principal pivot transform of this tableau has a positive definite matrix. In a positive definite matrix, every diagonal entry is positive. Thus, selection rule 1 can always be used. The only way the selection rule can fail is when the basic solution is feasible.

If the matrix of tableau 12.150 is merely positive semidefinite, as it can be even if M is positive definite, the selection rules can fail before a basic feasible solution is obtained. In that case we have to find an alternate way to proceed. It is easier to discuss how this can happen and what to do about it in the context of examples.

Before we go on, let us point out that if we can find a solution for the linear complementarity problem of tableau 12.150 then we can also obtain the optimal values of the objective variables for both quadratic programs. From 12.135 and 12.137 we see that the values of the objective variables are

$$f = cx - \frac{1}{2}x^T D^T Dx = f_0 - \frac{1}{2}r^T r \tag{12.152}$$

and

$$g = vb + \frac{1}{2}ss^T = g_0 + \frac{1}{2}ss^T \tag{12.153}$$

Since we take $s = -r^T$ we get

$$f + g = f_0 + g_0 \tag{12.154}$$

For an optimal solution we get $f = g$. Thus, each is

$$f = g = \frac{1}{2}(f_0 + g_0) \tag{12.155}$$

Now let us look at several examples. First consider the problem:

Maximize

$$f = 20x_1 - 10x_2 - (3x_1^2 + 2x_2^2) \tag{12.156}$$

subject to the constraints

$$\begin{aligned} 2x_1 - x_2 &\leq 6 \\ -x_1 + x_2 &\leq 10 \\ -2x_1 - 3x_2 &\leq -8 \end{aligned} \tag{12.157}$$

where x_1 and x_2 are nonnegative.

The tableau representing the associated linear complementary problem is given in 12.158.

(12.158)

x_1	6*	0	−2	1	2	20
x_2	0	4	1	−1	3	−10
x_1	2	−1	0	0	0	6
x_2	−1	1	0	0	0	10
v_1	−2	−3	0	0	0	−8
−1	20	−10	−6	−10	8	0
	$= u_1$	$= u_2$	$= y_1$	$= y_2$	$= y_3$	$= f + g$

All variables in tableau 12.158 are canonical. Although a phase 1 is described above, we are going to skip it. It would solve the problem too quickly. Since the leading principal symmetric submatrix and the trailing principal symmetric submatrix are both positive semidefinite, the entire 5×5 matrix is positive semidefinite, and the Bland selection rules can be used directly. We have indicated the first pivot entry by an asterisk. This leads to tableau 12.159.

(12.159)

u_1	1/6	0	−1/3	1/6	1/3	10/3
x_2	0	4	1	−1	3	−10
v_1	−1/3	−1	2/3*	−1/3	−2/3	−2/3
v_2	1/6	1	−1/3	1/6	1/3	40/3
v_3	1/3	−3	−2/3	1/3	2/3	−4/3
−1	−10/3	−10	2/3	−40/3	4/3	−200/3
	$= x_1$	$= u_2$	$= y_1$	$= y_2$	$= y_3$	$= f + g$

Rule 1 selects the third column and the diagonal entry of that column as the pivot entry. We get tableau 12.160.

(12.160)

u_1	0	−1/2	1/2	0	0	3
x_2	1/2	11/2	−3/2	−1/2	4*	−9
y_1	−1/2	−3/2	3/2	−1/2	−1	−1
v_2	0	1/2	1/2	0	0	13
v_3	0	−4*	1	0	0	−2
−1	−3	−9	−1	−13	2	−66
	$= x_1$	$= u_2$	$= v_1$	$= y_2$	$= y_3$	$= f + g$

In tableau 12.160 the selection rules select the y_3 column, but the diagonal entry is zero. Thus, rule 2 must be used. A second-order principal pivot exchange is made using the two entries with asterisks. We obtain tableau 12.161, for which the basic solution is feasible and an optimal solution.

(12.161)

u_1	0	$-1/8$	$3/8$	0	0	$13/4$
y_3	$1/8$	$11/32$	$-1/32$	$-1/8$	$1/4$	$-47/16$
y_1	$-3/8$	$-1/32$	$35/32$	$-5/8$	$1/4$	$-51/16$
v_2	0	$1/32$	$5/8$	0	0	$51/4$
u_2	0	$-1/4$	$-1/4$	0	0	$1/2$
-1	$-13/4$	$-47/16$	$-51/16$	$-51/4$	$-1/2$	$-445/8$
	$= x_1$	$= v_3$	$= v_1$	$= y_2$	$= x_2$	$= f + g$

Furthermore, the maximum value of the objective function is $f = 445/16$.

Now let us look at an example where the given selection rules fail. The matrix of tableau 12.162 is bisymmetric and positive semidefinite, but the principal symmetric submatrices are fragmented. One symmetric submatrix consists of rows and columns 1, 2, 3, and 5. The other is the intersection of row and column 4. The principal symmetric submatrices were fragmented to construct an example in which the selections rules would fail. The selection rules determine the first pivot exchange on the diagonal entry in the first column. The first three pivot entries by these rules are indicated with asterisks.

(12.162)

x_1	4*	2	-1	0	1	15
x_2	2	2*	-1	0	-3	9
x_3	-1	-1	1*	0	3	-4
x_4	0	0	0	0	-6	-20
x_5	1	-3	3	6	19	-4
-1	15	9	-4	20	-4	0
	$= u_1$	$= u_2$	$= u_3$	$= u_4$	$= u_5$	$= f + g$

When the first three pivot exchanges are complete, we have tableau 12.163. The selection rules would select the u_4 column. However, the diagonal entry is zero, and all entries in that column with smaller indices are also zero. Thus, the selection rules fail.

(12.163)

u_1	0.50	-0.50	0.00	0.00	2.00	3.00
u_2	-0.50	1.50	1.00	0.00	-2.00	2.00
u_3	0.00	1.00	2.00	0.00	3.00	1.00
x_4	0.00	0.00	0.00	0.00	-6.00*	-20.00
x_5	-2.00	2.00	-3.00	6.00	2.00	-4.00
-1	-3.00	-2.00	-1.00	20.00	-4.00	-59.00
	$= x_1$	$= x_2$	$= x_3$	$= u_4$	$= u_5$	$= f + g$

In general, the selected column must have a positive entry or else the program would be infeasible. Thus, there is a negative entry in the row of the dual variable, which we have indicated with an asterisk. Pivot on that entry. We obtain tableau 12.164. For the purpose of the following discussion, we assume that the tableau is nondegenerate. That is, that we will not get a zero entry in the c-row.

(12.164)

u_1	0.50	−0.50	0.00	0.00	0.33	−3.67
u_2	−0.50	1.50	1.00	0.00	−0.33	8.67
u_3	0.00	1.00	2.00	0.00	0.50	−9.00
u_5	0.00	0.00	0.00	0.00	−0.17	3.33
x_5	−2.00	2.00	−3.00	6.00	0.33	−10.67
-1	−3.00	−2.00	−1.00	20.00	−0.67	−45.67
	$= x_1$	$= x_2$	$= x_3$	$= u_4$	$= x_4$	$= f + g$

The last pivot exchange was not a principal pivot exchange and 12.164 is not in complementary form. There is a pair of nonbasic dual variables, u_5 and x_5, and there is a pair of basic dual variables, u_4 and x_4. Let us isolate the submatrix determined by these variables and examine the situation more closely.

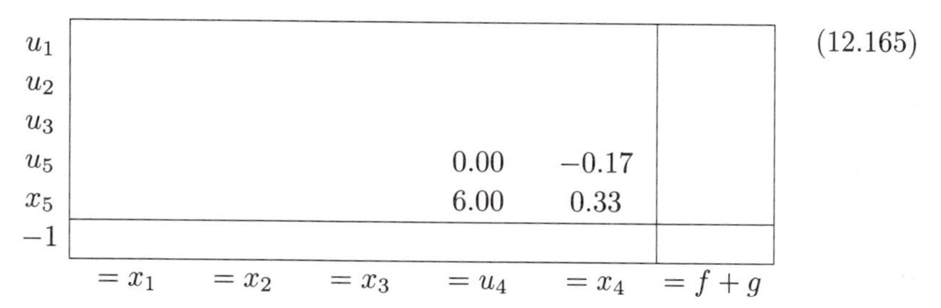

(12.165)

u_1						
u_2						
u_3						
u_5				0.00	−0.17	
x_5				6.00	0.33	
-1						
	$= x_1$	$= x_2$	$= x_3$	$= u_4$	$= x_4$	$= f + g$

This submatrix is call a *pair matrix.* All variables not involved in the pair matrix are complementary. A pivot exchange on any nonzero entry of the pair matrix would produce a tableau in complementary form. For this reason, a tableau with a single pair matrix, and all other dual pairs complementary, is said to be *almost-complementary.*

Since the tableau produced by a pivot exchange on one of these entries would produce a tableau with a bisymmetric positive semidefinite matrix, a number of important conclusions can be drawn. Consider tableau 12.166 in which the entries of the pair matrix are indicated with variables.

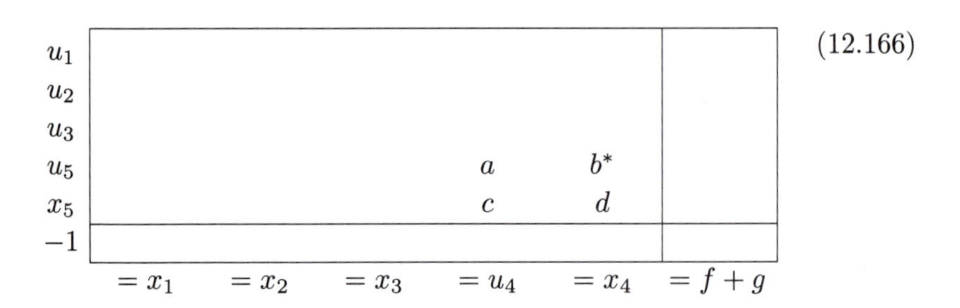

(12.166)

Suppose b is nonzero and we perform a pivot exchange on b. We obtain tableau 12.167.

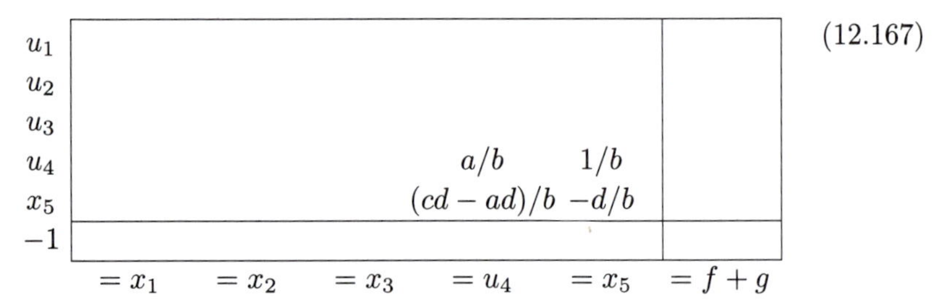

(12.167)

This submatrix, as a principal submatrix of a positive semidefinite matrix, is complementary, bisymmetric and positive semidefinite. Therefore, if a is nonzero, a and b have the same sign. If d is nonzero, d and b have opposite signs. Also, $cb - ad$ is ± 1. If $d = 0$, then $(cb - ad)/b = cb/b = -1/b$. That is, c and b have opposite signs.

At least one entry in a pair matrix is nonzero, since it was created by a pivot exchange on an entry in the pair matrix. No matter which one is nonzero, a pivot exchange on that entry will produce conclusions like the ones in the previous paragraph. Specifically, the determinant $cd - ad$ is nonzero, so that the pair matrix is invertible. In fact, $cd - ad$ is always ± 1. That means each row contains at least one nonzero entry, and each column contains at least one nonzero entry. Also, one row is nonnegative and the other row is nonpositive. If any entry is zero, the entries on the other diagonal are nonzero and opposite in sign.

We are now going to make a sequence of pivot exchanges, each producing another *almost-complementary* tableau, until a pivot exchange within a pair matrix restores complementarity. In tableau 12.164, the previous pivot exchange was in the row in which the entries of the pair matrix are nonpositive. We are going to make a pivot exchange in the other row of the nonbasic dual pair, the row in which the entries of the pair matrix are nonnegative.

In tableau 12.163, where we made our first nonprincipal pivot in the v_4 row, the first four entries in the pivot row were zeros and the first three entries in the c-row were nonpositive. The pivot exchange leaves all these entries unchanged. That is, we still have the c-entry of the selected column positive and all c-entries with smaller indices nonpositive (actually, negative because of the nondegeneracy assumption). The next pivot entry is chosen to preserve the nonpositivity of these c-entries. It is essentially a dual simplex pivot exchange. That is, we choose a negative entry a_{rj} for which the ratio c_j/a_{rj} is minimal. The pivot entry is indicated with an asterisk in tableau 12.168.

(12.168)

u_1	0.50	−0.50	0.00	0.00	0.33	−3.67
u_2	−0.50	1.50	1.00	0.00	−0.33	8.67
u_3	0.00	1.00	2.00	0.00	0.50	−9.00
u_5	0.00	0.00	0.00	0.00	−0.17	3.33
x_5	−2.00	2.00	−3.00*	6.00	0.33	−10.67
−1	−3.00	−2.00	−1.00	20.00	−0.67	−45.67
	$= x_1$	$= x_2$	$= x_3$	$= u_4$	$= x_4$	$= f + g$

When this pivot exchange is performed we have another almost-complementary tableau 12.169 with another nonbasic dual pair, u_3 and x_3. The next pivot entry, chosen in the row of the pair matrix in which the entries are nonnegative, is indicated with an asterisk in tableau 12.169.

(12.169)

u_1	0.50	−0.50	0.00	0.00	0.33	−3.67
u_2	−1.17	2.17	0.33	2.00	−0.22	5.11
u_3	−1.33*	2.33	0.67	4.00	0.72	−16.11
u_5	0.00	0.00	0.00	0.00	−0.17	3.33
x_3	0.67	−0.67	−0.33	−2.00	−0.11	3.56
−1	−2.33	−2.67	−0.33	18.00	−0.78	−42.11
	$= x_1$	$= x_2$	$= x_5$	$= u_4$	$= x_4$	$= f + g$

At each such almost-complementary pivot exchange, at least one of the entries in the basement row corresponding to the basic pair of dual variables decreases because of the nondegeneracy assumption and the fact that at least one entry of the pivot row within the pair matrix is positive. In fact, each of these two entries decreases at least as often as every other pivot exchange. We continue to make these almost-complementary pivot exchanges as long as at least one of these two entries is positive and as long as there are negative entries in the pivot row as candidates for the pivot entry. The entries within the pair matrix are not candidates during these steps because they are nonnegative. In tableau 12.170 the nonbasic dual pair is u_1 and x_1 and there are no negative entries in the u_1 row. In this case a pivot exchange on one of the entries in the pair matrix can be selected to make all five entries in the c-row nonpositive. If both c-entries are positive, select the pivot entry for which a_{rj} is positive and c_j/a_{rj} is maximal. The selected pivot entry is indicated with an asterisk in tableau 12.170.

(12.170)

u_1	0.38	0.38	0.25	1.50*	0.60	−9.71
u_2	−0.88	0.13	−0.25	−1.50	−0.85	19.21
x_1	−0.75	−1.75	−0.50	−3.00	−0.54	12.08
u_5	0.00	0.00	0.00	0.00	−0.17	3.33
x_3	0.50	0.50	0.00	0.00	0.25	−4.50
−1	−1.75	−6.75	−1.50	11.00	−2.04	−13.92
	$= u_3$	$= x_2$	$= x_5$	$= u_4$	$= x_4$	$= f + g$

This pivot exchange returns to complementary form and extends the row of nonpositive entries in the c-row. If fact, the basic solution in tableau 12.171 is feasible and we have an optimal solution.

(12.171)

u_4	0.25	0.25	0.17	0.67	0.40	−6.47
u_2	−0.50	0.50	0.00	1.00	−0.25	9.50
x_1	0.00	−1.00	0.00	2.00	0.67	−7.33
u_5	0.00	0.00	0.00	0.00	−0.17	3.33
x_3	0.50	0.50	0.00	0.00	0.25	−4.50
−1	−4.50	−9.50	−3.33	−7.33	−6.47	57.28
	$= u_3$	$= x_2$	$= x_5$	$= u_1$	$= x_4$	$= g$

The sequence of almost-complementary pivot exchanges results in a tableau in complementary form, but the tableau is not in principal form. Tableau 12.172 displays the rearrangement into principal form.

(12.172)

x_1	2.00	−1.00	0.00	0.67	0.00	−7.33
u_2	1.00	0.50	−0.50	−0.25	0.00	9.50
x_3	0.00	0.50	0.50	0.25	0.00	−4.50
u_4	0.67	0.25	0.25	0.40	0.17	−6.47
u_5	0.00	0.00	0.00	−0.17	0.00	3.33
−1	−7.33	−9.50	−4.50	−6.47	−3.33	57.28
	$= u_1$	$= x_2$	$= u_3$	$= x_4$	$= x_5$	$= g$

As the next example, consider the following tableau 12.173. It differs from the tableau of the previous example only in the fourth entry of the c-row.

(12.173)

x_1	4.00	2.00	−1.00	0.00	1.00	15.00
x_2	2.00	2.00	−1.00	0.00	−3.00	9.00
x_3	−1.00	−1.00	1.00	0.00	3.00	−4.00
x_4	0.00	0.00	0.00	0.00	−6.00	−8.00
x_5	1.00	−3.00	3.00	6.00	19.00	−4.00
−1	15.00	9.00	−4.00	8.00	−4.00	0.00
	$= u_1$	$= u_2$	$= u_3$	$= u_4$	$= u_5$	$= f + g$

The pivot exchanges for this example are the same as those for the previous example until we reach tableau 12.174, which corresponds to tableau 12.170. The important thing to observe here is not the nonnegative u_1-row. It is that both entries of the c-row for the basic dual pair have become negative. We select the x_1-row as the pivot row. This is the same row as the row in which the previous pivot exchange was performed. The previous pivot exchange was selected to preserve the nonpositivity of the first three entries of the c-row. The next pivot exchange is chosen to preserve the nonpositivity of those three entries and the nonpositivity of the two entries for the basic dual pair.

The new pivot entry will be within the pair matrix for the following reason. If the pivot entry were in one of the first three columns, it would have to be the same entry as the entry for the previous pivot exchange. That pivot exchange would exactly undo the previous pivot exchange and produce a positive entry in the c-row. In order for this argument to be valid, it is essential that this step be taken the first time that both c-entries for the basic dual variables become nonpositive. The selected pivot entry is indicated with an asterisk in tableau 12.174.

(12.174)

u_1	0.38	0.38	0.25	1.50	0.60	−2.46
u_2	−0.88	0.13	−0.25	−1.50	−0.85	8.96
x_1	−0.75	−1.75	−0.50	−3.00*	−0.54	5.58
u_5	0.00	0.00	0.00	0.00	−0.17	1.33
x_3	0.50	0.50	0.00	0.00	0.25	−1.50
−1	−1.75	−6.75	−1.50	−1.00	−2.04	−38.42
	$= u_3$	$= x_2$	$= x_5$	$= u_4$	$= x_4$	$= f + g$

When this pivot exchange is made we obtain tableau 12.175, for which the basic solution is feasible and complementary.

(12.175)

u_1	0.00	−0.50	0.00	0.50	0.33	0.33
u_2	−0.50	1.00	0.00	−0.50	−0.58	6.17
u_4	0.25	0.58	0.17	−0.33	0.18	−1.86
u_5	0.00	0.00	0.00	0.00	−0.17	1.33
x_3	0.50	0.50	0.00	0.00	0.25	−1.50
−1	−1.50	−6.17	−1.33	−0.33	−1.86	−40.28
	$= u_3$	$= x_2$	$= x_5$	$= x_1$	$= x_4$	$= f + g$

There is a possibility that the pivot exchange taken in tableau 12.170 is not possible because the entry in that position is zero. We will chase that possibility down, but for the moment we want to summarize what we have so far. Starting with tableau 12.163, which is in complementary form, we performed a pivot exchange to obtain an almost-complementary form. Then we performed several pivot exchanges leading from one almost-complementary form to another until we were able to restore complementary form. A similar sequence of events can be traced for the initial tableau 12.173. This is a principal pivot exchange of higher order than two, and with either kind of exit it leads to a tableau for which the positive basement entry with the least index is larger than it was before the exchange. We can continue using rule 1 or rule 2, or this more involved principal pivot exchange, which we will call *rule 3*.

Let us now consider what must be done if the exit rules for rule 3 fail. We might have a tableau that looks like tableau 12.176.

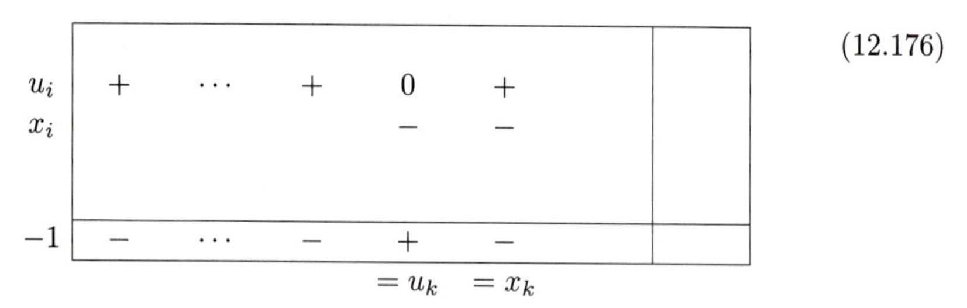

Here, u_i, x_i is the nonbasic pair and u_k, x_k is the basic pair. We are trying to maintain the negative entries in the first few entries of the basement row. The new pivot row should be the u_i-row. However, the nonnegative entries in that row corresponding to the negative entries in the basement row prevent a pivot exchange that will preserve the signs of those terms. The exit rule would then require a pivot exchange on a positive entry within the pair matrix in the column of the positive entry in the basement row. But that entry is zero.

If all the entries in the u_i-row were nonnegative, the problem would be infeasible. This is because the entries in the u_k-column would then have to be nonpositive, except for the basement entry. To see this, pivot on the positive entry in the pair matrix. This would restore complementary form without changing any entry in the u_k-column. Accordingly, let us assume that there is at least one negative entry in the u_i-row.

There are several strategies that can be used to delay or even avoid the next step that we wish to describe. However, it cannot be logically avoided and we might as well deal with it at this point.

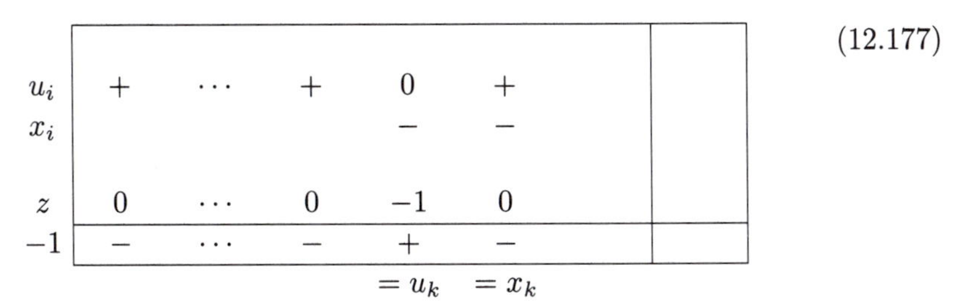

(12.177)

u_i	+	$\cdots$	+	0	+	
x_i				−	−	
z	0	$\cdots$	0	−1	0	
−1	−	$\cdots$	−	+	−	
				$=u_k$	$=x_k$	

We introduce an additional row and column. The nonbasic variable z is artificial, and the dual basic variable t is free. The t-column is used only to complement the artificial variable. We will make no use of the entries in that column, and we omit it from tableau 12.177. In the z-row, we enter a zero wherever the corresponding entry in the basement row in nonpositive. We enter -1 where shown. This could be the u_k-column or the x_k-column. It is the column with a positive entry in the basement row. Wherever else there is a positive entry in the basement row we enter $-N_j$ where N_j is anything larger than c_j/c_k.

If we were to pivot on the positive entry of the pair matrix, we would obtain a tableau in complementary form. We could then enter numbers in the t-column to make the matrix explicitly bisymmetric and positive semidefinite. Then we could pivot back on the same entry to obtain the entries that should be in the t-column. We do not have to take this step—it is sufficient to know that it could be done.

We now pivot on the -1 in the u_k-column. This makes all entries in the basement row negative, except for the positive entry in the u_k-column and the entry in the t-column, which doesn't matter.

The tableau we obtain is almost-complementary with u_i, x_i as the nonbasic pair and z, t as the basic pair. We select the u_i-row as the pivot row and make a dual simplex pivot in that row to preserve the negativity of the entries in the basement row.

We proceed exactly as we did with the pivoting rule 3 described above.

The positive entry in the basement row will be decreased as least as often as every other pivot exchange. The exit rules are also the same. Eventually we must see the entry in the z-column become positive or the pivot row will have no negative entry. In the first case we can pivot in the z-column to obtain a complementary form with a basic feasible solution. In the second case, if the entry in the z-column of the pair matrix is positive, we can pivot there to obtain a complementary form with a basic feasible solution. If that entry is zero, the problem is infeasible, as before.

We call this pivoting scheme *rule 4*. The way the description of the rules is formulated, it would appear that rule 1 is used until it fails, then rule 2 is used until it fails, etc. Actually, it is more likely that rules 1 and 2 will fail for small indices on the variables. If a column with a large index is selected, it is very likely that a row with a smaller index will have a nonzero entry. Experience seems to indicate that one must contrive special examples to obtain situations in which rules 1 and 2 do not suffice, once a tableau is obtained in which the smallest index for a positive basement entry is reasonable large.

Almost-complementary pivoting is an important part of several algorithms proposed for solving quadratic programs. In fact, Theorem 12.78 shows that it is not reasonable to expect an algorithm based on preserving a sign pattern in the c-row to make exclusive use of principal pivoting.

Of the many methods available for solving quadratic programs and linear complementarity problems, we wish to mention two, the Lemke algorithm by C. E. Lemke and the principal pivoting method by Richard W. Cottle and George B. Dantzig. Both use almost-complementary pivoting.

The *Lemke algorithm* consists of an initial pivot exchange that produces an almost-complementary form, and then all subsequent pivots are almost-complementary pivot exchanges until the algorithm terminates with a solution or in a dead end. No assumption other than nondegeneracy is required to assure that it will not cycle. Positive semidefiniteness is not required and it is not even necessary to assume that the linear complementarity problem represents a quadratic program. The Lemke algorithm is very robust in that it will work for quadratic programs and many other types of problems. When it is applied to a positive semidefinite quadratic program it can be shown that it terminates only with a solution or with a demonstration that the program is infeasible. For more general problems the dead-end termination is not always informative.

The algorithm described here more nearly resembles the *principal pivoting method* of Cottle and Dantzig than any other. The principal pivoting method does not use the steps we have described as rules 1 and 2. It uses rules 3 and 4, except that it provides for different techniques for initiating the sequence of almost-complementary pivot exchanges.

12.10 Noncanonical Quadratic Programs

Let us consider positive semidefinite quadratic programs for which some variables are noncanonical. We do require that dual pairs of variables satisfy dual feasibility specifications. If a solution is obtained for which the noncanonical variables are feasible, they are also complementary. That is, feasibility for noncanonical variables implies complementarity. For this reason, noncanonical feasibility specifications offer very little complication. It is sufficient to insert a step at the beginning to pivot all artificial variables to the positions of nonbasic variables, and pivot all free variables to the positions of basic variables.

Phase 0: Perform principal pivot exchanges, simple principal pivot exchanges if possible or second-order principal pivot exchanges if necessary, to move as many artificial variables as possible to the positions of nonbasic variables and as many free variables as possible to the positions of basic variables.

To do this, look among the basic variables for any artificial variables. If the diagonal entry for that column is positive, perform a simple principal pivot exchange on the diagonal entry. Since the dual nonbasic variable is free, this exchange will decrease the basic artificial variables by one.

If the diagonal entry is zero, look for a nonzero entry in that column for which the corresponding nonbasic variable is free or canonical. Perform a second-order principal pivot exchange on that entry and the entry that is symmetric to it with respect to the main diagonal. If the first pivot exchange exchanges a basic artificial variable and a nonbasic free variable, the second pivot exchange will exchange a basic artificial variable and a nonbasic free variables. If the first pivot exchange exchanges a basic artificial variable and a nonbasic canonical variable, the second pivot exchange will exchange a basic canonical variable and a nonbasic free variable. Either case will reduce the number of nonbasic artificial variables.

This process stops either when there are no more basic artificial variables or when the remaining basic artificial variables are in columns for which all nonzero entries are in the rows of nonbasic artificial variables. In the second case, if the basement entry in any of these columns is nonzero, the program is infeasible. If all are zero, or in the first case, we can proceed. This is the end of phase 0.

From this point on, we will not choose a pivot entry in the column of a free variable or in the row of an artificial variable. Let us look carefully at what this means if phase 0 terminated with basic artificial variables. The only nonzero entries in these columns are in the rows of artificial variables. Thus, we are not going to pivot in one of these columns and we are not going to pivot in any row that could change an entry in these columns. The rows of the nonbasic free variables dual to these basic artificial variables will also remain unchanged. The nonzero entries in these rows will be in the columns of basic free variables. We could put them aside until the rest of the pivoting runs its course. If an optimal solution is obtained, they would be used to obtain other solutions. The nonbasic variables would be parameters of the solution set.

Since the rest of the pivoting is restricted to rows and columns of canonical variables, the problem is reduced to a canonical problem and we can use the methods described in the previous section.

Let us consider several examples using the same numbers that were used in an example worked in the previous sections, but with different feasibility specifications. The initial tableau for the example is tableau 12.178, and the solution for the canonical problem is given in tableau 12.179.

O (12.178)

x_1	6.00	0.00	−2.00	1.00	2.00	20.00
x_2	0.00	4.00	1.00	−1.00	3.00	−10.00
v_1	2.00	−1.00	0.00	0.00	0.00	6.00
v_2	−1.00	1.00	0.00	0.00	0.00	10.00
v_3	−2.00	−3.00	0.00	0.00	0.00	−8.00
−1	20.00	−10.00	−6.00	−10.00	8.00	0.00
	$= u_1$	$= u_2$	$= y_1$	$= y_2$	$= y_3$	$= f + g$

F (12.179)

u_1	0.00	0.00	0.37	0.00	−0.12	3.25
u_2	0.00	0.00	−0.25	0.00	−0.25	0.50
y_1	−0.37	0.25	1.09	−0.62	−0.03	−3.19
v_2	0.00	0.00	0.63	0.00	0.13	12.75
y_3	0.12	0.25	−0.03	−0.12	0.34	−2.94
−1	−3.25	−0.50	−3.19	−12.75	−2.94	−55.62
	$= x_1$	$= x_2$	$= v_1$	$= y_2$	$= v_3$	$= f + g$

To see the effect of different feasibility specifications it is helpful to look at the graph of the components of the problem. The canonical feasible set is shaded in Figure 12.10. The optimal solution for the canonical problem is the point labeled F in Figure 12.10. The level curves for the objective function is a family of ellipses centered at $(3.91, -3.48)$. Three of these ellipses are shown in Figure 12.10.

Suppose, for example, that we change the feasibility specifications so that y_2 is artificial, x_1 is free, and the remaining variables are canonical.

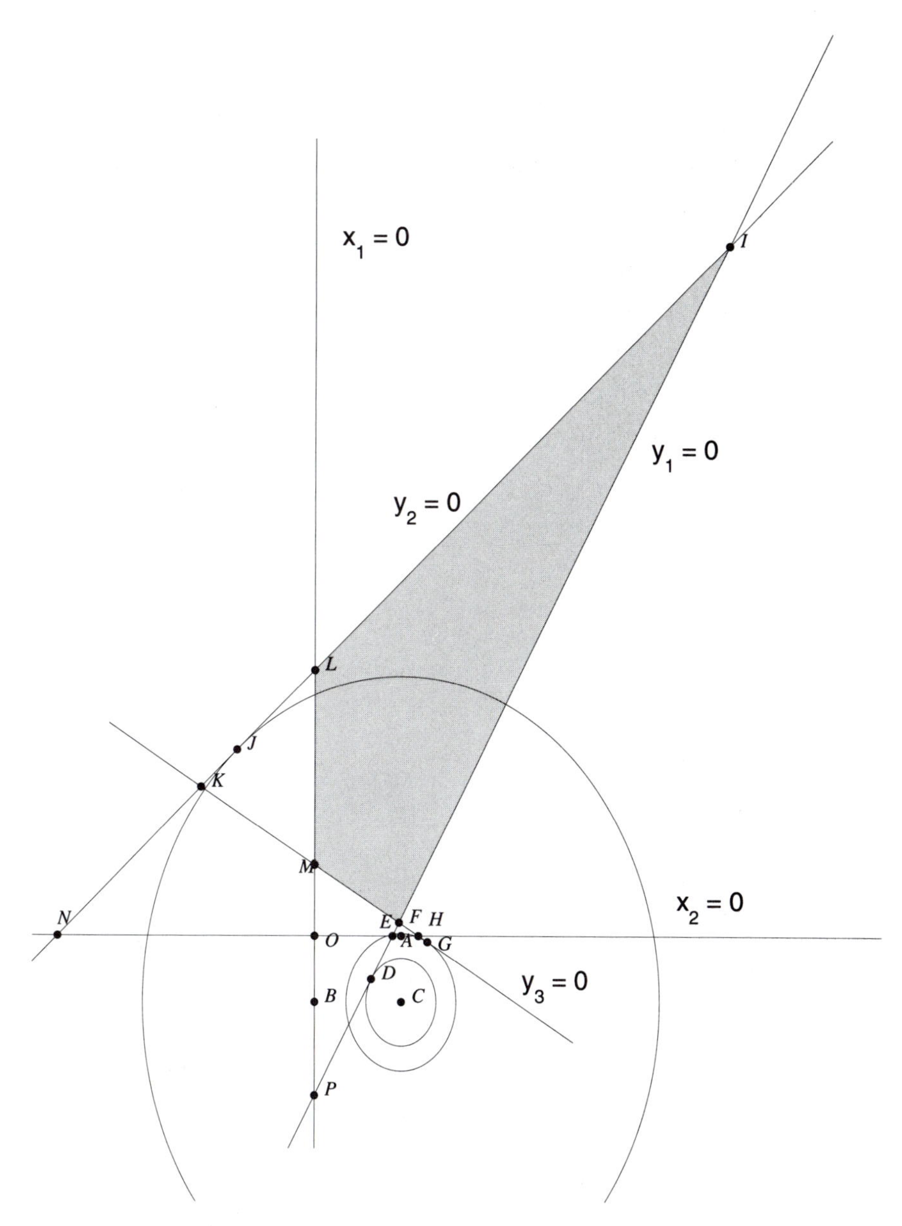

Figure 12.10: Optimal points for various feasibility specifications.

Of course, v_2, the variable dual to y_2, is free and u_1 is artificial. Starting with tableau 12.178, we pivot the artificial variables to the positions of nonbasic variables and then deal with the canonical variables. We obtain tableau 12.180 from which we see that $x_1 = -3$ and $x_2 = 7$. This is the point J in Figure 12.10. Notice that it is on the line $y_2 = 0$ and that one of the level curves is shown tangent to this line and passing though J.

J (12.180)

u_1	0.10	0.10	−0.10	−0.40	0.50	−3.00
u_2	0.10	0.10	−0.10	0.60	0.50	7.00
v_1	−0.10	−0.10	0.10	1.40	−0.50	19.00
y_2	0.40	−0.60	−1.40	2.40	−1.00	38.00
v_3	0.50	0.50	−0.50	1.00	2.50	7.00
−1	3.00	−7.00	−19.00	38.00	−7.00	510.00
	$= x_1$	$= x_2$	$= y_1$	$= v_2$	$= y_3$	$= f + g$

For every tableau in principal form that can be obtained it is possible to describe feasibility specifications for which the basic solution is feasible. It is sufficient to take any basic variables with nonnegative values to be canonical and any basic variables with negative values to be free. The following tableaux are all the remaining tableaux that can be obtained. Each is identified with a letter corresponding to a point shown in Figure 12.10. The reader should study each tableau and identify the feasibility specification for which the basic solution is optimal.

A (12.181)

u_1	0.17	0.00	−0.33	0.17	0.33	3.33
x_2	0.00	4.00	1.00	−1.00	3.00	−10.00
v_1	−0.33	−1.00	0.67	−0.33	−0.67	−0.67
v_2	0.17	1.00	−0.33	0.17	0.33	13.33
v_3	0.33	−3.00	−0.67	0.33	0.67	−1.33
−1	−3.33	−10.00	0.67	−13.33	1.33	−66.67
	$= x_1$	$= u_2$	$= y_1$	$= y_2$	$= y_3$	$= f + g$

B (12.182)

x_1	6.00	0.00	−2.00	1.00	2.00	20.00
u_2	0.00	0.25	0.25	−0.25	0.75	−2.50
v_1	2.00	0.25	0.25	−0.25	0.75	3.50
v_2	−1.00	−0.25	−0.25	0.25	−0.75	12.50
v_3	−2.00	0.75	0.75	−0.75	2.25	−15.50
−1	20.00	2.50	−3.50	−12.50	15.50	−25.00
	$= u_1$	$= x_2$	$= y_1$	$= y_2$	$= y_3$	$= f + g$

C (12.183)

u_1	0.17	0.00	−0.33	0.17	0.33	3.33
u_2	0.00	0.25	0.25	−0.25	0.75	−2.50
v_1	−0.33	0.25	0.92	−0.58	0.08	−3.17
v_2	0.17	−0.25	−0.58	0.42	−0.42	15.83
v_3	0.33	0.75	0.08	−0.42	2.92	−8.83
−1	−3.33	2.50	3.17	−15.83	8.83	−91.67
	$= x_1$	$= x_2$	$= y_1$	$= y_2$	$= y_3$	$= f + g$

D (12.184)

u_1	0.05	0.09	0.36	−0.05	0.36	2.18
u_2	0.09	0.18	−0.27	−0.09	0.73	−1.64
y_1	−0.36	0.27	1.09	−0.64	0.09	−3.45
v_2	−0.05	−0.09	0.64	0.05	−0.36	13.82
v_3	0.36	0.73	−0.09	−0.36	2.91	−8.55
−1	−2.18	1.64	−3.45	−13.82	8.55	−80.73
	$= x_1$	$= x_2$	$= v_1$	$= y_2$	$= y_3$	$= f + g$

E (12.185)

u_1	0.00	−0.50	0.50	0.00	0.00	3.00
x_2	0.50	5.50	−1.50	−0.50	4.00	−9.00
y_1	−0.50	−1.50	1.50	−0.50	−1.00	−1.00
v_2	0.00	0.50	0.50	0.00	0.00	13.00
v_3	0.00	−4.00	1.00	0.00	0.00	−2.00
−1	−3.00	−9.00	−1.00	−13.00	2.00	−66.00
	$= x_1$	$= u_2$	$= v_1$	$= y_2$	$= y_3$	$= f + g$

$$G \tag{12.186}$$

u_1	0.13	−0.09	−0.34	0.21	−0.11	4.34
u_2	−0.09	0.06	0.23	−0.14	−0.26	−0.23
v_1	−0.34	0.23	0.91	−0.57	−0.03	−2.91
v_2	0.21	−0.14	−0.57	0.36	0.14	14.57
y_3	0.11	0.26	0.03	−0.14	0.34	−3.03
−1	−4.34	0.23	2.91	−14.57	−3.03	−64.91
	$= x_1$	$= x_2$	$= y_1$	$= y_2$	$= v_3$	$= f + g$

$$H \tag{12.187}$$

u_1	0.00	1.50	0.00	0.00	−0.50	4.00
x_2	−1.50	17.50	4.00	−2.50	−4.50	−4.00
v_1	0.00	−4.00	0.00	0.00	1.00	−2.00
v_2	0.00	2.50	0.00	0.00	−0.50	14.00
y_3	0.50	−4.50	−1.00	0.50	1.50	−2.00
−1	−4.00	−4.00	2.00	−14.00	−2.00	−64.00
	$= x_1$	$= u_2$	$= y_1$	$= y_2$	$= v_3$	$= f + g$

$$I \tag{12.188}$$

u_1	0.00	0.00	1.00	1.00	0.00	16.00
u_2	0.00	0.00	1.00	2.00	0.00	26.00
y_1	−1.00	−1.00	10.00	14.00	−5.00	190.00
y_2	−1.00	−2.00	14.00	22.00	−8.00	304.00
v_3	0.00	0.00	5.00	8.00	0.00	102.00
−1	−16.00	−26.00	190.00	304.00	−102.00	4120.00
	$= x_1$	$= x_2$	$= v_1$	$= v_2$	$= y_3$	$= f + g$

$$K \tag{12.189}$$

u_1	0.00	0.00	0.00	−0.60	−0.20	−4.40
u_2	0.00	0.00	0.00	0.40	−0.20	5.60
v_1	0.00	0.00	0.00	1.60	0.20	20.40
y_2	0.60	−0.40	−1.60	2.80	0.40	40.80
y_3	0.20	0.20	−0.20	0.40	0.40	2.80
−1	4.40	−5.60	−20.40	40.80	2.80	529.60
	$= x_1$	$= x_2$	$= y_1$	$= v_2$	$= v_3$	$= f + g$

L (12.190)

x_1	10.00	1.00	−1.00	−4.00	5.00	−30.00
u_2	−1.00	0.00	0.00	1.00	0.00	10.00
v_1	1.00	0.00	0.00	1.00	0.00	16.00
y_2	−4.00	−1.00	−1.00	4.00	−3.00	50.00
v_3	−5.00	0.00	0.00	3.00	0.00	22.00
−1	−30.00	−10.00	−16.00	50.00	−22.00	600.00
	$= u_1$	$= x_2$	$= y_1$	$= v_2$	$= y_3$	$= f + g$

M (12.191)

x_1	7.78	−0.67	−2.67	1.67	−0.89	33.78
u_2	0.67	0.00	0.00	0.00	−0.33	2.67
v_1	2.67	0.00	0.00	0.00	−0.33	8.67
v_2	−1.67	0.00	0.00	0.00	0.33	7.33
y_3	−0.89	0.33	0.33	−0.33	0.44	−6.89
−1	33.78	−2.67	−8.67	−7.33	−6.89	81.78
	$= u_1$	$= x_2$	$= y_1$	$= y_2$	$= v_3$	$= f + g$

N (12.192)

u_1	0.00	−1.00	0.00	−1.00	0.00	−10.00
x_2	1.00	10.00	−1.00	6.00	5.00	70.00
v_1	0.00	1.00	0.00	2.00	0.00	26.00
y_2	1.00	6.00	−2.00	6.00	2.00	80.00
v_3	0.00	−5.00	0.00	−2.00	0.00	−28.00
−1	10.00	70.00	−26.00	80.00	28.00	1000.00
	$= x_1$	$= u_2$	$= y_1$	$= v_2$	$= y_3$	$= f + g$

P (12.193)

x_1	22.00	2.00	8.00	−1.00	8.00	48.00
u_2	−2.00	0.00	−1.00	0.00	0.00	−6.00
y_1	8.00	1.00	4.00	−1.00	3.00	14.00
v_2	1.00	0.00	1.00	0.00	0.00	16.00
v_3	−8.00	0.00	−3.00	0.00	0.00	−26.00
−1	48.00	6.00	14.00	−16.00	26.00	24.00
	$= u_1$	$= x_2$	$= v_1$	$= y_2$	$= y_3$	$= f + g$

Exercises

Nonlinear problems that are only slightly complex can be very difficult to solve. Since we are not interested in the difficulties of solving systems of nonlinear equations, the problems here are intended to be simple as far as equation solving is concerned. In many cases the solutions can be guessed. We are primarily concerned with formulating the feasibility specifications correctly, in determining and interpreting the Karush–Kuhn–Tucker conditions, and in finding the Karush–Kuhn–Tucker points. Then, finally, one should determine which Karush–Kuhn–Tucker points represent maxima or minima.

1. Find the maximum of $f = x_1^2 + x_2^2$, unconstrained.
2. Find the maximum of $f = x_1^2 + x_2^2$, constrained by

$$
\begin{aligned}
-1 \le x_1 \le 1 \\
-1 \le x_2 \le 1
\end{aligned}
$$

3. Find the maximum of $f = x_1^2 + x_2^2$, constrained by

$$
\begin{aligned}
0 \le x_1 \le 1 \\
0 \le x_2 \le 1
\end{aligned}
$$

4. Find the maximum of $f = x_1 x_2$, constrained by

$$
\begin{aligned}
-2 \le x_1 + x_2 \le 2 \\
-2 \le x_1 - x_2 \le 2
\end{aligned}
$$

5. Find the maximum of $f = x_1 x_2$, constrained by

$$
\begin{aligned}
x_1 + x_2 &\le 4 \\
(x_1 - 2)^2 + (x_2 - 2)^2 &\ge 2 \\
x_1 \ge 0, x_2 &\ge 0
\end{aligned}
$$

6. Find the maximum of $f = x_1 + x_2$, constrained by

$$
\begin{aligned}
x_1 + (x_2 - 2)^2 \le 4 \\
(x_1 - 1)^2 + (x_2 - 2)^2 \ge 1 \\
x_2 \le 2, x_1 \ge 0, x_2 \ge 0
\end{aligned}
$$

7. Find the maximum of $f = x_1 + x_2$, constrained by

$$
\begin{aligned}
2x_1^2 - 5x_1x_2 + 2x_2^2 + 2x_1 + 2x_2 \le 3 \\
x_1 + 3x_2 \le 8 \\
x_1 \ge 0, x_2 \ge 0
\end{aligned}
$$

8. Find the maximum of $f = x_1 + x_2$, constrained by

$$
\begin{aligned}
2x_1^2 - 5x_1x_2 + 2x_2^2 + 2x_1 + 2x_2 &\leq 4 \\
x_1 + 3x_2 &\leq 8 \\
x_1 \geq 0, x_2 &\geq 0
\end{aligned}
$$

9. Find the maximum of $f = -x_1x_2x_3$, constrained by

$$
\begin{aligned}
x_1^2 + x_2^2 + x_3^2 &= 3 \\
x_1 \geq 0, x_2 \geq 0, x_3 &\geq 0
\end{aligned}
$$

10. Find the maximum of $f = x_1x_2x_3$, constrained by

$$
\begin{aligned}
x_1^2 + x_2^2 + x_3^2 &= 6 \\
x_1 + x_2 - x_3 &= 0 \\
x_3 &\geq 0
\end{aligned}
$$

11. Find the "best" approximate linear function for the data in the following table.

t	y
1	0.3
2	−0.2
3	−0.1
4	0.1
5	1.1
6	2.1

12. Find the "best" approximate quadratic function for the data in the table for Exercise 11.

13. Find a point nearest the origin that satisfies the constraints

$$
\begin{aligned}
-s_1 + s_2 &\geq 1 \\
2s_1 - s_2 &\geq 3
\end{aligned}
$$

14. Find a point nearest the origin that satisfies the constraints

$$
\begin{aligned}
-s_1 + s_2 &\geq 1 \\
2s_1 - s_2 &\geq 3
\end{aligned}
$$

15. Find a point nearest the origin that satisfies the constraints

$$s_1 + s_2 + s_3 \geq 3$$
$$2s_1 - s_2 + 2s_3 \geq 4$$
$$2s_1 - s_2 - 2s_3 \leq 4$$

16. Find the minimum of $f = (x_1 + x_2 - 1)^2 + (x_1 + 3x_2 - 2)^2$, subject to the constraints

$$2x_1 + x_2 \geq 4$$
$$x_1 \geq 0, x_2 \geq 0$$

17. Find the minimum of the function given in Exercise 16. The constraints are the same except that x_2 is free.

18. Find the minimum of the function given in Exercise 16. The constraints are the same with the additional condition that

$$-x_1 + 2x_2 = 1$$

19. Find the maximum of the function

$$f = -4x_1^2 - 4x_2^2 - 4x_3^2 + x_1x_2 - x_2x_3 + x_1 + 2x_2 + 3x_3$$

subject to the constraints

$$\begin{aligned} 3x_1 - x_2 + 4x_3 &\leq 4 \\ -3x_1 + 6x_2 - 2x_3 &\leq -2 \\ x_1 \geq 0, x_2 \geq 0, x_3 \geq 0 \end{aligned}$$

20. Find the maximum for the problem given in Exercise 19, with the constraints

$$3x_1 - x_2 + 4x_3 \leq 4$$
$$-3x_1 + 6x_2 - 2x_3 = -2$$
$$x_1 \geq 0, x_2 \geq 0, x_3 \geq 0$$

Questions

For the following questions we assume all functions considered are differentiable everywhere on the feasible set.

Q1. A maximum nonlinear program always has a solution.

Q2. A maximum nonlinear program for which the feasible set is nonempty and for which the objective variable is bounded on the feasible set always has a solution.

Q3. For a maximum nonlinear program, at every point where a maximum value for the objective variable is taken on and at which the constraint qualification is satisfied, the Karush–Kuhn–Tucker conditions have a feasible solution.

Q4. For a maximum nonlinear program, for every point where the Karush–Kuhn–Tucker conditions are satisfied, there is either a local maximum or a global maximum.

Q5. For a maximum nonlinear program, for every point where the Karush–Kuhn–Tucker conditions are satisfied, there is either a local maximum or a local minimum.

Q6. For a maximum nonlinear program, there might be some points where the Karush–Kuhn–Tucker conditions are satisfied and at which there is a local minimum.

Q7. A quadratic program for a function with a negative semidefinite quadratic form always has a global maximum.

Q8. A quadratic program for a function with a negative semidefinite quadratic form and for which the feasible set is nonempty and bounded always has a global maximum.

Q9. For a quadratic program for a function with a negative semidefinite quadratic form, if the function has a local maximum it has a global maximum, and the global maximum is equal to the local maximum.

Q10. For a quadratic program for a function with a negative semidefinite quadratic form, if the function has a global maximum, the point where the maximum occurs is unique.

Q11. For a quadratic program for a function with a negative definite quadratic form, if the function has a global maximum, the point where the maximum occurs is unique.

Q12. For a least squares problem, the normal equations are always solvable.

Q13. For a least squares problem, the number of parameters is usually much smaller than the number of observations. If the rank of the matrix D in the defining relation $Dx - e = -r$ is of rank equal to the number of parameters, the optimal solution obtained by the methods described here is unique.

Q14. A feasible least distance problem always has an optimal solution.

Appendix A

Answers

Chapter 1

1. The **Furniture Maker's Problem.**

	Ddesk type					
	(1)	(2)	(3)	(4)	available	
carpentry time	4	9	7	10	6000	(hours)
finishing time	1	1	3	40	4000	(hours)
profit	12	20	28	40		(dollars)
		(per unit)			(per period)	

x_1	x_2	x_3	x_4	-1	
4	9	7	10	6000	$= -y_1$
1	1	3	40	4000	$= -y_2$
12	20	28	40		$= f$

2. The **President's Problem.**
Maximize

$$0.9x_1 + 0.6x_2$$

$$x_1 \geq 0, x_2 \geq 0$$

subject to

$$\begin{aligned} x_1 + x_2 &\leq 20 \\ x_1 &\leq 12 \\ x_2 &\leq 16 \end{aligned}$$

x_1	x_2	-1	
1	1	20	$= -y_1$
1		12	$= -y_2$
	1	16	$= -y_3$
0.9	0.6		$= f$

3. The **Wyndor Glass Co.**

	Doors made	Windows made	capacity available	
Plant 1	1		4	units Plant 1
Plant 2		2	12	units Plant 2
Plant 3	3	2	18	units Plant 3
unit profit	3	5		profit
	$\frac{\text{dollars}}{\text{door}}$	$\frac{\text{dollars}}{\text{window}}$		

x_1	x_2	-1	
1		4	$= -y_1$
	2	12	$= -y_2$
3	2	18	$= -y_3$
3	5		$= f$

4. The **Investment Manager's Problem.**

Many students have difficulties when they first encounter a linear program involving percentages. Not all such problems are handled the same way. Here it is reasonable to assume that all the money must be invested. Therefore, restricting each investment to \$4000 or less will yield the correct investment. However, if the conditions are different and it is not optimal to invest all the money available, this condition will not be correct. The problem is that $x_1 = \$4000$, $x_2 = \$4000$, and $x_3 = 0$ does not satisfy the verbal instructions given. These constraints do not properly define the feasible set.

A correct way to formulate the percentage constraint is in the form

$$x_1 \leq 0.4(x_1 + x_2 + x_3)$$

Thus, we formulate the problem in the following form:

Maximize

$$0.07x_1 + 0.08x_2 + 0.085x_3$$

subject to

$$x_1 + x_2 + x_3 \leq \$10,000$$

$$0.6x_1 - 0.4x2 - 0.4x3 \leq 0$$
$$-0.4x1 + 0.6x2 - 0.4x3 \leq 0$$
$$-0.4x1 - 0.4x2 + 0.6x3 \leq 0$$

$$x_1 \geq 0, x_2 \geq 0, x_3 \geq 0$$

x_1	x_2	x_3	-1	
1	1	1	10,000	$= -y_1$
0.6	−0.4	−0.4	0	$= -y_2$
−0.4	0.6	−0.4	0	$= -y_3$
−0.4	−0.4	0.6	0	$= -y_4$
0.07	0.08	0.085		$= f$

5. The **Investment Manager's Problem 2.** The additional condition is simple enough, $x_1 \geq 2500$. However, the inequalities for a max program must be in "less-than-or-equal" form. Inequalities for a min program must in "greater-than-or-equal" form. Thus, this inequality must be changed to $-x_1 \leq -2500$. The tableau becomes

x_1	x_2	x_3	-1	
1	1	1	10,000	$= -y_1$
−1			−2500	$= -y_2$
0.6	−0.4	−0.4	0	$= -y_3$
−0.4	0.6	−0.4	0	$= -y_4$
−0.4	−0.4	0.6	0	$= -y_5$
0.07	0.08	0.085		$= f$

6. The **Welfare Mother's Problem.** Minimize $20v_1 + 12v_2 + 16v_3$ subject to

$$v_1 + v_2 \geq 0.9$$

$$v_1 + v_3 \geq 0.6$$

$$v_1 \geq 0, v_2 \geq 0, v_3 \geq 0$$

v_1	1	1	20
v_2	1		12
v_3		1	16
-1	0.9	0.6	
	$= u_1$	$= u_2$	$= g$

7. The **Advertiser's Problem.**

	total	income $\geq$ \$8000	age 18 – 40	cost	
Magazine	8	3	4	40	(per ad)
Television	40	10	10	200	(per ad)
required	160	60	80		(per campaign)
		(millions)			

v_1	8	3	4	40
v_2	40	10	10	200
-1	160	60	80	
	$= u_1$	$= u_2$	$= u_3$	$= g$

8. The **MaxMin Problem.** Once you know how to express the value of the maximum of the minimum, this problem is routine. If x_1, x_2, x_3 are three variables, the minimum of these variables is the largest value of f for which $f \leq x_1$, $f \leq x_2$, $f \leq x_3$. Thus, we simply add these three inequalities to those given. The tableau is

x_1	x_2	x_3	f	-1	
1	2	1		16	$= -y_1$
4	1	3		30	$= -y_2$
1	4	5		40	$= -y_3$
-1			1	0	$= -y_4$
	-1		1	0	$= -y_5$
		-1	1	0	$= -y_6$
0	0	0	1		$= f$

9. The **MinMax Problem.** The maximum of v_1, v_2, v_3 is the smallest g for which $g \geq v_1$, $g \geq v_2$, $g \geq v_3$. Thus, we have

v_1	1	4	1	-1	0	0	0
v_2	2	1	4	0	-1	0	0
v_3	1	3	5	0	0	-1	0
g	0	0	0	1	1	1	1
-1	16	30	40	0	0	0	0
	$= u_1$	$= u_2$	$= u_3$	$= u_4$	$= u_5$	$= u_6$	$= g$

10. The **Maintenance Manager's Problem.** Let v_1, v_2, v_3, v_4, and v_5 be the portion of time the hanger is in each of the five configurations. Then

$$v_1 + v_2 + v_3 + v_4 + v_5 = 1$$

If a plane of type A arrives at a random time, the probability that it will not be accommodated immediately is $v_4 + v_5$, the portion of the time for which space is not available for a type A plane. Thus, the expected number of planes of type A that will have to wait is $9(v_4 + v_5)$. Similar expressions apply for each of the other types of planes. In particular, if g is the maximum of the number of planes that expect to wait then $g \geq 9(v_4 + v_5)$.

The previous problem gives us a model of how to find the minimum of the maximum. Thus, the tableau required is

v_1	0	0	−14	0	−15	1	0
v_2	0	−10	−14	0	0	1	0
v_3	0	−10	0	−12	0	1	0
v_4	−9	0	0	0	−15	1	0
v_5	−9	0	0	−12	0	1	0
g	1	1	1	1	1	0	1
−1	0	0	0	0	0	1	0
	$= u_1$	$= u_2$	$= u_3$	$= u_4$	$= u_5$	$= 0$	$= g$

11. The **Gardener's Problem.** The nitrogen and phosphorus constraints are inequality constraints, but the acidity constraint is an equality constraint. Thus, we have

v_1	0.25	0.10	2	0.20
v_2	0.10	0.05	−1	0.08
v_3	0.25	0.05	1	0.22
v_4	0	0	−10	0.02
−1	100	50	0	0
	$= u_1$	$= u_2$	$= 0$	$= g$

12. The **Metallurgist's Problem.** This is another problem involving percentages. The difference here is that the required percentages are exact. If A, B, ..., I are the fractions of each alloy in the blend, then the constraints are

$$A + B + C + D + E + F + G + H + I = 1$$

$$80A + 60B + 10C + 10D + 40E + 30F + 50G + 10H + 50I = 40$$

If the percentage of tin in the alloy were required to be between 30 and 40%, then we would have two constraints, both canonical.

$$80A + 60B + 10C + 10D + 40E + 30F + 50G + 10H + 50I \leq 40$$

$$80A + 60B + 10C + 10D + 40E + 30F + 50G + 10H + 50I \geq 30$$

This suggests that the given constraint could be formulated as

$$80A + 60B + 10C + 10D + 40E + 30F + 50G + 10H + 50I \leq 40$$

$$80A + 60B + 10C + 10D + 40E + 30F + 50G + 10H + 50I \geq 40$$

If a similar artifice is used for the other equality constraint we could obtain four inequality constraints. In fact, this is a commonly suggested device for converting equality constraints into inequality constraints. The disadvantage with such a device is that it increases the size of the problem to be solved. We handle equality constraints in Chapter 5 in a much more convenient way.

At any rate, the tableau for this problem is

A	1	80	\$4.10
B	1	60	\$4.30
C	1	10	\$5.80
D	1	10	\$6.00
E	1	40	\$7.60
F	1	30	\$7.50
G	1	50	\$7.30
H	1	10	\$6.90
I	1	50	\$7.30
-1	1	40	0
	$= 0$	$= 0$	$= g$

Note the zeros on the bottom margin to represent the equality constraints.

Chapter 2

1.

y_1	x_2	x_3	-1	
1	2	-3	4	$= -x_1$
-2	-1	7	-3	$= -y_2$
3	8	-6	12	$= f$

2.

x_1	x_2	y_1	-1	
$-1/3$	$-2/3$	$-1/3$	$-4/3$	$= -x_3$
$7/3$	$11/3$	$1/3$	$19/3$	$= -y_2$
-2	4	1	4	$= f$

3.

x_1	y_1	-1	
$-1/2$	$-1/2$	$3/2$	$= -x_2$
$9/2$	$3/2$	$-5/2$	$= -y_2$
$-1/2$	$-3/2$	$15/2$	$= -y_3$
$13/2$	$5/2$	$-7/2$	$= f$

4.

x_1	y_1	x_3	-1	
$-1/2$	$-1/2$	0	$3/2$	$= -x_2$
$9/2$	$3/2$	4	$-5/2$	$= -y_2$
$-1/2$	$-3/2$	5	$15/2$	$= -y_3$
5	0	-3	-1.00	$= -y_4$
$13/2$	$5/2$	2	$-7/2$	$= f$

5.

q	s	z	-1	
-5	-50	20	50	$= -p$
$1/5$	0	$3/5$	40	$= -x$
$-3/100$	-0.50	$13/50$	$3/2$	$= -r$
$-1/10$	5	$6/5$	25	$= -y$
$-3/2$	-25	$-13/2$	-525	$= f$

6.

v_1	$15/4$	$35/4$	$25/4$	$-1/4$	5000
u_4	$1/40$	$1/40$	$3/40$	$1/40$	100
-1	11	19	25	-1	-4000
	$= u_1$	$= u_2$	$= u_3$	$= v_2$	$= g$

7.

u_2	$-1/5$	$1/2$	23	$3/10$
v_2	-1	1	50	$4/5$
v_3	-1	$3/2$	110	$6/5$
u_1	$1/5$	$-2/5$	-17	$4/50$
-1	-10	-20	-500	-156
	$= v_4$	$= v_1$	$= u_3$	$= g$

8.

x_{11}	1	1	-1	-1	-1	1
z_2	1	0	-1	0	0	5
z_1	0	-1	0	1	0	2
x_{22}	-1	-1	1	0	0	1
x_{31}	0	1	-1	-1	0	2
y_1	1	0	-1	0	-1	2
-1	-30	-5	-10	-25	-20	-260
	$= x_{32}$	$= y_2$	$= y_3$	$= x_{21}$	$= x_{12}$	$= g$

9.

u_1	1	1	20
v_2	−1	−1	−8
v_3	0	1	16
−1	−0.90	−0.30	−18
	$= v_1$	$= u_2$	$= g$

10.

u_3	2	3/4	1/4	25/2
v_2	20	5/2	5/2	75
−1	0	0	−20	−1000
	$= u_1$	$= u_2$	$= v_1$	$= g$

11.

u_4	−1	−4	−1	−1	0	0	0
u_5	−2	−1	−4	0	−1	0	0
u_6	−1	−3	−5	0	0	−1	0
v_4	4	8	10	1	1	1	1
−1	16	30	40	0	0	0	0
	$= u_1$	$= u_2$	$= u_3$	$= v_1$	$= v_2$	$= v_3$	$= g$

12. Since this tableau has two rows, it is not possible to take more than two independent pivot exchanges. There are six nonzero entries in the tableau. To find those tableaux that are attainable with two independent pivot exchanges, we examine the 2×2 submatrices in the tableau. There are six 2×2 submatrices. Of these six submatrices, five are invertible and one is not invertible. Thus, there is one tableau attainable with zero pivot exchanges (the null pivot), six with one exchange, and five with two exchanges. The equivalence class contains 12 distinct tableaux. Since $(2+4)!/(2!4!) = 15$, there are three sets of variables that cannot be sets of basic variables. The sets $\{x_2, x_5\}$, $\{x_4, x_6\}$, and $\{x_1, x_3\}$ cannot be sets of basic variables.

13. Since $(3+3)!/(3!3!) = 20$, there are at most 20 tableaux in this equivalence class. The class contains the initial tableau and nine attainable in one pivot exchange. The 2×2 submatrices are easily counted by identifying each with a single entry—associate each entry with the submatrix obtained by deleting the row and column containing the entry. There are nine 2×2 submatrices. They are all invertible. The tableau itself is not invertible. Thus, the class of equivalent tableaux contains 19 tableaux.

14. Since $(4+4)!/(4!4!) = 70$, there could be as many as 70 tableaux in this equivalence class. The number is much smaller. There is the tableau itself and one attainable in four independent pivot exchanges. There are four attainable in one pivot exchange and four attainable in three independent pivot exchanges. Of the 36 2×2 submatrices, six are invertible. Thus,

there are 16 tableaux in this equivalence class.

15. Despite the large size of this tableau, it is easy to determine all the basic solutions. Every pivot exchange will leave the A-matrix and the b-column unchanged, and it will change only the sign of one entry in the c-row. Written in the form of Table 2.3, but with the rows and columns interchanged, they are:

	x_1 (u_1)	x_2 (u_2)	x_3 (u_3)	x_4 (u_4)	x_5 (u_5)	x_6 (u_6)	x_7 (u_7)	x_8 (u_8)
A	(−1)	(−1)	(−1)	(−1)	1	1	1	1
B	(−1)	1	(−1)	(−1)	(1)	1	1	1
C	(−1)	(−1)	(−1)	1	1	(1)	1	1
D	1	(−1)	(−1)	(−1)	1	1	(1)	1
E	(−1)	(−1)	1	(−1)	1	1	1	(1)
F	(−1)	1	(−1)	1	(1)	(1)	1	1
G	1	1	(−1)	(−1)	(1)	1	(1)	1
H	1	(−1)	(−1)	1	1	(1)	(1)	1
I	(−1)	1	1	(−1)	(1)	1	1	(1)
J	(−1)	(−1)	1	1	1	(1)	1	(1)
K	1	(−1)	1	(−1)	1	1	(1)	(1)
L	1	(−1)	1	1	1	(1)	(1)	(1)
M	1	1	1	(−1)	(1)	1	(1)	(1)
N	(−1)	1	1	1	(1)	(1)	1	(1)
O	1	1	(−1)	1	(1)	(1)	(1)	1
P	1	1	1	1	(1)	(1)	(1)	(1)

16. Of the 20 conceivable equivalent tableaux, only one is unattainable. If any tableau had two zero entries, there would be at least two unattainable tableaux. Thus, no tableau has more than one zero entry. Furthermore, the only tableaux that have a zero entry are those that are one pivot away from the unattainable configuration. Thus, nine tableaux (those that are two independent pivot exchanges away from the initial tableau) have one zero entry each.

17. The tableau is invertible and the A-matrix of the inverse tableau below is the inverse of the A-matrix of the initial tableau.

y_1	y_2	y_3	-1	
−1	−1	1	−1	$= -x_1$
8	7	−6	9	$= -x_2$
−3	−2	2	−3	$= -x_3$
−4	−4	3	−5	$= f$

18. This tableau is not invertible. Two independent pivot exchanges in the main diagonal will produce the following tableau. The zero in the main diagonal blocks the third pivot exchange.

y_1	y_2	x_3	-1	
1/3	0	−1/3	1/3	$= -x_1$
−2/3	1	8/3	1/3	$= -x_2$
−1	−1	0	−1	$= -y_3$
1/3	−1	−4/3	−2/3	$= f$

19. This tableau is invertible. Four independent pivot exchanges in any order will yield the following tableau.

	x_7	x_5	x_8	x_6	-1	
u_2	0	1	0	0	1	$= -x_2$
u_4	0	0	0	1	1	$= -x_4$
u_1	1	0	0	0	1	$= -x_1$
u_3	0	0	1	0	1	$= -x_3$
-1	−1	−1	−1	−1	0	$= f$
	$= u_7$	$= u_5$	$= u_8$	$= u_6$	$= g$	

This problem illustrates the difference between the inverse of a tableau and the inverse of the A-matrix. No sequence of pivot exchanges will yield a tableau in which the A-matrix is the inverse of the A-matrix of the initial tableau. We will have to rearrange the rows and columns. In the following tableau the A-matrix is the desired inverse matrix.

	x_5	x_6	x_7	x_8	-1	
u_1	0	0	1	0	1	$= -x_1$
u_2	1	0	0	0	1	$= -x_2$
u_3	0	0	0	1	1	$= -x_3$
u_4	0	1	0	0	1	$= -x_4$
-1	−1	−1	−1	−1	0	$= f$
	$= u_5$	$= u_6$	$= u_7$	$= u_8$	$= g$	

20. We cast the problem in tableau form.

	x_1	x_2	x_3	-1	
v_1	1*	−3	5	1	$= -y_1$
v_2	−2	5*	8	4	$= -y_2$
v_3	−3	7	1	6	$= -y_3$
	$= u_1$	$= u_2$	$= u_3$	$= g$	

We attempt to solve the system of equations by pivoting all the y's to make them nonbasic variables. If that is possible, the basic solution will yield the required values of the x's. We perform two pivot exchanges on the starred entries to obtain

	y_1	y_2	x_3	-1	
u_1	-5	-3	-19	-11	$= -x_1$
u_2	-2	-1	-8	-4	$= -x_2$
v_3	-31	-2	0	1	$= -y_3$
	$= v_1$	$= v_2$	$= u_3$	$= g$	

The zero in the third row blocks further pivoting. If the third entry in the b-column were zero, we would set $y_1 = y_2 = 0$ and obtain $y_3 = 0$, which would meet the requirements. Then x_3 would be a parameter for a one-dimensional family of solutions.

The fact that the third entry is nonzero means that the row system of equations does not have a solution. Most students in a course in linear algebra would say, "The rank of the augmented matrix is larger than the rank of the coefficient matrix. Therefore, the system of linear equations does not have a solution." What we are looking for here is the second alternative stated in Theorem 2.10.

The solution for the second alternative is obtained by setting $u_1 = u_2 = 0$ and $v_3 = 1$. This gives $u_1 = u_2 = u_3 = 0$ and $g = 1 \neq 0$, as required. This is not a petty, fine point. If the system of equations was a large system, the statement that rank of the augmented matrix was larger than the rank of coefficient matrix could only be checked by repeating all the calculations. The solution for the dual system can be checked very easily.

21. The application of the block pivot exchange should yield the following tableau.

y_1	y_2	x_3	-1	
2	-1	-4	1	$= -x_1$
-1	1	3	0	$= -x_2$
-9	4	18	-4	$= -y_3$
-1	0	2	-1	$= f$

22.

c_i		x_1	x_2	x_3	x_4	x_5	x_6	
0	x_4	1	1	-1	1	0	0	1
0	x_5	1	2	2	0	1	0	1
0	x_6	5	1	1	0	0	1	1
		1	1	1				0

23.

x_1	x_2	x_3	-1	
2	0	-1	2	$= x_4$
2	1	-2	-2	$= x_5$
5	1	1	3	$= x_6$
1	-2	4	0	$= x_7$

24.

c_i		x_1	x_2	x_3	x_4	x_5	x_6	
0	x_4	1	$-1/2$	0	1	$-1/2$	0	3
4	x_3	-1	$-1/2$	1	0	$-1/2$	0	1
0	x_6	6	$3/2$	0	0	$1/2$	1	2
		5	-0			2		-4

Questions

Q1. True. Theorem 2.12 says that one equivalent tableau can be obtained from the other by a sequence of independent pivot exchanges. The number of possible independent pivot exchanges is at most the minimum of the number of rows and the number of columns.

Q2. True. If any tableau has a zero entry, at least one division of the variables into basic variables and nonbasic variables is impossible.

Q3. True. Find the tableau for which the solution is the basic solution. The basic solution in that tableau for the dual problem is complementary.

Q4. False. In tableaux 2.45 tableaux B and C have the same basic solution for the row equations, but the basic solutions for the column equations are different. Both are basic solutions complementary to the same basic solutions to the row equations.

Q5. True. There is a more general principle at work here. If $L(x) = b$ is a linear equation with solutions S_1 and S_2, and c_1, c_2 are any constants for which $c_1 + c_2 = 1$, then

$$L(c_1S_1 + c_2S_2) = c_1L(S_1) + c_2L(S_2) = c_1b + c_2b = b$$

That is, $c_1S_1 + c_2S_2$ is also a solution. For an average, $c_1 = c_2 = 1/2$.

Q6. False. By Q4 a basic solution can have two different basic complementary solutions. By Q5 the average of two solutions is a solution. The average of two different basic complementary solutions is a complementary solution (not basic).

Q7. False. See the answer to Q6.

Q8. True. Consider any tableau. Assign any nonbasic variable the value 1 and all other nonbasic variables the value zero, and determine the values of the basic variables. Then, assign that same nonbasic variable the value -1 and all other nonbasic variables the value zero, and determine the values of the basic variables. The average of these two solutions assigns all nonbasic variables the value zero. That is, this average solution is a basic solution.

Q9. True. It takes a little care to find an example. We look for an example in two dimensions, since that context is easier to visualize. As a start, consider

x_1	x_2	-1	
1	1	2	$= -y_1$
?	?	?	$= -y_2$

By pivoting in the first row we can see parts of three basic solutions: $(0, 0, 2, ?)$, $(2, 0, 0, ?)$, and $(0, 2, 0, ?)$. The average of the last two basic solutions is $(1, 1, 0, ?)$. This suggests trying a line $y_2 = 0$ through the point $x_1 = 1$, $x_2 = 1$, $y_2 = 0$. So try

x_1	x_2	-1	
1	1	2	$= -y_1$
1	2	3	$= -y_2$

This works.

Q10. False. Consider the tableau

x_1	x_2	-1	
1	1	1	$= -y_1$
1	0	2	$= -y_2$

This tableau is invertible, but the tableau obtained by pivoting in the main diagonal will not be invertible.

Q11. True. For any solution, basic or not, when values are assigned to a set of variables that are nonbasic variables for some tableau in the equivalence class, the values of the other variables are uniquely determined. Consider the solution with the smaller set of variables that are zero. Since it is a basic solution there is a tableau in which some of the variables that are zero are nonbasic variables. Then the values of all other variables are determined and the other proposed solution cannot be a solution, basic or otherwise.

Chapter 3

1. It is possible to write down the dual program in inequality form in one step. However, to be clear we will go through the intermediate step of constructing the tableau with both dual programs shown.

	x_1	x_2	x_3	-1	
v_1	1	-2	3	4	$= -y_1$
v_2	4	5	-6	7	$= -y_2$
v_3	-7	8	9	10	$= -y_3$
-1	10	-11	12	13	$= f$
	$= u_1$	$= u_2$	$= u_3$	$= g$	

Then the min program is easily obtained.
Minimize $g = 4v_1 + 7v_2 + 10v_3 - 13$ subject to

$$\begin{aligned} v_1 + 4v_2 - 7v_3 &\geq 10 \\ -2v_1 + 5v_2 + 8v_3 &\geq -11 \\ 3v_1 - 6v_2 + 9v_3 &\geq 12 \end{aligned}$$

and $v_1 \geq 0$, $v_2 \geq 0$, $v_3 \geq 0$.

2. Both the max program and the min program are feasible. Both basic solutions are optimal.

3. The max program is feasible but unbounded. The min program is infeasible.

4. The max program is infeasible. The min program is feasible but unbounded.

5. Both the max program and the min program are infeasible.

6. Both programs are feasible and both basic solutions are optimal. The optimal solution for the min program is unique. A pivot exchange in the first column will yield a tableau with a different optimal basic solution for the max program.

7. Both programs are feasible. Both basic solutions are optimal. Both programs have multiple solutions. For the max program, x_1 can be increased (to 1) and yields an infinite number of optimal solutions. For the min program, u_5 can be increased (to 5) and yields an infinite number of optimal solutions.

8. Both programs are feasible. Both basic solutions are optimal. The min program has an infinite number of optimal solutions, found by increasing v_2. The basic solution for the max program is unique. Neither x_2 nor x_3 can be increased without decreasing the objective variable. Notice that x_1 cannot be increased without making x_4 negative. A further point of

interest in this problem is that the only optimal solutions for the min program that make u_1 nonzero are not basic solutions. Take u_4 large without bound.

9. Both programs are feasible. Both basic solutions are optimal. The max program has multiple optimal solutions, found by increasing x_1 (to 3). Notice that x_2 can be increased if x_1 is also increased to keep x_4 from becoming negative. The optimal solution for the min program is unique. Here, u_4 cannot be increased without making u_1 negative, and as long as u_4 cannot be increased neither can u_5.

10. The **Furniture Maker's Problem.** We will cast the constraints of both the given problem and its dual in a single tableau.

	x_1	x_2	x_2	x_4	-1	
v_1	4	9	7	10	6000	$= -y_1$
v_2	1	1	3	40	4000	$= -y_2$
-1	12	20	28	40		$= f$
	$= u_1$	$= u_2$	$= u_3$	$= u_4$	$= g$	

The variables of the primal problem have dimensions assigned to them. The dual is not formulated until appropriate dimensions are assigned to them. The Furniture Maker's Problem is similar to the Plastic Shop problem discussed in Section 1.1. Thus, the dual is also formulated in a similar way. The objective variable f is measured in dollars per six months. The objective function g for the dual problem must be in the same units. Since x_j is measured in desks per six months, u_j must be measured in dollars per desk. Since y_i is measured in hours per six months, v_i must be measured in dollars per hour.

11. The **President's Problem.**

	x_1	x_2	-1	
v_1	1	1	20	$= -y_1$
v_2	1		12	$= -y_2$
v_3		1	16	$= -y_3$
-1	0.9	0.6		$= f$
	$= u_1$	$= u_2$	$= g$	

The President's objective is to maximize the number of senators he can convince to support his policy. Let v_1 be a measure of a senator's ability to be convinced. To have a little more fun with this problem, let us refer to this factor as a senator's pliability. Then, v_2 is the extra contribution to his pliability if he is a Republican, and v_3 is the extra contribution to his pliability if he is a Democrat. Since a Republican is 90% pliable, we must have

$$v_1 + v_2 \geq 0.90$$

A Democrat is 60% pliable, so

$$v_1 + v_3 \geq 0.60$$

The Senate's desire is to minimize its total pliability. Thus, we want to minimize

$$g = 20v_1 + 12v_2 + 16v_3$$

12. The **Wyndor Problem** is a production problem similar to the Plastics Shop Problem and the Furniture Maker's Problem. We will give only the tableau part of the solution.

	x_1	x_2	-1	
v_1	1		4	$= -y_1$
v_2		2	12	$= -y_2$
v_3	3	2	18	$= -y_3$
-1	3	5		$= f$
	$= u_1$	$= u_2$	$= g$	

13. The **Financial Manager's Problem.**

	x_1	x_2	x_3	-1	
v_1	1	1	1	10,000	$= -y_1$
v_2	0.6	−0.4	−0.4	0	$= -y_2$
v_3	−0.4	0.6	−0.4	0	$= -y_3$
v_4	−0.4	−0.4	0.6	0	$= -y_4$
-1	0.07	0.08	0.085		$= f$
	$= u_1$	$= u_2$	$= u_3$	$= g$	

The primal problem could have been formulated in several different ways. For example, the variables could have been proportions instead of the amounts invested. In that case the chosen investment strategy would be scaled to any size investment. The point is that the choices made in the primal problem affect the formulation of the dual problem.

Percentages are dimensionless. All the variables for the max program are dimensioned in dollars and all variables for the min program are dimensioned in percentages.

14. Notice that the entries in the Welfare Mother's Problem are the same as the entries in the President's Problem. However, these two problems are not duals of each other since the dimensions of the variables do not match. This problem is a typical minimum requirements problem and is similar to the Feedlot Problem in Section 1.2.

15. The **MaxMin Problem.**

	x_1	x_2	x_3	f	-1	
v_1	1	2	1		16	$= -y_1$
v_2	4	1	3		30	$= -y_2$
v_3	1	4	5		40	$= -y_3$
v_4	-1			1	0	$= -y_4$
v_5		-1		1	0	$= -y_5$
v_6			-1	1	0	$= -y_6$
-1	0	0	0	1		$= f$
	$= u_1$	$= u_2$	$= u_d$	$= u_4$	$= g$	

The variables in the max program are dimensionless, and the variables in the min program are also dimensionless.

16. The **MinMax Problem.** The tableau for this problem is the transpose of the tableau for the MaxMin Problem. Since it is not the negative transpose, the MinMax Problem is not the dual of the MaxMin Problem.

17. The sequence of pivot exchanges called for in the feasibility algorithm is specific. We expect you work will determine whether the row system is feasible and obtain it with the same sequence of pivot exchanges. We list the pivot exchanges in order. (x_9, x_1), (x_{10}, x_3), (x_1, x_5), (x_{11}, x_6), (x_{12}, x_1), (x_3, x_4) $\Rightarrow$ Feasibility.

18. The sequence of pivot exchanges for this tableau is (x_9, x_2), (x_{11}, x_3), (x_2, x_4), (x_8, x_1), (x_3, x_7), (x_{12}, x_8), (x_{13}, x_9) $\Rightarrow$ The x_{14}-row is an infeasible row.

19. In the tableau for Problem 18, set $u_{14} = 1$. This gives $u_2 = 1$, $u_5 = 2$, $u_{13} = 1$, $g = -1$ and all other variables $= 0$.

Questions

Q1. False. The positive b-column implies that the max program is feasible, but the objective variable might be unbounded.

Q2. True. This is the content of Theorem 3.18.

Q3. True. This is the content of Theorem 3.17.

Q4. True. The two basic solutions for any tableau are always complementary. Therefore, because of the duality equation their objective variables are equal.

Q5. True. If the complementary solution were feasible, both solutions would be optimal.

Q6. True. If the complementary solution were optimal, the values of the objective variables would also be equal and optimal.

Q7. True. If the solutions were not complementary their values would be different.

Q8. True. There are only finitely many basic solutions, yet a linear program can have an infinite number of optimal solutions.

Q9. False. When a set is increased any function defined on both sets will have at least as large a range of values on the larger set as it has on the smaller set. The maximum value cannot be smaller. It might not be larger.

Q10. False. See the answer to Q9.

Q11. True. Regardless of the values of the entries in the c-row, the variables on the bottom margin can be made positive by taking the value of the nonbasic variable associated with the positive row sufficiently large.

Q12. True. The assumption means that the b-column is positive so that the row system is feasible. Also, there will be a positive row so that the statement in Q11 will apply.

Q13. True. This is merely the dual of the argument used for Q12.

Q14. True. The nonnegative b-column implies that the row system of equations is feasible.

Q15. True. By assigning all nonbasic variables the same positive value, the sum of the entries in each column of the A-matrix would be a multiple of the average and, therefore, positive. By taking this common value sufficiently large the values of all the variables can be made positive.

Q16. False. This would contradict Theorem 3.15.

Q17. False. The max program in the following tableau has only one optimal basic solution, but it has an infinite number of optimal solutions obtained by assigning positive values to x_1.

x_1	x_2	-1	
-1	1	1	$= -y_1$
0	-1	0	$= f$

Chapter 4

1. In the first column there is one simplex pivot—on the 2. In the second column there is one simplex pivot—on the 2. In the third column there are two simplex pivots. The 3 and the 9 are in a tie. In the fourth column there is one simplex pivot—on the 4.

3. The **Furniture Maker's Problem,**

x_1	x_2	y_1	x_4	-1	
0.47	1.29	0.14	1.43	857.14	$= -x_3$
−0.17	−2.86	−0.43	35.71	1428.57	$= -y_2$
−4	−16	−4	0	−24,000	$= f$

This answer gives the entries in decimal form. If you have done the arithmetic manually and have your answers in rational form, you can reconcile this answer with yours in the following way. Observe that this answer is obtained with a single pivot exchange on the entry 7. Since the entries in the initial tableau were all integers, the rational numbers in the tableau after the pivot exchange will all be multiples of 1/7. To find the rational forms, multiply each entry in the tableau by 7 and round the result to an integer. This integer will be the numerator of the rational entry. When we carry this step out we obtain the following equivalent answer in rational form.

x_1	x_2	y_1	x_4	-1	
4/7	9/7	1/7	10/7	6000/7	$= -x_3$
−5/7	−20/7	−3/7	250/7	10,000/7	$= -y_2$
−4	−16	−4	0	−24,000	$= f$

The optimal solution for the dual program is degenerate, so there are multiple solutions for the max program. We find the other basic optimal solutions by selecting simplex pivot exchanges in those columns with zeros in the c-row. In this way we obtain

x_1	x_2	y_1	y_2	-1	
3/5	7/5	7/50	−1/25	800	$= -x_3$
−1/50	−2/25	−3/250	7/250	40	$= -x_4$
−4	−16	−4	0	−24,000	$= f$

The Furniture Maker should make approximately 857 desks of type 3 (which will leave approximately 1429 units of time in the finishing shop unused), or 800 desks of type 3 and 40 desks of type 4 (which will use all the time available in both the carpentry shop and the finishing shop), or some mixture of these distributions. The optimal solution of the dual min program is unique. The solution of the dual program tells us that desks of type 1 are unprofitable by \$4.00 each, and the desks of type 2 are unprofitable by \$16.00 each. To make these desks profitable we would have to raise the price of each desk accordingly, or find a less expensive way to make the desks. An hour of time in the carpentry shop is worth \$4.00. That means that an additional hour of time available there would allow a

\$4.00 increase in the total profit. An additional hour in the finishing shop would not allow an increase in the profit.

4. The **President's Problem.**

y_2	y_1	-1	
-1	1	8	$= -x_2$
1		12	$= -x_1$
1	-1	8	$= -y_3$
0.3	0.6	-15.6	$= f$

The President should call 12 Republicans and 8 Democrats. He can expect to convince approximately 16 senators. Each senator can attribute a 0.6 pliability to being a senator, 0.3 pliability to being a Republican, and 0 pliability to being a Democrat.

5. The **Wyndor Glass Co. Problem,**

y_2	y_3	-1	
-0.33	0.33	2	$= -x_1$
0.33	-0.33	2	$= -y_1$
0.50	0	6	$= -x_2$
-1.50	-1	-36	$= f$

The Wyndor Glass Co. should make two windows and six doors with its capacity. All the capacity of Plants 2 and 3 would be utilized and 2 units of capacity at P1 would remain unused. An additional unit of capacity at Plant 2 would allow a \$1.50 increase in profit, and an additional unit at Plant 3 would allow a \$1 increase in profit.

6. The **Finance Manager's Problem.**

y_1	y_3	y_4	-1	
0.2	-1	-1	2000	$= -x_1$
0.2	1	1	2000	$= -y_2$
-0.4	1	0	4000	$= -x_2$
0.4	0	1	4000	$= -x_3$
0.08	0.01	0.015	-800	$= f$

There is no surprise. The investment manager should invest \$2000 in type 1 bonds, \$4000 in type 2 bonds, and \$4000 in type 3 bonds. He will get a return of \$800 on this investment. He invests all the money available and he could invest an additional \$2000 in type 1 bonds if it were more profitable. The profitability of an additional dollar available under these circumstances is \$0.08, which reflects the average return rate. If the restrictions were eased to allow more of each type of investment, nothing is gained by allowing more of type 2, an additional 1% is available by easing the restriction on type 3, and an additional 1.5% is available by easing the restriction on type 4.

7. The **Inverstment Manager's Problem 2.**

x_1	x_2	x_3	-1	
1	1	1	10,000	$= -y_1$
-1			-2500	$= -y_2$
0.6	-0.4	-0.4	0	$= -y_3$
-0.4	0.6	-0.4	0	$= -y_4$
-0.4	-0.4	0.6	0	$= -y_5$
0.07	0.08	0.085		$= f$

8. The **Welfare Mother's Problem.**

u_2	-1	1	8
u_1	1		12
v_3	1	-1	8
-1	0.9	0.6	-15.60
	$= v_2$	$= v_1$	$= g$

The Welfare Mother should buy 0.6 ounces of peanut butter, 0.9 small loafs of bread, and no milk. The carbohydrate requirement costs 12 cents a unit and the protein requirement costs 8 cents a unit. Milk is overpriced by 8 cents a cup. Her expenses come to about 16 cents.

9. The **Advertiser's Problem.**

u_2	8/3	1/3	4/3	40/3
v_2	40/3	$-10/3$	$-10/3$	100/3
-1	0	-20	0	-1000
	$= u_1$	$= v_1$	$= u_3$	$= g$

Only magazine ads are placed and the campaign will cost $1,000,000. The solution is unique, but the dual program has multiple optimal solutions. The dual program has four basic optimal solutions. They are obtained by making simplex pivot exchanges in the columns of the zeros in the c-row.

10. The **MaxMin Problem.**

y_4	y_5	y_6	y_2	-1	
-1	$9/2$	$-1/2$	$-1/2$	1	$= -y_1$
$1/2$	$1/8$	$3/8$	$1/8$	$15/4$	$= -f$
-4	$11/4$	$5/4$	$-5/4$	$5/2$	$= -y_3$
$-1/2$	$1/8$	$3/8$	$1/8$	$15/4$	$= -x_1$
$1/2$	$-7/8$	$3/8$	$1/8$	$15/4$	$= -x_2$
$1/2$	$1/8$	$-5/8$	$1/8$	$15/4$	$= -x_3$
$-1/2$	$-1/8$	$-3/8$	$-1/8$	$-15/4$	$= f$

The maximum of the minimum is $15/4$, and $x_1 = x_2 = x_3 = 15/4$. The outcome with the three variables equal to each other is something that experience with this kind of problem will lead one to expect. With this insight there is another approach to this problem. If this is expected, one can see that $y_4 = y_5 = y_6 = 0$. This suggests pivoting on the -1's in the last three rows. With that we obtain

y_4	y_5	y_6	f	-1	
1	2	1	4	16	$= -y_1$
4	1	3	8	30	$= -y_2$
1	4	5	10	40	$= -y_3$
-1	0	0	-1	0	$= -x_1$
0	-1	0	-1	0	$= -x_2$
0	0	-1	-1	0	$= -x_3$
0	0	0	1	0	$= f$

Notice that the last three rows do not constrain the variables. For any nonnegative values given to the nonbasic variables, x_1, x_2, and x_3 will be nonnegative. If one wishes, one could delete the last three rows. Furthermore, the nonbasic variables y_4, y_5, y_6 will be zero in the end, so the first three columns could also be ignored. Then, the optimal solution will be obtained with a single pivot exchange in the fourth column.

11. The **MinMax Problem.** With the insight gained in the previous problem, we can perform pivot exchanges on the -1's in the last three columns, and then perform a dual simplex pivot exchange in the last row. We then obtain

u_4	-0.6	-3.2	0.1	0.1	0.1	0.1	0.1
u_5	-0.4	2.2	0.4	0.4	-0.6	0.4	0.4
u_6	1	1	-0.5	0.5	0.5	-0.5	0.5
u_3	0.4	0.8	0.1	-0.1	-0.1	-0.1	0.1
-1	0	-2	-4	-4	-4	-4	-4
	$= u_1$	$= u_2$	$= g$	$= v_1$	$= v_2$	$= v_3$	$= g$

12. We can always do the pivoting to determine which of the four alternatives applies. However, we can see immediately that the column program is feasible because of the nonpositive entries in the c-row. For $x_1 = 1$, $x_2 = 2$, and $x_3 = x_4 = 0$, we get a negative sum in each row. By scaling these values arbitrarily large we can overwhelm the b-column and make all variables nonnegative. Thus, the row program is also feasible.

13. The given tableau and the following tableau are equivalent. We can see that the row system is feasible. The first column is an infeasible column, so the column system is infeasible.

	x_1	y_3	y_2	x_4	-1	
v_1	$-5/4$	$-1/2$	$-1/4$	$-7/4$	$5/4$	$= -y_1$
u_3	$-1/2$	0	$-1/2$	$-1/2$	$1/2$	$= -x_3$
u_2	$-1/4$	$-1/2$	$-1/4$	$-3/1$	$5/4$	$= -x_2$
v_4	$-1/4$	$-1/2$	$-5/4$	$-3/1$	$9/4$	$= -y_4$
-1	$1/4$	$-1/2$	$1/4$	$3/4$	$-5/4$	$= f$
	$= u_1$	$= v_3$	$= v_2$	$= u_4$	$= g$	

14. The given tableau and the following tableau are equivalent. The column system is feasible. The third row is an infeasible row, so the row system is infeasible.

	y_2	x_2	y_4	x_4	-1	
v_1	0	0	$-2/3$	$2/3$	3	$= -y_1$
u_1	-1	-1	$1/3$	$1/3$	2	$= -x_1$
v_3	0	2	$2/3$	$1/3$	-1	$= -y_3$
u_3	0	0	$1/3$	$2/3$	0	$= -x_3$
-1	-1	-2	$-1/3$	$-8/3$	0	$= f$
	$= v_2$	$= u_2$	$= v_4$	$= u_4$	$= g$	

15. The fourth row is an infeasible row, and the first column is an infeasible column. Thus, both systems are infeasible.

	x_1	x_2	y_2	x_4	-1	
v_1	-4	-3	-2	0	1	$= -y_1$
u_3	-1	-1	-1	0	1	$= -x_3$
v_3	-2	-4	-2	1	1	$= -y_3$
v_4	0	1	1	0	-2	$= -y_4$
-1	2	1	1	1	-3	$= f$
	$= u_1$	$= u_2$	$= v_2$	$= u_4$	$= g$	

Questions

Q1. True. This is the content of the theorem of the four alternatives and the existence-duality theorem. If the row program is infeasible, we must have case 3 or case 4 of the theorem of the four alternatives. If the column program is infeasible, we must have case 2 or case 4. Thus, with the information given we must have case 4.

Q2. True. If the row program is feasible and the column program is infeasible, we must have case 2.

Q3. True. If the column program is feasible and the row program is infeasible, we must have case 3.

Q4. True. If both the row program and the column programs are feasible, we must have case 1.

Q5. True. The existence of a positive row means the column program is feasible, and the existence of a negative column means the row program is feasible. (Of course, both of these cannot occur in the same tableau.) Thus, we must have case 1.

Q6. True. A simplex pivot exchange moves from one tableau with a nonnegative b-column to another tableau with a nonnegative b-column. Thus, the inverse pivot exchange also moves from a tableau with a nonnegative b-column to another tableau with a nonnegative b-column.

Q7. True. If we start with a tableau with a nonnegative b-column and the max program has an optimal solution, the simplex algorithm will lead to an optimal basic solution through a sequence of simplex pivot exchanges. Since the optimal solution is unique, every start from a tableau with a nonnegative b-column will lead to the same tableau. Thus, we can get from any tableau with a nonnegative b-column to any other by going through the tableau for the optimal solution.

Q8. False. The example in the following tableau shows a max program with an optimal solution, but the feasible values of f can be arbitrarily negative.

	x_1	x_2	-1	
v_1	-1	1	1	$= -y_1$
-1	0	-1	-1	$= f$
	$= u_1$	$= u_2$	$= g$	

Q9. True. This is stated in the existence-duality theorem. If both programs have optimal solutions, we must have case 1 of the theorem of the four alternatives. In that case, the tableau described there provides optimal solutions for both programs with the same values for the objective variables.

Q10. True. This is stated in the existence-duality theorem. If both programs have optimal solutions, we must have case 1 of the Theorem of the Four Alternatives. In that case, the tableau described there provides optimal solutions for both programs that are complementary.

Q11. True. A linear function is never unbounded on a bounded set. But if the objective variable is bounded, the max program must have an optimal solution. In that case the min program also has an optimal solution.

Q12. False. The example given in the answer to Q8 illustrates a problem for which the objective variable is unbounded below, but for which the min program has an optimal solution.

Q13. False. The following example shows that it can occur. The first pivot exchange is in the first column.

x_1	x_2	-1	
1	1	1	$= -y$
1	2	0	$= f$

Q14. True. After the first pivot exchange the entry in the c-row of the pivot column will be negative, and that column will not be selected as the pivot column for the second pivot exchange.

Chapter 5

1.

	x_1	x_2	x_3	x_4	x_5	t_6	-1	
v_1	0	0	-14	0	-15	1	0	$= -y_1$
v_2	0	-10	-14	0	0	1	0	$= -y_2$
v_3	0	-10	0	-12	0	1	0	$= -y_3$
v_4	-9	0	0	0	-15	1	0	$= -y_4$
v_5	-9	0	0	-12	0	1	0	$= -y_5$
g	1	1	1	1	1	0	1	$= 0$
-1	0	0	0	0	0	1	0	$= f$
	$= u_1$	$= u_2$	$= u_3$	$= u_4$	$= u_5$	$= 0$	$= g$	

The primal problem is the min program represented by the column system. The artificial basic variable in column 6 results from the fact that the decision variables are percentages. The objective variable, g, is certainly nonnegative since it represents the minimum of nonnegative quantities. However, the objective variable, which is equal to the nonbasic variable in row 6, is taken to be free, and this implies that the row system basic variable in row 6 is artificial.

2.

	x_1	x_2	t_3	-1	
v_1	18	-8	1	8	$= -y_1$
v_2	10	-16	1	6	$= -y_2$
-1	14	-10	1	0	$= f$
	$= u_1$	$= u_2$	$= z_3$	$= g$	

Note that t_3 is free and its dual variable, z_3, is artificial.

3.

	t_1	t_2	-1	
A	1	80	4.10	$= -y_1$
B	1	60	4.30	$= -y_2$
C	1	10	5.80	$= -y_3$
D	1	10	6.00	$= -y_4$
E	1	40	7.60	$= -y_5$
F	1	30	7.50	$= -y_6$
G	1	50	7.30	$= -y_7$
H	1	10	6.90	$= -y_8$
I	1	50	7.30	$= -y_9$
-1	1	40	0	$= f$
	$= 0$	$= 0$	$= g$	

All variables are canonical except for the artificial variables on the bottom margin and their dual variables on the top margin.

4. Using all four forms of dual pairs of variables we can write the dual in the form

Minimize $g = 200u + 30v + 10w - 10$ subject to

$$50u - 6v + 3w \geq 50$$

$$25u + 5v + 10w = 60$$

with $u \geq 0$, v free, and $w \leq 0$.

However, it is confusing to keep track of so many different kinds of side conditions. It is better to reverse the greater-than-or-equal-to constraint that makes w nonpositive and use the tableau

	x	t	-1	
u	50	25	200	$= -p$
s	-6	5	30	$= 0$
w	-3	-10	-10	$= -q$
-1	50	60	10	$= f$
	$= v$	$= 0$	$= g$	

All variables are canonical except s and t, which are free.

5.

	t_1	t_2	t_3	t_4	-1	
A	1	10	10	80	4.10	$= -y_1$
B	1	10	30	60	4.30	$= -y_2$
C	1	40	50	10	5.80	$= -y_3$
D	1	60	30	10	6.00	$= -y_4$
E	1	30	30	40	7.60	$= -y_5$
F	1	30	40	30	7.50	$= -y_6$
G	1	30	20	50	7.30	$= -y_7$
H	1	50	40	10	6.90	$= -y_8$
I	1	20	30	50	7.30	$= -y_9$
-1	1	30	30	40	0	$= f$
	$= 0$	$= 0$	$= 0$	$= 0$	$= g$	

The first column enforces the condition that the variables on the left margin are percentages.

6. The **Maintenance Manager's Problem.** The basic solutions in the following tableau are optimal.

	z_6	y_5	y_3	y_4	y_2	y_1	-1	
z_6	4.62	0.28	0.16	0.23	0.31	0.02	4.62	$= -t_6$
u_5	0.15	−0.02	−0.03	0.04	0.04	−0.03	0.15	$= -x_5$
u_3	0.17	0.05	0.04	−0.03	−0.02	−0.03	0.17	$= -x_3$
u_4	0.19	0.05	−0.04	−0.03	0.05	−0.04	0.19	$= -x_4$
u_2	0.23	−0.04	−0.04	0.06	−0.03	0.05	0.23	$= -x_2$
u_1	0.26	−0.04	0.06	−0.04	−0.04	0.06	0.26	$= -x_1$
-1	−4.62	−0.28	−0.16	−0.23	−0.31	−0.02	−4.62	$= f$
	$= t_6$	$= v_5$	$= v_3$	$= v_4$	$= v_2$	$= v_1$	$= g$	

The minimum for the maximum number of planes of any type that must wait is about 4.62. The units associated with the entries in the tableau are "planes per week." These are also the units associated with the slack variables and the objective variable. The percentages are dimensionless.

The initial tableau suggests that the dual program is a problem of the same type, except that the objective is to maximize a minimum. The units associated with the variables of the dual program are similar to those for the primal problem. We suggest the following narrative for the dual program.

Murphy (the same Murphy of Murphy's Law) wants to cause as much trouble as he can. He can control the probability of a plane's arrival for maintenance. That is, he controls the decision variables x_1 through x_5 for the max program. If the maintenance hanger is in the first configuration, the expected number of planes that will arrive and not find space available is

$14x_3 + 15x_5$. The minimum f for these expectations satisfies the inequality $14x_3 + 15x_5 \geq f$. Similar expressions hold for the other hanger configurations. Murphy's objective is to choose these probabilities to maximize f. His problem is the max program.

If you ask how Murphy can control these probabilities, we can only say that we don't know how he does his dastardly work. However, we can suggest a way this sort of thing can occur. There is a classical puzzle concerning a commuter, Kelly, who goes to the interurban station at random times to take the train south to work. He finds that about one-sixth of the time the next train is going his way, but that five-sixths of the time the train is going north. He suspects that his cousin, Murphy, is working against him. However, the explanation is simple. It is he who arrives at random times, not the trains. The trains are hourly and the schedule has the south-bound train arriving 10 minutes after the north-bound train. He is five times more likely to arrive before the north-bound train than before the south-bound train.

This interpretation is not just fun. There is a serious and valuable side to solving the dual program. By solving the dual program the maintenance manager can assure himself that the optimal schedule provided by solving his program is the best he can expect to do. If Murphy does his best (or worst), he could not do any better.

7. The **Gardener's Problem.** The initial tableau has a nonnegative b-column. The first pivot exchange is a simplex pivot exchange in the third column. Slik-Sod is overpriced by 12 cents per pound. Notice that limestone is overpriced by 12 cents per pound even though it costs only 2 cents per pound.

	x_1	y_2	y_1	-1	
z_3	0.01	−0.50	0.25	0.01	$= -t_3$
u_2	2.25	10.00	5.00	1.80	$= -x_2$
v_3	0.13	0.00	−0.50	0.12	$= -y_3$
v_4	0.13	−5.00	2.50	0.12	$= -y_4$
-1	−12.50	−500.00	−250.00	−90.00	$= f$
	$= u_1$	$= v_2$	$= v_1$	$= g$	

8. The **simpler Metallurgist's Problem.** Because the b-column is initially nonnegative and we have two artificial basic variables, we perform two simplex pivot exchanges, one in each column. Then we use phase II, remembering not to choose a pivot entry in the row on a nonbasic artificial variable.

	y_3	y_2	-1	
z_2	-0.02	0.02	-0.03	$= -t_2$
z_1	1.20	-0.20	6.10	$= -t_1$
v_1	0.40	-1.40	0.40	$= -y_1$
v_4	-1.00	0.00	0.20	$= -y_4$
v_5	-0.40	-0.60	2.70	$= -y_5$
v_6	-0.60	-0.40	2.30	$= -y_6$
v_7	-0.20	-0.80	2.70	$= -y_7$
v_8	-1.00	0.00	1.10	$= -y_8$
v_9	-0.20	-0.80	2.70	$= -y_9$
-1	-0.40	-0.60	-4.90	$= f$
	$= v_3$	$= v_2$	$= g$	

9. The more **complex Metallurgist's Problem.**

	t_1	y_2	y_3	y_4	-1	
v_1	0.00	-1.40	1.00	-0.60	0.28	$= -y_1$
z_2	0.01	0.02	-0.01	0.01	0.03	$= -t_4$
z_3	0.01	-0.00	0.03	-0.02	0.06	$= -t_3$
z_4	0.01	-0.00	-0.01	0.03	0.07	$= -t_2$
v_5	0.00	-0.60	0.00	-0.40	2.62	$= -y_5$
v_6	0.00	-0.40	-0.50	-0.10	2.28	$= -y_6$
v_7	0.00	-0.80	0.50	-0.70	2.56	$= -y_7$
v_8	0.00	0.00	-0.50	-0.5	1.00	$= -y_8$
v_9	0.00	-0.80	0.00	-0.20	2.66	$= -y_9$
-1	0.00	-0.60	0.00	-0.40	-4.98	$= f$
	$= z_1$	$= v_2$	$= v_3$	$= v_4$	$= g$	

The solution is degenerate but unique.

10. The exercises 10 through 21 illustrate the various possibilities represented in Table 5.1. Each exercise represents a possibility not contained in parentheses in that table.

	z_1	z_2	x_3	-1	
u_1	-0.33	1.67	0.67	2.00	$= -x_1$
u_2	-0.33	0.67	-0.33	1.00	$= -x_2$
s_3	-1.00	6.00	0.00	0.00	$= -z_3$
	$= s_1$	$= s_2$	$= u_3$	$= g$	

11. This problem is infeasible. Our intention is that you show the problem is infeasible by verifying the second alternative in Theorem 5.39. From the tableau below we can take $u_1 = u_2 = 0$, $u_3 = 1 \geq 0$, and obtain

a feasible solution of the column equations with $g < 0$.

	z_1	z_2	z_3	-1	
u_1	0.11	−0.22	0.00	2.00	$= -x_1$
u_2	0.67	−0.33	1.00	3.00	$= -x_2$
u_3	−0.33	0.33	−0.67	−1.00	$= -x_3$
	$= s_1$	$= s_2$	$= s_3$	$= g$	

12. The basic solution in the following tableau is feasible.

	y_3	x_2	x_3	-1	
v_1	1.00	3.00	3.00	6.00	$= -y_1$
v_2	−1.00	3.00	6.00	3.00	$= -y_2$
u_1	−0.33	0.33	1.00	2.00	$= -x_1$
	$= v_3$	$= u_2$	$= u_3$	$= g$	

13. The basic solution in the following tableau is feasible. The negative entry is acceptable since the t's are free.

	z_1	z_2	z_3	-1	
z_1	0.11	−0.22	0.00	2.00	$= -t_1$
z_2	0.67	−0.33	1.00	3.00	$= -t_2$
z_3	−0.33	0.33	−0.67	−1.00	$= -t_3$
	$= s_1$	$= s_2$	$= s_3$	$= g$	

14. By taking $v_1 = u_1 = 0$, $v_2 = 1$, we get a feasible solution for the column equations for which $g < 0$.

	y_3	x_2	x_3	-1	
v_1	−0.29	−0.29	2.43	1.29	$= -y_1$
v_2	0.29	0.29	1.57	−0.29	$= -y_2$
u_1	−0.14	0.86	0.71	0.14	$= -x_1$
	$= v_3$	$= u_2$	$= u_3$	$= g$	

15. The basic solution for the following tableau is feasible.

	z_1	z_2	z_3	-1	
z_1	1.00	−2.00	−1.00	2.00	$= -t_1$
z_2	−1.38	2.13	1.00	−2.38	$= -t_2$
z_3	0.25	0.25	0.00	0.25	$= -t_3$
	$= s_1$	$= s_2$	$= s_3$	$= g$	

16. The basic solution for the following tableau is feasible.

	y_1	y_2	x_3	-1	
u_1	-1.00	-1.00	-4.00	5.00	$= -x_1$
u_2	-1.00	-1.50	-5.50	6.00	$= -x_2$
v_3	-1.00	2.00	0.00	1.00	$= -y_3$
	$= v_1$	$= v_2$	$= u_3$	$= g$	

17. By taking $z_1 = z_2 = 0$, $s_3 = -1$, we get a feasible solution for the column equations for which $g < 0$.

	z_1	z_2	t_3	-1	
z_1	-1.00	-1.00	-4.00	5.00	$= -t_1$
z_2	-1.00	-1.50	-5.50	6.00	$= -t_2$
s_3	-1.00	2.00	0.00	1.00	$= -z_3$
	$= s_1$	$= s_2$	$= z_3$	$= g$	

18. By taking $u_1 = v_3 = 0$, $v_2 = 1$, we get a feasible solution for the column equations for which $g < 0$.

	y_1	x_2	x_3	-1	
u_1	-0.33	0.67	-0.33	1.67	$= -x_1$
v_2	0.67	0.67	3.67	-0.33	$= -y_2$
v_3	-2.33	-1.33	-7.33	1.67	$= -y_3$
	$= v_1$	$= u_2$	$= u_3$	$= g$	

19. By taking $z_1 = z_2 = 0$, $s_3 = -1$, we get a feasible solution for the column equations for which $g < 0$.

	z_1	z_2	t_3	-1	
z_1	-1.00	-1.00	-4.00	2.00	$= -t_1$
z_2	1.00	1.50	5.50	-0.50	$= -t_2$
s_3	-1.00	2.00	0.00	1.00	$= -z_3$
	$= s_1$	$= s_2$	$= z_3$	$= g$	

20. The basic solution for the following tableau is feasible.

	y_1	y_2	t_3	-1	
z_1	-1.00	-1.00	-4.00	2.00	$= -t_1$
z_2	1.00	1.50	5.50	-0.50	$= -t_2$
v_3	-1.00	2.00	0.00	1.00	$= -y_3$
	$= v_1$	$= v_2$	$= z_3$	$= g$	

21. By taking $z_1 = z_2 = 0$, $v_3 = 1$, we obtain a feasible solution for the column equations for which $g < 0$.

	y_1	y_2	t_3	-1	
z_1	-1.00	-1.00	-4.00	1.00	$= -t_1$
z_2	-1.00	-1.50	-5.50	-0.50	$= -t_2$
v_3	1.00	2.00	0.00	-1.00	$= -y_3$
	$= v_1$	$= v_2$	$= z_3$	$= g$	

22. One tableau with optimal basic solutions is

	x_1	y_1	x_3	-1	
u_2	2.67	0.33	1.33	16.67	$= -x_2$
v_2	13.33	-3.33	-3.33	33.33	$= -y_2$
-1	0.00	-20.00	0.00	-1000.00	$= f$
	$= u_1$	$= v_1$	$= u_3$	$= g$	

The solution for the min program is unique, but it is degenerate. The nonbasic variables, x_1 and x_3, are paired with zero dual variables. Perform a simplex pivot exchange in each of the columns of x_1 and x_3 to obtain

	y_2	y_1	x_2	-1	
u_3	-0.10	0.50	0.50	5.00	$= -x_3$
u_1	0.05	-0.13	0.13	3.75	$= -x_1$
-1	0.00	-20.00	0.00	-1000.00	$= f$
	$= v_2$	$= v_1$	$= u_2$	$= g$	

The optimal solution for the min program is the same, though it is a different basic solution. There are many choices for a pair of solutions that meet our requirements. None will include a basic solution for the max program. One possibility is to take the average of the two optimal solutions for the max program. We obtain $x_1 = 15/8$, $x_2 = 25/3$, $x_3 = 5/2$, $y_1 = 0$, $y_2 = 50/3$.

23. This problem can be worked by any combination of the methods described in this chapter. Even empirical pivoting will work. For a large problem we recommend the following procedure, which can be implemented effectively on a computer and which does not require human judgment.

First, insert a column headed by an artificial variable with -1's in the rows of any negative entries in the b-column.

	x_1	x_2	x_3	t_4	x_5	t_6	z_7	-1	
v_1	−2.00	0.00	0.00	−1.00	1.00	1.00	−1.00*	−2.00	$= -y_1$
v_2	−2.00	−1.00	−1.00	−1.00	−2.00	1.00	−1.00	−1.00	$= -y_2$
v_3	1.00	−1.00	0.00	1.00	−2.00	0.00	−1.00	−2.00	$= -y_3$
v_4	−2.00	−2.00	−1.00	0.00	0.00	−2.00	0.00	0.00	$= -y_4$
v_5	−1.00	−2.00	−1.00	0.00	−2.00	0.00	−1.00	−2.00	$= -y_5$
s_6	−1.00	−1.00	0.00	−1.00	0.00	1.00	−1.00	−2.00	$= -z_6$
-1	−2.00	−1.00	−2.00	0.00	−1.00	−1.00	0.00	−2.00	$= f$
	$= u_1$	$= u_2$	$= u_3$	$= z_4$	$= u_5$	$= z_6$	$= s_7$	$= g$	

We pivot on an entry in the new column to obtain a nonnegative b-column. We then insert a new row with the entries the sum of the entries in the rows of artificial variables.

	x_1	x_2	x_3	t_4	x_5	t_6	y_1	-1	
s_7	2.00	0.00	0.00	1.00	−1.00	−1.00	−1.00	2.00	$= -z_7$
v_2	0.00	−1.00	−1.00	0.00	−3.00	0.00	−1.00	1.00	$= -y_2$
v_3	3.00	−1.00	0.00	2.00	−3.00	−1.00	−1.00	0.00	$= -y_3$
v_4	−2.00	−2.00	−1.00	0.00	0.00	−2.00	0.00	0.00	$= -y_4$
v_5	1.00	−2.00	−1.00	1.00	−3.00	−1.00	−1.00	0.00	$= -y_5$
s_6	1.00	−1.00	0.00	0.00	−1.00	0.00	−1.00	0.00	$= -z_6$
v_7	3.00	−1.00	0.00	1.00	−2.00	−1.00	−2.00	2.00	$= -w$
-1	−2.00	−1.00	−2.00	0.00	−1.00	−1.00	0.00	−2.00	$= f$
	$= u_1$	$= u_2$	$= u_3$	$= z_4$	$= u_5$	$= z_6$	$= v_1$	$= g$	

We then perform a simplex pivot exchange in every column of a free variable. In this example, we pivot in the t_4-column and the t_6-column. We obtain

	x_1	x_2	x_3	y_5	x_5	y_3	y_1	-1	
s_7	1.00	2.00	1.00	−1.00	2.00	0.00	0.00	2.00	$= -z_7$
v_2	0.00	−1.00	−1.00	0.00	−3.00	0.00	−1.00	1.00	$= -y_2$
z_6	1.00	3.00	2.00	−2.00	3.00	1.00	1.00	0.00	$= -t_6$
v_4	0.00	4.00	3.00	−4.00	6.00	2.00	2.00	0.00	$= -y_4$
z_4	2.00	1.00	1.00	−1.00	0.00	1.00	0.00	0.00	$= -t_4$
s_6	1.00	−1.00	0.00	0.00	−1.00	0.00	−1.00	0.00	$= -z_6$
v_7	2.00	1.00	1.00	−1.00	1.00	0.00	−1.00	2.00	$= -w$
-1	−1.00	2.00	0.00	−2.00	2.00	1.00	1.00	−2.00	$= f$
	$= u_1$	$= u_2$	$= u_3$	$= v_5$	$= u_5$	$= v_3$	$= v_1$	$= g$	

We are now ready to begin phase I. For a large problem, the number of preceding steps is acceptably small. The number of pivot exchanges is not larger than the minimum of the number of free variables and the number of rows. In phase I we perform simplex pivot exchanges in the x_1-column, the x_2-column, and the y_5-column. We obtain

	z_6	y_4	x_3	z_7	x_5	y_3	y_1	-1	
v_5	-0.50	-0.38	-0.63	0.50	-0.75	-0.75	-0.25	1.00	$= -y_5$
v_2	-0.50	-0.13	-0.88	0.50	-2.25	-0.25	-0.75	2.00	$= -y_2$
z_6	0.00	-0.25	0.25	-1.00	-0.50	0.50	0.50	-2.00	$= -t_6$
u_2	-0.50	-0.13	0.13	0.50	0.75	-0.25	0.25	1.00	$= -x_2$
z_4	-1.00	0.00	0.00	-1.00	-1.00	1.00	1.00	-2.00	$= -t_4$
u_1	0.50	-0.13	0.13	0.50	-0.25	-0.25	-0.75	1.00	$= -x_1$
v_7	-1.00	0.00	0.00	-1.00	0.00	0.00	0.00	0.00	$= -w$
-1	0.50	-0.63	-1.38	0.50	-1.25	-0.25	-0.75	-1.00	$= f$
	$= s_6$	$= v_4$	$= u_3$	$= s_7$	$= u_5$	$= v_3$	$= v_1$	$= g$	

The minimum value of w is zero so the problem is feasible. The negative entries in the b-column are acceptable since they correspond to free variables, and the positive entry in the c-row is acceptable since it corresponds to an artificial variable. We can start phase II. Note that we do not select a pivot entry in the column of an artificial variable or the row of a free variable. Actually, no further work is required in phase II. Finally, we remove the artificial column and artificial row that were introduced.

	z_6	y_4	x_3	x_5	y_3	y_1	-1	
v_5	-0.50	-0.38	-0.63	-0.75	-0.75	-0.25	1.00	$= -y_5$
v_2	-0.50	-0.13	-0.88	-2.25	-0.25	-0.75	2.00	$= -y_2$
z_6	0.00	-0.25	0.25	-0.50	0.50	0.50	-2.00	$= -t_6$
u_2	-0.50	-0.13	0.13	0.75	-0.25	0.25	1.00	$= -x_2$
z_4	-1.00	0.00	0.00	-1.00	1.00	1.00	-2.00	$= -t_4$
u_1	0.50	-0.13	0.13	-0.25	-0.25	-0.75	1.00	$= -x_1$
-1	0.50	-0.63	-1.38	-1.25	-0.25	-0.75	-1.00	$= f$
	$= s_6$	$= v_4$	$= u_3$	$= u_5$	$= v_3$	$= v_1$	$= g$	

24. As with Exercise 23, we first introduce an artificial column. When we worked that exercise, we introduced the artificial row after the first pivot exchange. However, it is possible to introduce that row at the same time as we introduce the artificial column. In the following tableau, notice that the w-row is the sum of the two artificial rows, except in the z_7-column. There we have -3 instead of -2 since w is the sum of three artificial variables (don't forget z_7).

	x_1	x_2	x_3	x_4	x_5	t_6	z_7	-1	
v_1	−2.00	0.00	0.00	−1.00	1.00	1.00	−1.00*	−2.00	$= -y_1$
v_2	−2.00	−1.00	−1.00	−1.00	−2.00	1.00	−1.00	−1.00	$= -y_2$
v_3	1.00	−1.00	0.00	1.00	−2.00	0.00	−1.00	−2.00	$= -y_3$
v_4	−2.00	−2.00	−1.00	0.00	0.00	−2.00	0.00	0.00	$= -y_4$
s_5	−1.00	−2.00	−1.00	0.00	−2.00	0.00	−1.00	−2.00	$= -z_5$
s_6	−1.00	−1.00	0.00	−1.00	0.00	1.00	−1.00	−2.00	$= -z_6$
s	−2.00	−3.00	−1.00	−1.00	−2.00	1.00	−3.00	−4.00	$= -w$
-1	−2.00	−1.00	−2.00	0.00	−1.00	−1.00	0.00	−2.00	$= f$
	$= u_1$	$= u_2$	$= u_3$	$= u_4$	$= u_5$	$= z_6$	$= s_7$	$= g$	

A single pivot exchange in the artificial column gives us the following tableau.

	x_1	x_2	x_3	x_4	x_5	t_6	y_1	-1	
s_7	2.00	0.00	0.00	1.00	−1.00	−1.00	−1.00	2.00	$= -z_7$
v_2	0.00	−1.00	−1.00	0.00	−3.00	0.00	−1.00	1.00	$= -y_2$
v_3	3.00	−1.00	0.00	2.00	−3.00	−1.00*	−1.00	0.00	$= -y_3$
v_4	−2.00	−2.00	−1.00	0.00	0.00	−2.00	0.00	0.00	$= -y_4$
s_5	1.00	−2.00	−1.00	1.00	−3.00	−1.00	−1.00	0.00	$= -z_5$
s_6	1.00	−1.00	0.00	0.00	−1.00	0.00	−1.00	0.00	$= -z_6$
s	4.00	−3.00	−1.00	2.00	−5.00	−2.00	−3.00	2.00	$= -w$
-1	−2.00	−1.00	−2.00	0.00	−1.00	−1.00	0.00	−2.00	$= f$
	$= u_1$	$= u_2$	$= u_3$	$= u_4$	$= u_5$	$= z_6$	$= v_1$	$= g$	

There is only one free variable, so we perform a simplex pivot in the t_6-column.

	x_1	x_2	x_3	x_4	x_5	y_3	y_1	-1	
s_7	−1.00	1.00	0.00	−1.00	2.00	−1.00	0.00	2.00	$= -z_7$
v_2	0.00	−1.00	−1.00	0.00	−3.00	0.00	−1.00	1.00	$= -y_2$
z_6	−3.00	1.00	0.00	−2.00	3.00	−1.00	1.00	0.00	$= -t_6$
v_4	−8.00	0.00	−1.00	−4.00	6.00	−2.00	2.00	0.00	$= -y_4$
s_5	−2.00	−1.00	−1.00	−1.00	0.00	−1.00	0.00	0.00	$= -z_5$
s_6	1.00	−1.00	0.00	0.00	−1.00	0.00	−1.00	0.00	$= -z_6$
s	−2.00	−1.00	−1.00	−2.00	1.00	−2.00	−1.00	2.00	$= -w$
-1	−5.00	0.00	−2.00	−2.00	2.00	−1.00	1.00	−2.00	$= f$
	$= u_1$	$= u_2$	$= u_3$	$= u_4$	$= u_5$	$= v_3$	$= v_1$	$= g$	

We start phase I and obtain the following tableau.

	x_1	x_2	x_3	x_4	y_4	y_3	y_1	-1	
s_7	1.67	1.00	0.33	0.33	−0.33	−0.33	−0.67	2.00	$= -z_7$
v_2	−4.00	−1.00	−1.50	−2.00	0.50	−1.00	0.00	1.00	$= -y_2$
z_6	1.00	1.00	0.50	0.00	−0.50	0.00	0.00	0.00	$= -t_6$
u_5	−1.33	0.00	−0.17	−0.67	0.17	−0.33	0.33	0.00	$= -x_5$
s_5	−2.00	−1.00	−1.00	−1.00	0.00	−1.00	0.00	0.00	$= -z_5$
s_6	−0.33	−1.00	−0.17	−0.67	0.17	−0.33	−0.67	0.00	$= -z_6$
s	−0.67	−1.00	−0.83	−1.33	−0.17	−1.67	−1.33	2.00	$= -w$
-1	−2.33	0.00	−1.67	−0.67	−0.33	−0.33	0.33	−2.00	$= f$
	$= u_1$	$= u_2$	$= u_3$	$= u_4$	$= v_4$	$= v_3$	$= v_1$	$= g$	

Since the w-row contains nonpositive entries, we have a positive minimum for w. Thus, the original problem is infeasible.

Questions

Q1. False. Feasibility specifications are just constraints in a special form. Relaxing the feasibility specifications will enlarge the feasible set or leave it unchanged or, if the feasible set is empty, make the feasible set nonempty.

Q2. True. The feasible set might be larger.

Q3. False. The feasible set might be larger, but the additional points might not include a point where the objective variable is larger.

Q4. True. See the answer to Q3.

Q5. True. The feasible set might be smaller.

Q6. False. See the answer to Q5.

Q7. False. The feasible set might be smaller, but if the remaining set includes the point where the optimal value occurred there will be no change in the optimal value.

Q8. True. See the answer to Q7.

Q9. True. The existence-duality theorem is true for general linear programs as well as for canonical linear programs.

Q10. True. If the program does not have an optimal solution, the dual program is infeasible. If the dual program is infeasible, the primal program is unbounded. For those with a background in analysis a more general argument is available. The feasible set is closed since it is the intersection of a finite number of closed sets. A linear function is continuous. The continuous image of a closed set is closed. That is, the set of feasible values for the objective variable is a closed set. If it is bounded, it includes its least upper bound. This is the optimal value.

Q11. True. By the complementary slackness theorem, for each pair of canonical dual variables, there exists an optimal solution for one of the problems for which the variable in that pair is nonzero. Thus, one or the other of the problems in this case has an additional optimal solution.

Q12. True. Consider the following example.

	x	t	-1	
v	1	0	1	$= -y$
s	0	1	0	$= -0$
-1	-1	0	0	$= f$
	$= u$	$= 0$	$= g$	

Both basic solutions are optimal. The equation for s is $s = 0$, and the equation for t is $t = 0$.

Q13. True. See the answer to Q11.

Chapter 6

2.

$$C = \begin{bmatrix} 1 & 0 & 0 \\ -1 & 4 & 0 \\ 3 & -5 & -2 \end{bmatrix} \qquad R = \begin{bmatrix} 1 & 2 & 3 \\ 0 & 1 & 1 \\ 0 & 0 & 1 \end{bmatrix}$$

3.

$$C = \begin{bmatrix} 1 & 0 & 0 \\ -1 & 4 & 0 \\ 3 & -7 & 2 \end{bmatrix} \qquad R = \begin{bmatrix} 1 & 2 & 3 \\ 0 & 1 & 1 \\ 0 & 1 & 0 \end{bmatrix}$$

4.

$$C = \begin{bmatrix} 1 & 0 & 0 \\ -1 & 4 & -8/5 \\ 3 & -5 & 0 \end{bmatrix} \qquad R = \begin{bmatrix} 1 & 2 & 3 \\ 0 & 1 & 7/5 \\ 0 & 0 & 1 \end{bmatrix}$$

5.

x_1	x_2	x_3	x_4	x_5	-1	
1	2	-3	1	0	4	$= 0$
2	3	1	0	1	7	$= 0$
-3	2	3	0	0	0	$= f$

6. The first step is to introduce the bound as an extra constraint.

x_1	x_2	x_3	x_4	x_5	-1	
1	2	-3	1	0	4	$= 0$
2	3	1	0	1	7	$= 0$
1	1	1	1	1	50	$= -x_6$
-3	2	3	0	0	0	$= f$

Then we have to scale the coefficients to reduce the bound to 1.

x_1	x_2	x_3	x_4	x_5	-1	
50	100	-150	50	0	4	$= 0$
100	150	50	0	50	7	$= 0$
1	1	1	1	1	1	$= -x_6$
-150	100	150	0	0	0	$= f$

Then we pivot on the 1 in the last column.

x_1	x_2	x_3	x_4	x_5	x_6	
46	96	-154	46	-4	-4	$= 0$
93	143	43	-7	43	-7	$= 0$
1	1	1	1	1	1	$= 1$
-150	100	150	0	0	0	$= f$

Chapter 7

1. The linear program for this game is

	q_1	q_2	q_3	q_4	t_5	-1	
p_1	2	3	-1	2	1	0	$= -s_1$
p_2	1	0	3	2	1	0	$= -s_2$
p_3	-1	2	1	-1	1	0	$= -s_3$
t_4	1	1	1	1	0	1	$= -z_4$
-1	0	0	0	0	1	0	$= f$
	$= r_1$	$= r_2$	$= r_3$	$= r_4$	$= z_5$	$= g$	

The tableau for the optimal solutions is

	s_3	s_2	q_4	s_1	-1	
r_3	0.10	0.10	0.10	-0.20	0.40	$= -q_3$
r_2	0.25	-0.25	-0.25	0.00	0.50	$= -q_2$
r_1	-0.35	0.15	1.15	0.20	0.10	$= -q_1$
-1	-0.05	-0.55	-0.55	-0.40	1.30	$= f$
	$= p_3$	$= p_2$	$= r_4$	$= p_1$	$= g$	

The optimal strategy for the row player is $p = (0.4, 0.55, 0.05)$ and the optimal strategy for the column player is $q = (0.1, 0.5, 0.4, 0)$. The value of the game is 1.3 (to the row player.) The optimal strategies are unique since the tableau for the optimal solutions is nondegenerate.

2. The game matrix for this game was obtained from the game matrix for Exercise 1 by multiplying each entry by 2. This does not change the optimal strategies. The value of the game is multiplied by 2.

3. The game matrix for this game was obtained from the game matrix for Exercise 1 by adding 2 to each entry. This does not change the optimal strategies. The value of the game is 2 more than the value of the game for Exercise 1.

4. The game matrix is

	q_1	q_2	q_3	q_4	q_5	t_6	-1	
p_1	-1	0	-1	1	-1	1	0	$= -s_1$
p_2	2	-1	-1	0	1	1	0	$= -s_2$
p_3	-1	-1	1	0	1	1	0	$= -s_3$
p_4	0	2	-1	-2	0	1	0	$= -s_4$
p_5	0	1	1	0	-1	1	0	$= -s_5$
t_5	1	1	1	1	1	0	1	$= -z_4$
-1	0	0	0	0	0	1	0	$= f$
	$= r_1$	$= r_2$	$= r_3$	$= r_4$	$= r_5$	$= z_5$	$= g$	

A tableau for an optimal solution is

	z_6	s_2	s_4	s_3	s_5	s_1	-1	
z_6	0.00	0.21	0.10	0.24	0.26	0.19	0.00	$= -t_6$
r_2	0.33	-0.17	0.22	-0.11	-0.06	0.11	0.33	$= -q_2$
r_4	0.33	-0.02	-0.16	-0.06	-0.10	0.35	0.33	$= -q_4$
r_3	0.00	-0.07	-0.14	0.14	0.36	-0.29	0.00	$= -q_3$
r_5	0.33	-0.02	0.17	0.27	-0.44	0.02	0.33	$= -q_5$
r_1	0.00	0.29	-0.10	-0.24	0.24	-0.19	0.00	$= -q_1$
-1	0.00	-0.21	-0.10	-0.24	-0.26	-0.19	0.00	$= f$
	$= t_6$	$= p_2$	$= p_4$	$= p_3$	$= p_5$	$= p_1$	$= g$	

Since the solution for the row player is nondegenerate, the optimal solution for the column player is unique. However, the solution for the column player is degenerate and we must look for additional optimal solutions for the row player. The unique optimal solution for the column player is $q = (0, 1/3, 0, 1/3, 1/3)$. The optimal solution for the row player obtained from this tableau is $p = (8/42, 9/42, 10/42, 4/42, 11/42)$.

The other extreme point optimal solutions are found by performing dual simplex pivots in the rows of the zero entries in the b-column. We obtain $(0, 1/6, 1/3, 1, 1/2), (0, 1/2, 0, 0, 1/2)$, and $(1/3, 0, 1/12, 1/6, 1/12)$.

This game is an illustration of a fair game that is not symmetric. A fair game can be constructed by taking any game matrix and subtracting the value of the game from each entry. It is easy to construct a game matrix to achieve any value and optimal strategy that you want. Write the values of the desired optimal strategies on the margins of an empty matrix and fill in the matrix to achieve the desired sums, row by row and column by column.

5. **Scissors–Paper–Stone**. The game matrix is

	scissors	paper	stone
scissors	0	1	−1
paper	−1	0	1
stone	1	−1	0

It is easy to guess an optimal solution. The game is symmetric, so both players have the same optimal solutions and the value of the game is zero. The point of the exercise is that when you can guess a solution, you can verify it by trying it. The solutions are unique. There is no need to compute the solution.

	1/3	1/3	1/3	
1/3	0	1	−1	= 0
1/3	−1	0	1	= 0
1/3	1	−1	0	= 0
	= 0	= 0	= 0	

6. Strictly speaking, this game is not a finite game since there is no upper limit to the integer that a player can choose. However, there will be an upper limit to choices that will be played with a positive probability and the game can be solved as a finite matrix game. As soon as the probability of the preceding number (against which that choice will win 2) falls to less than one-half the sum of the probabilities of the preceding numbers (against which that choice will lose 1), it will be unprofitable to make the choice.

Use the fact that the game is symmetric and that the value of the game must be zero. Try several small size versions and test the optimal solution against a possible choice of a larger integer. It is not hard to obtain $(1/16, 5/16, 4/16, 5/16, 1/16, 0, 0, \ldots)$ as the optimal solution for both players.

7. **Morra.** We use ij to denote the choice of extending i fingers and guessing that the other player extends j fingers. Then the game matrix is

	11	12	13	21	22	23	31	32	33
11	0	2	2	−3	0	0	−4	0	0
12	−2	0	0	0	3	3	−4	0	0
13	−2	0	0	−3	0	0	0	4	4
21	3	0	3	0	−4	0	0	−5	0
22	0	−3	0	4	0	4	0	−5	0
23	0	−3	0	0	−4	0	5	0	5
31	4	4	0	0	0	−5	0	0	−6
32	0	0	−4	5	5	0	0	0	−6
33	0	0	−4	0	0	−5	6	6	0

To get the game matrix, place zeros down the main diagonal, since the game is symmetric. We divide the game matrix into zones in which the sum of the fingers extended is constant. Then the row player wins (or ties) in the first row of each zone in the first column, in the second row of each zone in the second column, and in the third row of each zone in the third column.

The following is a tableau for a pair of optimal solutions.

	s_4	s_5	s_7	s_2	32	11	s_3	33	12	−1	
r_5	−0.11	0.06	−0.06	0.08	1.28	0.04	0.03	0.04	0.77	0.32	$= -22$
r_6	0.00	−0.15	0.06	0.00	0.02	−0.77	0.09	1.23	−0.79	0.00	$= -23$
11	−0.70	0.00	0.70	−0.97	0.00	0.00	−0.03	0.00	0.00	0.11	$= -s_1$
r_4	0.00	0.09	0.11	0.00	−1.30	0.72	−0.19	−1.28	0.02	0.00	$= -21$
23	−0.03	0.00	−0.97	1.27	0.00	0.00	−1.27	0.00	0.00	0.08	$= -s_6$
r_3	0.19	0.00	−0.19	0.11	0.00	1.00	−0.11	0.00	1.00	0.43	$= -13$
32	1.30	−1.00	−1.30	0.03	0.00	0.00	−0.03	0.00	0.00	0.11	$= -s_8$
33	1.24	−1.00	−1.24	1.57	0.00	0.00	−1.57	0.00	0.00	0.27	$= -s_9$
r_7	−0.08	0.00	0.08	−0.19	1.00	0.00	0.19	1.00	0.00	0.24	$= -31$
−1	0.00	−0.26	−0.32	0.00	−0.11	−0.17	−0.43	−0.17	−0.06	0.00	$= f$
	$= 21$	$= 31$	$= 22$	$= 12$	$= r_8$	$= r_1$	$= 13$	$= r_9$	$= r_2$	$= g$	

This is confusing and not very informative. However, enough information can be gleaned to simplify things. Notice that for this optimal solution, r_1, r_2, r_8, and r_9 on the bottom margin are positive, and s_6 on the right margin is positive. This means the choices to which they are dual will be played with zero probabilities in every optimal strategy. And, since the game is symmetric the same choices will be avoided by both players. This allows us to start over with a much simpler game matrix and corresponding linear program. Consider the following game matrix.

	13	21	22	31
11	2	−3	0	−4
12	0	0	3	−4
13	0	−3	0	0
21	3	0	−4	0
22	0	4	0	0
23	0	0	−4	5
31	0	0	0	0
32	−4	5	5	0
33	−4	0	0	6

We have deleted the columns corresponding to choices that will be played with zero probability in an optimal strategy. We have not also deleted the corresponding rows since the optimal strategies for the column player must work against all choices for the row player. We can remove rows if they are redundant. For example, we can delete the 31-row.

When we construct the linear program and solve it we end up with

	21	s_4	s_2	s_5	-1	
11	−0.19	−0.70	−0.97	0.68	0.11	$= -s_1$
r_7	1.32	−0.08	−0.19	0.27	0.24	$= -31$
13	−7.00	0.00	0.00	−1.00	0.00	$= -s_3$
r_5	0.43	−0.11	0.08	0.03	0.32	$= -22$
23	−8.89	−0.03	1.27	−2.24	0.08	$= -s_6$
32	−4.19	1.30	0.03	−2.32	0.11	$= -s_8$
33	−7.03	0.76	0.43	−2.19	1.73	$= -s_9$
r_3	−0.76	0.19	0.11	−0.30	0.43	$= -13$
−1	−4.00	0.00	0.00	−1.00	0.00	$= f$
	$= r_4$	$= 21$	$= 12$	$= 22$	$= g$	

This doesn't seem to help much. But we note that there are multiple solutions, so we try them. We can't pivot in the first column since that is the column of an artificial variable. However, we can perform simplex pivots in the third and fourth columns to obtain three more extreme optimal solutions. Among them we get

	21	s_8	s_2	s_5	-1	
11	-2.46	0.54	-0.96	-0.58	0.17	$= -s_1$
r_7	1.06	0.06	-0.19	0.13	0.25	$= -31$
13	-7.00	0.00	0.00	-1.00	0.00	$= -s_3$
r_5	0.08	0.08	0.08	-0.17	0.33	$= -22$
23	-8.98	0.02	1.27	-2.29	0.08	$= -s_6$
21	-3.23	0.77	0.02	-1.79	0.08	$= -s_4$
33	-4.58	-0.58	0.42	-0.83	1.67	$= -s_9$
r_3	-0.15	-0.15	0.10	0.04	0.42	$= -13$
-1	-4.00	0.00	0.00	-1.00	0.00	$= f$
	$= r_4$	$= 32$	$= 12$	$= 22$	$= g$	

Here we obtain a relatively simple rational solution: $q = (0, 0,\ 5/12,\ 0,\ 1/3,\ 0,\ 1/4,\ 0,\ 0)$. This solution is optimal for both players.

As soon as it is established that only three choices will be played with positive probabilities, and both players obey that condition, it doesn't matter what those probabilities are. The 3×3 game matrix with just those strategies is a zero matrix. The players will consistently draw. However, if one player uses an optimal strategy and the other player tires of the draws and explores other choices, he will lose (on the average).

8. The **Bass and the Professor.** The game matrix is

	'tails	'flies	'bees
'tails	2	0	0
'flies	0	6	0
'bees	0	0	30

The optimal strategy for the professor is $(15/21, 5/21, 1/21)$ and the value of the game is $10/7$. This is not an absolute value since the game matrix is scaled. However, the game is clearly unfavorable for the bass.

9. The **Detective and the Mobster.** The game matrix for this problem is

	1	2	3	4
1	1	0	0	0
2	0	1	1/2	1/3
3	0	0	1	2/3
4	0	0	0	1

The entries in this matrix are the probabilities of success (to the detective), where the detective chooses the row and the mobster chooses the column. The only entries that might require an explanation are the fractions. The determining factor is the time that the tallest, among those who leave before the mobster, chooses to leave. If that person leaves before the time the detective chooses, no one will be followed until the mobster leaves. If he leaves at the chosen time or later, if no one who leaves earlier is followed he will be followed. In either case the mobster will get away. Set up the game matrix if the mobster has four cohorts.

When the optimal solutions are obtained we have $p = (6/17, 6/17, 3/17, 2/17)$ for the detective and $q = (6/17, 3/17, 2/17, 6/17)$ for the mobster. The value of the game is $6/17 = 0.35294$, a significant improvement from the 0.25 expectation without using this strategy.

Set up the game matrix if the mobster has four cohorts. The four-cohort game is the one discussed by Morton D. Davis in *Game Theory, a Nontechnical Introduction*, Harper Torchbooks, New York, 1970, p36. He gives a solution, but he does not discuss methods for reaching a solution.

10. The **Maintenance Manager's Problem.** You should obtain the same tableau for the associated linear program as that given as the answer to the Maintenance Manager's problem in Chapter 1.

11. **Simple Poker.** Notice that the optimal strategy uses (2,3) and (3,3) with positive probabilities. Both choices mean that the player will bet high on the low card. This is called "bluffing." Bluffing is usually regarded as a means of getting something for nothing. While there is some value in having a "poker face," the personality of the bettor has little to do with the matter.

The purpose of bluffing is to prevent systematic losses with poor cards. The crucial numbers are the 1's in the second column of the game matrix and the -1 and $+1$ in the third column corresponding to these choices. As the higher bet increases relative to the lower bet the probability of playing (2,3) decreases. The numbers in the second column become negative when the higher bet grows to more than twice the lower bet. At that point bluffing becomes too expensive to protect against those losses.

This discussion follows a much more extensive coverage of bluffing in *The Theory of Games and Economic Behavior*, by von Neumann and Morgenstern, second edition, Princeton University Press, Princeton (1947), pp 186–219. Although the version of poker discussed there is also simplified, they use the full deck.

Questions

Q1. True. Both players have feasible strategies, and a feasible strategy corresponds to a feasible solution for the associated linear program.

Q2. True. Solving a linear program involves only rational arithmetic. If the initial tableau has only integer entries, every equivalent tableau will have only rational entries.

Q3. True. Look at Exercise 4.

Q4. True. In the associated linear program, each canonical slack variable is at the opposite end of a row or column from a variable corresponding to a pure strategy. The extra row added corresponds to an artificial variable.

Q5. True. According to Q4, each probability corresponding to a pure strategy is a dual variable to a slack variable for the opponent's problem. By complementary slackness, a positive value for a slack variable in an optimal solution requires that the dual variable be zero in every optimal solution.

Q6. False. The following game has unique optimal pure strategies for both players and neither uses all pure strategies.

	H	T
H	1	2
T	0	1

Q7. True. Unlike linear programs in general, the feasible set for a matrix game is always bounded. Thus, there is never a ray of optimal solutions extending to infinity.

Q8. False. The classical matching pennies game is an example. If one player uses an optimal strategy, and announces his strategy by visibly using the penny, the expected value of the game is zero regardless of the strategy used by his opponent.

Q9. True. See Q8.

Q10. False. If, for example, the column player uses an optimal strategy that produces a positive slack variable for a choice for the row player, the row player must avoid that choice. He can, however, use any mixed strategy that assigns positive probabilities to the choices with zero slack variables for all optimal strategies of the column player.

Chapter 8

Generally, the value of the objective variable will be unique, but the assignment and the solution of the dual problem will not be unique. The only reliable check is to verify $u_i + v_j \leq c_{ij}$ and to compare the sum of the selected entries and the sum of the margin entries to see that they are equal.

1. Value of assignment = 21.

	6	1	2	0	8
−3	3*	14	18	2	5
2	14	9	19	2*	14
2	14	10	4*	17	12
1	7	2*	9	9	19
2	18	11	16	2	10*

2. Value of assignment = 25.

	5	0	1	1	0	7	0
−2	3*	14	18	14	2	5	12
2	14	9	19	3*	2	14	9
3	14	10	4*	13	17	12	8
2	7	2*	9	3	9	19	18
0	10	14	19	5	8	7*	20
2	18	11	16	8	2*	10	15
4	19	5	5	16	13	15	4*

3. Value of assignment = 34.

	5	6	0	2	1	1	0	7	0
−2	3*	4	14	17	18	14	2	5	12
−1	11	5	5	1*	1	19	13	17	2
2	14	19	9	15	19	3*	2	14	9
3	14	20	10	19	4*	13	1	12	8
2	11	20	2*	17	17	6	3	18	2
2	7	8*	2	12	9	3	9	19	18
0	10	11	14	2	19	5	8	7*	20
2	18	13	11	13	16	8	2*	10	15
4	19	15	5	7	5	16	13	15	4*

4. We need $u_i + v_j \geq c_{ij}$. The value of the maximum assignment is 87.

	4	0	4	1	4
14	3	14*	18	2	5
15	14	9	19*	2	14
16	14	10	4	17*	12
15	7	2	9	9	19*
14	18*	11	16	2	10

5. In the following table we show an assignment of seven applicants. The covered rows and columns are indicated with 1's in the margins. Since the cover contains seven lines, seven is the maximum of applicants that can be assigned jobs for which they are qualified.

	1	1	1		1		1		
	1	1	1		1		1*	1	
	1		1		1*				
1	1	1		1*					1
	1	1*			1		1		
	1	1	1*						
1			1	1		1*	1	1	
	1*				1				
	1		1						

6. In terms of systems of distinct representatives, we can think of each applicant as representing the jobs for which he is qualified, or the jobs representing applicants who are qualified for the job. Since there are eight applicants and nine jobs, part of the question is trivial. There are not enough applicants to represent all the jobs. Other, more interesting observations are possible. Since there are six uncovered rows and five vertical lines in the cover, there are six applicants who qualify for only five jobs collectively. Since there are four uncovered columns and only two horizontal lines in the cover, there are four jobs for which there are only two qualified applicants.

7. The assignment given is not maximal. The following shows a maximal assignment.

	A	B	C	D	E	F	G	H
a		1*						1
b	1			1*			1	
c			1*		1		1	
d			1			1		1*
e		1			1*			
f				1		1*		1
g	1						1*	
h	1*			1		1	1	

8. The "at least one" condition means that the variables y_i and z_j in equations 8.49 and 8.50 are nonnegative. Furthermore, any row or column with more than one entry has a positive value for the slack variable.

The dual variable corresponding to those nonzero slack variables must be zero. That makes the other variable on the margins easy to compute. The following table shows the optimal assignment, which has a value of 10.

	3	0	4
1	7	1	3
2	6	2	9
0	3	5	4

9. This is a nonsquare version of the previous problem. The observations about the feasibility specifications are the same. We start by selecting the smallest entry in each column. That doesn't produce an entry in each row. We replace the smallest entry in one column by an entry in the first row for which the penalty is least. But then we have to prove that the selection is optimal. The value of the optimal selection is 11.

	1	2	3	1	2
2	4	5	6	3*	5
0	1*	3	3*	2	2*
0	2	2*	4	1	3

10. We have $u_i + v_j \leq c_{ij}$ with the margin entries zero for any row or column with more than one entry selected. The optimal value is 15.

	1	4	2	0	1
2	4	7	5	2*	4
1	3	6	4	1*	3
0	1*	4*	2*	1	1*
4	6	9	7	4*	6

11. We have $u_i + v_j \geq c_{ij}$ with the margin entries zero for any row or column with less that one entry selected. The optimal value is 18.

	0	2	1	0	1
4	4*	6	5	2	4
3	3	5	4	1	4*
2	1	4*	2	1	1
5	5	6	6*	4	5

12. This is a maximization problem with at most one from each row and at most one from each column. We must have $u_i + v_j \geq r_{ij}$ with the margin variables zero for any column that does not contain a chosen entry. The optimal value is 64.

	7	0	0	4	0	4
11	18*	8	7	12	7	13
16	20	14	16*	20	15	20
12	16	11	10	16*	10	15
10	15	9	7	14	10	14*

13. We find a maximal matching and its associated minimal cover. The nodes in the cover are shown in the following graph. You will note that every potential pair has a darkened node at one end or the other. If these teams are visited, at least one team in every potential pairing will have been scouted.

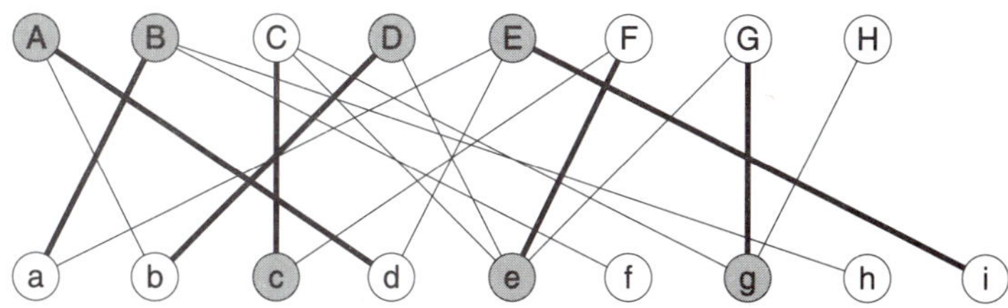

Figure A.1: Figure for Exercise 13.

14. This is an Egerváry problem. It is solved as though it were a maximum assignment problem, but we must make sure that the margin entries are nonnegative. The optimal value is 22.

	0	1	2	2
4	2	3	4	6*
2	1	3*	3	4
5	4	5	7*	7
6	6*	6	7	8

Questions

Q1. False. The example worked through in Section 8.1 provides a counter example.

Q2. False. The following example is a counter example.

1*	4
4	5*

Q3. True. There are only a finite number of different assignments.

Q4. True. There are only a finite number of different assignments.

Q5. True. The variables of the dual program on the margins are free variables. The w_{ij} are nonnegative, and the initialization step of the algorithm produces nonnegative w_{ij}.

Q6. False. Section 8.4 shows how an assignment problem can be represented as a linear program.

Q7. True. The linear program associated with an assignment problem has $2n$ basic variables. The assignment will produce only n nonzero x_{ij}.

Chapter 9

1. This initial feasible table is obtained by computing the maximum amount that can be shipped on one route in each column for the least cost in that column. The columns are scanned from left to right and within each column the rows are scanned from top to bottom. After each column has one entry, the columns are scanned again, from left to right, and the amount required in each column is fulfilled before going to the next column.

(15)	5	2	4	7
(1)	(4)	7	5	1
(0)	0	(4)	−1	−2
0	−1	0	(2)	−3
(0)	0	2	(3)	(2)

Note that if an assignment of an amount to a route exhausts both the supply and the demand, we assign (0) to the source. This preserves the tree structure of the solution. We do not make the assignment to a destination since we expect that a minimum-cost shipping schedule will not oversupply any destination (since every route has a positive cost.)

The cost associated with this schedule is 62.

2. Use the table obtained as an answer to Exercise 1 as a start. We increase the amount shipped on the first route (from top to bottom and left to right) with the maximum infeasibility. That entry is the −3 in the third row.

(15)	5	2	4	7
(1)	(4)	7	5	1
(0)	0	(4)	−1	−2
−3	4	3	3	(2)
(0)	0	2	(5)	(0)

This adjustment exhausts two entries in the loop. We remove the first one (in the ordering described for the Bland rule) and assign (0) to the other.

This time the maximum infeasibility is the −2 in the second row. We obtain

(15)	5	4	4	7
(1)	(4)	5	5	1
−2	2	(4)	1	(0)
−3	4	1	3	(2)
(0)	0	0	(5)	(0)

We obtain complementary feasible solutions. Thus, both are optimal. The minimum shipping cost is 56. Notice that both solutions are degenerate.

In most books, only balanced transportation problems are provided with a method for finding a solution. For an unbalanced transportation problem, like this one, a "dump" is provided as a zero-cost destination to balance the problem. The initial data table would then be the following.

(0)	(4)	(4)	(5)	(2)	(1)
(5)	5	9	9	8	0
(4)	5	2	3	5	0
(2)	6	2	4	4	0
(5)	5	4	4	7	0

The final table would then be

(16)	5	4	4	7	0
(0)	(4)	5	5	1	(1)
−2	2	(4)	1	(0)	2
−3	4	1	3	(2)	3
(0)	0	0	(5)	(0)	0

There is no practical difference between the two methods. Our left margin is equivalent to the dump. However, a dump is an unnecessary concept.

3.

(29)	7	5	8	6	2	6
−3	(7)	4	(1)	4	5	3
−2	0	(3)	0	(0)	3	(6)
−4	(0)	4	1	(4)	6	4
−7	4	6	(5)	6	8	6
(2)	−2	0	−3	6	(5)	(1)

4.

(29)	4	5	5	3	2	6
0	(7)	1	(1)	4	2	0
−2	0	(3)	3	3	3	(6)
−1	(0)	1	1	(4)	3	1
−4	4	3	(5)	6	5	3
(2)	1	0	(0)	3	(5)	(1)

Again, both optimal solutions are degenerate.

5.

(21)	4	5	5	4
−1	(4)	(4)	1	1
(0)	3	(1)	(5)	2
(3)	2	2	3	(7)

6.

(21)	4	7	5	4
−2	1	1	2	(7)
−2	4	(4)	2	4
0	1	(1)	3	(4)
−1	(4)	(1)	(5)	2
(4)	0	0	1	(0)

7.

(27)	4	5	4	5	8
−1	2	3	(5)	4	2
−2	1	1	3	4	(3)
0	(4)	1	(0)	5	0
(1)	2	(2)	1	(3)	2
(1)	1	0	(3)	1	(1)
−2	6	4	6	4	(6)

8.

(4)	5	4	4	7
(0)	(1)	5	5	1
−2	2	(1)	1	(0)
−3	4	1	3	(1)
(0)	0	0	(1)	(0)

This is an assignment problem represented as a transportation problem. Work the problem as an assignment problem and see how much faster the Hungarian algorithm is.

9.

(21)	4	7	7	4
−1	(7)	(−4)	−1	−1
(4)	2	−2	−2	(0)
1	(0)	−1	(5)	−1
−1	0	(9)	−2	2
0	(−3)	0	−1	(7)

Neither basic solution is feasible, but that is not the question. For any choice of a spanning tree, the complementary basic solutions for both problems are determined.

10.

(0)	(4)	(3)	(5)
(6)	3	4	7
(5)	3	5	6
(2)	5	5	4

11. Even for this small table this is an elaborate process. First, construct a blank tableau and fill in the basic and nonbasic variables appearing

in the table. We can also insert the entries for the b-column, the c-row, and the d-corner.

	w_{11}	w_{12}	w_{22}	w_{23}	w_{33}	u_2	0	-1	
x_{13}								2	$= -w_{13}$
x_{21}								-1	$= -w_{21}$
x_{31}								3	$= -w_{31}$
x_{32}								2	$= -w_{32}$
y_1								1	$= -u_1$
y_3								2	$= -u_3$
z_1								4	$= -v_1$
z_2								5	$= -v_2$
z_3								6	$= -v_3$
-1	-4	-2	-1	-3	-2	-1	-12	-51	$= f$
	$= x_{11}$	$= x_{12}$	$= x_{22}$	$= x_{23}$	$= x_{33}$	$= y_2$	$= t$	$= g$	

This much is easy. Filling in the A-matrix is more involved. The process is described in general terms in Section 9.7. We will give more detail here. Consider the graph representing a three-source, three-destination transportation problem, shown below.

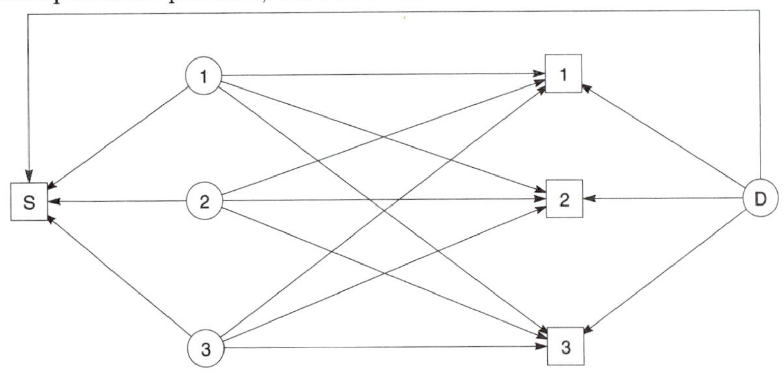

Figure A.2: Three-source three-destination transportation network.

Notice that the graph for a transportation problem is bipartite. Furthermore, to get the correct signs in the tableau corresponding to the entries in the transportation table, all arcs are oriented and all point from a circle node to a square node.

The tree representing the table in the exercise is shown in Figure A.3.

The easiest (least complicated) way to determine the entries in the A-matrix is to deal with the A-matrix row by row. For example, consider the row corresponding to the nonbasic variable x_{31}. This row corresponds to

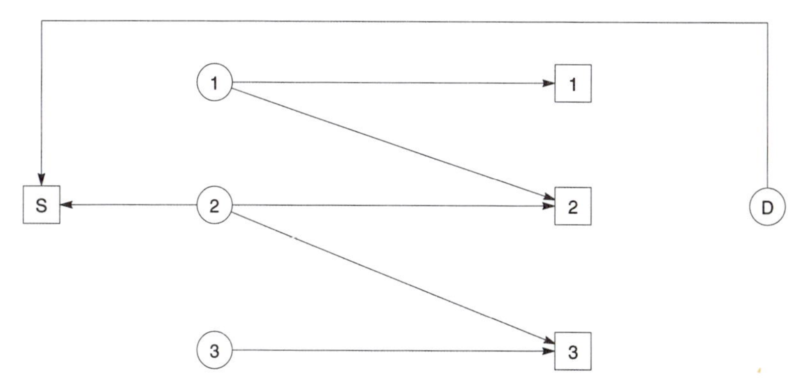

Figure A.3: Basic solution tree.

the dotted arc and the loop containing that arc in the Figure A.4.

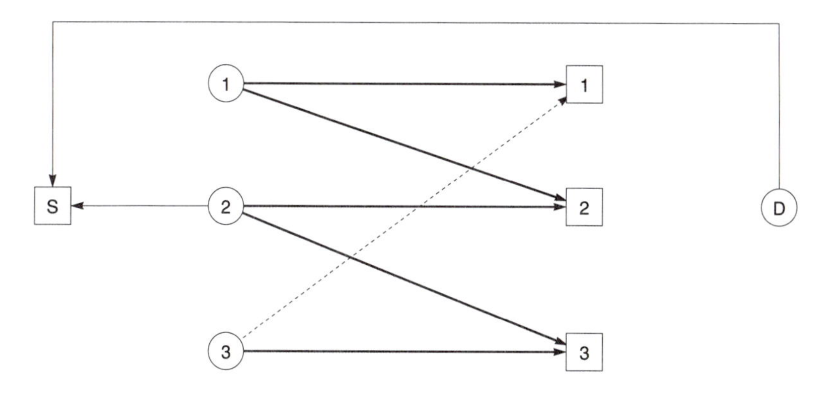

Figure A.4: Loop created in a network.

Trace the loop that contains the arc corresponding to x_{31}, starting in the direction indicated by that arc. As you pass through each arc in the loop, assign 1 if that arc is oriented in the direction of transit and -1 if that arc is oriented in the opposite direction. When the x_{31}-row is filled in we have

	w_{11}	w_{12}	w_{22}	w_{23}	w_{33}	u_2	0	-1	
x_{13}								2	$= -w_{13}$
x_{21}								-1	$= -w_{21}$
x_{31}	-1	1	-1	1	-1			3	$= -w_{31}$
x_{32}								2	$= -w_{32}$
y_1								1	$= -u_1$
y_3								2	$= -u_3$
z_1								4	$= -v_1$
z_2								5	$= -v_2$
z_3								6	$= -v_3$
-1	-4	-2	-1	-3	-2	-1	-12	-51	$= f$
	$= x_{11}$	$= x_{12}$	$= x_{22}$	$= x_{23}$	$= x_{33}$	$= y_2$	$= t$	$= g$	

You can fill in each row without drawing the network. For each variable, x_{31} for example, locate its position in the transportation table. Then trace through the loop in the table that it completes. As you pass through the positions in the loop, enter a -1 for the odd (negative) positions in the loop into the tableau for the corresponding variables, and a $+1$ for the even (positive) positions in the loop. Finally, the entire tableau is

	w_{11}	w_{12}	w_{22}	w_{23}	w_{33}	u_2	0	-1	
x_{13}		-1	1	-1				2	$= -w_{13}$
x_{21}	-1	1	-1					-1	$= -w_{21}$
x_{31}	-1	1	-1	1	-1			3	$= -w_{31}$
x_{32}			-1	1	-1			2	$= -w_{32}$
y_1		-1	1			1		1	$= -u_1$
y_3				1	-1	1		2	$= -u_3$
z_1	1	-1	1			1	-1	4	$= -v_1$
z_2			1			1	-1	5	$= -v_2$
z_3				1		1	-1	6	$= -v_3$
-1	-4	-2	-1	-3	-2	-1	-12	-51	$= f$
	$= x_{11}$	$= x_{12}$	$= x_{22}$	$= x_{23}$	$= x_{33}$	$= y_2$	$= t$	$= g$	

12. The new table is

(12)	3	4	5
0	(3)	(3)	2
(1)	(1)	1	(3)
-1	3	2	(2)

The tableau equivalent to the transportation table is shown below. You should perform the pivot exchange in the tableau in the answer to Exercise 6 independently to see that the transportation table faithfully represents the tableau.

	w_{11}	w_{12}	w_{22}	w_{23}	w_{33}	u_2	0	-1	
x_{13}	-1		1	-1				1	$= -w_{13}$
x_{21}	1	-1	-1					1	$= -w_{21}$
x_{31}			-1	1	-1			4	$= -w_{31}$
x_{32}	1	-1	-1	1	-1			3	$= -w_{32}$
y_1	-1		1			1		0	$= -u_1$
y_3				1	-1	1		2	$= -u_3$
z_1			1			1	-1	3	$= -v_1$
z_2	-1	1	1			1	-1	4	$= -v_2$
z_3				1		1	-1	6	$= -v_3$
-1	-3	-3	-1	-3	-2	-1	-12	-50	$= f$
	$= x_{11}$	$= x_{12}$	$= x_{22}$	$= x_{23}$	$= x_{33}$	$= y_2$	$= t$	$= g$	

13. The solution for the transportation problem is nondegenerate. Therefore, the solution for the dual max program is unique. The solution for the max program is degenerate, and that means there must be multiple solutions for the min program. Other solutions are obtained by increasing the shipment for either route with a zero residual cost. For example, if we increase the amount shipped x_{41}, we obtain

(17)	5	4	4	7
(2)	(3)	5	5	1
-2	2	(4)	1	(1)
-3	4	1	3	(2)
0	(1)	0	(5)	(1)

There are several other optimal basic solutions.

14. This time the optimal solution to the max program is nondegenerate. Thus, the optimal solution to the min program is unique. There are other solutions to the max program, but they are harder to find than other optimal solutions for the min program. To preserve the feasibility of the min program we perform a dual simplex pivot exchange in a row of a tableau. To preserve the feasibility of the solution for the max program we perform a simplex pivot exchange in a column of a tableau. A column is more difficult to identify in a transportation table.

To identify a row, we select an entry corresponding to a variable in the max program and trace the loop it generates. The eligible entries correspond to odd positions in the loop. To identify a column, we select an entry corresponding to a variable in the min program and trace all loops through that entry. The eligible entries correspond to even positions in the loop.

In this exercise, select the (0) in the x_{44} position. In the following table we have identified all the entries in some loop through that entry. We have labeled with a $-$ sign those in odd positions, and we have labeled with a $+$ sign those in even positions. Those in even positions are the eligible entries. The minimum is in the x_{23} position.

(15)	5	4^-	4	7^-
(1)	(4)	5^-	5	1^-
-2^+	2^+	(4)	1^+	(0)
-3^+	4^+	1	3^+	(2)
(1)	1	1^-	(5)	(0)

When the required pivot exchange is performed we have

(15)	5	3	4	6
(1)	(4)	6	5	2
-1	1	(4)	(0)	(0)
-2	3	1	2	(2)
(1)	1	2	(5)	1

This gives us a new optimal solution for the max program. There are several others.

15. Both optimal solutions are degenerate. At first glance, it might seem that we can perform the kinds of pivot exchanges described in the previous two exercises and find multiple optimal solutions for both programs. If you try that, you will find another optimal solution for the min program to be elusive. There are, however, other optimal solutions for the max program. Among them we find the equivalent transportation table below.

(15)	5	6	7	8
(1)	(4)	3	2	(0)
-4	1	(4)	(0)	1
-4	5	(0)	1	(2)
-3	3	1	(5)	2

In this table, the optimal solution for the max program is nondegenerate. Therefore, the optimal solution for the min program is unique. It is the same solution as that occurring in the table given in the exercise. There are different basic variables, but the values of the variables are the same.

16.

2	2	4	3	3	3
2	2	4	3	3	3
2	2	4	3	3	3
3	3	2	5	5	5
3	3	2	5	5	5
4	4	3	4	4	4

The first three rows represent the availability of 3 units of the commodity at the first source. The first two columns represent the demand of 2 units at the first destination. Although the Hungarian algorithm is much more efficient than the transportation algorithm, the increase in the size of the table that occurs with this transformation negates the advantage.

17.

(0)	(1)	(1)	(1)	(1)
(1)	2	4	1	3
(1)	5	3	5	4
(1)	2	4	3	5
(1)	4	5	4	3

Converting an assignment problem into a transportation problem is not recommended as a computational procedure because the transportation algorithm is not as efficient as the Hungarian algorithm. However, we can gain some insight. The optimal solution, obtained by the transportation algorithm, will have nine ($2n+1$ in general) basic variables. However, only four (n in general) of them will be nonzero. Thus, the optimal solution to an assignment problem is always degenerate.

18. If the possibilities not voided contain a loop, the solution is not unique. The set of unvoided possibilities should be a tree. Any spanning tree satisfies the requirements.

19. We assign a cost of 100 to each forbidden route. When the resulting transportation problem is solved we obtain

(21)	5	95	98	100
−93	90	1	1	(3)
−95	94	4	(3)	(1)
−93	94	(5)	(2)	97
(3)	(4)	5	2	(3)

Since the optimal solution must use one of the forbidden routes, the problem is infeasible.

20. We assign a cost of 100 to each forbidden route. When the resulting transportation problem is solved we obtain

(21)	5	5	8	1^-
−3	(0)	1	1	(3)
−5	4	4	(2)	(1)
−3	4	(5)	(3)	93
(6)	(4)	95	92	90

The solution of this transportation problem is also a solution of the given problem.

Questions

Q1. False. The problem will not be feasible if the total supply is insufficient to meet the total demand. If the total supply is at least as large as the total demand, the transportation problem is feasible. A feasible solution can always be found by the "Northwest corner rule." Start in the upper-left corner. Assign enough to that route to exhaust either the source or the destination, or both. If the source is exhausted, move down one cell. If the destination is exhausted, move right one cell. Continue in that fashion until a feasible solution is obtained. This methods always works, though it may not produce as good a start as some other methods.

For a feasible transportation problem, the total amount shipped is bounded by the total supply. Since a transportation problem is a special case of a linear program, the fact that the feasible set is bounded is sufficient to show that an optimal solution exists.

Q2. False. Consider the following transportation problem.

(0)	(1)	(1)
(1)	1	2
(1)	2	4

Q3. False. Consider the following transportation problem.

(0)	(1)	(1)
(1)	1	3
(1)	3	4

Q4. True. Phase II of the Dantzig transportation algorithm is identical to phase II of the dual simplex algorithm. The feasibility of the primal problem is preserved by the rules for selecting the pivot exchanges.

Q5. True. See Section 9.3.

Q6. True. See Section 9.3.

Q7. True. When the transportation network is expanded to include a super source and a super sink, the network has $m = n+2$ nodes. The spanning tree for that network has $m+n+1$ arcs, but one of those arcs connects the sink and the source. Some of the remaining $m + n$ arcs may connect the super source with some of the individual sources. The remaining arcs connect the sources with the destinations.

Q8. False. See the principal theorem about spanning trees.

Chapter 10

1. The network in Figure A.5 has two additional nodes. F is the super source, and G is the super sink. The arcs between these additional nodes and the rest of the network must have capacities large enough so that they do not impose restrictions more stringent than those imposed by the original network.

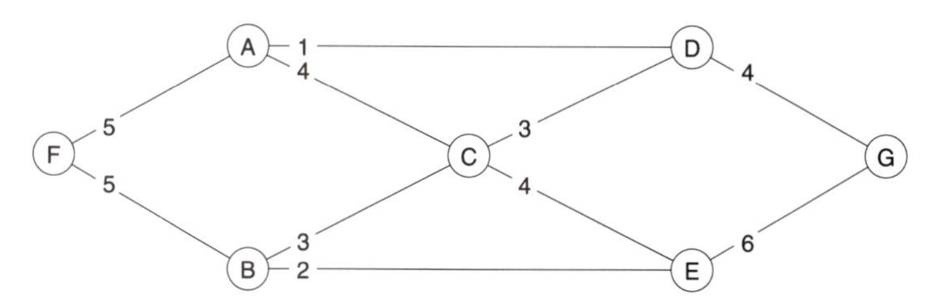

Figure A.5: Expanded flow network.

2.

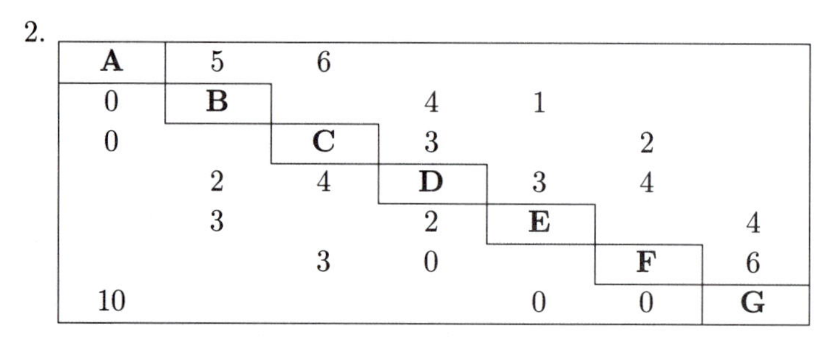

A	5	6				
0	**B**		4	1		
0		**C**	3		2	
	2	4	**D**	3	4	
	3		2	**E**		4
		3	0		**F**	6
10				0	0	**G**

3. Equivalent networks can look quite different. Figure A.6 shows one way to draw the network.

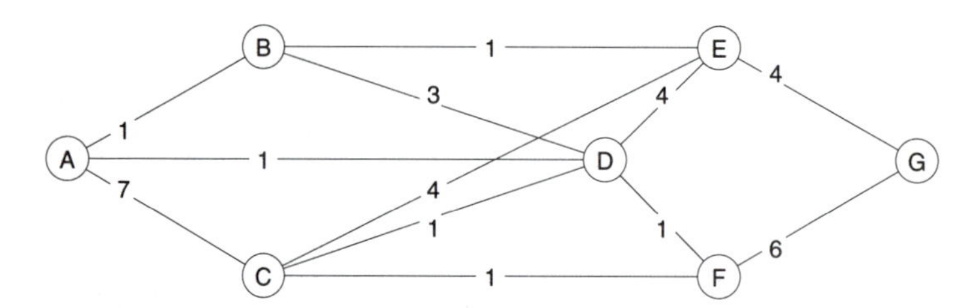

Figure A.6: Figure for Exercise 3.

4. Figure A.7 shows a cut with value 10. If the network is connected, almost any way of slashing the network into two pieces, one containing the source and one containing the sink, will produce a cut. The min cut may be considerably more complex, though any cut can be converted into a slash by rearranging the nodes. You can make your own example of a min cut by drawing any connected network and selecting any combination of arcs that divide the network into two connected pieces. Assign capacities to those arcs in any way you please. Then assign capacities to the other arcs that are sufficiently large that they will not impose a restriction on the flow.

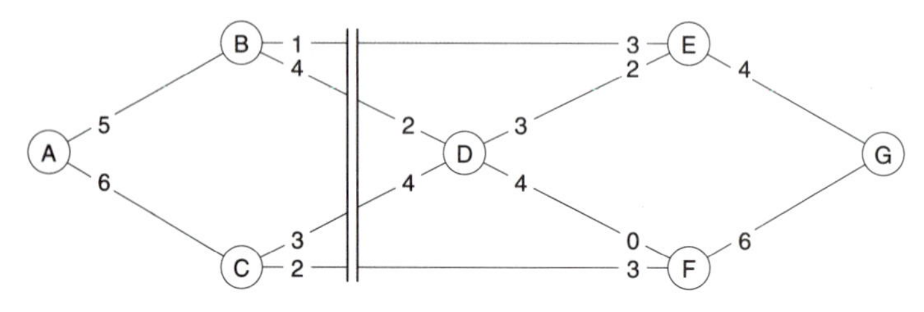

Figure A.7: Figure for Exercise 4.

5. Start by covering the source with a vertical line, and then nodes with an entry in the row of the source. We can extend the covering to node E through the 7 in the row of node B. Then cover the sink with a horizontal line, and then cover nodes with an entry in the column of the sink, with exceptions of those nodes already covered.

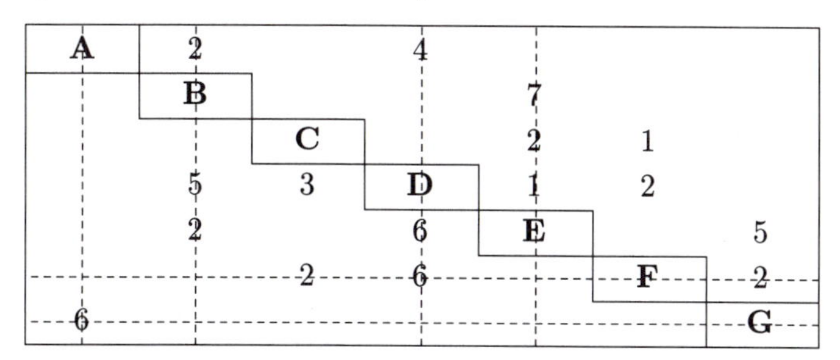

We continue to extend the vertical and horizontal lines to cover more nodes. We never cover a node previously covered. A node can be covered with a vertical line if there is an entry in its column in a row of a node previously covered vertically. A node can be covered with a horizontal line if there is an entry in its row in a column of a node previously covered horizontally.

By this rule, the node C could be covered by a vertical line through the 3 in its column or covered by a horizontal line through the 1 in its row. The choice is arbitrary.

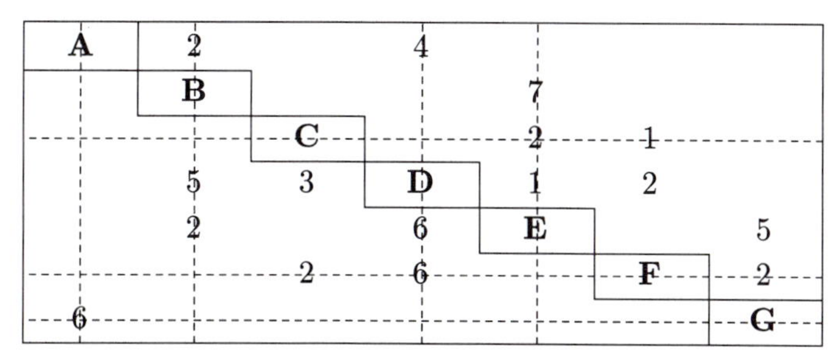

We get a cut with value 10. We could just draw a vertical line through the source column and a horizontal line through all other rows. In this case, that would produce a cut with a value of 6. This is certainly a better bound. However, such a casual method will not show whether the sink set is connected. If the source and sink are not connected, the method described will produce a cut with a value of 0. To see how this occurs, apply this method of producing a cut to the residual capacity table in the answer to Exercise 6.

6. When we work through the Ford–Fulkerson algorithm, we obtain

A	2	0	0		6
10	**B**		0	1	0
2		**C**	1	2	
2	2	5	**D**		2
	5	0		**E**	0
2	2		10	2	**F**

The source set is (*ABE*) and the sink set is (*CDF*). The max flow, indicated by the entry in the upper-right corner, is 6. The following table shows the min cut.

A	6	1	1		
6	**B**		1	3	1
1		**C**	3	1	
1	1	3	**D**		6
	3	1		**E**	1
8	1		6	1	**F**

7. When we work through the Ford–Fulkerson algorithm, we obtain

A			2			12
8	**B**		7			
4		**C**	4	1	2	
14	9		**D**		1	
	6	11	2	**E**		2
			7		**F**	
2			2	20	4	**G**

The source set is (*ABDF*) and the sink set is (*CEG*). The max flow, indicated by the entry in the upper-right corner, is 12. The following table shows the min cut.

A	4	2				
4	**B**		8	3		
2		**C**	2	6	1	
8	8	2	**D**	1	4	1
	3	6	1	**E**		11
		1	4		**F**	2
14			1	11	2	**G**

8. When we work through the Ford–Fulkerson algorithm, we obtain

A	2	1				12
5	**B**			1		
5		**C**				
2	3	7	**D**	6	1	
	7			**E**		
		7	9		**F**	
			5	2	5	**G**

The source set is $(ABCE)$ and the sink set is (DFG). The max flow, indicated by the entry in the upper-right corner, is 12. The following table shows the min cut.

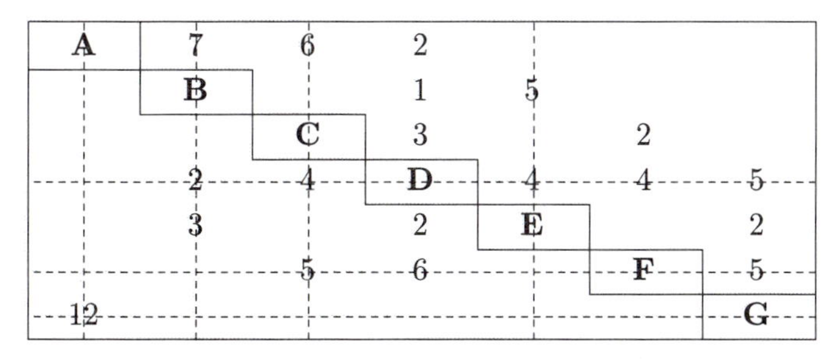

A	7	6	2			
	B		1	5		
		C	3		2	
	2	4	**D**	4	4	5
	3		2	**E**		2
		5	6		**F**	5
12						**G**

9. When we work through the Ford–Fulkerson algorithm, we obtain

A	2		1			12
10	**B**		7			
4		**C**	4	1	2	
13	9		**D**		1	
	6	11	2	**E**		2
			7		**F**	
1			2	11	4	**G**

The source set is $(ABDF)$ and the sink set is (CEG). The max flow, indicated by the entry in the upper-right corner, is 12. The following table shows the min cut.

A	6	2				
6	**B**		8	3		
2		**C**	2	6	1	
7	8	2	**D**	1	4	1
	3	6	1	**E**		11
		1	4		**F**	2
13			1	11	2	**G**

10. When we work through the Ford–Fulkerson algorithm, we obtain

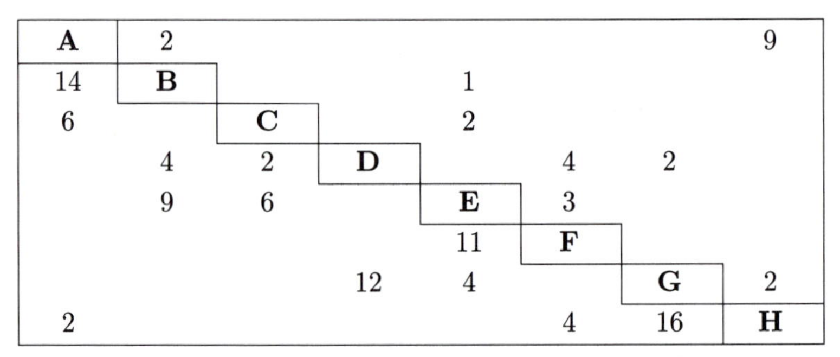

A	2						9
14	**B**			1			
6		**C**		2			
	4	2	**D**		4	2	
	9	6		**E**	3		
				11	**F**		
			12	4		**G**	2
2					4	16	**H**

The source set is $(ABCEF)$ and the sink set is (DGH). The max flow, indicated by the entry in the upper-right corner, is 9. The following table shows the min cut.

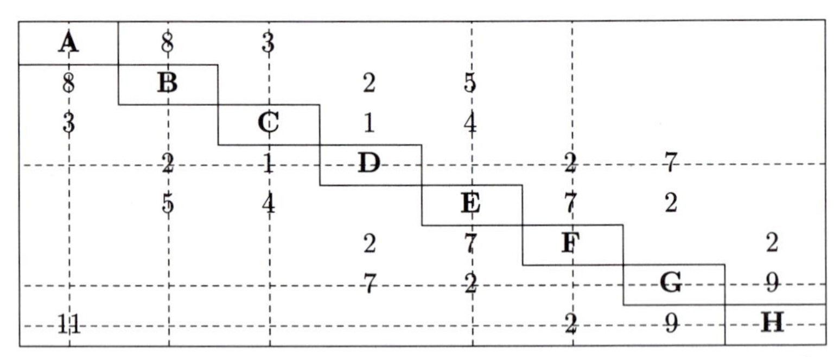

A	8	3					
8	**B**		2	5			
3		**C**	1	4			
	2	1	**D**		2	7	
	5	4		**E**	7	2	
			2	7	**F**		2
			7	2		**G**	9
11					2	9	**H**

11. If we accept the assertion that every 1 represents an arc in the cut, no line in the cover can cover a 1. Thus, wherever a 1 occurs, we draw a vertical line through the node in its row and a horizontal line through the node in its column. It turns out that with a vertical line through the source and a horizontal line through the sink, every node will be covered.

In order for this covering to produce a cut, it is necessary that each portion be connected. Establishing that is usually not difficult, but here it is even easier. You can trace a path from the source through arcs with capacity 5 to every node in the source set, and a path from the sink through arcs with capacity 5 to every node in the sink set.

Finally, we face the problem of showing that the cut is a min cut. There are two ways to go about this. One would be to compare the value of the cut with all other cuts, and the other would be to find one flow with the same value as the value of the cut. It turns out that there is a path from the source to the sink through each arc with capacity 1. These paths do not use any common arc. Thus, there is a flow with value 4.

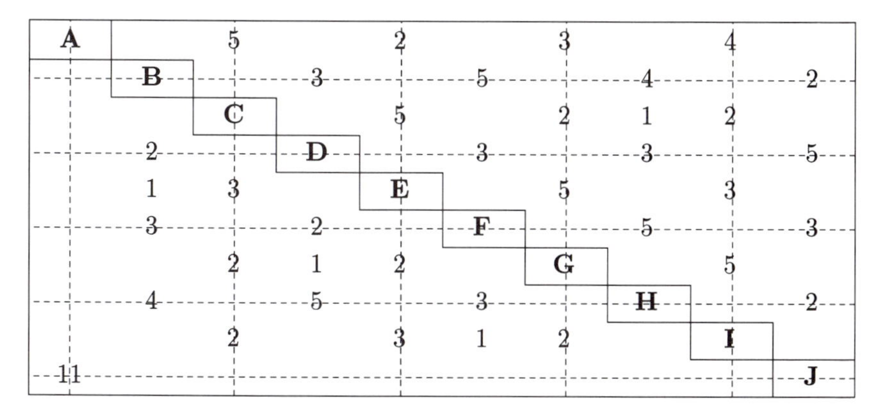

A		5		2		3		4	
	B		3		5		4		2
		C		5		2	1	2	
	2		**D**		3		3		5
	1	3		**E**		5		3	
	3		2		**F**		5		3
		2	1	2		**G**		5	
	4		5		3		**H**		2
		2		3	1	2		**I**	
11									**J**

Questions

Q1. True. If a network is not connected, there is no feasible solution. However, if the network is connected, any path from source to sink will produce a feasible flow, though that flow might have value zero. Since the flow is bounded by any cut, there is an optimal solution.

Q2. True. Since the network is connected and each part is connected, there is at least one arc connecting the two parts. The sum of the capacities in these arcs is the value of the cut.

Q3. False. The source set and the sink set must be connected in order for the covering to produce a cut. If the proposed cut set for the following table is *(ABD/CE)*, the source set would not be connected.

A	4	2		
	B	1		3
	2	**C**	2	5
	2	2	**D**	4
				E

From an abstract point of view we could allow either the source set or the sink set to be disconnected. However, if a residual capacity network could be divided into three connected pieces with the source in one set, the sink in another, and another floating piece, the algorithm might not identify the min cut because the labeling might not extend to the min cut.

Chapter 11

1.

(0)						(1)	(4)
(2)	(0)	3	4		2		
(3)	3	(0)	2	5			
(0)	4	2	(0)			2	3
(0)		5		(0)	2	4	2
(0)	2			2	(0)		5
			2	4		(0)	
			3	2	5		(0)

2. The following table gives the distances from A to each node.

A	B	C	D	E	F	G	H	I
0	15	4	29	16	11	27	10	41

3. The following table gives the distances from A to each node.

A	B	C	D	E	F	G	H	I
0	3	4	9	8	13	12	15	16

4.

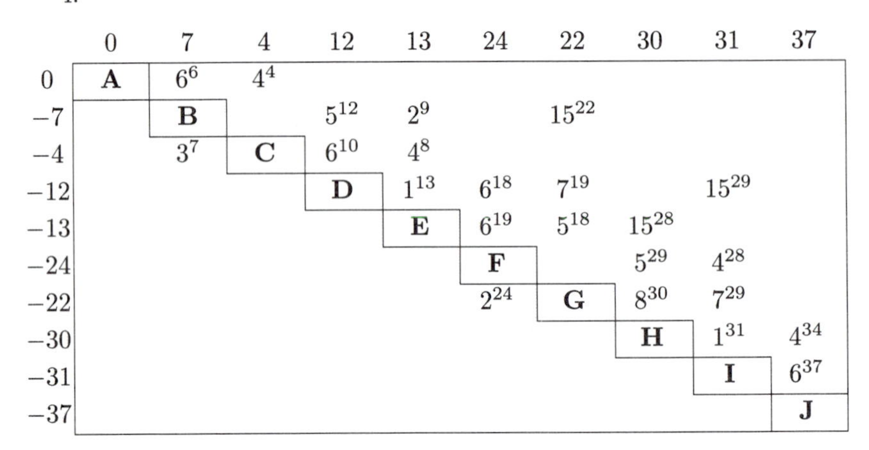

	0	7	4	12	13	24	22	30	31	37
0	**A**	6^{6}	4^{4}							
−7		**B**		5^{12}	2^{9}		15^{22}			
−4		3^{7}	**C**	6^{10}	4^{8}					
−12				**D**	1^{13}	6^{18}	7^{19}		15^{29}	
−13					**E**	6^{19}	5^{18}	15^{28}		
−24						**F**		5^{29}	4^{28}	
−22						2^{24}	**G**	8^{30}	7^{29}	
−30								**H**	1^{31}	4^{34}
−31									**I**	6^{37}
−37										**J**

5. It is important to follow the steps of the algorithm carefully with this exercise. We will give every other step, omitting the steps in which the cover is drawn. Note that the sums of the basic variables in each row and in each column are preserved. Note also that the basic variables consistently form a tree.

	0	0	0	0
(0)			(1)	(2)
(2)	(0)		2	
(1)		(0)	1*	2
0	2	1	(0)	
0		2		(0)

	0	0	1	1
(1)				(2)
(2)	(0)		1	
(0)		(0)	(1)	1*
-1	3	2	(0)	
-1		3		(0)

	0	0	1	2
(1)				(2)
(2)	(0)		1*	
0		(0)	(1)	(0)
-1	3	2	(0)	
-2		4		(0)

	0	1	2	3
(2)				(1)
(1)	(0)		(1)	
-1		(0)		(1)
-2	4	2*	(0)	
-3		4		(0)

	0	3	2	5
(3)				
(0)	(0)		(2)	
−3		(−1)	2	(2)
−2	4	(1)	(−1)	
−5		4		(0)

6.

	0	1	3	4	2	5	6
(5)							
(0)	(0)	2	1		(2)		
−1	4	(0)	(3)	2			
−3	7	4	(−3)			(1)	(2)
−4		8		(−2)	4	3	(2)
−2	4			(2)	(−2)		1
−5			4	5		(0)	
−6			6	4	9		(0)

7. There are several different ways to set this problem in network form. It is, in fact, possible to pose this problem as a transportation problem with blocked routes. In this form there are four destinations, the days on which napkins must be supplied. The store is one source. The soiled napkins available at the end of each day is another. Since the napkins from the last day are unavailable, there are four sources. The following transportation table represents this problem.

(0)	(100)	(60)	(60)	(90)
(260)	2	2	2	2
(100)		0.75	0.25	0.25
(60)			0.75	0.25
(60)				0.75

The question of what to do with the napkins soiled on the last day is vaguely stated in the problem. The expectation in the table above is that they will remain soiled until they are laundered for a subsequent requirement. If we wish to impose a condition that all remaining napkins are laundered, we can create an additional node, the linen closet, to which all napkins are finally sent. We should also add another source, the napkins from the last day. The slack variables for the variables on the left margin, except for the store, should be made artificial to ensure that everything is laundered.

Chapter 12

1. The Karush–Kuhn–Tucker (KKT) tableau for this problem is

	dx_1	dx_2	
-1	$2x_1$	$2x_2$	$= df$
	$= 0$	$= 0$	

There is a feasible solution to the column equations: $x_1 = x_2 = 0$. This is a KKT point, but it is a minimum, not a maximum point.

2. The KKT tableau for this problem is

	dx_1	dx_2	
v_1	1	0	$= -dy_1$
v_2	-1	0	$= -dy_2$
v_3	0	1	$= -dy_3$
v_4	0	-1	$= -dy_4$
-1	$2x_1$	$2x_2$	$= df$
	$= u_1$	$= u_2$	

The feasible set is a square centered at the origin, 2 units on each side. The feasible set must be subdivided into nine regions in which the feasibility specifications are different. We recommend setting up a table for the feasibility specifications in these regions.

	x_1	x_2	y_1	y_2	y_3	y_4	dx_1	dx_2	dy_1	dy_2	dy_3	dy_4
a	free	free	> 0	> 0	> 0	> 0	free	free	free	free	free	free
b	free	free	0	> 0	> 0	> 0	free	free	≥ 0	free	free	free
c	free	free	0	> 0	0	> 0	free	free	≥ 0	free	≥ 0	free
d	free	free	> 0	> 0	0	> 0	free	free	free	free	≥ 0	free
e	free	free	> 0	0	0	> 0	free	free	free	≥ 0	≥ 0	free
f	free	free	> 0	0	> 0	> 0	free	free	free	≥ 0	free	free
g	free	free	> 0	0	> 0	0	free	free	free	≥ 0	free	≥ 0
h	free	free	> 0	> 0	> 0	0	free	free	free	free	free	≥ 0
i	free	free	0	> 0	> 0	0	free	free	≥ 0	free	free	≥ 0

One may wonder why x_1 and x_2 are free throughout. The restrictions on these variables are exercised through the slack variables. Consider, for example, the situation where the boundary lines are not parallel to the coordinate axes. The interior of the square is defined in row a. The remaining rows define the four edges and the four corners.

We use this table to determine the dual feasibility specifications.

	dx_1	dx_2	dy_1	dy_2	dy_3	dy_4	u_1	u_2	v_1	v_2	v_3	v_4
a	free	free	free	free	free	free	0	0	0	0	0	0
b	free	free	≥ 0	free	free	free	0	0	≥ 0	0	0	0
c	free	free	≥ 0	free	≥ 0	free	0	0	≥ 0	0	≥ 0	0
d	free	free	free	free	≥ 0	free	0	0	0	0	≥ 0	0
e	free	free	free	≥ 0	≥ 0	free	0	0	0	≥ 0	≥ 0	0
f	free	free	free	≥ 0	free	free	0	0	0	≥ 0	0	0
g	free	free	free	≥ 0	free	≥ 0	0	0	0	≥ 0	0	≥ 0
h	free	free	free	free	free	≥ 0	0	0	0	0	0	≥ 0
i	free	free	≥ 0	free	free	≥ 0	0	0	≥ 0	0	0	≥ 0

Because of symmetry, we can cover all cases by looking at three regions.

a. The interior. The KKT tableau reduces to the tableau of Exercise 1. The origin is a KKT point, but it is a minimum, not a maximum.

b. The right boundary. The dual problem of the KKT tableau reduces to the following.

	dx_1	dx_2	
v_1	1	0	$= -dy_1$
-1	2	$2x_2$	$= df$
	$= 0$	$= 0$	

We get $x_1 = 1$, $x_2 = 0$, $v_1 = 2$, $v_2 = v_3 = v_4 = 0$ as a KKT point. It is a saddle point. The objective variable decreases towards the interior of a circle about the origin tangent to the border at that point. It increases in directions tangent to the circle.

c. The upper-right corner. The problem of the KKT tableau reduces to the following.

	dx_1	dx_2	
v_1	1	0	$= -dy_1$
v_3	0	1	$= -dy_2$
-1	2	2	$= df$
	$= 0$	$= 0$	

We get $x_1 = 1$, $x_2 = 1$, $v_1 = 2$, $v_3 = 2$, $v_2 = v_4 = 0$, as a KKT point. It is a point where a maximum is taken on.

3. The KKT tableau is

	dx_1	dx_2	
v_1	1	0	$= -y_1$
v_2	0	1	$= -y_2$
-1	$2x_1$	$2x_2$	$= df$
	$= u_1$	$= u_2$	

The feasible set is a square in the first quadrant. It can be divided into nine subregions in a way similar to that used in Exercise 2. However, the objective function is not quite as symmetric, so we must examine five subregions.

The origin is a KKT point, but it is a minimum. The upper-right corner is a KKT point and it is a maximum. The other two corners are KKT points, but they are saddle points. There are no other KKT points.

4. The KKT tableau is

	dx_1	dx_2	
v_1	1	1	$= -y_1$
v_2	-1	-1	$= -y_2$
v_3	1	-1	$= -y_3$
v_4	-1	1	$= -y_4$
-1	x_2	x_1	$= df$
	$= u_1$	$= u_2$	

The feasible set is a square, similar to the square in the previous exercise except that it is rotated 45 degrees. Because of symmetry it is sufficient to examine four of the nine subregions. The same table of cases that we used in Exercise 2 can be used here.

a. The interior. The origin is a KKT point, but it is a saddle point.

b. The boundary in the first quadrant. The KKT tableau is

	dx_1	dx_2	
v_1	1	1	$= -y_1$
-1	x_2	x_1	$= df$
	$= 0$	$= 0$	

We see from the column equations that we have $x_1 = x_2$. That combined with the condition $x_1 + x_2 = 2$ on that boundary gives us $x_1 = 1$, $x_2 = 1$. The KKT tableau has a feasible solution and we have a KKT point. It is also a point where the function takes on its maximum.

c. The right corner. The KKT tableau is

	dx_1	dx_2	
v_1	1	1	$= -y_1$
v_3	1	-1	$= -y_3$
-1	0	2	$= df$
	$= u_1$	$= u_2$	

This does not yield a KKT point.

d. The boundary in the fourth quadrant. The KKT tableau is

	dx_1	dx_2	
v_3	1	-1	$= -y_1$
-1	x_2	x_1	$= df$
	$= 0$	$= 0$	

The column equations require $x_1 + x_2 = 0$, which, with $x_1 + x_2 = 2$, gives $x_1 = -1$, $x_2 = 1$. This is a KKT point, but it is a minimum.

5. The KKT tableau is

	dx_1	dx_2	
v_1	1	1	$= -dy_1$
v_2	$-2(x_1 - 2)$	$-2(x_2 - 2)$	$= -dy_2$
-1	x_2	x_1	$= df$
	$= u_1$	$= u_2$	

The feasible set is a right isosceles triangle with a semicircle cut out of the hypotenuse. The point $x_1 = 1$, $x_2 = 1$ is a KKT point, but it is a saddlepoint. Both the intersections of the circle with the hypotenuse, $(3, 1)$ and $(1, 3)$, are KKT points, and both are maxima.

6. The KKT tableau is

	dx_1	dx_2	
v_1	$2x_1$	$2(x_2 - 2)$	$= -dy_1$
v_2	$-2(x_1 - 1)$	$-2(x_2 - 2)$	$= -dy_2$
v_3	0	1	$= -dy_3$
-1	1	1	$= df$
	$= u_1$	$= u_2$	

The feasible set is the interior of the larger circle, outside the smaller circle, below the line $x_2 = 2$, and in the first quadrant. There are three KKT points. $(0, 2)$ is a KKT point and a local maximum. $(2, 2)$ is a KKT point and a global maximum. $(1 - 1/\sqrt{2}, 2 - 1/\sqrt{2})$ is KKT point, but it is a saddle point.

7. The KKT tableau is

	dx_1	dx_2	
v_1	$4x_1 - 5x_2 + 2$	$-5x_1 + 4x_2 + 2$	$= -dy_1$
v_2	1	3	$= -dy_2$
-1	1	1	$= df$
	$= u_1$	$= u_2$	

The conic section is a hyperbola. The only purpose for the linear constraint is to select one branch of the hyperbola. The linear constraint is otherwise not effective. Since $y_2 > 0$ throughout the feasible set, $v_2 = 0$. That is, the KKT tableau always reduces to

	dx_1	dx_2	
v_1	$4x_1 - 5x_2 + 2$	$-5x_1 + 4x_2 + 2$	$= -dy_1$
-1	1	1	$= df$
	$= u_1$	$= u_2$	

Since $v_1 = 0$ except on the hyperbola, the only KKT point is on the hyperbola. The maximum occurs at $(1, 1)$ and that point is a KKT point.

8. The KKT tableau is

	dx_1	dx_2	
v_1	$4x_1 - 5x_2 + 2$	$-5x_1 + 4x_2 + 2$	$= -dy_1$
v_2	1	3	$= -dy_2$
-1	1	1	$= df$
	$= u_1$	$= u_2$	

Again, the linear constraint merely selects part of the region defined by the first quadratic constraint. This time the hyperbola degenerates into two intersecting straight lines. The maximum occurs at this intersection point, $(2, 2)$. However, the constraint qualification is not satisfied and this is not a KKT point. The KKT tableau for this point is

	dx_1	dx_2	
v_1	0	0	$= -dy_1$
v_2	1	3	$= -dy_2$
-1	1	1	$= df$
	$= 0$	$= 0$	

9. The KKT tableau is

$$
\begin{array}{r|ccc|l}
 & dx_1 & dx_2 & dx_3 & \\
\hline
v_1 & 2x_1 & 2x_2 & 2x_3 & = 0 \\
\hline
-1 & -x_2x_3 & -x_1x_3 & -x_1x_2 & = df \\
\hline
 & = u_1 & = u_2 & = u_3 &
\end{array}
$$

Note that the constraint is an equality. Therefore, the variable on the right margin is artificial and v_1 is free.

The feasible set is the part of the surface of a sphere in the first octant. The objective function is negative throughout the interior of the feasible set and it is zero on the boundary. Thus, all boundary points are maxima. These points are also KKT points. There is one KKT point in the interior, where the objective function is a minimum.

10. The KKT tableau is

$$
\begin{array}{r|ccc|l}
 & dx_1 & dx_2 & dx_3 & \\
\hline
v_1 & 2x_1 & 2x_2 & 2x_3 & = 0 \\
v_2 & 1 & 1 & -1 & = 0 \\
\hline
-1 & x_2x_3 & x_1x_3 & x_1x_2 & = df \\
\hline
 & = 0 & = 0 & = u_3 &
\end{array}
$$

The feasible set is a semicircle lying on the sphere. A frontal assault requires solving the following system of equations (except for the endpoints, where $u_3 \geq 0$).

$$
\begin{aligned}
2v_1x_1 + v_2 - x_2x_3 &= 0 \\
2v_1x_2 + v_2 - x_1x_3 &= 0 \\
2v_1x_3 - v_2 - x_1x_2 &= 0 \\
x_{12} + x_{22} + x_{32} &= 6 \\
x_1 + x_2 - x_3 &= 0
\end{aligned}
$$

This is certainly daunting. Because of symmetry it is possible to guess that the point $x_1 = x_2$ is a point with special properties. At that point the KKT tableau reduces to

$$
\begin{array}{r|ccc|l}
 & dx_1 & dx_2 & dx_3 & \\
\hline
v_1 & 2 & 2 & 4 & = 0 \\
v_2 & 1 & 1 & -1 & = 0 \\
\hline
-1 & 2 & 2 & 1 & = df \\
\hline
 & = 0 & = 0 & = 0 &
\end{array}
$$

$v_1 = 1/2$, $v_2 = 1$ provides a feasible solution for the column equations. Thus, this point is a KKT point.

The endpoints are $(\sqrt{3}, -\sqrt{3}, 0)$ and $(-\sqrt{3}, \sqrt{3}, 0)$. Both are KKT points. The objective variable is 0 there and negative otherwise near these points. We have two local maxima and one global maximum.

11. A linear function is of the form $f = a + bt$. The parameters are a and b. The tableau giving the constraints is

a	b	-1	
1	1	0.3	$= -r_1$
1	2	−0.2	$= -r_2$
1	3	−0.2	$= -r_3$
1	4	0.1	$= -r_4$
1	5	1.1	$= -r_5$
1	6	2.1	$= -r_6$

We compute $D^T D$ and $D^T e$.

$$D^T D = \begin{bmatrix} 6 & 21 \\ 21 & 91 \end{bmatrix}, \quad D^T e = \begin{bmatrix} 3.3 \\ 18.1 \end{bmatrix}$$

Solving the normal equation, we get $f = -0.76 + 0.37t$.

12. A quadratic function is of the form $f = a + bt + ct^2$. The tableau giving the constraints is

a	b	c	-1	
1	1	1	0.3	$= -r_1$
1	2	4	−0.2	$= -r_2$
1	3	9	−0.2	$= -r_3$
1	4	16	0.1	$= -r_4$
1	5	25	1.1	$= -r_5$
1	6	36	2.1	$= -r_6$

We compute $D^T D$ and $D^T e$.

$$D^T D = \begin{bmatrix} 6 & 21 & 91 \\ 21 & 91 & 441 \\ 91 & 441 & 2275 \end{bmatrix}, \quad D^T e = \begin{bmatrix} 3.3 \\ 18.1 \\ 103.3 \end{bmatrix}$$

When we solve the normal equation we get

$$f = 1.09 - 1.01t + 0.2t^2$$

13. This is a least distance problem. The appropriate tableau is

	x_1	x_2	
s_1	-1	2	$= -r_1$
s_2	1	-1	$= -r_2$
-1	1	3	$= f^0$
	$= u_1$	$= u_2$	

We require x and u to be canonical, and r and s to be free.
The associated linear complementarity problem is

x_1	2	-3
x_2	-3	5
-1	1	3
	$= u_1$	$= u_2$

We try to find a complementary feasible solution by principal pivoting. There are not many possibilities and we obtain

u_1	5	3
u_2	3	2
-1	-14	-9
	$= x_1$	$= x_2$

From $x_1 = 14$, $x_2 = 9$ we obtain $s_1 = -r_1 = 4$ and $s_2 = -r_2 = 5$.
14. This is another least distance problem. The appropriate tableau is

	x_1	x_2	
s_1	1	2	$= -r_1$
s_2	-1	-1	$= -r_2$
-1	-1	3	$= f_0$
	$= u_1$	$= u_2$	

We require x and u to be canonical, and r and s to be free.
Using the same methods used for Exercise 13, we obtain $u_1 = 14/5$, $x_2 = 3/5$, $s_1 = 6/5$, $s_2 = -3/5$.
15. The appropriate starting tableau is

	x_1	x_2	x_3	
s_1	1	2	-2	$= -r_1$
s_2	1	-1	1	$= -r_2$
s_3	1	2	-2	$= -r_3$
-1	3	4	-4	$= f_0$
	$= u_1$	$= u_2$	$= u_3$	

The solution is $s_1 = 7/6$, $s_2 = 2/3$, $s_3 = 7/6$.

16. This is a quadratic program, but we must convert it to a maximum program and expand the parentheses to make it conform to the format required for the methods provided here. We maximize

$$\begin{aligned} f &= 2x_1^2 - 8x_1x_2 - 10x_2^2 + 6x_1 + 14x_2 - 5 \\ &= \frac{1}{2}(4x_1^2 - 16x_1x_2 - 20x_2^2) + 6x_1 + 14x_2 - 5 \end{aligned}$$

subject to

$$-2x_1 - x_2 \geq -4$$

and $x_1 \geq 0$, $x_2 \geq 0$.

The initial tableau is

x_1	4	8	2
x_2	8	20	1
v_1	-2	-1	0
-1	6	14	4
	$= u_1$	$= u_2$	$= y_1$

All variables are canonical. Not many principal pivot exchanges must be considered. Principal pivot exchanges in the main diagonal yield

u_1	0	$1/2$	$-1/2$
x_2	$-1/2$	13	3
y_1	$1/2$	3	1
-1	-2	-1	-1
	$= x_1$	$= u_2$	$= v_1$

The optimal point is $x_1 = 2$, $x_2 = 0$. The maximum value is -1.

17. The initial tableau is the same. Consider the final tableau obtained in Exercise 16. Since u_2 is artificial, an additional principal pivot exchange is required in that column. We obtain

u_1	$1/52$	$-1/26$	$-8/13$
u_2	$-1/26$	$1/13$	$3/13$
y_1	$8/13$	$-3/13$	$4/13$
-1	$-53/13$	$1/13$	$-10/13$
	$= x_1$	$= x_2$	$= v_1$

The optimal point is $x_1 = 53/13$, $x_2 = -1/13$. The maximal value is $-12/13$. Making x_2 free increased the value of the objective variable slightly.

18. The initial tableau is

x_1	2	4	2	1
x_2	4	10	1	−2
v_1	−2	−1	0	0
v_2	−1	2	0	0
−1	6	14	4	−1
	$= u_1$	$= u_2$	$= y_1$	$= y_2$

All the variables are canonical except v_2, which is free, and y_2, which is artificial. Principal pivoting yields

u_1	0	0	−2/5	−1/5
u_2	0	0	−1/5	2/5
y_1	2/5	1/5	2.72	−2.24
y_2	1/5	−2/5	−2.24	2.08
−1	−7/5	−6/5	−7.92	6.64
	$= x_1$	$= x_2$	$= v_1$	$= v_2$

Since v_2 is free, the positive entry in the c-row is acceptable. The optimal point is $x_1 = 7/5$, $x_2 = 6/5$. Actually, this point is the intersection of the two lines. The maximum value is −11.56.

19. The initial tableau is

x_1	8.00	−1.00	0.00	−3.00	3.00
x_2	−1.00	8.00	−1.00	1.00	−6.00
x_3	0.00	−1.00	8.00	−4.00	2.00
v_1	3.00	−1.00	4.00	0.00	0.00
v_2	−3.00	6.00	−2.00	0.00	0.00
−1	1.00	2.00	3.00	−4.00	2.00
	$= u_1$	$= u_2$	$= u_3$	$= y_1$	$= y_2$

Three principal pivot exchanges yield the following tableau, from which we can obtain the optimal solution.

u_1	0.04	−1.37	−0.06	0.12	−0.23
x_2	1.37	25.53	0.95	−6.90	3.31
u_3	−0.06	−0.95	0.09	−0.17	−0.15
v_1	0.12	6.90	−0.17	0.35	1.31
u_2	0.23	3.31	0.15	−1.31	0.62
−1	−0.33	−0.39	−0.51	−0.98	−0.54
	$= x_1$	$= u_2$	$= x_3$	$= y_1$	$= v_2$

$x_1 = 0.33$, $x_2 = 0$, $x_3 = 0.51$ is point where the maximum value occurs.

20. Additional principal pivot exchanges must be made to move the free variable v_1 to the bottom margin. We obtain

u_1	0.08	0.02	−0.06	0.11	−0.01
u_2	0.02	0.01	−0.02	0.12	0.18
u_3	−0.06	−0.02	0.04	0.19	0.05
u_1	−0.11	−0.12	−0.19	0.45	0.19
u_2	0.01	−0.18	−0.05	−0.19	0.26
−1	−0.43	−0.12	−0.71	0.49	−0.28
	$= x_1$	$= x_2$	$= x_3$	$= v_1$	$= v_2$

Questions

Q1. False. The feasible set might be empty. If the feasible set is nonempty, the objective variable might be unbounded. If the objective variable is bounded, it might not take on its least upper bound.

Q2. False. The objective variable might not take on its least upper bound. Consider $f = 1 - e^{-x}$ on the positive real axis.

Q3. True. This is the content of the Karush–Kuhn–Tucker theorem.

Q4. False. There are several exercises above that illustrate points that satisfy the KKT conditions but are either minima or saddle points.

Q5. False. A KKT point might be a saddle point.

Q6. True. Exercise 1 illustrates such a case.

Q7. False. The feasible set might be empty.

Q8. True. The function will be bounded. Since the feasible set is closed and bounded, the function will take on its least upper bound.

Q9. True. Since the constraints are linear, a negative semidefinite quadratic function cannot have two isolated local maxima. Actually, this assertion is true if the feasible set is merely convex. If you draw a straight line between two points on the surface for a negative semidefinite quadratic function, the straight line lies below or on the surface. It can have more than one local maximum, but they will be connected. Imagine a sheet of paper in the form of an arch.

Q10. False. Imagine a sheet of paper in the form of an arch.

Q11. True. The sheet-of-paper example is semidefinite but not definite. There can still be complications from the boundary of the feasible set. Consider the function $f = -x_{12} - x_{22}$ constrained by $x_{12} + x_{22} \geq 1$. However, as we have defined a quadratic program, the constraints are linear. For a negative definite quadratic function, a line joining two distinct points on the surface lies below the surface. The feasible set is convex and the maximum point is unique.

Q12. True. This was proved in connection with tableau 12.47.

Q13. True. In the normal equation $D^T Dx = D^T e$, the rank of $D^T D$ is equal to the rank of D. The assumptions imply that $D^T D$ is invertible.

Q14. True Since the problem is feasible, we can draw a circle about the origin that will contain the point at which the least distance must occur. Since the distance is always nonnegative and the feasible set is closed, there is a minimum distance.

Selected Bibliography

Some citations in this bibliography are of historic interest. Some are sources for topics that we include in this book or for treatments that we present. Some extend or complement what we cover here.

BALINSKI, M. L. and A. W. TUCKER (1969). "Duality theory of linear programs: A constructive approach with applications," *SIAM Review, 11*: 347–377.

BEALE, E. M. L. (1955). "Cycling in the dual simplex algorithm," *Naval Research Logistics Quarterly 2*: 269–275.

BLAND, R. G. (1977). "New finite pivoting rules for the simplex method." *Mathematics of Operations Research 2*: 103–107.

BOREL, E. (1924). "Sur les jeux oú interviennent l'hasard et l'habilité des joueurs," in *Theorie des Probabilitiés*, Paris: Librarie scientifique, Hermann. [English translation by L. J. Savage: *Econometrica 21*, (1953): 97–100.

BRADLEY, G. H., G. G. BROWN, and G. W. GRAVES (1977). "Design and implementation of large scale primal transshipment algorithms," *Management Science 24*: 1–34.

COTTLE, R. W. (1968). "The principal pivoting method of quadratic programming," in *Mathematics of the Decision Sciences, Part I*, G. B. Dantzig and A. F. Veinott, Jr., eds. American Mathematical Society, Providence, RI, 144–162.

COTTLE, R. W. and G. B. DANTZIG (1968). "Complementary pivot theory of mathematical programming," in *Mathematics of the Decision Sciences, Part I*, G. B. Dantzig and A. F. Veinott, Jr., eds. American Mathematical Society, Providence, RI, 115–136.

COTTLE, R. W., J-S PANG, and R. E. STONE (1992). *The Linear Complementarity Problem*, Boston, Massachusetts: Academic Press.

CHVÁTAL, V. (1975). *Linear Programming*, New York, NY: Springer-Verlag.

DANTZIG, G. B. (1951). "Application of the simplex method to a transportation problem," in *Activity Analysis of Production and Allocation*, T. C. Koopmans, ed. New York, NY: John Wiley and Sons, 359–373.

DANTZIG, G. B. (1963). *Linear Programming and Extensions*, Princeton, New Jersey: Princeton University Press.

DAVIS, M. D. (1970). *Game Theory: A Nontechnical Introduction*, New York, NY: Basic Books.

DIJKSTRA, E. (1959). "A note on two problems in connexion with graphs," *Numerische Mathematik 1*: 269–271.

DORFMAN, R., P. A. SAMUELSON, and R. M. SOLOW (1958), *"Linear Programming and Economic Anlysis*, New York, NY: McGray-Hill.

EGERVÁRY, J. (1931). "Matrixok kombinatorikus tulajdonságairól," *Mathematikai és Fizidai Lápok 38*: 16–28.

FARKAS, J. (1902). "Theorie der einfachen Ungleichungen," *Journal für die reine und angewandte Mathematik 124*: 1–27.

FORD, L. R., JR., and D. R. FULKERSON (1956). "Maximal flow through a network," *Candian Journal of Mathematics 8*: 399-404.

FORD, L. R., JR., and D. R. FULKERSON (1962). *Flows in Networks.* Princeton, New Jersey, Princeton University Press.

FORSYTHE. G., and C. B. MOLER (1967). *Computer Solution of Algebraic Systems*, Englewood Cliffs, New Jersey: Prentice-Hall.

GALE, D. (1960). *The Theory of Linear Economic Models*, New York, NY: McGraw-Hill.

GALE, D., H. W. KUHN, and A. W. TUCKER (1951). "Linear programming and the theory of games," in *Activity Analysis of Production and Allocation*, T. C. Koopmans, ed. New York, NY: John Wiley and Sons, 317–329.

GASS, S. I. (1969, 1985). *Linear Programming: Methods and Applications*, Fifth Edition. New York, NY: McGraw-Hill.

GOOD, R. A. (1959). "Systems of Linear Relations," *Siam Review 1*: 1–31.

HALL, P. (1935). "On representatives of subsets," *Journal of the London Mathematical Society 10*: 26–30.

HITCHCOCK, J. L. (1941). "The distribution of a produce from several sources to numerous localities," *Journal of Mathematical Physics 20*: 224–230.

HOFFMAN, A. J. (1953). "Cycling in the simplex algorithm," National Bureau of Standards, Report 2974.

JACOBS, W. (1954). "The caterer problem," *Naval Research Logistics Quarterly 1*: 154–165.

KANTOROVICH, L. V. (1939). "Mathematical methods in the organization and planning of production," [English translation: *Management Science 6*, (1960): 366–422.]

KARMARKAR, N. (1984). "A new polynomial-time algorithm for linear programming," *Combinatorica 4*: 373–395.

KARUSH, W. (1939). "Minima of functions of several variables with inequalities as side conditions," Masters Thesis, Department of Mathematics, University of Chicago.

KHACHIYAN, L. G. (1979). "A polynomial algorithm in linear programming" (in Russion), *Doklady Akademia Nauk SSSR 244*: 1093–1096. [English translation: *Soviet Mathematics Doklady 20*: 191–194.]

KLEE, V., and G. J. MINTY (1972). "How good is the simplex algorithm?" in *Inequalities–III*, O, Shisha, ed. New York: Academic Press, 159–175.

KÖNIG, D. (1931). "Graphen und Matrizen," *Matematikai eś Fizikai Lápok 38*: 116–119.

KUHN, H. W. (1955). "The Hungarian method for the assignment problem," *Naval Research Logistics Quarterly 2*: 83–97.

KUHN, H. W. (1976). Nonlinear programming: A historical view," in *Nonlinear Programming. SIAM-AMS Proceedings*, Vol 9, American Mathematical Society, Providence, RI, 1-26.

KUHN, H. W. and A. W. TUCKER (1951). "Nonlinear programming," in *Proceeding of the Second Berkeley Symposium on Mathematical Statistics and Probability*, J. Neyman, ed. University of California, Berkeley, CA, 481–492; also in *Econometrica, 19*: 50–51 (abstract).

KUHN, H. W. and A. W. TUCKER (1956). eds. *Linear Inequalities and Related Systems*, Annals of Mathematics Studies, 38, Princeton, NJ: Princeton University Press.

LEMKE, C. E. (1965). "Bimatrix equilibrium points and mathematical programming," *Management Science 11*, 681–689.

LEMKE, C. E. (1968). "On complementary pivot theory," in *Mathematics for the Decision Sciences, Part 1*, G. B. Dantzig and A. F. Veinott, Jr., eds. Americal Mathematical Society, Providence, RI, 95–114.

LEMKE, C. E. and J. T. HOWSON (1964). "Equilibrium points of bimatrix games," *SIAM Journal of Applied Mathematics 12*, 413–423.

LUCE, W. D., and H. RAIFFA (1957). *Games and Decisions*, New York, NY: John Wiley and Sons.

LUSTIG, I. J., R. E. MARSTEN, and D. F. SHANNO (1991). "Computational experience with a primal–dual interior point method for linear programming," *Linear Algebra and its Applicaitions 152*: 191–222.

NERING, E. D., and A. W. TUCKER (1979). "Priority pivoting," *Tenth International Symposium on Mathematical Programming*, Montreal.

ORDEN, A. (1956). "The transshipment problem," *Management Science 2*: 276–285.

PARSONS, T. D., and A. W. TUCKER (1971). "Hybrid programs: linear and least distance," *Mathematical Programming 1*: 153–167.

SCHRIJVER, A. (1986). *Theory of Linear and Integer Programming*, New York, NY: John Wiley and Sons.

SINGLETON, R. R., and W. F. TYNDALL (1974). *Games and Programs—Mathematics for Modeling*, W. H. Freeman, San Francisco.

STRAYER, J. K. (1989). *Linear Programming and its Applications*, New York: Springer-Verlag.

TUCKER, A. W. (1956). "Dual systems of homogeneous linear equations," in *Linear Inequalities and Related Systems*, H. W. Kuhn and A. W. Tucker, eds. *Annals of Mathematics Studies 38*: 3–18.

TUCKER, A. W. (1957). " Linear and nonlinear programming," *Operations Research 5*: 244–257.

TUCKER, A. W. (1960). "Simplex method and theory," in *Mathematical Optimization Techniques*, Richard Bellman, ed. University of California Press, Berkeley, California 213–231.

TUCKER, A. W. (1960). "Solving a matrix game by linear programming," *IBM Journal of Research and Development 4*: 507–517.

TUCKER, A. W. (1967). "Pivotal algebra," Seminar notes (taken by T. D. Parsons), Department of Mathematics, Princeton University, Princeton, NJ.

TUCKER, A. W. (1976). "Least squares extension of linear programming," *Survey of Mathematical Programming* A. Prékopa, ed. Proceedings of 9th International Mathematical Programming Symposium. Budapest.

TUCKER, A. W. (1982). "Constructive LP thoery—with extensions." *Nordic Symposium on Linear Complementarity Problems*, Linköping, Sweden.

VON NEUMANN, J., and O. MORGENSTERN (1980). *The Theory of Games*, Princeton, New Jersey: Princeton University Press.

WILLIAMS, J. D. (1954). *The Compleat Strategyst: Being a Primer on the Theory of Games of Strategy*, New York, NY: McGraw-Hill; Mineola, NY: Dover.

Index

ISBN 0-12-515440-2